FOUNDATIONS OF AERODYNAMICS:
Bases of Aerodynamic Design

FOUNDATIONS OF AERODYNAMICS:
Bases of Aerodynamic Design,

Third Edition

Arnold M. Kuethe
Department of Aerospace Engineering,
University of Michigan

Chuen-Yen Chow
Department of Aerospace Engineering Sciences,
University of Colorado

John Wiley & Sons
New York ● Santa Barbara ● London ● Sydney ● Toronto

Library of Congress Cataloging in Publication Data

Kuethe, Arnold Martin, 1905–
    Foundations of aerodynamics.

    Bibliography
    Includes index.
    1. Aerodynamics.   I. Chow, Chuen-Yen, 1932–
Joint author.   II. Title.
TL570.K76 1977   629.132′3   76-20761
ISBN 0-471-50953-1

Preface

The first two editions of this book, on which J. D. Schetzer and I collaborated, appeared in 1950 and 1959. Professor Schetzer resigned from the University in 1960 to enter Industry, and he decided not to collaborate on the present edition. I was pleased when Professor Chuen-Yen Chow agreed to take over the co-authorship; his broad knowledge of the field and especially of computational methods were significant factors in the revision.

The collaboration with J. D. Schetzer was for me a stimulating and fruitful experience. His reputation as a most effective teacher and as a gifted expositor are reflected in many of the sections that were carried over with little change from the second edition.

Arnold M. Kuethe

In this third edition we include, at the level consistent with the prerequisites chosen, treatments of the most significant advances since 1959, the publication date of the second edition. Furthermore, we focus on those concepts that relate most closely to current methods of aerodynamics design of aircraft and, to some extent, of fluid machinery in general.

The use of SI Metric units throughout is in keeping with the general standardization throughout the world. Conversion factors between SI and English units are given in Table 1 and on overleafs of the front and back covers.

Our objective continues to be the same as it was for the previous editions, that is, to provide a textbook for courses in aerodynamics at the junior-senior level in engineering and physics. Many of the topics discussed are applicable as well to other specialized aspects of fluid mechanics. The prerequisites we aim at are one year each of calculus and physics. In addition, a course in advanced calculus or differential equations should be a pre- or corequisite. A course in elementary fluid mechanics with laboratory experiments and demonstrations would help the student to visualize the physical concepts.*

With each new edition the choice of what material to add and what to delete becomes, of course, most difficult. The main additions here are the panel method, the area rule, and the boundary layer control methods, and some of their ramifications in design syntheses. Specific deletions are the appendices on "Aerodynamic

---

*A very useful and inexpensive companion volume would be, *Illustrated Experiments in Fluid Mechanics* (preface by Asher Shapiro) M.I.T. Press, Cambridge, Mass. The book describes experiments and results illustrating many aspects of fluid mechanics; it comprises mainly descriptions and explanatory material on many half hour films and more than 100 four-minute film loops (available from Encyclopedia Britannica, Chicago, Ill. 60611) that effectively illustrate the details of many flow phenomena and their dependence on Reynolds number and a myriad of other similarity parameters.

Characteristics of Wings," and on "Real Fluid Effects in High Speed Flight"; however, some of the treatments in these appendices are expanded and incorporated in the pertinent chapters of the text. The net result is that the present edition is appreciably longer than the previous ones, especially since the rewriting of many sections and the addition of more worked-out examples and new problems expands even the older material.

In Chapter 4 we introduce the panel method, one of the most significant recent contributions toward narrowing the gap between theory and experiment. Conceived by A. M. O. Smith, its utility has been expanded by him and his colleagues at Douglas Aircraft and by other research and design groups. By means of this method, the flow fields of sources, doublets, and vortices are added to a uniform flow field to generate the velocity and pressure distributions for flows about complicated shapes; the properties of an airfoil, given in Fig. 5.23, and of a wing-fuselage combination, shown in Fig. 6.19, are prime examples. Later chapters describe the area rule, by which compressibility effects can be incorporated, and in the last chapters the material on boundary layers is expanded and the utilization of their properties in design are indicated.

We are indebted to many of our friends and colleagues for their help in the preparation of this edition. We thank them for their suggestions, copies of photographs and illustrations, and problems. Professor M. V. Morkovin of the Illinois Institute of Technology made many suggestions, especially concerning stability and boundary layer transition and flow separation. Flow photographs were kindly supplied by Professor Emeritus F. N. M. Brown of the University of Notre Dame, Professor H. W. Liepmann of the California Institute of Technology, Dr. L. B. Gratzer and G. E. Saaris of The Boeing Company, Dr. G. R. Seddon of the U.S. Army Ballistics Research Laboratory, and Dr. Julianna Chow of the National Center for Atmospheric Research. Professors Harm Buning, V. C. Liu, A. F. Messiter, Jr., and Martin Sichel of the University of Michigan, and L. C. Garby and M. S. Uberoi of the University of Colorado read parts of the manuscript and suggested problems and topics. Among those who also made useful suggestions are A. M. O. Smith and Dr. T. Cebeci of the Douglas Aircraft Company, Dr. P. E. Rubbert, and G. R. Saaris of The Boeing Company, Dr. W. P. Rodden, Consulting Engineer, and L. Hasel of the NASA Langley Research Center. Prof. D. M. Rao of Texas A & M University reviewed the first six chapters and Prof. J. C. Williams III of North Carolina State University reviewed the entire manuscript; their suggestions are greatly appreciated. Publications of the NASA and The British Aeronautical Research Council have proved most useful. We are also indebted to Gloria Lyons, Margaret Fillion, and Mary Pudim for their excellent secretarial services.

Arnold M. Kuethe
Chuen-Yen Chow

Contents

CHAPTER 1

The Fluid Medium

1.1 Introduction

The science of aerodynamics deals with the determination of the characteristics of the flow through passages and past bodies of various shapes. Once the flow pattern has been established, the aerodynamic forces and moments acting on the body can be calculated.

This book attempts to build up from first principles a background of concepts that may be utilized in applying aerodynamics and fluid dynamics in general to problems in aeronautics and other fields of engineering. A clear understanding of these concepts is necessary since, owing to mathematical complexities and often hypothetical physical premises, we are constantly dealing with approximations to the actual problems we are attempting to solve. Therefore, many of the more difficult problems involve in their solution an intuitive approach. This approach must be disciplined by a comprehension of those concepts that have been shown by experiment to be valid.

This chapter deals with the properties of the fluid medium, which we define as a material that is at rest *only* when all forces acting on it are in equilibrium. Although the concepts treated are of primary interest in aerospace engineering, applications in other fields will be pointed out. These applications will generally be considered through analogies and illustrative examples.

The fluid properties we discuss here are the pressure, temperature, density, elasticity, and transport properties, especially viscosity; they are related to the molecular structure of the fluid. Numerical data are given for both air and water. The chapter closes with brief descriptions of some aspects of dimensional analysis and a discussion of the subdivisions of aerodynamics according to altitude and speed of flight.

1.2 Units

The SI (Systéme International) system of units is used throughout this book. In this system, the unit of force, the newton, is defined by the equation

$$1 \text{ newton (N)} = 1 \text{ kilogram mass (kg)} \times 1 \text{ m/sec}^2$$

The British system, used in the previous editions, is based on the definition of one pound as the unit of force given by

$$1 \text{ pound (lb)} = 1 \text{ slug} \times 1 \text{ ft/sec}^2$$

If the first equation above is multiplied by the acceleration of gravity, 9.807 m/sec^2, we see that a mass of 1 kg weighs 9.807 newtons or 1 kg-force in the "gravitational" MKS system; in nontechnical fields, the gravitational system is commonly used. Similarly, in the English system (g = 32.174 ft/sec^2), 1 slug weighs 32.174 lb.

The only other fundamental unit we need is that of absolute temperature T, expressed in "degrees Rankine" or in "degrees Kelvin"; the latter is used in this book. In terms of the Fahrenheit and "Celcius" (degrees Centigrade) scales, these units are defined by

$$T(^{\circ}R) = {}^{\circ}F + 460$$
$$T(^{\circ}K) = {}^{\circ}C + 273$$
$${}^{\circ}C = ({}^{\circ}F - 32)/1.8$$

Conversion factors relating the English and SI units are given in Table 1 at the end of the book and on overleafs of the front and back covers.

1.3   Properties of Gases at Rest

A gas consists of a large number of molecules moving in a random fashion relative to one another. The "number density" of molecules is determined by **Avogadro's law,** which states that a gas contains 6.025 $\times$ 10^{26} molecules/kg-mole* (2.732 $\times$ 10^{26}/slug-mole). For air under standard conditions (see Section 1.4) the number density of air is 2.7 $\times$ 10^{19} molecules/cc (4.4 $\times$ 10^{20}/cu in). The **ideal gas** is one in which intermolecular forces are negligible. Its bulk properties, which closely approximate those of real gases, except at very low and very high temperatures and densities, can be expressed in terms of its molecular properties: the **mass** m of the molecule, the **average random speed** c of the molecule, and the mean distance the molecule travels between collisions with other molecules, namely, the **mean free path** λ.

1. *Density.*

The density of matter is defined as the mass per unit volume; it is thus the total mass of the molecules per unit volume. The dimensions of density are then force $\cdot$ sec^2/(length)^4; it is designated by ρ and has dimensions of kilograms per cubic meter (kg/m^3) in SI units and slugs per cubic foot (slugs/ft^3) in FPS units. Table 2 gives its variation with temperature for air at sea level pressure and for water; in Table 3, values of the density in SI units are given for the "Standard Atmosphere."

---

*One kg-mole is the number of kilograms of gas numerically equal to the atomic weight. Thus, one kg-mole of air has a mass of 28.97 kg. Avogadro's number has the same value for all gases.

2. *Pressure.*

When molecules strike a surface they rebound, and, by Newton's second law, a force is exerted on the surface equal to the time rate of change of momentum of the rebounding molecules; that is, the force is equal to the sum of the changes in momentum experienced by all the molecules striking and rebounding from the surface per second. **Pressure** is defined as the force per unit area exerted on a surface immersed in the fluid and at rest relative to the fluid; it is expressed in newtons per square meter [N/m^2 (pascals)] or pounds per square foot (lb/ft^2). Experiment indicates that the collisions among molecules and with surfaces are elastic so that the mean change in momentum is a vector normal to the surface regardless of the angle of incidence of the collision; therefore, we conclude that *fluid pressure acts normal to a surface.*

We show now that the fluid pressure is proportional to the kinetic energy of molecules of the gas. We compute the force exerted on the walls of a unit cube of gas (Fig. 1) and, since we wish only to identify the combination of gas properties that determine the pressure, we adopt the following simplified model of the molecular motion: All of the N (N is the number density) molecules in the unit cube are assumed to have identical masses m and identical speeds c. They are assumed to travel parallel to the coordinate axes, $N/3$ parallel to and $N/6$ in the positive direction of each axis. The $(N/6)\Delta x$ molecules in the thin layer shown in Fig. 1 will strike the right x face in time $\Delta t = \Delta x/c$. The collisions are assumed to be elastic so that the momentum of each molecule is changed by $2mc$. Newton's second law then predicts that a force equal to the product of $2mc$ and the number of molecules striking the surface per second will be exerted on the right face. The number striking per second will be $(N/6)c$ and since $Nm = \rho$, the fluid density, the force acting on the right face, in fact on each face, is given by the formula

$$p = (N/6)c \cdot 2mc = \frac{\rho c^2}{3} \tag{1}$$

The pressure p has the dimensions of force/area. Physically, Eq. 1 states that *the pressure is proportional to the kinetic energy of the molecular motion.* Since the

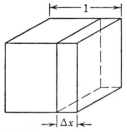

Fig. 1 Model for interpretation of pressure.

pressure is equal on all faces, the cube is at rest relative to the surrounding fluid; that is, either the flow speed is zero or the cube is moving at the flow speed. Also, since speed of the molecules varies, the pressure is actually proportional to the mean of the square of the speed rather than to the square of the mean speed (see Problem 1, Section 1.3). This approximation, however, affects only the magnitude of the proportionality factor in Eq. 1.

3. *Temperature.*

According to the kinetic theory of gases, *the absolute temperature is proportional to the mean translational kinetic energy of the molecules.* It can be interpreted, in terms of the **equation of state** for an ideal gas

$$p = \rho RT \tag{2}$$

where T is the absolute temperature and R, the gas constant, has a specific value depending only on the composition of a gas. For air $R = 287 \text{m}^2/\text{sec}^2\ {}^\circ\text{K}$. For systems in which the mass per unit volume remains constant, any process, such as the addition of heat, that increases the kinetic energy of the random motion will increase the temperature and pressure by proportional amounts.

4. *Elasticity.*

When a pressure is applied to a gas, its volume per unit mass changes. **Elasticity** *is defined as the change in pressure per unit change in specific volume.*

$$E = \frac{-dp}{d\rho^{-1}/\rho^{-1}} = \rho\,\frac{dp}{d\rho} \tag{3}$$

It will be shown later that $dp/d\rho$ is the square of the speed of sound through the medium. Therefore, the density and speed of sound define the elasticity.

1.4 Fluidstatics—The Standard Atmosphere

In order to establish uniformity in the presentation of data, standard atmospheric conditions have been adopted and are in general use. Commonly referred to as sea level conditions, these are

$$p = 1.013 \times 10^5 \text{ N/m}^2 \text{ or pascals } (2116 \text{ lb/ft}^2)$$
$$\rho = 1.23 \text{ kg/m}^3 \ (0.002378 \text{ slugs/ft}^3)$$
$$T = 273 + 15^\circ\text{C} = 288^\circ \text{ Kelvin } (520^\circ \text{ Rankine})$$

Under these standard conditions, the speed of sound a is 340 m/sec.

The temperature, pressure, and density in the atmosphere vary with altitude; their magnitudes up to 20 km above sea level are plotted in Fig. 2, and the properties extended to a much higher level are tabulated in Table 1 at the end of the

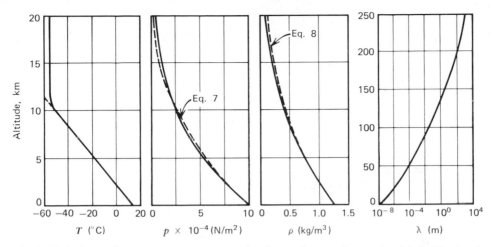

Fig. 2 Variations of temperature, pressure, density, and mean free path with height in the standard atmosphere. The approximate relations, Eqs. 6, 7, and 8 are plotted as dashed lines.

book. The mean free path of the air molecules is also plotted in Fig. 2, up to an altitude of 250 km.

We show now that the variations of pressure and density shown in Fig. 2 are consistent with the measured temperature distribution under the hypothesis that no net force acts on any element of the fluid; the atmosphere is then in **static equilibrium**. Figure 3 shows a cube of fluid of volume $\Delta x \Delta y \Delta z$ oriented with its base $\Delta x \Delta y$ at a height z above an arbitrary datum, for example, sea level. The pressures are equal on all vertical faces of the cube so the pressure p varies only with z. Under the assumed equilibrium conditions, the weight of the element $\bar{\rho} g \Delta x \Delta y \Delta z$, where $\bar{\rho}$ is the average density of the cube, is balanced by the difference between the pressure forces on the lower and upper faces. The net force acting on the cube is in the z direction, and is

$$p\Delta x \Delta y - \left[p + \left(\frac{dp}{dz} \right) \Delta z \right] \Delta x \Delta y = \bar{\rho} g \Delta x \Delta y \Delta z$$

In the limit as the volume of the cube approaches zero $\bar{\rho} \to \rho$, the density at z, we obtain the **equation of aerostatics**

$$\frac{dp}{dz} = -\rho g \qquad (4)$$

We may take g as constant, but both p and ρ vary with z, as shown in Fig. 2. Then the pressure variation as expressed by the integral of Eq. 4 is

$$p = p_0 - \int_0^z \rho g \, dz \qquad (5)$$

where p_0 is the pressure at $z = 0$.

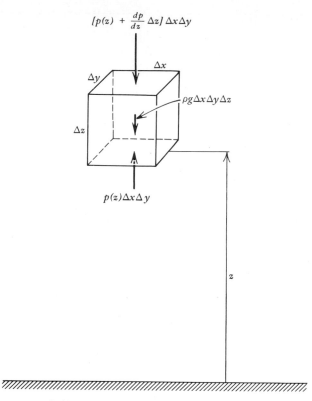

Fig. 3 Force balance on a fluid element.

If we use Eq. 2 and express $\rho = p/RT$, Eq. 4 becomes

$$\frac{dp}{p} = -g\,\frac{dz}{RT}$$

We can derive an analytic expression for the pressure in the lower 10 km of the atmosphere by observing (Fig. 2) that over this interval the temperature decreases approximately linearly with height. Thus we write

$$T = T_0 - \alpha z \tag{6}$$

where T_0 is the temperature at $z = 0$ (sea level) and α is the "temperature lapse rate"; its value is 6.50 °C/km in this region. Substitution into the above equation yields the differential equation

$$\frac{dp}{p} = -g\,\frac{dz}{R(T_0 - \alpha z)}$$

which integrates to

$$p(z) = p_0 \left(1 - \frac{\alpha z}{T_0}\right)^{g/R\alpha} \tag{7}$$

and, for the density,

$$\rho(z) = \rho_0 \left(1 - \frac{\alpha z}{T_0}\right)^{-1 + g/R\alpha} \tag{8}$$

in which p_0 and ρ_0 are the pressure and density, respectively, at $z = 0$. Equation 7 and 8 are plotted in Fig. 2; we see that the calculated and measured values agree very well even up to the 20 km level.

1.5 Buoyancy Force on a Body

The buoyancy force acting on a body submerged in a fluid, for example, on a balloon in the atmosphere, depends on **Archimedes principle,** according to which, *a body submerged in a stagnant fluid is subjected to force equal to the weight of the fluid it displaces.*

To verify this principle, we reason as follows: if the body is replaced by fluid and the fluid volume is subdivided into infinitesimal elements, Eq. 4 (Fig. 3) shows that the pressure force (downward) on the upper surface of each element is less than that (upward) on the lower surface by the weight of the fluid element. Then summation of the net upward force over all of the elements replacing the body shows that the fluid volume experiences a buoyancy (or gravity) force equal to its weight. It follows that a submerged body experiences an upward force equal to the weight of the fluid it displaces.

Example 1.

A hot-air balloon uses a blowtorch to heat the air to the temperature necessary to hover at a given altitude. The total weight of the balloon, equipment, and occupants is 2000 N and its volume is 1000 m^3. For standard conditions at sea level, find the balloon air temperature necessary to hover at sea level and at 5 km.

Solution:

At sea level, the weight of 1000 m^3 of air is $1.23 \times 9.81 \times 10^3 = 12,070$ N. The weight of the air in the balloon must be reduced by 2000 N by heating so it can support the balloon. The internal pressure will be unchanged so, by Eq. 2,

$$T = (12{,}070/10{,}070) \cdot 288° \text{ K} = 345° \text{ K} = 72° \text{ C}$$

or, in lighter-than-air craft terminology, the air must be "superheated" $72° - 15° = 57°$ C.

At 5 km altitude, Fig. 2 gives ambient values of ρ and T, $\rho_a = .737$ kg/m^3, $T_a = 256°$ K, and we find that hovering would require superheating of $99°$ C.

Example 2.

Grebes (an example is the common "hell-diver") are among the birds that hunt their prey underwater. Unlike ducks and other surface water birds, whose feathers are completely water repellent, the outer two-thirds of the grebes' body feathers are wettable. However, like the duck, they require the buoyancy as well as the thermal insulation of the air trapped by a thick mantle of feathers when they are on the surface. In order to facilitate the underwater maneuverability required to catch its prey, the grebe increases its specific gravity to near that of the water by drawing its feathers close to its body (each feather has eight muscles); their partial wettability assists in expelling most of the air, leaving only a thin layer at the skin surface for thermal insulation.

We approximate the bird's body by a sphere of radius $r = 5$ cm and assume it has a specific gravity of 1.1 and find the thickness Δr of the layer of air under sea level conditions required to bring the specific gravity of the combination to unity.

Solution:

The density of water is 1000 kg/m^3 and that of air is 1.23 kg/m^3, so that the specific gravity of air is about 1/800. Then the equation to be solved for Δr is

$$\frac{4}{3}\pi r^3 \cdot (1.1) + \frac{\frac{4}{3}\pi[(r + \Delta r)^3 - r^3]}{800} = \frac{4}{3}\pi(r + \Delta r)^3$$

If $\Delta r \ll r$, we may neglect quadratic and cubic terms in $\Delta r/r$. The result is $\Delta r = 0.0017$ m $= 1.7$ mm. Note that $\Delta r/r = .03$, so that the neglect of its square and cube is justified.

1.6 Hydrostatic Forces—Center of Pressure

Consider the tank 2 m wide with the cross section of Fig. 4 filled to a depth of 2 m with water ($\rho_w = 1000$ kg/m^3; $\rho_w g = 9807$ N/m^3). The gate AB of length 1.5 m of negligible weight, is hinged at A and inclined at an angle of 30° to the floor of the tank. We will calculate the total force, which acts normal to AB, the moment acting on the hinge A, the force at B, and the "center of pressure" of the force on AB.

Since the atmospheric pressure p_a acts on both sides of AB, the net resultant force $F = F_A + F_B$ is given by

$$F = \int_A^B (p - p_a)\,dS = \int_0^{1.5} \rho_w g[(2 - 1.5\sin 30°) + x\sin 30°]\,2\,dx$$

$$= 47{,}800 \text{ N}$$

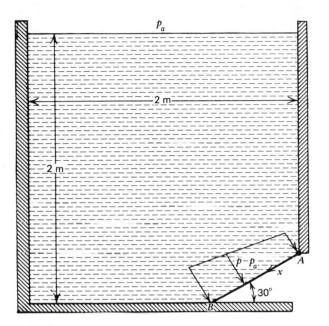

Fig. 4 Pressure distribution on a gate.

The moment acting on the hinge A is

$$M_A = \int_A^B (p - p_a)\, 2x\, dx = \int_0^{1.5} 9807 \left(1.25 + \frac{x}{2} \right) 2x\, dx = 38{,}600 \text{ Nm}$$

The force at B is

$$F_B = \frac{38{,}600}{1.5} = 25{,}700 \text{ N}$$

The **center of pressure** $x_{C.P.}$ is the point of application of a single force which will produce the same moment about the given point as does the force distribution. Then

$$x_{C.P.} = \frac{M_A}{F} = 0.81 \text{ m}$$

Throughout aircraft design, the location of the center of pressure relative to the center of gravity is of vital concern since it determines (1) the magnitude of the aerodynamic moment acting if the craft is disturbed (as when it enters a gust), and (2) whether the craft is stable, in other words, whether the sign of the aerodynamic moment in the disturbed orientation is in the direction to right the craft automatically.

1.7   Fluids in Motion

PRESSURE

When a fluid is in motion, the surface on which pressure is exerted is assumed to move with the fluid; for definiteness when there is a chance for misinterpretation, the pressure given by Eq. 1 is termed the **static pressure**. In Chapter 3, we designate $\frac{1}{2}\rho V^2$, where V is the fluid speed, as the **dynamic pressure** and the sum of the static and dynamic pressures at a point, $p + \frac{1}{2}\rho V^2$, as the **stagnation pressure** for incompressible flow; for example, if a symmetrical body is held stationary with its axis along the direction of flow, the pressure exerted at the nose is the stagnation pressure.

VISCOSITY

The instantaneous velocity of a molecule in a fluid in motion is the vector sum **V + c**, where **V** is the fluid velocity and **c** is the instantaneous velocity of the molecule, measured by an instrument that moves with the fluid. Since the molecular velocity in the fluid at rest is random in magnitude and direction, its mean value is zero, so that the fluid velocity at a given point is the mean vector sum of the velocities of the molecules passing that point.

If the flow velocities are different on two layers aligned with the flow, the exchange of molecules between them tends to equalize their velocities; that is, the random molecular motion effects a transfer of downstream momentum between them. The process of momentum transfer by the molecular motions is termed **viscosity**. The viscous force per unit area, termed the **shearing stress,** is defined as the rate at which the molecules accomplish the cross-stream transfer of downstream momentum per unit area. The stress is thus a force per unit area and is characterized by an equal and opposite reaction, in that positive momentum is transferred from the higher to the lower speed layer and vice versa.

A consequence of the existence of fluid viscosity is the "no-slip condition" at the solid surface, as illustrated in Fig. 5 for the flow between coaxial cylinders, one of which is rotating with angular velocity **ω**. The velocity distribution, designated schematically by the vectors, is established as a result of the random motions of the molecules with the no-slip condition as a constraint at each surface; that is, the

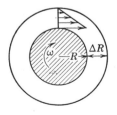

Fig. 5 Flow in an annulus.

monomolecular layer adjacent to a surface has zero velocity relative to it. Mole-cules rebounding from the outer (stationary) surface will have zero average *ordered* velocity, while those rebounding from the inner (moving) surface will have an aver-age *ordered* tangential velocity of magnitude ωR. The mean free path between collisions, as shown in Fig. 2 for air, will, under the great majority of practical conditions, be much smaller than the gap ΔR, so that the molecules will on the average carry their ordered velocities only a very short distance before they collide with molecules of slightly different ordered velocities. The large number of colli-sions constitutes an effective mixing process, resulting in the continuous velocity distribution shown in Fig. 5. If the gap $\Delta R \ll R$, a small segment of the flow will approximate flow between parallel planes in relative motion, as shown in Fig. 6.

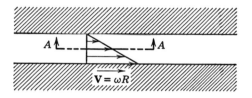

Fig. 6 Shearing stress in a fluid.

The lower plane will have a speed $V = \omega R$, and the transfer of momentum across the intervening space will tend to drag the upper plane to the right. For $\Delta R \ll R$, experiment shows that the stress, τ, exerted on the upper plane varies directly as the relative speed V and inversely as the distance between the planes. Thus

$$\tau = \mu \frac{V}{\Delta R}$$

where μ is the **coefficient of viscosity** of the fluid. Returning momentarily to the coaxial flow of Fig. 5, the torque on the outer cylinder will be $2\pi (R + \Delta R)^2 \tau$ per unit length. The above formula may be generalized by considering the shearing stress exerted between two adjacent layers of fluid, say the layers immediately above and immediately below the section AA in Fig. 6. For this example, the rela-tive speed of the layers is an infinitesimal dV, and the distance between layers is also an infinitesimal ds. Then we may write

$$\tau = \mu \frac{dV}{ds} \qquad (9)$$

It is understood that the derivative in Eq. 9 is always in a direction perpendicular to the plane on which the shearing stress is being computed. When air is flowing with a velocity u over a solid surface, the mixing by the random molecular motion re-sults in the formation of a thin boundary layer, as identified by Ludwig Prandtl (1875–1953) and shown schematically in Fig. 7. Each infinitesimal section of the

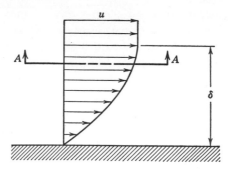

Fig. 7 Boundary layer.

boundary layer can be visualized as two planes in relative motion, and so the shearing stress at any point is given by Eq. 9. At the surface, the shearing stress is given by

$$\tau_0 = \mu \left(\frac{du}{dy}\right)_{y=0} \tag{10}$$

The velocity gradient du/dy is large at the surface and becomes substantially zero at $y = \delta$, where δ is defined as the thickness of the boundary layer.

The coefficient of viscosity may be interpreted in terms of the molecular properties of a fluid by evaluating the shearing stress associated with it. The shearing stress exerted by the fluid below AA in Fig. 7 on that above AA is a retarding effect; it is equal to the rate of loss of momentum of the fluid above AA resulting from the exchange of molecules across unit area of the plane AA.

To calculate the shearing stress, we evaluate the rate of transport of ordered momentum through unit area of AA by the random motion of the molecules. If our frame of reference moves with the fluid velocity at AA and we restrict our consideration to the immediate vicinity of AA, we may postulate the linear velocity distribution shown in the blowup in Fig. 8. Thus the velocities above AA are posi-

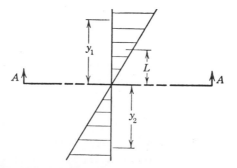

Fig. 8 Kinetic interpretation of coefficient of viscosity.

tive and those below are negative. A molecule originating at y_1 and moving downward through AA will carry with it a positive momentum $m(du/dy)y_1$. Similarly,
a molecule moving upward through AA and originating at y_2 will carry with it a
negative momentum $m(du/dy)y_2$. Both these excursions represent a transfer of
ordered momentum across AA, and it is the sum of such losses that occur in 1 sec
through unit area of AA that equals the shear stress τ.

As in the pressure calculation of Section 1.3, the random molecular motion is
assumed to be split equally among the three coordinate directions. Then, if there
are N molecules per unit volume, if their average speed is c, and if one-third of them
have a motion perpendicular to AA, $Nc/3$ molecules will pass through AA each
second. Each of these molecules will carry with it a momentum corresponding
to the position y at which it originates. The sum of these momenta is the shear
stress. There is an effective height at which all the molecules could originate with
the same resulting shear stress. If we call this height L, the shear stress becomes

$$\tau = \left(\frac{1}{3} Nc\right) m \left(\frac{du}{dy}\right) L$$

The product Nm is the density ρ; therefore

$$\tau = \frac{1}{3} \rho c L \frac{du}{dy} \qquad (11)$$

A comparison of Eqs. 9 and 11 indicates that

$$\mu = \tfrac{1}{3} \rho c L$$

The effective height L is related to the mean free path λ, and more accurate calculations show that

$$\mu = 0.49 \, \rho c \lambda \qquad (12)$$

As the density of a gas decreases, the mean free path increases such that the product $\rho\lambda$ remains nearly constant. Then μ is proportional to c, which in turn is proportional to the square root of the absolute temperature. Thus *the coefficient of
viscosity depends only on the temperature and is independent of the pressure.* This
deduction is a good first approximation for gases; deviations from experiment are
discussed in Chapter 16.

1.8 Analogy Between Viscosity and Other Transport Properties

The random molecular motion may transport properties of the gas other than the
ordered momentum. For example, if a gradient exists in the temperature, the temperature will become equalized by the random process. The transport of high-

temperature fluid to regions of lower temperature is interpreted as heat conduction. In general, let Q be a property of the gas. Then, by an argument similar to that leading to Eq. 11, we can write

$$\frac{d(Q/A)}{dt} = \frac{1}{3} \rho c L \frac{d(Q/M)}{ds}$$

where $d(Q/A)/dt$ is the property transported through unit area each second and Q/M is the property per unit mass. From the above formula, it is clear that all properties of the fluid that are subject to transport by the random molecular motion have a time rate of transport that is related to the viscosity by the equation

$$\frac{d(Q/A)}{dt} = \mu \frac{d(Q/M)}{ds} \tag{13}$$

For the case of heat conduction, let Q/M be the heat content of the gas per unit mass which may be written in terms of the specific heat c_p and temperature as*

$$\frac{Q}{M} = c_p T \tag{14}$$

From Fourier's law of heat conduction through a continuum we have

$$\frac{d(Q/A)}{dt} = k \frac{dT}{ds} \tag{15}$$

where dT/ds is the temperature gradient and k is the thermal conductivity. k is the proportionality constant that relates heat transfer to temperature gradient in the same manner that μ relates shearing stress to velocity gradient. Equation 15 is comparable to Eq. 9 and Figs. 7 and 8 illustrate Fourier's law if the velocity distributions are considered to be temperature distributions instead.

A comparison of Eqs. 13 and 15 leads to the result

$$\mu \frac{d(Q/M)}{ds} = k \frac{dT}{ds}$$

After using the value for Q/M from Eq. 14 and considering c_p to be constant, the following relation is obtained

$$\mu c_p = k \tag{16}$$

The measured value of the viscosity coefficient at standard temperature ($15°C$) is 1.78×10^{-5} N sec/m² (kg/msec).

Another parameter of importance in determining the flow characteristics is the **kinematic viscosity** $\nu = \mu/\rho$. Because the viscous forces are proportional to μ and the inertia forces are proportional to ρ, the kinematic viscosity may be interpreted

*The specific heats of a gas are defined in Chapter 8.

as a measure of the relative importance of viscous and inertia forces. That is, for two flows in which the velocity patterns are identical, the viscosity plays a relatively greater role in the fluid that has the greater kinematic viscosity. Under standard conditions the kinematic viscosity of air has a value of 1.45×10^{-5} m^2/sec. The variations of density, coefficient of viscosity, and kinematic viscosity with temperature for air and water are shown in Table 2.

1.9 Force on a Body Moving Through a Fluid

The force on a body arising from the motion of the body through a fluid depends on the properties of the body, the properties of the fluid, and the relative velocity between body and fluid. The size, shape, and orientation of the body are of consequence in determining the force arising from the relative motion. The fluid properties of importance are the density, viscosity, and elasticity, the last being determined by the density ρ and speed of sound a.

For bodies of given shape, we may describe the size by specifying a characteristic dimension l. Then, for a body of given shape and orientation moving through a flow with speed V, the force experienced may be written in the functional form

$$F = f(\rho, V, l, \mu, a)$$

or alternatively,

$$g(F, \rho, V, l, \mu, a) = 0 \qquad (17)$$

Equation 17 states a relation among physical quantities, and therefore its form is partially dictated by the dimensions of the parameters involved. The method of **dimensional analysis*** shows that Eq. 17 can always be written in the equivalent form

$$f_1 \left(\frac{F}{\rho V^2 l^2}, \frac{\rho V l}{\mu}, \frac{V}{a} \right) = 0$$

where each of the three combinations of parameters is a dimensionless quantity. The above equation may be solved for the first dimensionless combination, and then we have the fundamental relation

$$\frac{F}{\rho V^2 l^2} = g_1 \left(\frac{\rho V l}{\mu}, \frac{V}{a} \right) \qquad (18)$$

To interpret Eq. 18, consider a body of a given shape and orientation in motion in a fluid such that the quantities $F/(\rho V^2 l^2)$, $\rho V l/\mu$, and V/a have certain definite values. Then, if a geometrically similar body with the same orientation is moved

*The method of dimensional analysis and its application to the present problem may be found in Appendix A.

through the same or another fluid such that $\rho Vl/\mu$ and V/a have the same values as for the first body, then $F/(\rho V^2 l^2)$ will also have the same value.

Assume for the moment that μ and a in Eq. 17 have no influence on the force F. Then an application of dimensional analysis will lead to the result

$$f_1\left(\frac{F}{\rho V^2 l^2}\right) = 0$$

the solution of which is

$$F = C_F \rho V^2 l^2 \qquad (19)$$

where C_F is a dimensionless constant. Equation 19 states that, for a body of given orientation and shape in motion through a fluid, the force experienced is proportional to the *kinetic energy* per unit volume of the relative motion of the fluid $\rho V^2/2$ and to a characteristic area l^2. For example, if the force on an airplane of given shape, orientation, and size is known at a given flight speed and altitude, the force on another airplane of geometrically similar shape at the same orientation and flying at a different speed and altitude can be predicted from Eq. 19. This result was given by Isaac Newton (1642–1727), and the dimensionless constant C_F is sometimes referred to as the Newtonian coefficient. C_F is a dimensionless quantity that characterizes the force and in the following chapters will be called the **force coefficient**. The force coefficient is of great importance in experimental aerodynamics, for it makes possible the prediction of forces on full-scale airplanes at various altitudes and flight speeds from data obtained on models tested in wind tunnels.

Newton's result, Eq. 19, is an approximation that is accurate only under specialized conditions to be described later. The viscosity and elasticity of the fluid are important in general, and Eq. 18 shows that the force coefficient is not a constant for a body of given shape and orientation. It is a function of the combinations $\rho Vl/\mu$ and V/a, which bears the names **Reynolds number** and **Mach number**, respectively, after Osborne Reynolds (1842–1912) and Ernst Mach (1838–1916) who investigated the effects of these parameters in flow problems.

Dimensional analysis has shown that the force coefficient for a body of given orientation and shape is a function of the Reynolds number and the Mach number. The accuracy of this result depends entirely on the correctness with which the parameters governing the force are initially chosen. *If important properties of the flow are omitted from the initial choice of parameters, the method of dimensional analysis will not expose this fact.* For example, properties that were neglected and that influence the force on the body are surface roughness, turbulence of the stream, the presence of other bodies in the vicinity, and so forth. In applying data from model tests to the full-scale airplane, these facts must be considered.

Finally, if the geometries of the two flows are similar (geometric similarity) and if the Mach numbers are equal and the Reynolds numbers are equal, the flows are said to be *dynamically similar*. Dynamically similar flows have *equal* force coefficients. The Mach and Reynolds numbers are called **similarity parameters.**

1.10 The Approximate Formulation of Flow Problems

Strictly speaking, a gas is a compressible, viscous, inhomogeneous substance, and the physical principles underlying its behavior are not understood well enough to permit us to formulate, exactly, *any* flow problem. Even if this were possible, the resulting equations would, at present, be too difficult to solve. Therefore, all methods used are approximations, and the answers they yield must be tested by experiment.

In order to render the problems of aerodynamics tractable, we consider three different fluids, each of which provides a good approximation for airflow problems of particular types.

1. *Perfect Fluid.*

This fluid is homogeneous (not composed of discrete particles), inelastic, and inviscid, corresponding to zero Mach number and infinite Reynolds number. The assumption of a perfect fluid gives good agreement with experiment for flows outside of the boundary layer and wake of well-streamlined bodies moving with velocities of less than about 400 km/hr (111 m/sec) at altitudes under about 30 km. The scale effect is neglected for problems treated by perfect-fluid theory, and the force coefficient given by Eq. 19 is constant.

2. *Compressible, Inviscid Fluid.*

This fluid differs from the perfect fluid in that the elasticity, characterized by the speed of sound, is taken into account (nonzero Mach number, infinite Reynolds number). It provides a good approximation for problems involving the flow outside the boundary layer and wake of bodies at high Reynolds numbers at altitudes below about 30 km.

3. *Viscous, Compressible Fluid.*

This fluid differs from that described under (2) in that the viscosity is taken into account (nonzero Mach number, finite Reynolds number). Although it is not feasible to treat the entire flow around a body, that within the boundary layer and wake is amenable to accurate analysis, provided the flow is *laminar; turbulent* flow has so far yielded only to semiempirical analyses. The agreement of the analyses with experiment is good for all speeds at altitudes below about 30 km.

At altitudes above about 60 km, the mean free path of the molecules will generally not be small compared with a significant dimension of the body. Then, as the altitude increases further, the characterization of air as a fluid becomes more and more approximate. Finally, at altitudes above approximately 150 km, the flow (if it can be termed a flow) consists simply of the collision of the body with those molecules directly in its path.

1.11   Plan for the Following Chapters

The objective as stated in Section 1.1 may, in view of the intervening discussion, be rephrased as follows: to provide a background of concepts for finding approximate solutions of problems in the flow of a compressible, viscous, inhomogeneous gas. The approximations that may be made depend on the particular aspect of the flow being investigated. It has, for instance, been abundantly demonstrated experimentally that the **lift** and **moment** acting on aircraft at flight speeds under about 400 km/hr (250 mph) are very slightly affected by compressibility and viscosity. Therefore, the first six chapters of this book are devoted to a study of the flow of a perfect fluid and to the application of perfect-fluid theory to the prediction of the lifting characteristics of wings.

Chapters 7 to 13 deal with the compressible inviscid fluid and its application to the flow through channels and about wings. Both the subsonic and supersonic flow problems are formulated, and the basic differences between the two regimes are discussed in physical terms. In some cases, exact solutions of the equations are given; in others, useful approximations that neglect the nonlinear terms are introduced.

The effects of viscosity on the flow of incompressible and compressible fluids are taken up in Chapters 14 through 19. Here we are concerned with the flow in boundary layers and in tubes. The main objectives are to give an understanding of the approximations that have been made and to analyze the problems of viscous drag and flow separation.

Two appendices, tables, problems and a bibliography appear at the end of the book. In Appendix A, a brief treatment of dimensional analysis is given. The equations of motion and energy for a viscous compressible fluid are given in Appendix B and the reduction to the boundary layer equations is carried out.

CHAPTER 2

Kinematics of a Flow Field

2.1 Introduction—Fields

In this and the following chapter we develop equations that describe the properties of incompressible, inviscid flows as regards (1) **field properties,** that is, the velocities, pressures, temperatures, and the like, at any point in space, and (2) **particle properties,** that is, the variations of those properties experienced by a fluid element as it moves along its path in the flow. In Chapter 2 we restrict ourselves to the *kinematic* properties, that is, those that follow from the indestructibility of matter. The *dynamic* properties, those that follow from the application of Newton's laws to the individual (tagged) fluid elements, are discussed in Chapter 3; the *dynamics* and *thermodynamic* properties of compressible viscous and inviscid flows are reserved for later chapters.

The term "**field**" denotes a region throughout which a given quantity is a function of the coordinates within the region and of time. Examples are pressure, density, and temperaure fields, within which these properties can be represented as functions of x, y, z, and t. These are **scalar fields,** in that the magnitude of the quantity at any point at a given instant of time is represented completely by a single number. **Vector fields,** on the other hand, such as velocity, momentum, or force fields, each require at every point three numbers for their complete description; that is, the vector field results from the superposition of three scalar fields. If the quantity is independent of time, the field is *steady* or *stationary*.

The empirical laws on which fluid mechanics is based are applicable to **fluid elements** of fixed identity, that is, to small regions that still comprise an inordinately large number of *tagged* molecules. As any given element moves through a flow field, along a path determined by the force field, the principle of conservation of mass states that its mass remains constant. This indestructibility of matter is thus a **particle property.** Other conservation laws, such as momentum and energy, taken up later also apply to particle properties. Then, the application of these conservation laws to the particle properties defines a flow field in which each fluid element moves in conformity with these laws.

In most problems, rather than solve for the particle properties, it is more convenient to derive the **field properties,** that is, the flow properties as functions of the space coordinates at any given instant. When the conservation laws are applied, we show that the particle properties so derived determine the field properties.

2.2 Scalar Field, Directional Derivative, Gradient

In the previous section a scalar field was defined as one in which a property, such as temperature, pressure, and the like, at a given point in space at a given instant is represented completely by a single number. We will abbreviate the analyses by postulating that the field does not vary with time, that is, it is *steady*. Then the property Q is defined in Cartesian coordinates by

$$Q = Q(x, y, z)$$

We thus describe scalars as single-valued functions of position. They and their derivatives are also generally continuous so that a Taylor expansion can be used to relate magnitude of Q at neighboring points A and B in terms of Q_A, the value at A. Thus

$$Q_B = Q_A + \left(\frac{\partial Q}{\partial x}\right)_A (x_B - x_A) + \left(\frac{\partial Q}{\partial y}\right)_A (y_B - y_A) + \left(\frac{\partial Q}{\partial z}\right)_A (z_B - z_A)$$

$$+ \left(\frac{\partial^2 Q}{\partial x^2}\right)_A \frac{(x_B - x_A)^2}{2!} + \dots$$

Let $B \to A$, then $(Q_B - Q_A) \to dQ$, $(x_B - x_A) \to dx$, and so on, and in the limit the terms multiplying the higher derivatives vanish and the equation becomes

$$dQ = \left(\frac{\partial Q}{\partial x}\right) dx + \left(\frac{\partial Q}{\partial y}\right) dy + \left(\frac{\partial Q}{\partial z}\right) dz \tag{1}$$

This equation can now be extended to define the *directional derivative*, that is, the rate of change of Q with respect to distance in a specified direction. We may, for instance, specify this direction as that along a tangent to a curve in space at any given point. We define a variable s as the distance measured along the curve from an arbitrary point. Then the coordinates x, y, and z and their differentials at a point on the curve are given by

$$x = x(s) \quad \text{and} \quad dx = \left(\frac{dx}{ds}\right) ds$$

$$y = y(s) \quad \text{and} \quad dy = \left(\frac{dy}{ds}\right) ds$$

$$z = z(s) \quad \text{and} \quad dz = \left(\frac{dz}{ds}\right) ds$$

Since Q is a function of x, y, and z and therefore of the new variable s, substitution into Eq. 1 yields the derivative of Q along the tangent to the curve at any s:

$$\frac{dQ}{ds} = \frac{\partial Q}{\partial x}\frac{dx}{ds} + \frac{\partial Q}{\partial y}\frac{dy}{ds} + \frac{\partial Q}{\partial z}\frac{dz}{ds} \tag{2}$$

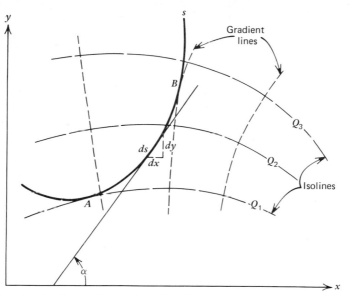

Fig. 1 Directional derivative; isolines and gradient lines.

For geometrical simplicity the problem is reduced in two dimensions, as illustrated in Fig. 1. The curve labeled s is shown, along with *isolines*, that is, curves on which Q is constant, and *gradient lines* which are everywhere in the direction of the greatest rate of change of Q. The geometrical derivatives in Eq. 2 are, by reference to Fig. 1,

$$\frac{dx}{ds} = \cos \alpha; \qquad \frac{dy}{ds} = \sin \alpha$$

so that

$$\frac{dQ}{ds} = \frac{\partial Q}{\partial x} \cos \alpha + \frac{\partial Q}{\partial y} \sin \alpha \tag{3}$$

We can see from Fig. 1 that the distribution of a scalar throughout a two-dimensional region is defined by a family of isolines or by a family of gradient lines. It is therefore important that we establish the relationship between them. At the point A the tangent to s is along an isoline, so that $(dQ/ds)_A = 0$ and, designating the direction of the isoline by α_i, Eq. 3 gives

$$\tan \alpha_i = - \frac{\partial Q/\partial x}{\partial Q/\partial y} \tag{4}$$

To find α_g, the directions of the gradient line, we observe in Fig. 1 that at B the tangent to s is along a gradient line and hence dQ/ds is a maximum at that point. Therefore α_g is the value of α for which the derivative of Eq. 3 with respect to α

vanishes. Thus

$$\frac{d}{d\alpha}\left(\frac{dQ}{ds}\right) = 0 = -\frac{\partial Q}{\partial x}\sin \alpha_g + \frac{\partial Q}{\partial y}\cos \alpha_g$$

$$\tan \alpha_g = +\frac{\partial Q/\partial y}{\partial Q/\partial x} \qquad (5)$$

Since the right-hand sides of Eqs. 4 and 5 are negative reciprocals, *gradient lines and isolines are everywhere orthogonal*.

Example 1

The equations of the two families of lines are related by Eqs. 4 and 5. For instance, assume that the scalar is distributed according to the equation

$$Q = xy \qquad (A)$$

That is, the isolines are members of a family of rectangular hyperbolas asymptotic to the x and y axes, each one corresponding to a different value of Q. Then $\partial Q/\partial x = y$ and $\partial Q/\partial y = x$, so that, by Eq. 5, the slope of the gradient line is given by

$$\tan \alpha_g = \frac{dy}{dx} = \frac{x}{y}$$

and the equation of the gradient line is the solution of the differential equation

$$ydy - xdx = 0$$

which integrates to

$$y^2 - x^2 = \text{const.} \qquad (B)$$

This equation also represents a family of rectangular hyperbolas, orthogonal to those of Eq. A, and asymptotic to axes at $45°$ to the x and y axes.

If we were given instead the equation for the gradient lines, Eq. B, we would differentiate and take the negative reciprocal to find the differential equation

$$xdy + ydx = 0$$

for the isolines. The integral is Eq. A. We show later that the hyperbolas of Eq. A trace the approximate paths of fluid elements in the immediate neighborhood of a "stagnation point," that is, for instance, the point on an airfoil where the incident flow divides.

Example 2

Consider a long heated wire along the z axis in a stagnant fluid and neglect gravity forces. Then convection currents would be absent and the temperature T would be constant on circular cylindrical surfaces. We show later that conservation of energy requires that

$$T = \frac{A}{r} = \frac{A}{\sqrt{x^2 + y^2}} \tag{A}$$

where A = constant. Then $\partial T/\partial x = -Ax/(x^2 + y^2)^{3/2}$ and $\partial T/\partial y = -Ay/(x^2 + y^2)^{3/2}$ and, by Eq. 5, the differential equation for the gradient lines is $dy/y = dx/x$, the solution of which is $y = kx$ where k is a constant. The gradient lines are therefore radial lines through the origin, orthogonal to the circular isolines.

It follows from Eqs. 4 and 5, and is evident in the two examples, that since the two families are orthogonal, *their designation as isolines or gradient lines are interchangeable*. Thus, in Example 1, if Eq. B describes the isolines, Eq. A describes the gradient lines; in Example 2, if the planes $y = kx$ are specified as isotherms the gradient lines are concentric circles.

2.3 Vector Field—Method of Description

It was shown in the previous section that the gradient of a scalar at a given point is a vector, tangent to the gradient line at that point. Its magnitude is the absolute value of the maximum directional derivative at the point. Thus, to be specific, let us consider Q in Eqs. 3 and 5 to be the pressure p in a flow field and substitute Eq. 5 into Eq. 3 to find

$$\left(\frac{dp}{ds}\right)_{max} = \sqrt{\left(\frac{\partial p}{\partial x}\right)^2 + \left(\frac{\partial p}{\partial y}\right)^2}$$

The vector is designated as grad p or ∇p and, written in vector notation in three dimensions, is

$$\text{grad } p = \mathbf{i}\frac{\partial p}{\partial x} + \mathbf{j}\frac{\partial p}{\partial y} + \mathbf{k}\frac{\partial p}{\partial z} \tag{6}$$

where $\mathbf{i}, \mathbf{j}, \mathbf{k}$ are, respectively, the unit vectors in the x, y, z directions; the vector has the magnitude

$$|\text{grad } p| = \sqrt{\left(\frac{\partial p}{\partial x}\right)^2 + \left(\frac{\partial p}{\partial y}\right)^2 + \left(\frac{\partial p}{\partial z}\right)^2} \tag{7}$$

A velocity field, in which u, v, w are, respectively, the x, y, z components of the velocity is represented in vector notation by

$$V = u\mathbf{i} + v\mathbf{j} + w\mathbf{k}$$

where u, v, and w are each scalar functions of x, y, z and time t. They are field properties; that is, they represent the velocity components of the fluid element passing through a given point (x, y, z) at a given instant of time; at a later instant the element will have moved to a new position and its velocity will correspond to the field position it occupies at the later instant. References to Eq. 6 indicate that we can calculate the gradients of u and v for specific flows.

In two-dimensional polar coordinates we may write

$$V = \mathbf{e_r}\,u_r + \mathbf{e_\theta}\,u_\theta$$

$$\text{grad } u_r = \mathbf{e_r}\,\frac{\partial u_r}{\partial_r} + \mathbf{e_\theta}\,\frac{1}{r}\frac{\partial u_r}{\partial\theta}; \qquad \text{grad } u_\theta = \mathbf{e_r}\,\frac{\partial u_r}{\partial r} + \mathbf{e_\theta}\,\frac{1}{r}\frac{\partial u_\theta}{\partial\theta}$$

where $\mathbf{e_r}$, u_r and $\mathbf{e_\theta}$, u_θ are, respectively, the unit vectors and velocity components in the r and θ directions.

A streamline is defined as a path whose tangent at any point is in the direction of the velocity vector at that point. A velocity field may then alternatively be described in terms of equations of the streamlines and the absolute value of the velocity. Letting the absolute value of V be denoted by $|V|$, and representing a streamline by two intersecting surfaces, we may write the alternative description as

$$|V| = f(x, y, z); \qquad f_1(x, y, z) = 0; \qquad f_2(x, y, z) = 0$$

In two dimensions, two equations suffice to describe the field by either representation.

Example 1

Frequently, in the following material, something will be known about the streamlines and the magnitude of the velocity. For mathematical operations, however, the components are needed as functions of the field coordinates, and a conversion from the second method of representation to the first is necessary. In treating the two-dimensional case, the conversion is readily performed by remembering that v/u is the slope of a streamline and the magnitude of the velocity vector is $\sqrt{u^2 + v^2}$. These values provide two equations in two unknowns:

$$\frac{dy}{dx} = \frac{v}{u}; \qquad |V| = \sqrt{u^2 + v^2}$$

For example, assume it is known that the streamlines are circular and the magnitude of the velocity is constant on a given streamline. Then the equation of the streamline is

$$x^2 + y^2 = \text{constant}$$

and the magnitude of the velocity is

$$|\mathbf{V}| = f(x^2 + y^2)$$

From the first equation above,

$$\frac{dy}{dx} = -\frac{x}{y} = \frac{v}{u}$$

and from the second,

$$\sqrt{u^2 + v^2} = u \sqrt{1 + \left(\frac{v}{u}\right)^2} = f(x^2 + y^2)$$

Substituting $-x/y$ for v/u and solving for u,

$$u = -\frac{yf(x^2 + y^2)}{\sqrt{x^2 + y^2}} = -\frac{yf(r)}{r}$$

where $x^2 + y^2 = r^2$. Similarly,

$$v = +\frac{xf(x^2 + y^2)}{\sqrt{x^2 + y^2}} = +\frac{xf(r)}{r}$$

The symbol r in the above equations and in the material that follows represents the magnitude of the radius vector in polar coordinates. The signs on u and v must be opposite in order to satisfy the condition $v/u = -x/y$. The sense of $\mathbf{V}$, that is, the sign of $f(r)$, determines the component that bears the negative sign. Thus, either $u > 0$, $v < 0$, or $u < 0$, $v > 0$ satisfy the conditions postulated for the flow, that is, that the streamlines are circular and $|\mathbf{V}| = f(r)$.

This solution becomes particularly concise in polar coordinates, where the velocity components are, respectively, the radial and circumferential components u_r and u_θ. Since the streamlines are circles, their equation is $r = $ constant and therefore the radial velocity component vanishes; it follows that the flow is defined completely by $u_r = 0$, $u_\theta = f(r)$. The Cartesian components given above are simply the components of u_θ in the x and y directions.

Each component of the velocity is a continuous single-valued function of the field coordinates. As in Eq. 1, the values of the components at any point can be expressed in terms of their values at a neighboring point and nine first-order derivatives.

The properties of the flow field are thus determined by the values of nine derivatives throughout the field. These are:

$$\partial u/\partial x \qquad \partial u/\partial y \qquad \partial u/\partial z$$
$$\partial v/\partial x \qquad \partial v/\partial y \qquad \partial v/\partial z$$
$$\partial w/\partial x \qquad \partial w/\partial y \qquad \partial w/\partial z$$

It will be shown in the following sections that the nine derivatives are not generally independent; that is, the physics of the problems require definite relations among some of the derivatives.

2.4   Divergence of a Vector, Theorem of Gauss

The three derivatives on the diagonal extending from the upper left-hand corner to the lower right-hand corner of the matrix have a property in common. Each represents the rate of change of a component of the velocity in the direction of that component. The sum of these extension derivatives is called the *divergence of the velocity* and is written

$$\text{div } \mathbf{V} = \frac{\partial u}{\partial x} + \frac{\partial v}{\partial y} + \frac{\partial w}{\partial z} \tag{8}$$

The divergence of the velocity vector can be interpreted physically as the efflux per unit volume from a point. To see this, consider the velocity vector $\mathbf{V}$, which is a field property, and let $\hat{R}$ (Fig. 2) be a fixed region in the field, called a *control volume*, and $\hat{S}$ an imaginary *control surface* enclosing the control volume. At any instant of time, a fluid particle in the control volume has the field velocity at the point it occupies at that instant. In Fig. 2, the control volume $\hat{R}$ has been oriented

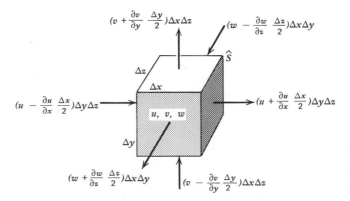

Fig. 2 Velocity components normal to faces of a cubical fluid element.

for convenience with its edges parallel to the x, y, z axes; the velocity components at its center are indicated by u, v, and w. To find the volume of fluid flowing per second, or the volume flux, through the control surface enclosing $\hat{R}$, field values of the velocity components normal to each of the six faces of the control surface have been indicated on the figure; it is evident that only those velocity components normal to the surface contribute to a flux through the surface. The magnitudes of the normal components have been derived from their values at the center of the control volume by the argument of Section 2.2. For example, the value of the x component of the velocity on an x face of the cube, that is, on one of the faces normal to the x axis, differs from that at the center of the control volume by an increment $(\partial u/\partial x)\,(\Delta x/2)$; it will be remembered from the discussion on Section 2.2 that this is an exact statement when Δx is vanishingly small. Consequently, for a control volume of infinitesimal proportions, the average value of the velocity component on the right-hand x face is $u + (\partial u/\partial x)\,(\Delta x/2)$. Note that $\partial u/\partial x$ has the dimensions of t^{-1} and, since the area of an x face is $\Delta y \Delta z$, the rate of volume flow *out* of the right-hand x face is $[u + (\partial u/\partial x)\,\Delta x/2]\,\Delta y\Delta x$; that *into* the left-hand face is $[u - (\partial u/\partial x)\,\Delta x/2]\,\Delta y\Delta x$.   Thus the net outflow from the x faces is $(\partial u/\partial x)\,\Delta x\Delta y\Delta z$ per unit time.

The flux of volume from a point is, by definition,

$$\lim_{\hat{R} \to 0} \frac{\text{volume outflow/sec} - \text{volume inflow/sec}}{\hat{R}}$$

With the aid of Fig. 2, the above expression becomes $\partial u/\partial x + \partial v/\partial y + \partial w/\partial z$, for the net outflow per second from the three pairs of faces of the cube; this is precisely the definition of div $\mathbf{V}$ given in Eq. 8. It is a scalar and its magnitude is the rate at which fluid volume is leaving a point per unit volume, that is, the *volume flux from the point*.

In a similar manner the expression for divergence of a vector can be derived in other orthogonal coordinate systems. Figure 3 shows a control volume in cylindrical polar coordinates, r, θ, z.

It can be shown (Problem 1) that

$$\text{div } \mathbf{V} = \frac{\partial u_r}{\partial r} + \frac{1}{r}\left(u_r + \frac{\partial u_\theta}{\partial \theta}\right) + \frac{\partial u_z}{\partial z} \tag{9}$$

where u_r, u_θ, and u_z are, respectively, the radial, circumferential, and axial velocity components.

We return to the consideration of flow through a finite control volume for the purpose of deriving an important integral relation and thus gaining further insight into the meaning of divergence. We consider a finite control volume $\hat{R}$ as shown in Fig. 4. The region $\hat{R}$ is divided into elementary cubes and div $\mathbf{V}$ is found for each cube. The flux through common faces of adjacent cubes cancel because the inflow through one face is equal to the outflow through the other. Now, if we sum the

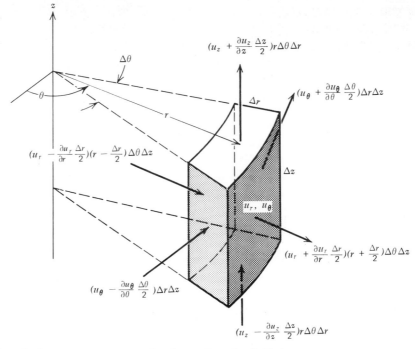

Fig. 3 Velocity components normal to faces of a polar fluid element.

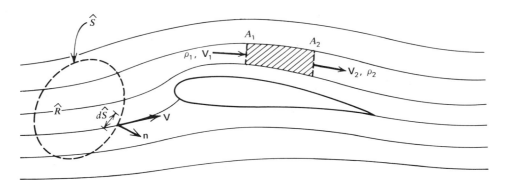

Fig. 4 Control volumes.

fluxes for all cubes, only the flow rate through the faces of the surface enclosing the entire region will contribute to the summation. In the limit, therefore, as the cubes approach zero volume the integral of div **V** over $\hat{R}$ is equal to the net flux through $\hat{S}$, the surface enclosing $\hat{R}$, that is,

$$\iiint_{\hat{R}} \text{div } \mathbf{V} \, d\hat{R} = \iint_{\hat{S}} \mathbf{V} \cdot \mathbf{n} d\hat{S} \tag{10}$$

This equation is a specialized form of an integral theorem applying to vector fields in general. It is useful in many fields of technology and is referred to as the *divergence theorem*; it is generally attributed to Johann Gauss (1777–1855).

In the next section we treat the conservation of matter applied to fluid flows. All the analysis by which we identified div **V** in the present section as the flux of volume at a point carries over to the flux of mass by substituting ρ**V** for **V** in Eq. 10, which then when set equal to zero simply states that there can be no accumulation of mass in a steady flow.

2.5   Conservation of Mass and the Equation of Continuity

One of the empirical laws that forms the basis of "Newtonian mechanics" states that mass can neither be created nor destroyed. This conservation principle applies to fluid elements of fixed identity, and, in order to express the principle in terms of field properties of a flow, we apply the following physical reasoning.

Consider a control volume $\hat{R}$ fixed in the field, as shown in Fig. 4. If more mass flows out of than into $\hat{R}$ each second, the mass within $\hat{R}$ must be decreasing. Specifically, *the net efflux of mass through $\hat{S}$ is equal to the time rate of decrease of mass within $\hat{R}$.* Through any incremental area $d\hat{S}$ of the control surface the mass flow per unit time is ρ**V** · **n**$d\hat{S}$ where **n** is the unit vector normal to $d\hat{S}$; *and* **n** *is counted positive when directed outward.* Then the mass flux through the control surface is, by application of Eq. 10,

$$\iint_{\hat{S}} \rho \mathbf{V} \cdot \mathbf{n} d\hat{S} = \iiint_{\hat{R}} \text{div } \rho \mathbf{V} d\hat{R} \tag{10a}$$

where ρ and **V** are field properties. The total mass within $\hat{R}$ is

$$\iiint_{\hat{R}} \rho d\hat{R}$$

where $\rho d\hat{R}$ is the infinitesimal contribution of the mass at a given point to the total within the region $\hat{R}$.

The conservation of mass principle stated in terms of field properties is then

$$\iiint_{\hat{R}} \text{div } \rho \mathbf{V} d\hat{R} = - \iiint_{\hat{R}} \frac{\partial \rho}{\partial t} d\hat{R} \tag{11}$$

This equation states simply that the net rate at which mass flows through the boundaries of a closed region must equal the rate at which the mass within the region is decreasing.

As a simple application of the above principle, consider the steady flow of a fluid through the shaded stream-tube shown in Fig. 4. The control volume is contained within the two neighboring streamlines and the dotted lines. There is no flux through the streamlines because they are everywhere tangent to the local fluid velocities, and it is assumed that ρ_1 and V_1 are average values across the area A_1, and ρ_2 and V_2 are average values across A_2. Because the flow is steady, the right-hand side of Eq. 11 is zero and the conservation of mass principle becomes

$$\rho_2 V_2 A_2 - \rho_1 V_1 A_1 = 0$$

The "equation of continuity" is a statement of the conservation of mass principle in terms of field properties at a point. It follows simply from Eq. 11 written in the form

$$\iiint_{\hat{R}} \left(\text{div } \rho\mathbf{V} + \frac{\partial \rho}{\partial t} \right) d\hat{R} = 0 \tag{12}$$

The interchange of the order of differentiation and integration in Eq. 11 is permissible because the limits of the space integration are independent of time.

Equation 12 must hold for all control volumes regardless of size, and therefore the integrand must be identically zero. Then the *equation of continuity*, applied to conditions at a point, becomes

$$\text{div } \rho\mathbf{V} = - \frac{\partial \rho}{\partial t} \tag{13}$$

This equation is satisfied in a *physically possible flow*, that is, a flow in which mass is conserved. When the flow is *steady* the field properties are not functions of time, and continuity reduces to

$$\text{div } \rho\mathbf{V} = 0 \tag{14}$$

If the fluid is incompressible, ρ is a constant and may be divided out of the equation. Continuity becomes

$$\text{div } \mathbf{V} = 0 \tag{15}$$

A given velocity field is *physically possible* only if continuity, as defined by the one of these equations appropriate to the flow being considered, is satisfied.

The most elementary example of a physically possible flow is one in which the fluid velocity is constant everywhere. There, regardless of the orientation of the x, y, z axes, the separate terms in div $\mathbf{V}$ vanish identically, and the continuity is satisfied for uniform incompressible flow.

Another example occurs in the incompressible flow in which the streamlines radiate from the origin; for simplicity we assume that the radial velocity u_r depends only on the distance r from the origin, and we determine its variation such that continuity is satisfied.

For the two-dimensional problem the formula for div $\mathbf{V}$ is given in Eq. 9, and since $u_\theta = 0$ everywhere the condition for continuity, Eq. 15, becomes

$$\frac{du_r}{dr} + \frac{u_r}{r} = 0 \tag{16}$$

and the solution is $ru_r = $ constant $\equiv k$, that is,

$$u_r = \frac{k}{r}$$

The constant k is evaluated by writing $\Lambda = 2\pi r u_r = 2\pi k$ for the rate of volume flow through the circle of radius r. Then $k = \Lambda/2\pi$ and the two-dimensional flow radiating uniformly in all directions from the origin and satisfying continuity is described by

$$u_r = \frac{\Lambda}{2\pi r}; \qquad u_\theta = 0 \tag{17}$$

These equations describe a *source flow of strength* Λ, $\Lambda > 0$ for flow outward, $\Lambda < 0$ for flow inward. Λ has dimensions of length^2/time.

The same result will be obtained by use of Eq. 11 for a steady flow if we choose $\hat{R}$ as the annular region shown in Fig. 5, bounded by circular cylinders of radii r_1 and r_2, with $r_2 > r_1$. For flow outward, fluid volume enters unit length of cylinder r_1 at the rate $2\pi r_1 u_{r_1}$ and leaves unit length of cylinder r_2 at the rate $2\pi r_2 u_{r_2}$. The surface $\hat{S}$ enclosing $\hat{R}$ comprises the surfaces of the two cylinders and a connecting "corridor" of infinitesimal width, as indicated in Fig. 5. We arbitrarily choose the positive direction of integration so that the region $\hat{R}$ is always on the left. Then, since the outward drawn normal is positive, $\mathbf{V} \cdot \mathbf{n} = -u_{r_1}$ and $+u_{r_2}$, respectively, on the inner and outer surfaces and $\mathbf{V} \cdot \mathbf{n}$ vanishes on the boundaries of the corridor. Then Eq. 11 becomes

$$2\pi r_2 u_{r_2} - 2\pi r_1 u_{r_1} = 0 \tag{18}$$

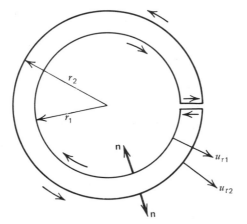

Fig. 5 Control surface for volume flux from a source.

Since r_1 and r_2 are arbitrary, we may let r_2 approach r_1. Then $r_2 = r_1 + \Delta r$ and $u_{r_2} = u_{r_1} + \Delta u_r$ and the equation becomes

$$(r_1 + \Delta r)(u_{r_1} + \Delta u_r) - r_1 u_{r_1} = 0$$

as $\Delta r \to 0$, $\Delta u_r \to 0$, and we obtain Eq. 16, which integrates to Eqs. 17.

We note that in the preceding analyses the region $\hat{R}$ excluded the origin. Now we take the region $\hat{R}$ as *including* the origin and proceed to evaluate div **V** at $r = 0$, where Eqs. 17 indicate that u_r becomes infinite. To this end: since div **V** vanishes everywhere except at $r = 0$, the volume integral on the left-hand side of Eq. 10 becomes in the limit the product of div **V** and a volume $\Delta \hat{R}$ which approaches zero as $r \to 0$; therefore, since both integrals in the equation must be equal to the constant efflux represented by the right-hand side of Eq. 10,

$$(\text{div } \mathbf{V})_{r=0} = \lim_{\Delta \hat{R} \to 0} \frac{\oint \mathbf{V} \cdot \mathbf{n} d\hat{S}}{\Delta \hat{R}} = \infty \tag{19}$$

A source flow is thus physically possible everywhere except at the center where div **V** $= \infty$.

We discuss separately these elementary flows, such as uniform flow, sources, sinks, and vortices (see Section 2.9), preliminary to demonstrating that they are "building blocks" for more complicated flows. We show later the methods by which the details of the flow around, and the fluid pressures acting on, a body of a given shape in motion through a fluid can be determined by adding the contributions from these building blocks, whose respective strengths can be determined from the body shape.

In the following sections, quantitative methods are developed for describing these flows to facilitate their application to practical problems.

2.6   Stream Function in Two-Dimensional Incompressible Flow

As a consequence of the conservation of mass, it is possible to define uniquely a function of the field coordinates from which the velocity in both magnitude and direction is derivable for two-dimensional incompressible flow. In Fig. 6, let *ab* and *cd* represent streamlines in a two-dimensional incompressible flow.* *A streamline is a curve whose tangent at every point conicides with the direction of the velocity vector.* Therefore, no fluid can cross *cd* or *ab*. Incompressible flow has been assumed, the density of the fluid within the boundary *efgh* does not vary with time, and, therefore, the same mass of fluid must cross *ef* per unit time as crosses

---

*Such a function can be found in three dimensions for flows with axial symmetry, for example, the flow about a body of revolution (see Karamcheti, 1966). For a two-dimensional compressible flow, the stream function is derived in Section 7.2.

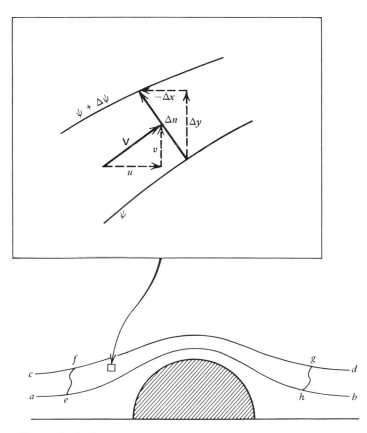

Fig. 6 Streamlines and velocity components.

hg, or any other path connecting the streamlines. If the streamline *ab* is arbitrarily chosen as a base, every other streamline in the field can be identified by assigning to it a number ψ, called the *stream function; its magnitude is the volume of fluid passing per second between an arbitrary base streamline and the one passing through the given point, per unit distance normal to the plane of the motion.* Isolines of ψ are the streamlines; the isoline $\psi = 0$ is chosen as the base streamline.

The fluid velocity can be derived from a stream function from the following considerations. Let Δn in Fig. 6 be the incremental length of a normal between two neighboring streamlines ψ and $\psi + \Delta\psi$. The streamline and therefore **V** are normal to Δn and, according to the definition of stream function, the rate of volume flow is $\Delta\psi \simeq V\Delta n$, where V is the magnitude of **V**. Then

$$V = \lim_{\Delta n \to 0} \frac{\Delta\psi}{\Delta n} = \frac{\partial\psi}{\partial n} \tag{20}$$

It states that the magnitude of the fluid velocity at a point is obtained by differentiating the stream function in the direction normal to the velocity.

If **V** is decomposed into u and v, and Δn into $-\Delta x$ and Δy as shown in Fig. 6, we may write

$$\Delta \psi = u\Delta y - v\Delta x$$

or, after taking the limit,

$$d\psi = udy - vdx$$

Mathematically, we may also write

$$d\psi = \frac{\partial \psi}{\partial x} dx + \frac{\partial \psi}{\partial y} dy$$

Since these two equations are identical everywhere in the flow field, the coefficients of dx and dy must respectively be identical in the two equations. It therefore follows that

$$u = \frac{\partial \psi}{\partial y} ; \qquad v = - \frac{\partial \psi}{\partial x} \tag{21}$$

From Eqs. 20 and 21, a general rule can be drawn: *the velocity component in any direction is found by differentiating the stream function in the direction at right angles to the left of that direction.* It follows from Eqs. 21 that the dimensions of ψ are length^2/time.

In polar coordinates, in which the counterclockwise direction is positive for θ, the application of the above general rule gives

$$u_r = \frac{1}{r} \frac{\partial \psi}{\partial \theta} ; \qquad u_\theta = - \frac{\partial \psi}{\partial r} \tag{22}$$

The use of the stream function in describing a two-dimensional incompressible fluid flow becomes clear if we substitute Eqs. 21 or 22 for the terms in the continuity equation (div **V** in polar coordinates is given in Eq. 9)

$$\text{div } \mathbf{V} = \frac{\partial u}{\partial x} + \frac{\partial v}{\partial y} = \frac{\partial u_r}{\partial r} + \frac{1}{r} \left(u_r + \frac{\partial u_\theta}{\partial \theta} \right) = 0$$

We find that the equations are satisfied identically and therefore that *continuity is identically satisfied by the "existence" of a stream function*, that is, by the existence of an analytic expression ψ from which the velocity components can be found by partial differentiation according to the above rule.

Introduction of the stream function simplifies the mathematical description of a two-dimensional flow because the two velocity components can thereby be expressed in terms of the single variable ψ.

Example 1

For a uniform flow, Eqs. 21 give

$$u = \frac{\partial \psi}{\partial y} = A ; \qquad v = - \frac{\partial \psi}{\partial x} = B$$

where A and B are constants. Since the derivatives are partials with respect to y and x, respectively, we find two expressions for ψ by integrating the first equation, keeping x constant, and the second keeping y constant. Thus, except for integration constants that vanish in the derivatives,

$$\psi = Ay + f(x); \qquad \psi = -Bx + g(y)$$

where $f(x)$ is a function only of x and $g(y)$ only of y; they are evaluated from the condition that the two equations must give identical expressions for ψ. This identity condition requires that $f(x) = -Bx$ and $g(y) = Ay$. Thus

$$\psi = Ay - Bx$$

describes a uniform flow with velocity $|\mathbf{V}| = \sqrt{A^2 + B^2}$, inclined at an angle $\tan^{-1}(B/A)$ to the x axis.

In the same way, by the integration of Eqs. 22 for the source flow of Section 2.5 $[u_r = f(r), u_\theta = 0]$ we find

$$\psi = \frac{\Lambda}{2\pi}\theta$$

for the stream function; differentiation according to Eqs. 22 yields the velocity components of Eqs. 17:

$$u_r = \frac{\Lambda}{2\pi r}, \; u_\theta = 0$$

It follows that *the existence of a stream function is a necessary condition for a physically possible flow*, that is, one that satisfies mass conservation.

Example 2

A flow for which a stream function does *not* exist is described by the components $u = A$, $v = By$, where A and B are constants. If this flow is physically possible, we must find a function ψ, such that, in accordance with Eqs. 21, $\partial\psi/\partial x = -By$ and $\partial\psi/\partial y = A$. If such a function exists the integral of $\partial\psi/\partial x$ with y constant must be *identically* equal to the integral of $\partial\psi/\partial y$ with x constant. Thus

$$\psi = Ay + f(x) \equiv -Bxy + g(y)$$

Since no choice of $f(x)$ and $g(y)$ will satisfy this identity, we must conclude that the flow cannot be described by a stream function as defined in Eqs. 21; in other words, a stream function does not exist for the flow. Furthermore, it follows from

$$\mathrm{div}\; \mathbf{V} = \frac{\partial u}{\partial x} + \frac{\partial v}{\partial y} = B$$

that the flow would require the creation of B cubic meters of fluid per cubic meter of the flow field per second at every point of the field. Thus the assumed flow field is not physically possible and, as was shown above, it cannot be described by a stream function.

2.7 The Shear Derivatives—Rotation and Strain

We have so far identified the sum of the derivatives along the diagonal of the matrix in Section 2.3 as the divergence of the vector **V** and, physically, as the flux of fluid volume from a point. The six remaining derivatives, those *off* the diagonal, are distinguished by the fact that they represent rates of change of the velocity components in directions *normal* respectively to these components. These are called the *cross* or *shear derivatives;* this section describes the roles they play in identifying further physical properties of flows.

The derivative $\partial u/\partial y$, for instance, measures the rate at which the x component of the velocity changes with position normal to the xz plane. Consider the distortion of an element as a function of time as it transverses a flow in which u is the total velocity and $\partial u/\partial y$ is the only nonzero shear derivative. Figure 7 shows schematically the changes in shape of the element with time. These changes can be described in terms of (1) the rate of deformation or strain defined in terms of the rate of change of the included angles, and (2) the angular velocity or rate of rotation of the diagonal.

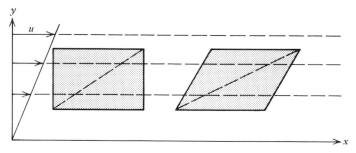

Fig. 7 Fluid element in a shear flow.

In a general three-dimensional flow the *coplanar pairs of shear derivatives* in each of three orthogonal planes determine three components of rate of strain and of angular velocity of an element. These coplanar sets are: $\partial u/\partial y$ and $\partial v/\partial x$ for the xy plane, $\partial u/\partial z$ and $\partial w/\partial x$ for the xz plane, $\partial v/\partial z$ and $\partial w/\partial y$ for the yz plane. Rate of strain and angular velocity are thus identified as vectors; the three components are designated respectively by subscripts indicating the axis normal to the planes on which they act.

For purposes of illustration we choose the xy plane and thus derive expressions

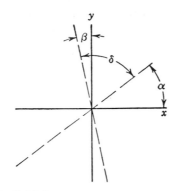

Fig. 8 Strain and rotation of a fluid element.

for the z components of rate of strain and angular velocity, respectively γ_z and ϵ_z; the corresponding vectors are designated $\boldsymbol{\gamma}$ and $\boldsymbol{\epsilon}$.

At a given instant, consider two needles of fluid particles intersecting orthogonally at a point in the field; these represent adjacent sides of an element as shown in Fig. 8. Attach the coordinate system to the point of intersection of the needles, and after an interval of time observe the positions of the two needles. In general, each will have rotated as shown by the dotted lines on Fig. 8. Counterclockwise rotation is taken as positive. The immediate neighborhood of a point is being considered, and, consequently, the needles in their final position have been shown on Fig. 8 as straight lines; it follows that the other two sides of the fluid element will rotate such that the cross section will remain a parallelogram.

The components of the rate of strain γ_z and the angular velocity ϵ_z are defined by the following equations:

$$\gamma_z = -\frac{d\delta}{dt} = -\frac{d}{dt}(90° - \alpha + \beta) = -\frac{d}{dt}(\beta - \alpha)$$

$$\epsilon_z = \frac{d}{dt}\left(\frac{\alpha + \beta}{2}\right) = \frac{1}{2}\frac{d}{dt}(\alpha + \beta)$$

(23)

The derivative $d\beta/dt$ is equal to $-\partial u/\partial y$. This may be seen from Fig. 9. The solid vertical line represents the needle at time t_1 and the dotted line represents the needle at time t_2. In the time interval $(t_2 - t_1)$ a fluid particle situated at a distance y from the intersection will have traveled a distance $-(\partial u/\partial y)y\,(t_2 - t_1)$, and the angle β is given by $-\tan^{-1}[(\partial u/\partial y)\cdot(t_2 - t_1)]$. The negative sign is needed because the positive sense for angular displacement is taken counterclockwise. The derivative becomes

$$\frac{d\beta}{dt} = \lim_{(t_2 - t_1)\to 0}\left(\frac{\beta}{t_2 - t_1}\right) = \lim_{(t_2 - t_1)\to 0}\frac{-\tan^{-1}[(\partial u/\partial y)\cdot(t_2 - t_1)]}{t_2 - t_1} = -\frac{\partial u}{\partial y}$$

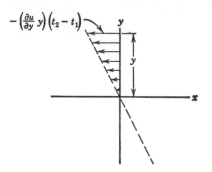

Fig. 9 Rate of angular deformation in terms of shearing derivative.

In a similar manner it can be shown that $d\alpha/dt = + \partial v/\partial x$. If these relations are used in Eqs. 23, strain and angular velocity in the xy plane become

$$\gamma_z = \frac{\partial v}{\partial x} + \frac{\partial u}{\partial y}$$

$$\epsilon_z = \frac{1}{2}\left(\frac{\partial v}{\partial x} - \frac{\partial u}{\partial y}\right) \equiv \frac{\omega_z}{2}$$

(24)

where ω_z is defined as the component of the **vorticity** normal to the x, y plane. The magnitude of the vorticity vector, **ω**, is twice that of the angular velocity of the fluid element, and it is directed along its axis of rotation. The factor of 2 is introduced for convenience in the analyses.

Example 1

These two flow properties can be illustrated in terms of the flow $u = ky$, $v = 0$, shown in Fig. 7. Here $\partial v/\partial x = 0$ so that $\gamma_z = \partial u/\partial y = k$ and $\omega_z = -\partial u/\partial y = -k$.

We make use in later chapters of the fact that the *shearing stress* in a two-dimensional viscous flow, such as a "boundary layer," is $\mu\gamma_z$, where μ is the coefficient of viscosity of the fluid. For inviscid flows, the quantity ϵ_z is more important, it represents the angular velocity of the element, specifically that of the diagonal.

It is instructive to consider the angular velocity as defined by Eqs. 24 in terms of "natural coordinates," that is, along and normal to the streamlines. Let s be a coordinate in the direction of the tangent to a streamline, and let n be a coordinate normal to s and lying in a plane of the streamline at the point in question. See Fig. 10.

In the interval $(t_2 - t_1)$, the tangent needle rotates by the amount θ, that is, by $(t_2 - t_1)$ (V/R). The amount the radial needle rotates depends on the difference in velocity between its upper and lower ends, that is, $(\partial V/\partial n)\Delta n$. The difference in displacement of the upper and lower ends is then $(\partial V/\partial n)\Delta n (t_2 - t_1)$. This ex-

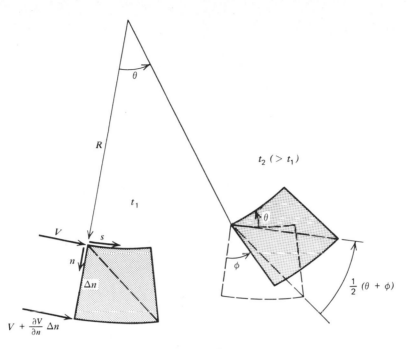

Fig. 10 Angular velocity in terms of natural coordinates.

pression divided by Δn gives the angular displacement ϕ. The mean angular displacement of the fluid element in the interval $(t_2 - t_1)$ is $\frac{1}{2}(\theta + \phi)$ or $\frac{1}{2}(V/R + \partial V/\partial n)(t_2 - t_1)$. Thus the vorticity of the element for small $(t_2 - t_1)$ is given by

$$\omega_z = \frac{V}{R} + \frac{\partial V}{\partial n} \tag{25}$$

where R and n have the same direction. Outward along a radius and counterclockwise are considered positive.

Similarly it can be shown that the rate of strain is given by

$$\gamma_z = \frac{\partial V}{\partial n} - \frac{V}{R} \tag{26}$$

Example 2

If the streamlines of a flow are concentric circles about the origin, Eqs. 25 and 26 can be expressed in polar coordinates by replacing V, R, and ∂n by u_θ, r, and ∂r, respectively. Figure 11 shows an example of such a flow in which the two intersecting edges of an element rotate at the same rate but in opposite directions. Thus the diagonal of the element has vanishing angular velocity but is subjected to a *pure strain*. This flow is called a two-

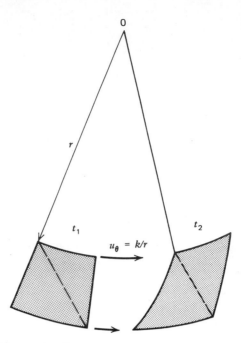

Fig. 11 Fluid element in a vortex flow.

dimensional *vortex*, whose velocity field u_θ is derived in Section 2.9 as k/r, where k is a constant. After substituting in Eqs. 25 and 26 and using the method of Section 2.6 to find the stream function ψ, the flow field of a two-dimensional vortex has the following properties (see Eqs. 22)

$$u_r = 0; \quad u_\theta = \frac{k}{r}; \quad \therefore \psi = -k \ln r$$

$$\omega_z = 0; \quad \gamma_z = -\frac{2k}{r^2} \text{ (at } r \neq 0)$$

(27)

Example 3

Figure 12 shows a flow in *solid-body rotation* about a point O. The two intersecting edges rotate at the same angular velocity, causing a pure rotation of the element without any deformation. In this case u_θ is given by Ωr, where Ω is the constant angular speed of the flow. Using Eqs. 25 and 26, and the method of Section 2.6 to find the stream function ψ, the flow field of a fluid in solid-body rotation has the properties:

$$u_r = 0; \quad u_\theta = \Omega r; \quad \therefore \psi = -\frac{\Omega r^2}{2}$$

$$\omega_z = 2\Omega; \quad \gamma_z = 0$$

(28)

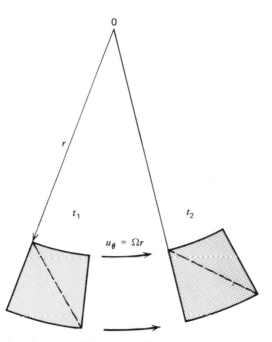

Fig. 12 Fluid element in a flow in solid-body rotation.

Since $|\boldsymbol{\omega}|^2$ is proportional to the kinetic energy of rotation of the fluid element, it is an important variable in flows that involve transfer between translational and rotational energies; examples are boundary layers and some flows involving shock waves.

2.8 Circulation, Curl of a Vector

We have in the last section defined the field properties, rate of strain and vorticity. The *circulation*, Γ, a property related to the net vorticity over a region of the flow, is defined by

$$\Gamma = -\oint \mathbf{V} \cdot d\mathbf{s} \qquad (29)$$

where $\mathbf{V} \cdot d\mathbf{s}$, the dot product of the velocity vector and the vector element of length along the path of integration, is, as indicated in Fig. 13, the scalar of magnitude $|\mathbf{V}| \cos \theta \, |d\mathbf{s}| \equiv V_s ds$; the circle through the integral sign indicates that the circulation is defined as the line integral around a *closed* path of the product of V_s, the velocity component *parallel* to the path, and ds. The integral is carried out in the *counterclockwise* sense; the clockwise direction of integration for circulation,

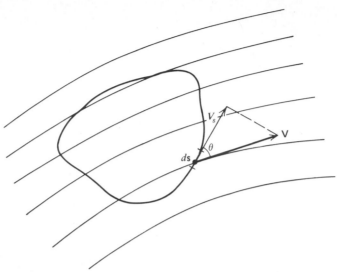

Fig. 13 $\mathbf{V} \cdot d\mathbf{s} = V_s ds$.

indicated by the minum sign in Eq. 29, is chosen in aerodynamics for convenience in the analyses of airfoil and wing theory.

The integrand of Eq. 29 may be expressed in two-dimensional Cartesian notation (see Fig. 13) as $\mathbf{V} \cdot d\mathbf{s} = V \cos \theta \, ds = udx + vdy$ and, in three dimensions,

$$\Gamma = - \oint \mathbf{V} \cdot d\mathbf{s} = - \oint (udx + vdy + wdz) \tag{30}$$

Line integrals such as that defining the circulation have various physical interpretations in the different fields of technology; in fluid mechanics we will show later that the circulation around a contour in a flow enclosing a body such as an airfoil is proportional to the lift developed by the body.

At this point we show the relation between the concepts of circulation and vorticity by means of the vector designated curl $\mathbf{V}$, defined by

$$\text{curl } \mathbf{V} \equiv \left(\frac{\partial w}{\partial y} - \frac{\partial v}{\partial z} \right) \mathbf{i} + \left(\frac{\partial u}{\partial z} - \frac{\partial w}{\partial x} \right) \mathbf{j} + \left(\frac{\partial v}{\partial x} - \frac{\partial u}{\partial y} \right) \mathbf{k}$$

The respective components are seen to be exactly the components of the vorticity, that is,

$$\text{curl}_x \mathbf{V} = \frac{\partial w}{\partial y} - \frac{\partial v}{\partial z} = \omega_x$$

$$\text{curl}_y \mathbf{V} = \frac{\partial u}{\partial z} - \frac{\partial w}{\partial x} = \omega_y \tag{31}$$

$$\text{curl}_z \mathbf{V} = \frac{\partial v}{\partial x} - \frac{\partial u}{\partial y} = \omega_z$$

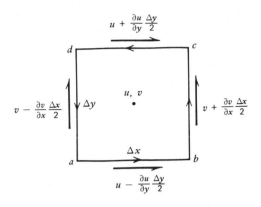

Fig. 14 V_s on the sides of an element.

In vector notation

$$\boldsymbol{\omega} = \text{curl } \mathbf{V} \tag{32}$$

The vector is directed along the axis of rotation of the fluid element.

The components of $\boldsymbol{\omega}$ at a given point are identified with the limit of the circulation around a contour enclosing the point. For example, by carrying out the line integral of Eq. 30 around the element of Fig. 14, one can verify that

$$\omega_z = \lim_{\Delta \hat{S} \to 0} \left\{ \frac{\oint (u\,dx + v\,dy)}{\Delta \hat{S}} \right\} = \lim_{\Delta \hat{S} \to 0} \left\{ \frac{-\Delta \Gamma}{\Delta \hat{S}} \right\} \tag{33}$$

where $\Delta \hat{S}$ is the area enclosed by the path about which the line integral is taken; x and y are measured along the surface and become rectilinear axes in the limit. The directions of ω_x, ω_y, and ω_z are determined according to the "right-hand rule."

In cylindrical polar coordinates (r, θ, z) (Problem 1, Section 2.8), the vorticity component in the z direction is found to be

$$\omega_z = \frac{\partial u_\theta}{\partial r} + \frac{1}{r} \left(u_\theta - \frac{\partial u_r}{\partial \theta} \right) \tag{34}$$

2.9 Irrotational Flow

In 1904, Ludwig Prandtl hypothesized that the flow of gases and "watery liquids" around streamlined bodies can be assumed to have zero vorticity, that is, be *irrotational*, except in a thin boundary layer and a narrow wake behind the body. In other words, at the Reynolds numbers of flight of aircraft or for the motion of many devices and animals through fluids, a good approximation to the flow field, outside of the immediate neighborhood of the body, can be found by assuming irrotationality. In terms of Eqs. 31, an irrotational flow is characterized by

$$\omega_x = \omega_y = \omega_z = 0$$

that is,

$$\frac{\partial v}{\partial x} = \frac{\partial u}{\partial y} \; ; \qquad \frac{\partial w}{\partial y} = \frac{\partial v}{\partial z} \; ; \qquad \frac{\partial u}{\partial z} = \frac{\partial w}{\partial x} \tag{35}$$

In cylindrical polar coordinates, a two-dimensional irrotational flow (Eq. 34) is expressed by

$$r \frac{\partial u_\theta}{\partial r} + u_\theta \equiv \frac{\partial}{\partial r} (r u_\theta) = \frac{\partial u_r}{\partial \theta} \tag{36}$$

Example 1

The simplest and most obvious irrotational flow is of course a uniform flow. In Section 2.6 we identified the flow $\psi = Ay - Bx$ as a uniform flow inclined at the angle $\tan^{-1} (B/A)$ to the x axis. By Eqs. 21, $u = A$, $v = B$, and, by Eqs. 31, $\omega = \omega_z k = 0$, so the flow is irrotational.

Example 2

Consider the flow described by the stream function

$$\psi = Axy \tag{A}$$

This example was used to identify iso- and gradient lines in Section 2.2, where it was pointed out that the rectangular hyperbolas described by ψ = constant represent the approximate paths of fluid elements, that is, the streamlines, in the neighborhood of a "stagnation point." By Eqs. 21, $u = Ax$ and $v = - Ay$. Both $\partial v/\partial x$ and $\partial u/\partial y$ vanish so that, by Eqs. 31, the flow is irrotational.

In general, the vorticity has a definite magnitude and direction, expressed by Eq. 32, at each point in the flow field. We see from Eqs. 31 that any component of the vorticity vanishes if the shear derivatives in the normal plane vanish (as in the illustrative examples above) or if coplanar pairs are equal to each other. This latter condition, which seems highly restrictive, is actually fulfilled by a large class of flow fields of importance in fluid mechanics.

Of special importance is the two-dimensional irrotational flow in which the streamlines are circular and the velocity is constant on each streamline. We will solve for the velocity distribution as a function of the radius. The flow is described in polar coordinates by

$$\psi = f(r); \qquad u_\theta = - \frac{df}{dr}; \qquad u_r = \frac{1}{r} \frac{\partial f}{\partial \theta} = 0$$

Since u_θ is a function only of r, the condition for irrotationality, Eq. 36, may be written as an ordinary differential equation:

$$\frac{d(ru_\theta)}{dr} = 0$$

and the integral is

$$u_\theta = \frac{k}{r} \tag{37}$$

where k is a constant. This result was referred to in Section 2.7 (Eqs. 27) where it was shown that the stream function for this flow is $\psi = -k \ln r$ and that the vorticity vanishes but the strain does not.

As was mentioned, this flow is termed a "vortex flow," and we have shown above that the flow is irrotational everywhere *except* at $r = 0$ where the derivative $\partial u_\theta / \partial r$ becomes infinite and Eq. 34 indicates that the vorticity at that point must be evaluated by other means; this is carried out in the next section by the application of Stokes' theorem.

2.10   Theorem of Stokes

A theorem connecting the integral over a surface with the line integral around it is indicated by the definition of vorticity, or curl, given in Eq. 33. In terms of the vorticity component normal to an arbitrarily oriented surface, the definition may be expressed as

$$\omega_n = \operatorname{curl}_n \mathbf{V} = \lim_{\Delta \hat{S} \to 0} \left\{ \frac{\oint \mathbf{V} \cdot d\mathbf{s}}{\Delta \hat{S}} \right\}$$

where the subscript n designates the component of $\boldsymbol{\omega}$ normal to $\Delta \hat{S}$.

In a given flow, a continuous surface of arbitrary shape and orientation, shown schematically in Fig. 15, is chosen and is subdivided into k segments. If the flow is known, the component of the vorticity normal to each segment can be determined by resolution of the components defined by Eqs. 31. The above relation, applied to segment 1 of Fig. 15, reads

$$(\omega_n \Delta \hat{S})_1 \simeq \left(\oint \mathbf{V} \cdot d\mathbf{s} \right)_1$$

and, if we sum over the k segments,

$$\sum_{i=1}^{k} (\omega_n \Delta \hat{S})_i \simeq \sum_{i=1}^{k} \left(\oint \mathbf{V} \cdot d\mathbf{s} \right)_i \tag{38}$$

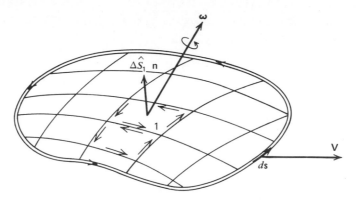

Fig. 15 Stokes' theorem.

In summing the line integrals on the right-hand side of the equation, it will be observed that the integrals over paths that are common to any two segments can make no contribution to the summation. Therefore, the sum of the line integrals about the segments $\Delta\hat{S}_i$ is just equal to the line integral around the entire surface $\hat{S}$. Equation 38 becomes a better and better approximation as each $\Delta\hat{S}_i$ approaches zero, and it is exact in the limit. Under these conditions, the summation becomes a surface integral and Eq. 38 may be written

$$\iint_{\hat{S}} \omega_n \, d\hat{S} = \oint \mathbf{V} \cdot d\mathbf{s} = -\Gamma \qquad (39)$$

The relation between the double integral and line integral in Eq. 39 bears the name of Sir George Stokes (1819–1903). The negative sign on Γ arises because circulation is counted positive clockwise but the positive sense for curl is counterclockwise.

It is evident from Eq. 39 that the circulation around any contour is zero if the flow is irrotational within the contour of integration. However, for a vortex flow described by $u_\theta = k/r$ (Eq. 37) the flow was shown to be irrotational everywhere except at $r = 0$, so that Eq. 39 will not vanish unless the origin $r = 0$ is outside of the contour of integration. If the origin is *within* the contour, however, the value of $-\Gamma$ will be the limiting value of $\omega_n \Delta\hat{S}$ as $\Delta\hat{S}$ approaches zero, and hence $\omega_n \to \infty$ as $r \to 0$. Thus the vortex flow $u_\theta = k/r$ is irrotational everywhere except at the origin where the vorticity is infinite.

We evaluate k in terms of the circulation by integrating Eq. 39 along a streamline, that is, along a circle $r = $ constant. The product $\mathbf{V} \cdot d\mathbf{s}$ is simply $(k/r)r\, d\theta$ and the circulation becomes

$$\Gamma = -\oint \mathbf{V} \cdot d\mathbf{s} = -\int_0^{2\pi} (k/r)r \, d\theta = -2\pi k$$

If k is replaced with its equivalent $-\Gamma/2\pi$, the equation that describes the vortex flow becomes

$$u_\theta = \frac{-\Gamma}{2\pi r}$$

$$\psi = \left(\frac{\Gamma}{2\pi}\right) \ln r \qquad (40)$$

Thus the intensity or strength of a vortex is measured by the circulation around it; the sense of the velocity is clockwise when Γ is a positive number.

In summary, vortex flow is irrotational everywhere except at the center of the vortex itself, where the vorticity is infinite, it satisfies the equation of continuity everywhere and therefore represents a physically possible flow.

This important flow is discussed at greater length in Section 2.12.

2.11 Velocity Potential

From Stokes' theorem, it is apparent that the line integral of $\mathbf{V} \cdot d\mathbf{s}$ around a closed path in an irrotational field is zero. Then, if A and B of Fig. 16 represent two points on a closed path in an irrotational field, the line integral from A to B along either branch must be the same. Since an infinite number of closed paths may be drawn through A and B, around each one of which the line integral is zero, it necessarily follows that the line integral along any path connecting A and B is independent of the path; its value is therefore a function only of its integration

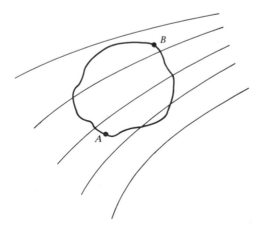

Fig. 16 Integration paths between points in a flow.

limits, that is,

$$\oint_A^B \mathbf{V} \cdot ds = \oint_A^B (udx + vdy + wdz) = f(A, B) \tag{41}$$

However, a line integral can be independent of the path of integration only if the integrand is an exact differential. Therefore, Eq. 41 requires that $\mathbf{V} \cdot ds$ be an exact differential of some function of the field coordinates. This function is given the symbol $\phi(x, y, z)$ and called the *velocity potential*.

We accordingly expand the total differential $d\phi$ and set it equal to the integrand $\mathbf{V} \cdot ds$ in Eq. 41:

$$udx + vdy + wdz = d\phi \equiv \frac{\partial \phi}{\partial x} dx + \frac{\partial \phi}{\partial y} dy + \frac{\partial \phi}{\partial z} dz$$

Since these expressions must be identical the coefficients of dx, dy, and dz must respectively be equal, that is,

$$u = \frac{\partial \phi}{\partial x}; \quad v = \frac{\partial \phi}{\partial y}; \quad w = \frac{\partial \phi}{\partial z} \tag{42}$$

In cylindrical polar coordinates (r, θ, z),

$$u_r = \frac{\partial \phi}{\partial r}, \quad u_\theta = \frac{1}{r}\frac{\partial \phi}{\partial \theta}, \quad u_z = \frac{\partial \phi}{\partial z} \tag{43}$$

and, in vector notation,

$$\mathbf{V} = \text{grad } \phi \equiv \nabla \phi$$

We note that, unlike the stream function, the velocity potential is defined in three dimensions; its dimensions are length^2/time.

Thus in a flow that is irrotational everywhere (containing neither isolated vortices nor regions of distributed vorticity), since the circulation around *any* contour vanishes, the integrand of Eq. 41 must be everywhere a perfect differential, which we call $d\phi$. Then, for this flow, a velocity potential $\phi(x, y, z) = \int d\phi$ "exists," and ϕ *has the property its partial derivative in any direction is the velocity component in that direction.* It follows that *the existence of ϕ is the sole criterion for irrotationality*, since when we substitute the expressions for u, v, w in Eqs. 42 into Eqs. 31, we find that each component of the vorticity, ω_x, ω_y, or ω_z, vanishes identically. Thus, *an irrotational flow is termed a potential flow.* The usefulness of the velocity potential in flows of practical significance derives from the circumstance that, for a body in relative motion in an originally irrotational flow, the circulation vanishes around any contour that does not include the body or does not intersect the boundary layer or the wake, therefore, a velocity potential can be found to describe the flow everywhere outside of the boundary layer and the wake.

The stream function and the velocity potential have some analogous features: the existence of the velocity potential implies *irrotationality*, and in a three-dimensional flow the three velocity components can be derived from a single variable, $\phi(x, y, z)$. The existence of the stream function, on the other hand, implies *continuity* in a two-dimensional flow and the two velocity components can be derived from a single variable, $\psi(x, y)$. Both ϕ and ψ have the dimensions length^2/time.

In Section 2.7, it was shown that the components of the velocity determine the stream functions for uniform flow, sources, and sinks. We can use the same method to find the velocity potential. For instance, by the use of Eqs. 43, Eqs. 40 give, for the potential vortex,

$$u_r = \frac{\partial \phi}{\partial r} = 0; \qquad u_\theta = \frac{1}{r} \frac{\partial \phi}{\partial \theta} = \frac{-\Gamma}{2\pi r}$$

Keeping in mind that θ is constant in the first equation and r is constant in the second, and, setting the integration constants equal to zero, the equations integrate to

$$\phi = f(\theta) \qquad \text{and} \qquad \phi = -\left(\frac{\Gamma}{2\pi}\right)\theta + g(r)$$

These expressions can be reconciled everywhere only if $g(r) = 0$ and $f(\theta) = -(\Gamma/2\pi)\theta$, so that $\phi = -(\Gamma/2\pi)\theta$ for a potential vortex. The "equipotentials" are thus straight lines radiating from the origin. If the same method is applied to a *rotational flow* such as the plane Couette flow of Fig. 7 in which $u = Ay$, $v = 0$ ($\psi = \frac{1}{2} Ay^2$), we can easily prove the *nonexistence* of a velocity potential. We do this by first assuming a function $\phi(x, y)$ exists, such that

$$u = \frac{\partial \phi}{\partial x} = Ay; \qquad v = \frac{\partial \phi}{\partial y} = 0$$

We then integrate the first equation keeping y constant, the second keeping x constant, and let the integration constants vanish. The two solutions are

$$\phi = Axy + f(y); \qquad \phi = g(x)$$

The function $g(x)$ is an arbitrary function of x only, so the term Axy in the first expression cannot possibly occur in the second. The two expressions for ϕ are therefore irreconcilable. Thus, we conclude that, contrary to the original assumption, a velocity potential does *not* exist for the flow.

Application of the same method to the other irrotational flows treated earlier yields the velocity potential functions given along with the stream functions in the first four entries in Eqs. 44. For the rotational flows, the method shows, as indicated in the last two entries, that velocity potentials do not exist.

Flow	ψ	ϕ
Vortex	$(\Gamma/2\pi)\ln r$	$-(\Gamma/2\pi)\theta$
Source	$(\Lambda/2\pi)\theta$	$(\Lambda/2\pi)\ln r$
Uniform flow	$Ay - Bx$	$Ax + By$
90° Corner flow	Axy	$\frac{1}{2}A(x^2 - y^2)$
Solid body rotation	$\frac{1}{2}\Omega r^2$	Does not exist
Plane Couette flow	$\frac{1}{2}Ay^2$	Does not exist

$$(44)$$

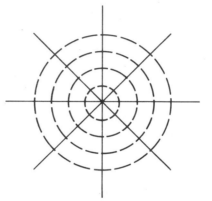

Fig. 17 Flow net for source or vortex.

The first two entries in Eqs. 44 indicate that the circles in Fig. 17 represent streamlines of a vortex or equipotentials of a source, while the radial lines represent equipotentials of a vortex or streamlines of a source. Uniform flow is represented by a network of orthogonal straight lines. The streamlines and equipotentials of the flow in 90° corners is shown in Fig. 18; they are orthogonal families of hyperbolas and are the isolines and gradient lines of Example 1, Section 2.2. The curves in each quadrant describe the flow in a corner and those in adjacent quadrants describe the flow against a plane surface or in the immediate neighborhood of a stagnation point. Since the two families are orthogonal, except at the stagnation point, their designations are interchangeable; that is, the streamlines for the flow in one quadrant are the equipotentials for the flow in the quadrant at 45°.

We show that this orthogonality condition is general for irrotational flows by reference to the formulas we have derived for the velocity components. In Cartesian and polar coordinates, these are

$$u = \frac{\partial \phi}{\partial x} = \frac{\partial \psi}{\partial y}; \qquad u_r = \frac{\partial \phi}{\partial r} = \frac{1}{r}\frac{\partial \psi}{\partial \theta}$$

$$v = \frac{\partial \phi}{\partial y} = -\frac{\partial \psi}{\partial x}; \qquad u_\theta = \frac{1}{r}\frac{\partial \phi}{\partial \theta} = -\frac{\partial \psi}{\partial r}$$

$$(45)$$

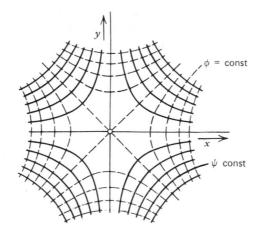

Fig. 18 Flow net for 90° corner flow, dashed curves are equipotentials.

Note that the slope of a streamline is given by

$$\frac{v}{u} = \frac{\partial\phi/\partial y}{\partial\phi/\partial x} = -\frac{\partial\psi/\partial x}{\partial\psi/\partial y} \tag{46}$$

Since the ratios of the partial differentiations are negative reciprocals of each other, it follows from Eq. 46 that *the streamlines, ψ = constant, and the equipotentials, ϕ = constant, are orthogonal except at stagnation points, that is, where the velocity components vanish simultaneously.*

2.12 Point Vortex, Vortex Filament, Law of Biot and Savart

The vortex flow described by Eqs. 40 is of great importance in aerodynamics. The equations describe a flow pattern in a nonviscous, incompressible fluid, and therefore it cannot be expected that a pure example of this pattern is to be found in nature. However, examples of actual vortices described in the next section show that, while the action of viscosity diffuses the peak of infinite vorticity and velocity at the center, outside of a central core the flow field conforms closely with that of Eqs. 40.

We therefore treat the properties and motion of a theoretical vortex as an irrotational flow with infinite vorticity at the center. *The center point is referred to as a point vortex whose strength is defined as the circulation about that point.* It is customary to speak of the point vortex as *inducing* a flow in the surrounding region. However, it should be remembered that the point vortex and the flow in the surrounding region simply coexist. One is not actually the cause of the other.

The two-dimensional velocity field described by Eqs. 40 is a vortex flow in the *xy* plane with the point vortex at the origin of coordinates. Two-dimensional flow means the pattern is identical in all planes parallel to the *xy* plane. This point

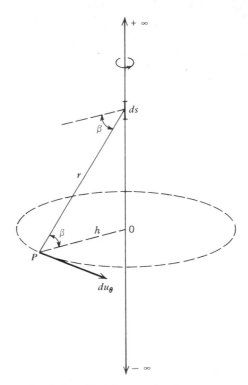

Fig. 19 Velocity increment du_θ induced by element ds of vortex.

vortex, then, must be duplicated in every parallel plane, the configuration of the points forming a straight line perpendicular to the xy plane and extending from $-\infty$ to $+\infty$. In three dimensions, such a line is called a vortex filament and may be defined as a line coinciding with the axis of rotation of successive fluid elements.

It is convenient to speak of the velocity induced in the region surrounding a vortex filament by an element ds of the filament shown in Fig. 19. The increment of velocity induced at P by element ds of the filament is given by the *Biot-Savart law*:

$$du_\theta = \frac{\Gamma}{4\pi} \frac{\cos\beta\ ds}{r^2} \tag{47}$$

where Γ is the strength of the vortex filament, r is the distance between the filament and the point P, and β is the angle between the length r and the normal to the filament. The direction of the velocity increment induced by ds is perpendicular to a plane containing the length ds and the point P. (With Γ the direct current and u_θ the magnetic field, Eq. 47 expresses the proportionality between the strength of a magnetic field induced in a region surrounding a wire that is carrying a current.) The velocity at P induced by the entire filament is obtained by integration.

$$u_\theta = \frac{\Gamma}{4\pi} \int_{-\infty}^{\infty} \frac{\cos \beta \, ds}{r^2} \tag{48}$$

$$r = h \sec \beta; \qquad s = h \tan \beta \tag{49}$$

The h in Eqs. 49 is the perpendicular distance between the filament and the point P. If Eqs. 49 are substituted in Eq. 48 and if the integration with respect to β between the limits $\pm \pi/2$ is performed,

$$u_\theta = \frac{\Gamma}{2\pi h} \tag{50}$$

This is precisely the velocity induced at P, arising from a point vortex at the origin in a two-dimensional flow.

The concept of the vortex filament is extended by lifting the restriction that it must be a straight line perpendicular to the plane of the two-dimensional point vortex. The picture now is one of a line curving arbitrarily in space in such a manner that it coincides with the axis of rotation of successive fluid particles. The law of Biot and Savart as given by Eq. 47 is applicable to the curved filament.

2.13 Helmholtz' Vortex Theorems

It is readily shown by the following method that the strength of a vortex filament is constant along its length. Enclose a segment of the vortex filament with a sheath from which a slit has been removed, as pictured in Fig. 20. Since the vorticity at every point on the curved surface enclosed by the perimeter of the split sheath is zero, it follows, from Stokes' theorem, that the line integral of the velocity along the perimeter is also zero. In traversing the perimeter, the contributions to the

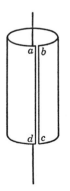

Fig. 20 Split sheath control surface around vortex filament.

total circulation of the line integrals from *b* to *c* and *d* to *a* will be of equal magnitude and opposite sign, providing the slit is very narrow. In order for the total circulation to be zero, therefore, the line integrals around the sheath from *a* to *b* and from *c* to *d* must be of equal magnitude and opposite sign. This means that the vorticity enclosed by the top and bottom perimeters of the cylinder must be identical, which demonstrates the truth of Helmholtz' first theorem:

The strength of a vortex filament is constant along its length.

Carrying the demonstration a step further, presume the split sheath so placed that the filament ends midway between the top and bottom edges. The line integrals around the sheath from *a* to *b* and from *c* to *d* can no longer be of equal magnitude, and the condition that the circulation around the perimeter be zero is therefore violated. Hence, the vortex filament cannot end in space; this conclusion is embodied in Helmholtz' second theorem:

A vortex filament cannot end in a fluid; it must extend to the boundaries of the fluid or form a closed path.

A third Helmholtz theorem follows by use of the fact that in an inviscid, incompressible flow, only pressure forces are exerted on a fluid element; since pressure forces act normal to all surfaces of the element, it can be shown that their resultant passes through the centroid of the element and therefore can cause no rotation, that is, no vorticity. In mathematical terms, the curl of the force vanishes and the law is stated as follows.

In the absence of rotational external forces, a fluid that is initially irrotational remains irrotational.

From the theorem of Stokes, a corollary may be written immediately.

In the absence of rotational external forces, if the circulation around a path enclosing a definite group of particles is initially zero, it will remain zero.

A further corollary may be written.

In the absence of rotational external forces the circulation around a path that encloses a tagged group of elements is invariant.

It follows from these laws that an element is set in rotation only by a contiguous rotating element; in other words, while pressure changes propagate throughout the flow field with the speed of sound, vorticity propagates only through contiguous fluid elements.

It should be noted that the vortex theorems are proved for inviscid fluid flows; they are therefore valid to a good approximation in viscous-fluid flows in regions where the viscosity may be neglected.

2.14 Vortices in Viscous Fluids

We have shown that the potential vortex $u_\theta = -\Gamma/2\pi r$ (Eqs. 40) is a physically possible flow field of an inviscid incompressible fluid with a point of infinite vorticity at $r = 0$. In a real fluid, however, viscosity causes diffusion of the vorticity to form instead a core within which the vorticity is approximately constant; in Section 2.7 (Eqs. 28), we showed that if the vorticity is constant, the fluid is in solid body rotation with $u_\theta = \Omega r$. As time goes on, this core spreads; that is, the vortex decays from within. The process is illustrated in Fig. 21 where theoretical

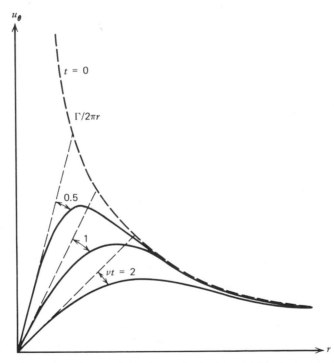

Fig. 21 Decay of a vortex filament in a viscous fluid. At the initial instant $t = 0$, $u_\theta = -\Gamma/2\pi r$. Prandtl (1943) shows that the important nondimensional parameter is $r^2/\nu t$, where ν is the kinematic viscosity (dimensions l^2/t). Velocity distributions for three values of νt are shown along with linear distribution of u_θ corresponding roughly to cores of solid body rotation (dashed lines) with radii proportional to $\sqrt{\nu t}$.

Fig. 22 (*a*) Funnel of a tornado. (Courtesy of National Severe Storms Laboratory, National Oceanic and Atmospheric Administration.) (*b*) Satellite photograph of hurricane Hilda, October 1, 1964 (grid is about 300 × 300 km). (Courtesy of T. N. Krishnamurti, Florida State University.)

velocity distributions are shown for three values of the parameter vt, along with those for solid body rotation with the same total circulation. The calculated curves are all asymptotic to that for $t = 0$; they indicate further that the vortex may be represented approximately by a core of solid body rotation joined to the theoretical distribution for a vortex with circulation given by $-\Gamma = \pi r_1^2 \omega_1$, where r_1 and ω_1 are, respectively, the radius of and the vorticity everywhere in the core.

A tornado in the atmosphere (Fig. 22*a*) is an example of a vortex in nature. Because of viscous effects, a funnel-shaped core is formed whose diameter increases upward. The vortex originates near the ground so, as the air within it moves upward, the circulation around the core remains constant while the radius increases. Thus, velocities near the ground will be much higher than they are aloft; the resulting high pressure variations associated with the flow around structures is one of the main causes of damage to structures. A hurricane (Fig. 22*b*), because its genesis involves convection over a much larger sea area than does a tornado over land, has a much more extensive core, termed the "eye of the storm"; its passing is marked

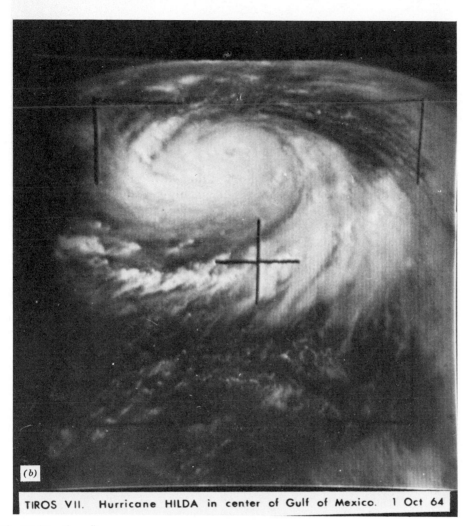

TIROS VII.   Hurricane HILDA in center of Gulf of Mexico.   1 Oct 64

Fig. 22 (*Continued*)

by an interval of relative calm, after which the wind speed again increases but at 180° to its former direction.

Many other examples of vortices are commonly observed.   Behind an airplane there are two trailing vortices of opposite directions of circulation, as shown in Fig. 23.   The formation of such a vortex system is described in Chapter 6.   The body of a streamline-shaped automobile is aerodynamically a finite wing.   The

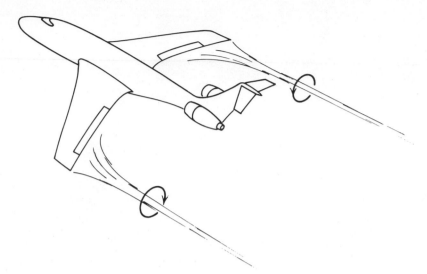

Fig. 23 Trailing vortices behind aircraft in flight.

vortex pair behind such a vehicle in motion is sometimes traced out by the engine exhaust or by snow, dust, or fog. The commonly observed smoke rings illustrate a vortex filament that closes on itself.

CHAPTER 3

Dynamics of Flow Fields

3.1. Introduction

In the last chapter, methods for describing simple flow fields were developed without regard to the forces generating them. The basic flow fields treated were those of uniform flow, sources, and vortices. We found in general that the continuity and irrotationality conditions uniquely determine a velocity field by identifying a network of orthogonal families of curves, representing, respectively, the streamlines and equipotentials of the flow.

This description is, however, incomplete in that the pressure field required to calculate the fluid dynamic force on a body in relative motion in the fluid is still not known.

The dynamics of the flow field determines this pressure distribution, for which the governing equations are dictated by Newton's law expressing conservation of momentum. This chapter is devoted to deriving the equations expressing momentum conservation, which, together with mass conservation (continuity), determines all the details of an incompressible inviscid flow field. The analysis will first be carried out for flow along a streamline and then generalized to give the equations describing the motion throughout the flow field.

3.2 Dynamics of Flow Along a Streamline

Before deriving the *general* equations of flow dynamics, we analyze the inviscid flow along a streamline by applying Newton's law of conservation of linear momentum to a "tagged" elementary mass, termed a fluid element or particle. The element is defined such that its dimensions are small compared with a characteristic dimension of the flow but large enough to contain an inordinately large number of molecules. Figure 1 shows schematically the streamlines of a flow in which the motion of an element will be analyzed in "natural coordinates," that is, in terms of orthogonal directions s and n, respectively, parallel to and normal to the tangent to the streamline at a given point. The blowup of the element is shown in order to indicate more clearly the gravity force on the element and the pressure forces acting on the faces; for simplicity the flow shown is two-dimensional though the result is also valid for a three-dimensional flow. The dimensions of the element are Δs by Δn by unit length normal to the plane of Fig. 1 and it moves along the streamline a distance δs in time δt. Under the action of the forces the element

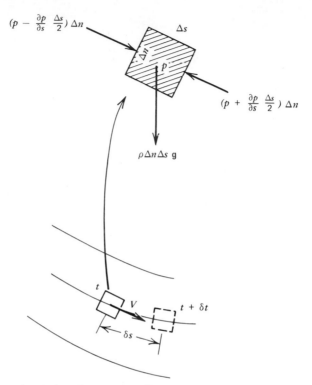

$$(p - \frac{\partial p}{\partial s} \frac{\Delta s}{2}) \, \Delta n$$

Δs

p'

$$(p + \frac{\partial p}{\partial s} \frac{\Delta s}{2}) \, \Delta n$$

$\rho \Delta n \Delta s \, g$

t

V

$t + \delta t$

δs

Fig. 1 Fluid element in motion along a streamline.

accelerates and its change in speed over the distance δs is

$$\delta V = \frac{\partial V}{\partial s} \delta s + \frac{\partial V}{\partial t} \delta t \tag{1}$$

where the speed V is the scalar magnitude of the velocity. The first term in Eq. 1 represents the change in velocity in a **steady** flow; the second measures the change in the velocity if the flow field is time-dependent, such as occurs at a point near a maneuvering or accelerating aircraft.

After dividing Eq. 1 by δt, and letting $\delta t \to 0$, Eq. 1 describes the acceleration of the fluid element as it moves along the streamline. The acceleration is then given by what we term the **particle** or **substantial** derivative of the flow speed. It is designated by the script $\mathcal{D}$ as defined by

$$\frac{\mathcal{D}V}{\mathcal{D}t} = \frac{\partial V}{\partial t} + \frac{\partial V}{\partial s} \lim_{\delta t \to 0} \left(\frac{\delta s}{\delta t} \right)$$

In the limit $\delta s / \delta t$ becomes V and we write

$$\frac{\mathcal{D}V}{\mathcal{D}t} = \frac{\partial V}{\partial t} + V \frac{\partial V}{\partial s} \tag{2}$$

where

$$V \frac{\partial V}{\partial s} = \frac{\partial}{\partial s} \left(\frac{1}{2} V^2 \right) \tag{3}$$

is the streamwise component of the acceleration in a steady flow; it is termed the **convective acceleration**; $\partial V/\partial t$ is termed the **local acceleration**.

Example

   To illustrate that Eq. 3 represents the acceleration of an element, consider a liquid issuing from a horizontal pipe under the influence of gravity. Viewed from a station fixed relative to the pipe, the flow is steady and, since no forces are acting on the fluid in the horizontal direction, the acceleration of a fluid element is given by

$$\frac{d^2 z}{dt^2} = g \tag{A}$$

where z is the vertical coordinate of the tagged fluid element, positive downward. Two integrations give, respectively,

$$V = \frac{dz}{dt} = gt + A$$

$$z = \tfrac{1}{2} gt^2 + At + B$$

Assume that at time $t = 0$, $V = z = 0$, so that $A = B = 0$,

$$V = gt = \sqrt{2gz}, \quad \frac{dV}{dz} = \sqrt{\frac{g}{2z}}$$

and

$$V \frac{dV}{dz} = g \tag{B}$$

   The identity of Eqs. A and B illustrates that Eq. 3 is the acceleration of a particle in a steady flow.
   We now proceed with the derivation of the equation of motion for the fluid element. To do so, we apply *Newton's law*, which states that *the rate of change of momentum of a fluid element tangent to a streamline is equal to the sum of the components of the pressure and gravity forces in that direction.* By Fig. 1, the net pressure force acting on the faces normal to the s direction is that on the upstream-face minus that on the downstream-face and has the value $-(\partial p/\partial s) \Delta s \Delta n$. The rate of change of momentum of the element is $\mathcal{D}(\rho \Delta s \Delta n V)/\mathcal{D}t$, which must, by Newton's law, equal the sum of the pressure and gravity forces in the stream

direction:

$$\frac{\mathcal{D}}{\mathcal{D}t}(\rho \Delta s \Delta n V) = -\frac{\partial p}{\partial s}\Delta s \Delta n + \rho \Delta s \Delta n (\mathbf{g} \cdot \mathbf{s}) \tag{4}$$

where $\mathbf{g}$ is the acceleration of gravity, $\mathbf{s}$ is the unit vector in the downstream direction, so that $\mathbf{g} \cdot \mathbf{s}$ is the component of the acceleration tangent to the streamline. Regardless of whether ρ is constant, the mass of the element, $\rho \Delta s \Delta n$, is constant and may therefore be carried through the differential sign. After cancelling $\Delta s \Delta n$, the equation describing the dynamics of the flow in the stream direction is

$$\rho \frac{\mathcal{D}V}{\mathcal{D}t} = -\frac{\partial p}{\partial s} + \rho \mathbf{g} \cdot \mathbf{s} \tag{5}$$

If we restrict our analysis to *steady flow* along a streamline, we may write Eq. 4 in the form

$$\rho V \frac{dV}{ds} = \rho \frac{d(\frac{1}{2}V^2)}{ds} = -\frac{dp}{ds} + \rho \mathbf{g} \cdot \mathbf{s} \tag{5a}$$

The last term is the streamwise component of the gravity force per unit volume.

It is convenient for the integration of this equation to define a **gravitational potential** U analogous to the velocity potential ϕ defined in section 2.11; just as $\partial \phi / \partial l$ is the velocity component in the l direction so is $\partial U / \partial l$ the component of the gravitational acceleration in the l direction, that is, $\mathbf{g} \cdot \mathbf{s} = \partial U / \partial s$. In Cartesian coordinates, if z is the vertical coordinate, *positive downward*, $U = gz$, so that $dU/dz = g$. Then, the equation above becomes

$$\rho \frac{d(\frac{1}{2}V^2)}{ds} + \frac{dp}{ds} - \rho \frac{dU}{ds} = 0$$

If this equation is multiplied by ds, after dividing by ρ the equation may be written for flow along a given streamline,

$$d\left(\frac{V^2}{2}\right) + \frac{dp}{\rho} - d(gz) = 0$$

which integrates to

$$\frac{V^2}{2} + \int \frac{dp}{\rho} = \text{constant} + gz \tag{6}$$

For flow of an ideal compressible gas $\rho = \rho(p)$ must be a known function (Chapter 8) to evaluate the integral. In an incompressible flow ρ is constant and Eq. 6 integrates to

$$\frac{1}{2}\rho V^2 + p - \rho g \Delta z = p_0 \tag{7}$$

where, for convenience in the physical applications to follow, the integration constant has been written in the form $p_0 - \rho g z_0$, and $\Delta z = z - z_0$. Equation 7 is a special form of **Bernoulli's equation**, Eq. 6, named after Daniel Bernoulli (1700–1782). The equation is applicable to steady, inviscid, incompressible flow along a streamline; however, applications described later will show that these restrictions are, for many applications, not nearly as stringent as they seem.

The terms in Eq. 7 have the dimensions of pressure. Each is associated with a particular aspect of the flow, as follows: $p - \rho g \Delta z$, the **static pressure**, is the pressure measured, relative to that at a reference height z_0, by an infinitesimal barometer attached to a given fluid element in the flow; $\frac{1}{2} \rho V^2$, called the **dynamic pressure**, is the pressure identified with the fluid motion; then p_0, the sum of the static and dynamic pressures, is the **total pressure** at every point in the flow field. If $V = 0$, at a point in the flow, that point is a **stagnation point** of the flow and the pressure there is the **stagnation pressure**, equal to the sum of the static and dynamic pressures at every other point. In line with the above discussion, $\Delta z = 0$ if p and p_0 are measured at the same altitude z. As will be shown later, for gases ρ is small enough so that $\rho g \Delta z$ is negligible in most practical applications.

Equation 7 states the law of conservation of mechanical energy for an element as it moves along a streamline in a steady, inviscid incompressible flow. The terms $\frac{1}{2} \rho V^2$ and $p - \rho g \Delta z$ are, respectively, the kinetic and potential energies of flow per unit volume, and their sum is the total energy, p_0, which is a constant for the flow on the given streamline. At a given z, the static pressure measured at the element then increases as its speed decreases and vice versa such that Eq. 7 is always satisfied, in a *steady* flow. In an unsteady flow field such as that around an accelerating or maneuvering aircraft, the term $\rho \partial V / \partial t$, neglected in Eq. 5, becomes significant. For simplicity, this "transient term" was not included in the above analysis; it is included in the general derivation of Section 3.5.

3.3 Application of Bernoulli's Equation in Incompressible Flows

The **pitot-static tube** shown schematically in Fig. 2, is a device that depends on Bernoulli's equation for its usefulness. The tube is of cylindrical cross section, and the U tube manometer attached to it measures the difference between the **stagnation pressure**, p_0, at the point where the streamline intersects the body and the pressure at the station behind the nose where the pressure, termed the **static pressure**, p, is equal to that in the undisturbed flow. In order to avoid errors resulting from small misalignment, static pressure orifices are distributed around the circumference. For the tube shown, known as the "Prandtl tube," the nose is hemispherical in diameter and the static pressure orifices are three diameters behind the nose; calibration shows that at or behind this station, the pressure at the tube surface is very nearly equal to the static pressure far from the body with $\Delta z = 0$; in other

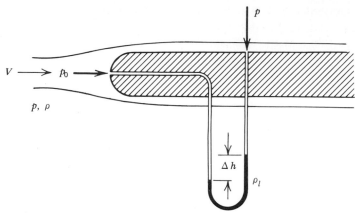

Fig. 2 Pitot-static tube.

words, the flow interference due to the nose extends about three diameters along the body.

The U tube manometer indicates a difference in level Δh and thus the pressure difference, which is equal to the weight of the column of the manometer liquid of unit cross section and height Δh gives

$$p_0 - p = \rho_l g \Delta h \tag{8}$$

where ρ_l is the mass density of the manometer liquid. Then, after substituting for the left side from Eq. 7 with $\Delta z = 0$ and solving for V, Eq. 8 becomes

$$V = \sqrt{\frac{2 \rho_l g \Delta h}{\rho}} \tag{9}$$

The **venturi tube** is another device for measuring the flow speed; it is shown schematically in Fig. 3. We assume the area change is gradual enough so that the

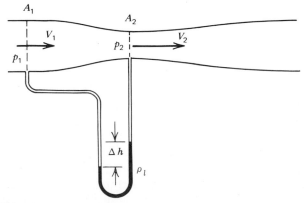

Fig. 3 Venturi tube.

speeds are constant across the cross sections at stations 1 and 2, and their magnitudes are V_1 and V_2. We assume also that $\rho_1 = \rho_2 = \rho$. The flow is assumed irrotational and $\Delta z = 0$, so that, by Eq. 7,

$$p_1 - p_2 = \tfrac{1}{2}\rho(V_2^2 - V_1^2) = \rho_l g \Delta h$$

Since mass is conserved and the density is constant,

$$V_2 A_2 = V_1 A_1$$

we can solve, for instance, for V_2:

$$V_2 = \sqrt{\frac{2\,\rho_l g \Delta h}{\rho}\;\frac{1}{1-(A_2/A_1)^2}} \tag{10}$$

The accuracy of Eqs. 9 and 10 depends on the assumption that compressibility of the fluid can be neglected, that is, on the relation

$$\frac{p_0 - p}{\tfrac{1}{2}\rho V^2} = 1$$

In Chapter 9, we find that the accuracy of this relation depends on the Mach number of the flow. For instance at a Mach number of 0.4, that is, a flow velocity of 136 m/sec, under standard conditions the ratio is about 1.04; it follows that the velocity calculated by Eq. 9; that is, assuming constant density, would be about 2 percent high. Restrictions on the use of these formulas in flows around aircraft, where the flow speeds at the location of the pitot tube differ appreciably from the flight speed, are discussed in Chapter 4.

Furthermore, in gas flows the variation in height of the streamline relative to the reference level is not large; the variation of the term $\rho g z$ along a streamline in a gas flow will be small compared with p_0 so that its magnitude may be absorbed in the constant p_0. For instance, for air under standard conditions $\rho g z = 1.23 \times 9.81 = 12.07$ N/m^2 per meter change in z; and for $V = 30$ m/sec, $\tfrac{1}{2}\rho V^2 = 553$ N/m^2. Then if there is 1 m difference in altitude between the points where p_0 and p are measured, the error involved in neglecting $\rho g z$ at a flow speed of 30 m/sec will be about 2 percent in $\tfrac{1}{2}\rho V^2$ or 1 percent in the velocity. The error will decrease as the reciprocal square of the flow velocity and is therefore neglected in aircraft aerodynamics.

In practical measurements on flow around bodies data are generally presented in terms of the **pressure coefficient,** C_p. To define C_p, we assume irrotational flow so that p_0 is constant everywhere and we identify p_∞ and V_∞ as values far from the body; then Eq. 7 yields

$$p_\infty + \tfrac{1}{2}\rho V_\infty^2 = p + \tfrac{1}{2}\rho V^2$$

Using what has become a standard abbreviation, $q = \frac{1}{2}\rho V^2$, we define the **pressure coefficient**

$$C_p = \frac{p - p_\infty}{q_\infty} = 1 - \left(\frac{V}{V_\infty}\right)^2 \tag{11}$$

where $q_\infty = \frac{1}{2}\rho V_\infty^2$. Then, in an incompressible flow, $C_p = 1$ at a stagnation point, that is, where $V = 0$, and $C_p = 0$ far from the body where $V = V_\infty$.

It is to be noted that for flight through the atmosphere p_∞ is the barometric pressure at a given altitude and remains constant during flight at the same level, while the stagnation pressure p_0 increases with flight speed. On the other hand, when we are considering flow in a tube of varying cross section, the stagnation pressure p_0 is constant at every station in the tube, regardless of the speed, while the static pressure varies as indicated by Eq. 7.

So far we have been considering flow along specific streamlines. In the next section, we generalize the analysis and show that Eq. 7 with p_0 constant is valid throughout an irrotational flow, but in a rotational flow the value of p_0 will vary from streamline to streamline by an amount depending on the velocity and the vorticity.

3.4   Euler's Equation

Newton's law of motion, in general terms, is: **the rate of change of momentum of a tagged element in a flow is equal in both magnitude and direction to the resultant force acting on the element.** Applied to an element of mass $\rho\Delta x\Delta y\Delta z$, it gives

$$\mathbf{F}\Delta x\Delta y\Delta z = \frac{\mathfrak{D}}{\mathfrak{D}t}(\rho\Delta x\Delta y\Delta z\mathbf{V})$$

where $\mathbf{F}$ is the resultant of the forces acting per unit volume on the element; they arise through viscous stresses, pressure gradients, gravity forces, and, for a conducting fluid, electromagnetic stresses. The right-hand side is the substantial rate of change of momentum of the element. As with Eq. 4, since the mass of the element $\rho\Delta x\Delta y\Delta z$ is fixed, the vector equation reduces to

$$\mathbf{F} = \rho\frac{\mathfrak{D}\mathbf{V}}{\mathfrak{D}t} \tag{12}$$

which is then valid for compressible or incompressible flow.

The force on an element arising from pressure gradients in the flow is obtained from a consideration of Fig. 4. It is to be noted that the element shown is in motion along a streamline of the flow, and the pressures designated are *static pressures*; that is, they are the barometric pressures indicated by an instrument moving with the flow. The pressure at the center of the element is p and the pressures acting on the six faces of the cube are designated in the figure. The resultant force in the

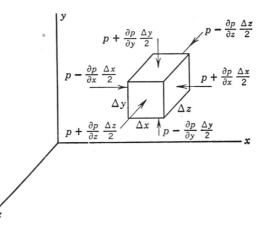

Fig. 4 Dynamic equilibrium of a fluid element.

positive direction of x, y, and z axes are, respectively, $-(\partial p/\partial x)\Delta x \cdot \Delta y \Delta z$, $-(\partial p/\partial y)\Delta y \cdot \Delta x \Delta z$, and $-(\partial p/\partial z)\Delta z \cdot \Delta x \Delta y$, where $\Delta y \Delta z$ and so forth are the areas of the appropriate faces of the cube.   In vector notation, the pressure force may be written

$$-\left(\mathbf{i}\frac{\partial p}{\partial x} + \mathbf{j}\frac{\partial p}{\partial y} + \mathbf{k}\frac{\partial p}{\partial z}\right)\Delta x \Delta y \Delta z = -\boldsymbol{\nabla}p\Delta x \Delta y \Delta z$$

and, thus, $-\operatorname{grad} p \equiv -\boldsymbol{\nabla}p$ is, in magnitude and direction, the pressure force per unit volume acting on a fluid element.   Let $\mathbf{g}$ represent the acceleration due to gravity. The weight of the fluid element is $(\rho \Delta x \Delta y \Delta z)\mathbf{g}$.   Then the total external force on the element is

$$\mathbf{F}\Delta x \Delta y \Delta z = (\rho \mathbf{g} - \boldsymbol{\nabla}p)\Delta x \Delta y \Delta z$$

and Eq. 12 becomes

$$\rho \mathbf{g} - \boldsymbol{\nabla}p = \rho \frac{\mathfrak{D}\mathbf{V}}{\mathfrak{D}t} \tag{13}$$

This equation, known as Euler's equation of motion, is named after Leonard Euler (1703–1783) who was responsible for its formulation.   It is applicable to compressible or incompressible flows.   In Cartesian form, in which the substantial derivative is expanded to show the local and convective accelerations (see Eqs. 2 and 3), the three component equations are

$$g_x - \frac{1}{\rho}\frac{\partial p}{\partial x} = \frac{\partial u}{\partial t} + u\frac{\partial u}{\partial x} + v\frac{\partial u}{\partial y} + w\frac{\partial u}{\partial z}$$

$$g_y - \frac{1}{\rho}\frac{\partial p}{\partial y} = \frac{\partial v}{\partial t} + u\frac{\partial v}{\partial x} + v\frac{\partial v}{\partial y} + w\frac{\partial v}{\partial z} \tag{14}$$

$$g_z - \frac{1}{\rho}\frac{\partial p}{\partial z} = \frac{\partial w}{\partial t} + u\frac{\partial w}{\partial x} + v\frac{\partial w}{\partial y} + w\frac{\partial w}{\partial z}$$

where g_x, g_y, and g_z are the three components of the gravitational acceleration. The development of Euler's equation has assumed the fluid to be nonviscous. For cases in which viscous shearing stresses cannot be neglected, further terms must be added to the force **F** of Eq. 12. This subject is treated in Chapter 14 and Appendix B.

We may expand the right side of Eq. 13 to obtain

$$\rho \mathbf{g} - \nabla p = \rho \left[\frac{\partial \mathbf{V}}{\partial t} + (\mathbf{V} \cdot \nabla) \, \mathbf{V} \right] \tag{15}$$

where the "operator"

$$\mathbf{V} \cdot \nabla = (\mathbf{i}u + \mathbf{j}v + \mathbf{k}w) \cdot \left(\mathbf{i}\frac{\partial}{\partial x} + \mathbf{j}\frac{\partial}{\partial y} + \mathbf{k}\frac{\partial}{\partial z} \right) = u\frac{\partial}{\partial x} + v\frac{\partial}{\partial y} + w\frac{\partial}{\partial z}$$

since $\mathbf{i} \cdot \mathbf{i} = \mathbf{j} \cdot \mathbf{j} = \mathbf{k} \cdot \mathbf{k} = 1$ and all "cross" dot products $\mathbf{i} \cdot \mathbf{j}$ and the like vanish. We recognize from the last expression that $(\mathbf{V} \cdot \nabla) \, \mathbf{V}$ in Eqs. 15 is the convective derivative of the velocity.

3.5. Integration of Euler's Equation

In spite of their seeming complexity, Euler's equations can be integrated (1) throughout an irrotational flow, and (2) along a streamline even if the flow is rotational. These integrations are sketched below.

IRROTATIONAL FLOW

We multiply the first equation in Eqs. 14 by dx, the second by dy, the third by dz, or, equivalently, form the dot products of the terms of Eq. 13 or 15 with $d\mathbf{s}$, an incremental distance with arbitrary direction, and by so doing express work done in direction $d\mathbf{s}$:

$$\rho \mathbf{g} \cdot d\mathbf{s} - \nabla p \cdot d\mathbf{s} = \rho \frac{\mathcal{D} \mathbf{V}}{\mathcal{D} t} \cdot d\mathbf{s} \tag{16}$$

Now we impose irrotationality by introducing Eqs. 2.35:

$$\frac{\partial u}{\partial y} = \frac{\partial v}{\partial x} ; \quad \frac{\partial u}{\partial z} = \frac{\partial w}{\partial x} ; \quad \frac{\partial v}{\partial z} = \frac{\partial w}{\partial y}$$

and obtain for the first term on the right in Eq. 16,

$$\rho \frac{\mathcal{D} u}{\mathcal{D} t} \, dx = \rho \left[\frac{\partial u}{\partial t} + \frac{\partial}{\partial x}\left(\frac{u^2}{2}\right) + \frac{\partial}{\partial x}\left(\frac{v^2}{2}\right) + \frac{\partial}{\partial x}\left(\frac{w^2}{2}\right) \right] dx$$

with corresponding expressions for the remaining two terms. The velocity potential (Eqs. 2.42) is then introduced, so that, with

$$\frac{\partial u}{\partial t} = \frac{\partial}{\partial t}\frac{\partial \phi}{\partial x} = \frac{\partial}{\partial x}\frac{\partial \phi}{\partial t}$$

the right side of Eq. 16 becomes

$$\rho \frac{\mathcal{D}\mathbf{V}}{\mathcal{D}t} \cdot d\mathbf{s} = \rho \left[d\left(\frac{\partial \phi}{\partial t}\right) + d\left(\frac{V^2}{2}\right) \right] \tag{17}$$

where $V^2 = u^2 + v^2 + w^2$. Next, the gravitational potential, U, defined in Section 3.2, is introduced, so that the left side of Eq. 16 becomes

$$\rho \mathbf{g} \cdot d\mathbf{s} - \boldsymbol{\nabla} p \cdot d\mathbf{s} = \rho dU - dp \tag{18}$$

and Eq. 16, after using Eqs. 17 and 18, becomes

$$d\left(\frac{\partial \phi}{\partial t} + \frac{V^2}{2}\right) = dU - \frac{dp}{\rho} \tag{19}$$

If the density is assumed constant Eq. 19 integrates to

$$\frac{\partial \phi}{\partial t} + \frac{V^2}{2} + \frac{p}{\rho} - U = f(t) \tag{20}$$

Since $U = gz$, Eqs. 19 and 20 are identical, respectively, with Eqs. 6 and 7 except for the inclusion of the unsteady term. Thus, if the flow is steady and irrotational, and the gravity term is negligible or constant, there is a unique relation between the pressure and velocity; for unsteady flow this relation is affected throughout by the rate of change of the velocity potential as a function of the coordinates and time.

FLOW ALONG A STREAMLINE

Instead of making use of the irrotationality conditions to integrate Eq. 16, as we did above, the terms in the equations are expressed as perfect differentials by means of the relations that restrict their application to flow properties on a given stream-line. These are the relations that express the slopes of the streamlines in terms of the velocity components; these are

$$\frac{v}{u} = \frac{dy}{dx} ; \quad \frac{w}{u} = \frac{dz}{dx} ; \quad \frac{w}{v} = \frac{dz}{dy} \tag{21}$$

By use of these relations the first term on the right-hand side of Eq. 16 becomes

$$\rho \frac{\mathcal{D}u}{\mathcal{D}t} dx = \rho \left[\frac{\partial u}{\partial t} dx + d\left(\frac{u^2}{2}\right) \right]$$

Now, however, since we do not assume irrotationality, a velocity potential does not exist, and we are therefore not able to reduce the unsteady term to a perfect differential. We therefore restrict the analysis to a steady flow. Then, Eqs. 21 enable us to reduce Eq. 16 to Eq. 19, minus the unsteady term.

The conclusion we draw from these analyses is: **In steady inviscid flow, the stagnation pressure, p_0, occurring in Eqs. 6 and 7, is constant throughout an irrotational flow; in a rotational flow p_0 is constant along any streamline but changes from streamline to streamline.** The analysis of the next section gives the magnitude of the change of p_0.

3.6.   Bernoulli's Equation for Rotational Flow

To get an expression for the variation of p_0 normal to the streamlines in a steady flow, it is necessary to consider the equilibrium of the fluid in a direction normal to the streamlines. If the streamlines are curved, there will be an inertia force per unit mass V^2/R, where R is the local radius of curvature of the streamline, directed normal to the velocity, and, if gravity force is neglected, there will exist an equal and opposite pressure force. The force balance normal to a streamline is shown in Fig. 5. The equilibrium equation is

$$\left(\frac{\partial p}{\partial n}\right) \Delta n \Delta s \Delta z - \left(\frac{\rho V^2}{R}\right) \Delta n \Delta s \Delta z = 0 \tag{22}$$

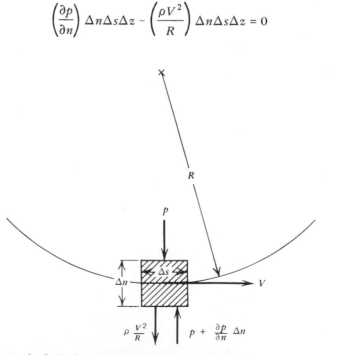

Fig. 5 Equilibrium of a fluid element perpendicular to a streamline.

where Δz is the dimension perpendicular to the plane of the paper. From Bernoulli's equation, Eq. 7 (neglecting the term $\rho g \Delta z$) becomes

$$\frac{\partial p}{\partial n} = \frac{\partial p_0}{\partial n} - \rho V \frac{\partial V}{\partial n} \tag{23}$$

Substituting Eq. 23 in Eq. 22 and solving for $(\partial p_0/\partial n)$,

$$\frac{\partial p_0}{\partial n} = \rho V \left(\frac{V}{R} + \frac{\partial V}{\partial n} \right)$$

According to Fig. 2.10 and Eq. 2.25 $(V/R + \partial V/\partial n)$ is the vorticity component $\omega_\perp$ normal to the plane of the flow. We have, therefore,

$$\frac{\partial p_0}{\partial n} = \rho V \omega_\perp \tag{24}$$

Example 1. Solid Body Rotation

In Section 2.7, solid body rotation was characterized by (Eqs. 2.28) $u_r = 0$, $u_\theta = \Omega r$ and by constant angular velocity $\epsilon_z = \Omega$; therefore, the vorticity $\omega_\perp = 2 \epsilon_z = 2\Omega$. Then, by Eq. 24, $\partial p_0/\partial n \equiv dp_0/dr = 2\rho u_\theta \Omega = 2\rho\Omega^2 r$, so that

$$p_0 = \rho\Omega^2 r^2 + p_0^0 = \rho u_\theta^2 + p_0^0$$

where $p_0^0 = p_0$ at $r = 0$. Then Bernoulli's equation becomes

$$p + \tfrac{1}{2} \rho u_\theta^2 = p_0^0 + \rho u_\theta^2$$

or the static pressure is given by

$$p = p_0^0 + \tfrac{1}{2} \rho u_\theta^2$$

and the pressure force per unit volume on a fluid element is in the r direction. Thus, since $u_\theta = \Omega r$, we find

$$|\nabla p| = \frac{dp}{dr} = \frac{\rho u_\theta^2}{r} = \rho\Omega^2 r$$

Thus the pressure gradient is radial and is equal and opposite to the centrifugal force per unit volume.

Example 2. Couette Flow

In Fig. 2.7, the flow $u = ky$, $v = 0$, termed "plane Couette flow," is shown. By Eqs. 2.31, $\omega_z = \partial v/\partial x - \partial u/\partial y = -k$, so that $\partial p_0/\partial n \equiv -dp_0/dy = -\rho k^2 y$ and

$$p_0 = \tfrac{1}{2} \rho k^2 y^2 + p_0^0 = \tfrac{1}{2} \rho u^2 + p$$

where $p_0^0 = p_0$ at $y = 0$, and thus from Bernoulli's equation $p = p_0^0$ is constant. This result is predicted by the conditions of the problem, since the streamlines are rectilinear; for, if p were to vary with y, $\partial p/\partial y$ would not vanish and the resulting force on the fluid elements would cause the streamlines to curve in order to generate a balancing centrifugal force.

3.7 The Momentum Theorem of Fluid Mechanics

Euler's equation, developed in the last section, is a statement of the conditions that must be fulfilled at each point in the field if the fluid is to be in dynamic equilibrium. Frequently, in aerodynamics the details of a flow field are too complicated to deal with and a gross relation involving a group of field points is desired. The momentum theorem of fluid mechanics provides this relation. It is an extension of Newton's law of motion used in the foregoing section and may be stated: **the time rate of change of momentum of a group of particles of fixed identity is equal in both magnitude and direction to the force acting on the particles.**

The conservation of momentum principle is stated in terms of particle properties. Its use in field theory requires a conversion to field properties, which may be made by considering the control surface $\hat{S}$ in Fig. 6. At time t, control surface $\hat{S}$, which is fixed in the fluid, encloses a region $\hat{R}$ containing a tagged set of fluid elements. At time t_1, these elements will have moved to the region enclosed by the dotted surface $\hat{S}_1$.

Let $\mathbf{M}_A$, $\mathbf{M}_B$, and $\mathbf{M}_C$ be the momentum of the fluid within the regions A, B, and C. For instance, $\mathbf{M}_A = \iiint_A \rho \mathbf{V} dA$. Then at time t the fluid within $\hat{S}$ has the momentum $\mathbf{M}_A(t) + \mathbf{M}_B(t)$, and at time t_1, the same collection of particles has momentum $\mathbf{M}_B(t_1) + \mathbf{M}_C(t_1)$. The momentum change during the interval $t_1 - t$ is the difference between these two expressions; it may be written in the form

$$[\mathbf{M}_B(t_1) - \mathbf{M}_B(t)] + [\mathbf{M}_C(t_1) - \mathbf{M}_A(t)]$$

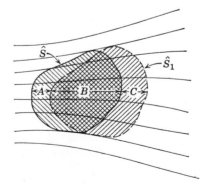

Fig. 6 Conversion to field properties.

We divide by $t_1 - t$ and let $t_1 \to t$; then in the limit $\hat{S}_1$ and $\hat{S}$ coincide and the rate of change of momentum of the fluid within $\hat{S}$ is given by

$$\lim_{t_1 \to t} \left[\frac{\mathbf{M}_B(t_1) - \mathbf{M}_B(t)}{t_1 - t} + \frac{\mathbf{M}_C(t_1) - \mathbf{M}_A(t)}{t_1 - t} \right] \tag{25}$$

In the limit, the first term represents the time rate of change of momentum in the region $\hat{R}$, enclosed by the surface $\hat{S}$, which, as was stated above, is fixed in space. The first term in (25) may therefore be written

$$\frac{d}{dt} \iiint_{\hat{R}} \rho \mathbf{V} d\hat{R} = \iiint_{\hat{R}} \frac{\partial}{\partial t} (\rho \mathbf{V}) \, d\hat{R} \tag{26}$$

in which d/dt is an ordinary differentiation. The two expressions are equal only if $\hat{S}$, enclosing the tagged elements comprising the region $\hat{R}$, is fixed in space. In the following equations we use the first form of Eq. 26. The second term in (25) represents in the limit the rate at which momentum leaves $\hat{S}$ minus the rate at which it enters; it thus represents the momentum flux through $\hat{S}$, outflow being positive. Therefore, if we define the outward drawn normal $\mathbf{n}$ as positive, this term is written:

$$\iint_{\hat{S}} \rho \mathbf{V} (\mathbf{V} \cdot \mathbf{n}) \, d\hat{S} \tag{27}$$

The momentum theorem states that the sum of the terms (26) and (27) is equal to the force acting on the fluid. Since most of the practical applications of the momentum theorem involve finding the force acting on a body within a control volume, it is convenient to break this force $\mathbf{F}$ into three parts: a force $-\mathbf{F}_e$ exerted by the body on the fluid, a pressure force $-\iint_{\hat{S}} p\mathbf{n} d\hat{S}$ exerted on the control surface by the fluid outside the surface, and a body force (gravity or electromagnetic) acting on the fluid within $\hat{R}$. If the body force is gravitational its magnitude is $\iiint_{\hat{R}} \rho \mathbf{g} d\hat{R}$. Then

$$\mathbf{F} = -\mathbf{F}_e - \iint_{\hat{S}} p\mathbf{n} \, d\hat{S} + \iiint_{\hat{R}} \rho \mathbf{g} \, d\hat{R} \tag{28}$$

and, for the mathematical formulation of the momentum theorem we combine the terms (26) and (27) with Eq. 28 to get

$$\frac{d}{dt} \iiint_{\hat{R}} \rho \mathbf{V} \, d\hat{R} + \iint_{\hat{S}} \rho \mathbf{V} (\mathbf{V} \cdot \mathbf{n}) \, d\hat{S} = -\mathbf{F}_e - \iint_{\hat{S}} p\mathbf{n} \, d\hat{S} + \iiint_{\hat{R}} \rho \mathbf{g} \, d\hat{R} \tag{29}$$

The two terms on the left measure the momentum gained per second by the fluid streaming through the region; the three terms on the right measure the force that was exerted to cause that increase in momentum. There are two contributions to the momentum increase: the first term is the rate of increase of momentum of the fluid within $\hat{R}$ at a given instant and will vanish in a steady flow (it would be *nonzero* for instance for the flow near an accelerating or maneuvering aircraft); the second term is the steady flow contribution and measures the flux of momentum through $\hat{S}$, that is, the amount by which the momentum leaving $\hat{S}$ per second is greater than that entering. The first term on the right, $-\mathbf{F}_e$, has a nonzero value if there is a body being held within $\hat{S}$ against a force $+\mathbf{F}_e$ exerted by the fluid on the body, as would be the case for instance if the body experienced a drag, $\mathbf{D}$; then the force on the fluid would be upstream and the first term on the right would be $-\mathbf{D}$. The second term on the right is the resultant contribution of pressure on the control surface to the force on the fluid; the sign is negative because the unit normal vector is positive outward; this term vanishes if the pressure is constant on $\hat{S}$. The third term on the right represents a net body force on the fluid within $\hat{R}$.

An important aspect of the momentum theorem, and one that makes it so extremely useful in engineering computations, is that one needs to know only the flow properties at the control surface boundary and the magnitude and direction of any external and body forces acting across the boundary.

In the applications treated here, we neglect gravity forces. Then the three Cartesian components of Eq. 29 are

$$\frac{d}{dt}\iiint_{\hat{R}} \rho u \, d\hat{R} + \iint_{\hat{S}} \rho u (\mathbf{V} \cdot \mathbf{n}) \, d\hat{S} = -F_x - \iint_{\hat{S}} p \cos (\mathbf{n}, \mathbf{i}) \, d\hat{S}$$

$$\frac{d}{dt}\iiint_{\hat{R}} \rho v \, d\hat{R} + \iint_{\hat{S}} \rho v (\mathbf{V} \cdot \mathbf{n}) \, d\hat{S} = -F_y - \iint_{\hat{S}} p \cos (\mathbf{n}, \mathbf{j}) \, d\hat{S} \qquad (30)$$

$$\frac{d}{dt}\iiint_{\hat{R}} \rho w \, d\hat{R} + \iint_{\hat{S}} \rho w (\mathbf{V} \cdot \mathbf{n}) \, d\hat{S} = -F_z - \iint_{\hat{S}} p \cos (\mathbf{n}, \mathbf{k}) \, d\hat{S}$$

where cos $(\mathbf{n}, \mathbf{i})$ is the cosine of the angle between the unit normal and the unit vector $\mathbf{i}$, and so forth, and F_x, F_y, and F_z are respectively the magnitudes of the components of $\mathbf{F}_e$ in the direction of x, y, and z axes.

Because of the importance of the momentum theorem, three examples of applications of Eq. 29 for steady flows are given below. Other examples are given in later chapters.

Example 1. Drag of a Cylindrical Body.

A practical means of determining the drag of a body from velocity measurements in its wake is used as the first illustration of the momentum theorem. Although the method involves a number of approximations, good results can be obtained.

Because of viscosity effects, a wake of retarded flow exists behind the cyl-

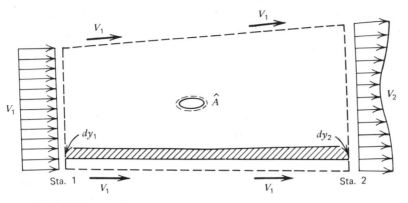

Fig. 7 Drag of a body from wake measurements.

inder of Fig. 7. In the wake region, the velocity is less than the upstream value, as illustrated by the profile at the right.

The control volume is bounded by the streamlines far from the body and the lines normal to the streamlines labeled station 1 and station 2, respectively, far upstream and far downstream of the body. Then the pressure on the control surface is constant and the pressure integral in Eq. 29 vanishes. We assume steady incompressible flow and negligible gravity forces. Then the force exerted by the fluid on the body, follows from Eq. 29:

$$\mathbf{F}_e = - \iint_{\widehat{S}} \rho \mathbf{V} (\mathbf{V} \cdot \mathbf{n}) \, d\widehat{S} = -\mathbf{i} \int_{\text{sta.}\,2} V_2 (\rho V_2 \, dy_2) + \mathbf{i} \int_{\text{sta.}\,1} V_1 (\rho V_1 \, dy_1)$$

where y_1 and y_2 are the coordinates measured along the normals to the streamlines respectively at stations 1 and 2. Further, $\rho V_1 \, dy_1$ is the rate at which fluid enters the control volume through dy_1 and $\rho V_2 \, dy_2$ is the rate at which fluid leaves dy_2; the signs in the equation are determined by the convention that the outward drawn normal is positive. $\mathbf{F}_e$ is the drag of the cylinder per unit length, $\mathbf{D}'$.

The practical use of the equation is facilitated by expressing the two integrals in terms of the single variable y_2. Consideration of the cross-hatched section of the stream tube in Fig. 7, bounded at its ends by dy_1 and dy_2, indicates that, for every stream tube, mass conservation requires that

$$\rho V_1 \, dy_1 = \rho V_2 \, dy_2$$

and when this relation is introduced the above equation may be written

$$D' = \int_{\text{sta.}\,2} \rho V_2 (V_1 - V_2) \, dy_2$$

The physical meaning of this equation becomes clear when we note that $\rho V_2 \, dy_2$ is the mass of fluid leaving dy_2 in one second and during its flow around the body its velocity decreased from V_1 to V_2. The integrand is therefore the momentum lost by the fluid leaving the control volume through dy_2 per second. In the absence of pressure forces on the control surface, the integral is the loss of momentum suffered by the fluid passing through the downstream plane per second, which, by the momentum theorem, is exactly equal to the drag per unit length of the cylinder.

The equation is used practically for instance to determine the contribution at a particular spanwise station to the total drag of a wing; then V_1 is the flight speed and V_2 is the measured velocity as a function of position along a line normal to the trailing edge. The location of the measurement plane behind the trailing edge so that the contribution of the pressure integral in Eq. 29 can be neglected must be determined by experiment; measurements behind airfoils at low angles of attack (see Goldstein, 1938, p. 262) show that if the measurement plane location is 0.12 chord behind the trailing edge, pressure variation is small enough so that the above integral gives the drag to within a few percent of its correct value.

Example 2. Force on a Pipe Bend.

We calculate the force required to cause a change in flow direction by applying the momentum theorem to the flow in a pipe bend as shown in Fig. 8.

Fluid enters the pipe bend with pressure p_1, density ρ_1, and velocity $\mathbf{V}_1$ parallel to the x axis, and leaves with pressure p_2, density ρ_2, and velocity

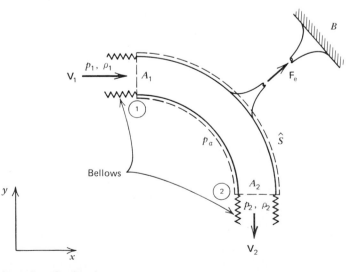

Fig. 8 Flow through a pipe bend.

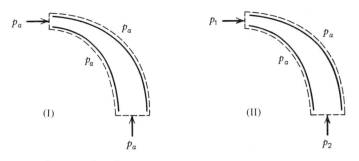

Fig. 9 Pressures on the control surface.

V_2 parallel to the y axis. The bend is connected with the rest of the piping through perfectly flexible bellows so that the external force $\mathbf{F}_e$ is exerted only at the support B. A control surface $\hat{S}$ is drawn as shown, coincident with the outer surface of the pipe and normal to the flow at the stations one and two. Then, since there are no velocity gradients normal to $\hat{S}$, no shearing stresses exist at the boundary, and only pressure forces act on the fluid at the control surface. The external pressure is p_a.

We evaluate the pressure integral in Eq. 29 by comparing cases I and II in Fig. 9, the only difference between them being that in I, with constant pressure p_a everywhere on $\hat{S}$, no resultant pressure force can act on the fluid within, whereas under the *actual* conditions II the pressures are p_1 and p_2 over A_1 and A_2, respectively, and p_a over the remainder of $\hat{S}$. It follows that the resultant pressure force on the fluid within $\hat{S}$ can be found by subtracting pressure forces in I from those in II. Accordingly (with $\mathbf{n} = -\mathbf{i}$ over A_1, and $\mathbf{n} = -\mathbf{j}$ over A_2),

$$-\iint_{\hat{S}} p\mathbf{n} \, d\hat{S} = \mathbf{i}(p_1 - p_a) A_1 + \mathbf{j}(p_2 - p_a) A_2$$

The momentum flux, with the $\mathbf{n}$ directions shown above, becomes

$$\iint_{\hat{S}} \rho\mathbf{V}(\mathbf{V} \cdot \mathbf{n}) \, d\hat{S} = -\rho_2 \mathbf{j} V_2 [(-\mathbf{j} V_2) \cdot (-\mathbf{j})] A_2 + \rho_1 \mathbf{i} V_1 [(\mathbf{i} V_1) \cdot (-\mathbf{i})] A_1$$

or, since $\mathbf{i} \cdot \mathbf{i} = \mathbf{j} \cdot \mathbf{j} = 1$,

$$\iint_{\hat{S}} \rho\mathbf{V}(\mathbf{V} \cdot \mathbf{n}) \, d\hat{S} = -\rho_2 V_2^2 A_2 \mathbf{j} - \rho_1 V_1^2 A_1 \mathbf{i}$$

Finally, if we postulate steady flow, neglect gravity forces, and use the above relations in Eq. 29, we get

$$-\rho_2 V_2^2 A_2 \mathbf{j} - \rho_1 V_1^2 A_1 \mathbf{i} = -\mathbf{F}_e + \mathbf{i}(p_1 - p_a) A_1 + \mathbf{j}(p_2 - p_a) A_2$$

and the force $\mathbf{F}_e$ on the support becomes

$$\mathbf{F}_e = \mathbf{i}[\rho_1 A_1 V_1^2 + (p_1 - p_a) A_1] + \mathbf{j}[\rho_2 A_2 V_2^2 + (p_2 - p_a) A_2]$$

It will be noted that the viscosity of the fluid is not neglected in the above analysis. However, the viscosity will affect the pressures and densities as well as the uniformity of the velocities over A_1 and A_2, and therefore will in general have a significant effect on $\mathbf{F}_e$.

Mass conservation provides an additional relation between the properties at stations one and two, that is,

$$\rho_1 V_1 A_1 = \rho_2 V_2 A_2 = \dot{m}$$

Then the equation for the external force may be written

$$\mathbf{F}_e = \mathbf{i}[\dot{m} V_1 + (p_1 - p_a) A_1] + \mathbf{j}[\dot{m} V_2 + (p_2 - p_a) A_2]$$

Example 3. Deflection Through Vanes.

The two-dimensional cascade of vanes shown in Fig. 10 deflects the flow through an angle and changes the pressure from p_1 to p_2. F_x and F_y are the force components acting on the vane normal to and parallel to the plane of the cascade.

The cascade is taken to comprise an infinite number of blades with constant gap h, so that all corresponding streamlines, such as those designated AB and CD, are identical. The control surface $\hat{S}$ is made up of the streamlines AB and CD and the surfaces AC and BD. Assume steady incompressible flow with p_1, V_1 and p_2, V_2, respectively, constant along AC and BD. The

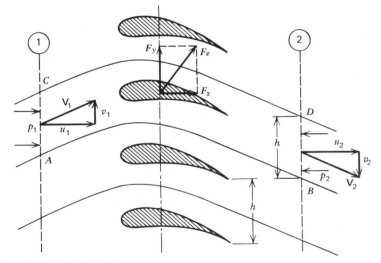

Fig. 10 Flow through cascade of vanes.

first term in Eq. 29 vanishes since the flow is steady, and, since AB and CD are streamlines, continuity requires that the mass flow rate $\rho u_1 h = \rho u_2 h \equiv \dot{m}$. Therefore, $u_1 = u_2$ and, if we neglect gravity, the remaining terms of Eq. 29 are

$$\iint_{\hat{S}} \rho V(V \cdot n)\, d\hat{S} = i \cdot 0 + j \rho u_2 h (v_2 - v_1)$$

$$\iint_{\hat{S}} pn\, d\hat{S} = i(p_2 - p_1)\, h$$

and the force exerted by the fluid on the vane is

$$F_e = i(p_1 - p_2)\, h + j \dot{m}(v_1 - v_2)$$

The flow is irrotational so that $p_0 = $ constant, and

$$p_1 + \tfrac{1}{2}\rho V_1^2 = p_2 + \tfrac{1}{2}\rho V_2^2$$

But, since $u_1 = u_2$ and $V_2^2 - V_1^2 = v_2^2 - v_1^2$,

$$F_e = iF_x + jF_y = i\,\tfrac{1}{2}\,\rho h\,(v_2^2 - v_1^2) + j\dot{m}(v_1 - v_2)$$

Thus,

$$F_x = \tfrac{1}{2}\rho h(v_2^2 - v_1^2) = (p_1 - p_2)\, h; \qquad F_y = \dot{m}(v_1 - v_2)$$

It should be pointed out that in this example v_1 is positive and v_2 is negative according to their directions shown in Fig. 10.

The flow through the cascade resembles roughly that through a turbine wheel far from the axis of rotation if the flow is viewed from the moving blade. Then F_x and F_y are analogous respectively to the thrust and torque per unit span of the blade; they are functions of the fluid density, the spacing h (determined by the number of blades in the wheel), the pressure increase, the mass flow rate $\dot{m}$, and the change in peripheral speed $(v_1 - v_2)$ through the wheel.

CHAPTER
4

Flow About a Body

4.1 Introduction

The current chapter describes the synthesis of the steady irrotational flow about bodies in an incompressible inviscid fluid through the addition (superposition) of the simple flows treated in Chapter 2. These are uniform flows, sources, sinks, and vortices. It was shown there that these flows satisfy continuity, and are thus physically possible, and that they are irrotational. We show in this chapter that a given flow field resulting from suitable superposition of these simple flows represents the flow around a body that has accelerated from rest to a steady velocity in a stationary fluid.

Once the flow field is known, we can, by Bernoulli's equation derived in Chapter 3, calculate the pressure distribution on the surface of the body; the forces and moments acting follow from integration of pressures and moments over the surface.

We describe the flow about an airfoil and the manner in which the viscosity, by causing the fluid to flow smoothly past the sharp trailing edge, is responsible for generating a circulation which is, in turn, directly proportional to the lift. After this condition of smooth flow at the trailing edge, called the *Kutta condition*, is imposed, the methods of inviscid fluid analysis enable calculation of the aerodynamic characteristics of airfoils; however, this calculation and comparison with experiment is reserved for Chapter 5.

Approximate methods that utilize computer techniques for representing the flow around complicated bodies are described for two- and three-dimensional configurations, and are shown to be in excellent agreement with experiment.

4.2 The Governing Equations; Boundary Conditions

The analyses of the present chapter apply to two- and three-dimensional flows that satisfy the equations for continuity and irrotationality. The equation for continuity, that is, conservation of mass,

$$\text{div } \mathbf{V} = 0 \tag{1}$$

for incompressible flow is identically satisfied in a two-dimensional flow by the existence of a stream function ψ (Section 2.6). The vector equation expressing irrotationality, that is, that the three components of the vorticity (Eqs. 2.31 and

2.35) vanish, is

$$\text{curl } \mathbf{V} = 0 \tag{2}$$

for compressible *or* incompressible flow; it is satisfied identically by the existence of a velocity potential, ϕ.

The equations for the velocity components in a two-dimensional flow are (Eqs. 2.45)

$$u = \frac{\partial \phi}{\partial x} = \frac{\partial \psi}{\partial y} \; ; \qquad u_r = \frac{\partial \phi}{\partial r} = \frac{1}{r}\frac{\partial \psi}{\partial \theta}$$

$$v = \frac{\partial \phi}{\partial y} = -\frac{\partial \psi}{\partial x} \; ; \qquad u_\theta = \frac{1}{r}\frac{\partial \phi}{\partial \theta} = -\frac{\partial \psi}{\partial r} \tag{3}$$

Substituting for u and v from Eqs. 3, Eqs. 1 and 2 become, respectively, for two-dimensional flows,

$$\nabla^2 \phi = 0$$
$$\nabla^2 \psi = 0 \tag{4}$$

In these equations the operator

$$\nabla^2 \equiv \frac{\partial^2}{\partial x^2} + \frac{\partial^2}{\partial y^2} \equiv \frac{\partial^2}{\partial r^2} + \frac{1}{r}\frac{\partial}{\partial r} + \frac{1}{r^2}\frac{\partial^2}{\partial \theta^2} \tag{5}$$

termed the Laplacian (named for Pierre de Laplace 1749–1827), is important in electrical, as well as in fluid flow field theory and other fields of physics.

It follows from the foregoing that either of the following two pairs of conditions

$$\phi \quad \text{exists and} \quad \nabla^2 \phi = 0$$
$$\psi \quad \text{exists and} \quad \nabla^2 \psi = 0 \tag{6}$$

gives the identical information, that is, that the flows they describe satisfy continuity and irrotationality. Also (Section 2.11) the curves $\phi = $ constant and $\psi = $ constant are orthogonal families. Which of the two pairs of Eqs. 6 are used to solve a specific flow problem is a matter of which pair enables the equations to be put in the most convenient form for solution.

In addition to the Eqs. 6 we must have suitable **boundary conditions** to enable us to choose the solution that describes the particular problem. For instance, to find the flow field about a specific body moving at speed V_∞ in the minus x direction through a stagnant fluid we must impose the conditions that the surface of the body is a streamline of the flow, that is,

$$\psi = \text{constant} \quad \text{or} \quad \frac{\partial \phi}{\partial n} = 0 \tag{7}$$

at the surface, on which n is the direction of the normal, and that at a great distance the velocity of the fluid *relative to the body* is $V_\infty \mathbf{i}$, that is, in Cartesian coordinates, the velocity components far from the body are

$$u = \frac{\partial \phi}{\partial x} = \frac{\partial \psi}{\partial y} = V_\infty$$

$$v = \frac{\partial \phi}{\partial y} = -\frac{\partial \psi}{\partial x} = 0$$

(8)

It can be shown that a solution of either of Eqs. 4 that satisfies Eq. 7 at the surface of a given body about which a given circulation exists, and Eqs. 8 at infinity is unique; in other words, uniqueness of the irrotational flow about a given body is established if the boundary conditions at infinity and at the body are satisfied and, *in addition, if the magnitude of the circulation around the body is specified*.

Thus the kinematical problem of finding the flow pattern of an incompressible inviscid flow about a body is reduced to the purely mathematical one of finding a suitable particular solution of Laplace's equation. The incompressible, inviscid, irrotational flows derived in Chapter 2 are summarized in first three entries of Eqs. 2.44. These are

Flow	ψ	ϕ	
Vortex	$(\Gamma/2\pi) \ln r$	$-(\Gamma/2\pi)\theta$	
Source	$(\Lambda/2\pi)\theta$	$(\Lambda/2\pi) \ln r$	(9)
Uniform flow in x direction	$V_\infty y$	$V_\infty x$	

Each of the stream functions and velocity potentials in Eqs. 9 is expressed in the coordinate system in which it takes its most concise form; the system used to solve a specific problem is chosen on the basis of conciseness and tractability.

In the next section we show how these simple flows may be superimposed to describe the flow about bodies of arbitrary shape.

4.3   Superposition of Flows

We show in this section that since Eqs. 4 are *linear* we can "superimpose" flows in the sense that, given two or more flows with stream functions $\psi_1, \psi_2, \ldots, \psi_n$ such that $\nabla^2 \psi_i = 0$ for each, the resulting flow described by $\psi = \psi_1 + \psi_2 + \cdots + \psi_n$ also satisfies $\nabla^2 \psi = 0$. Thus the flow resulting from the superposition of incompressible, irrotational flows is also incompressible and irrotational. Furthermore, the resulting flow automatically satisfies the boundary conditions, Eqs. 7 and 8, given by summing those of the component flows, $\psi_1 \cdots \psi_n$.

Since the first of Eqs. 4 is also linear the velocity potentials are additive under the same rules as for the stream functions.

That the linearity of Eqs. 4 in ϕ and ψ is the key to the validity of the superposition process, can easily be seen from the nature of the equation. Consider a representative term of $\nabla^2(\psi_1 + \psi_2)$ as

$$\frac{\partial^2}{\partial x^2}(\psi_1 + \psi_2) = \frac{\partial^2 \psi_1}{\partial x^2} + \frac{\partial^2 \psi_2}{\partial x^2}$$

and a similar term for the y derivative. This expansion shows that a point-by-point addition of ψ_1 and ψ_2 describes a new flow that is also incompressible and irrotational. This superposition principle can obviously be extended to the addition of any number of flows.

Since Eqs. 1 and 2 are also linear, the velocity components and therefore the velocity vectors, given by the derivatives of ϕ and ψ (Eqs. 3), are also additive, that is,

$$\mathbf{V} = \mathbf{V}_1 + \mathbf{V}_2 = (u_1 + u_2)\mathbf{i} + (v_1 + v_2)\mathbf{j}$$

$$u = u_1 + u_2; \qquad v = v_1 + v_2 \tag{10}$$

where u and v are the components of $\mathbf{V}$.

It is important to note that the pressures of the component flows *cannot* be superimposed, because they are *non-linear* (in fact, quadratic) functions of the velocity (see for instance Eqs. 3.7 and 3.14).

In the following sections we superpose elementary flows and in this way synthesize the flow fields about specific bodies.

4.4 Source in a Uniform Stream

The stream function for the uniform flow of a fluid with velocity V_∞ in the direction of the negative x axis is given by the expression

$$\psi_1 = -V_\infty y$$

If, at the origin, a source of strength Λ with stream function $\psi_2 = (\Lambda/2\pi)\theta$ is superimposed on the uniform flow the resultant stream function is

$$\psi_3 = \psi_1 + \psi_2 = -V_\infty y + \left(\frac{\Lambda}{2\pi}\right)\theta$$

$$= V_\infty\left(\frac{h\theta}{\pi} - y\right) \tag{11}$$

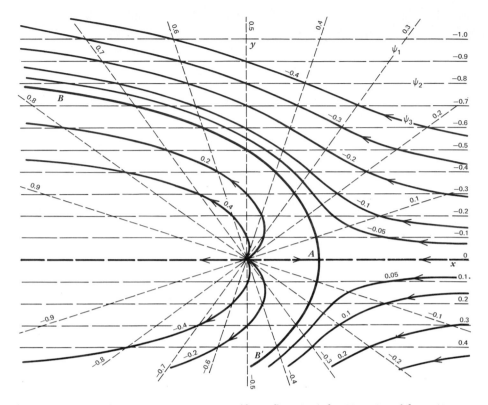

Fig. 1 Streamlines for ψ_3 = source (ψ_1) + uniform flow (ψ_2) for $V_\infty = 1$ and $h = \Lambda/2V_\infty = 1$. Note for instance that the $\psi_3 = 0$ streamline (OA and BAB') is the locus of the intersections of the streamlines $\psi_1 = -\psi_2$. Similarly for the $\psi_3 = 0.1$ streamline $\psi_1 + \psi_2 = 0.1$, and so forth.

where $h = \Lambda/2V_\infty$ is a "characteristic length" of the combined flow. Some streamlines plotted for $V_\infty = h = 1$ are shown in Fig. 1. If the region enclosed by the curve BAB of the streamline $\psi_3 = 0$ is considered as a solid surface which encloses the source, the flow exterior to the surface satisfies continuity everywhere and is irrotational. The flow field may be visualized as that of a horizontal wind past a cliff, whose shape (y_0, θ) is described by the equation $\psi = 0$, that is,

$$y_0 = r_0 \sin \theta = \frac{h\theta}{\pi} \qquad 0 \leqslant \theta \leqslant \pi \tag{12}$$

Thus h represents the height which the cliff approaches asymptotically when θ tends to π.

The velocity components, obtained by differentiating the stream function given in Eq. 11 with the substitution $\theta = \tan^{-1}(y/x)$, have the expressions

$$u = \frac{\partial \psi}{\partial y} = -V_\infty + \frac{V_\infty h}{\pi} \frac{x}{x^2 + y^2}$$

$$v = -\frac{\partial \psi}{\partial x} = \frac{V_\infty h}{\pi} \frac{y}{x^2 + y^2}$$

We see from these equations that the velocity vanishes ($u = v = 0$) at the point $(h/\pi, 0)$, that is, at $(\Lambda/2\pi V_\infty, 0)$; in other words, the velocity vanishes at the point on the x axis where the velocity from the source, $\Lambda/2\pi x$, just cancels V_∞.

Since the flow is irrotational the term p_0 in Bernoulli's equation $p = p_0 - \frac{1}{2}\rho V^2$ (see Eq. 3.7 with $\Delta z = 0$) is constant everywhere in the field so that p is a maximum where $V = 0$, that is, at the stagnation point; thus p_0 is termed the stagnation pressure. It follows from Eq. 3.11 that the pressure coefficient $C_p = 1$ at a stagnation point.

Example 1

To remain aloft in a light wind a glider will seek the point in the flow field where the vertical wind speed is a maximum. To locate this position, the vertical wind speed on the hill is written as

$$v_0 = \frac{h \sin \theta}{\pi r_0} V_\infty$$

On the cliff $\psi = 0$, whose shape is given by Eq. 12,

$$r_0 = \frac{\theta}{\pi \sin \theta} h$$

After eliminating r_0 from the above two equations, we obtain

$$v_0 = \frac{\sin^2 \theta}{\theta} V_\infty$$

and the derivative

$$\frac{dv_0}{d\theta} = \frac{\sin \theta \cos \theta (2\theta - \tan \theta)}{\theta^2} V_\infty$$

$dv_0/d\theta = 0$ at $\theta = 66.8°$ so that v_0 is a maximum at that point on the body. That maximum value of v_0 is $0.725 V_\infty$. From Eq. 12, the height there is $0.37 h$.

4.5   Flow Pattern of a Source and Sink of Equal Strength–Doublet

As a first step in the synthesis of the flow pattern for a doublet, the case of a source of strength Λ at $(-x_0, 0)$ and a sink of strength $-\Lambda$ at $(x_0, 0)$ is considered. See Fig. 2.

The angles θ_1 and θ_2 are measured from the positive x axis, and the x axis is also chosen as the zero streamline of each flow. The stream function of the combined flow may be written (Eqs. 9)

$$\psi = -\frac{\Lambda}{2\pi}(\theta_1 - \theta_2) = -\frac{\Lambda}{2\pi}\left(\tan^{-1}\frac{y}{x - x_0} - \tan^{-1}\frac{y}{x + x_0}\right)$$

and, by use of the trigonometric relation

$$\tan^{-1}\alpha - \tan^{-1}\beta = \tan^{-1}\left(\frac{\alpha - \beta}{1 + \alpha\beta}\right),$$

$$\psi = -\frac{\Lambda}{2\pi}\tan^{-1}\frac{y/(x - x_0) - y/(x + x_0)}{1 + [y^2/(x - x_0)(x + x_0)]}$$

which simplifies to

$$\psi = -\frac{\Lambda}{2\pi}\tan^{-1}\frac{2x_0 y}{x^2 + y^2 - x_0^2} \tag{13}$$

The flow pattern represented by this stream function has a simple geometrical form. The equation of a streamline is given by ψ = constant. It is put in a recognizable form in the following manner:

$$-\tan\frac{2\pi\psi}{\Lambda} = \frac{2x_0 y}{x^2 + y^2 - x_0^2}$$

$$x^2 + y^2 + 2x_0 y \cot\frac{2\pi\psi}{\Lambda} = x_0^2$$

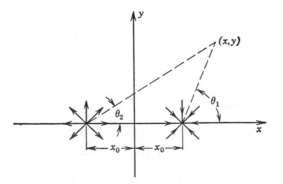

Fig. 2 Source-sink pair.

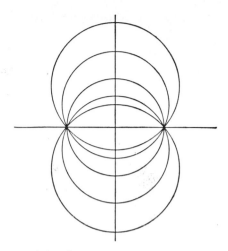

Fig. 3 Streamlines of a source-sink pair.

Completing the square on the left-hand side,

$$x^2 + \left(y + x_0 \cot \frac{2\pi\psi}{\Lambda}\right)^2 = x_0^2 + x_0^2 \cot^2 \left(\frac{2\pi\psi}{\Lambda}\right) \tag{14}$$

The equation is seen to represent a family of circles with centers on the y axis. When $y = 0$, $x = \pm x_0$ for all values of ψ. The flow pattern is shown in Fig. 3.

An especially useful flow pattern results when the distance between a source and a sink of equal strengths approaches zero while their strengths approach infinity in such a way that their product $2x_0\Lambda$ remains constant. In the limit the resulting flow is called a *doublet of strength* κ, where

$$\kappa = 2x_0\Lambda$$

The stream function of the doublet is given by the limit of Eq. 13 as x_0 approaches zero.

$$\psi = \lim_{x_0 \to 0} \left[-\frac{\kappa}{4\pi x_0} \tan^{-1} \frac{2x_0 y}{x^2 + y^2 - x_0^2} \right]$$

As x_0 approaches zero, $2x_0 y/(x^2 + y^2 - x_0^2)$ becomes small, and we can write

$$-\frac{\kappa}{4\pi x_0} \tan^{-1} \frac{2x_0 y}{x^2 + y^2 - x_0^2} = -\frac{\kappa}{4\pi x_0} \frac{2x_0 y}{x^2 + y^2 - x_0^2}$$

In the limit, the stream function and velocity potential are

$$\psi = -\frac{\kappa}{2\pi} \frac{y}{x^2 + y^2} = -\frac{\kappa}{2\pi} \frac{\sin\theta}{r}$$

$$\phi = +\frac{\kappa}{2\pi} \frac{\cos\theta}{r} \tag{15}$$

The equation for ϕ can be verified by using Eqs. 3 to show that the velocity components calculated from both functions are identical.

The streamlines (lines of constant ψ), given by the first of Eqs. 15 are circles, as is readily seen if it is rearranged in the following manner:

$$x^2 + y^2 + \frac{\kappa y}{2\pi\psi} = 0$$

$$x^2 + \left(y + \frac{\kappa}{4\pi\psi}\right)^2 = \left(\frac{\kappa}{4\pi\psi}\right)^2$$

This flow is illustrated in Fig. 4. The streamlines are a series of circles that pass through the origin. The centers lie on the y axis. In the same way the equipotentials can be shown to be circles with centers on the x axis.

It is preferable in some applications to use "vortex doublets," instead of source-sink doublets. For example we note from Eqs. 9 that the stream function for a source is identical with the negative of the velocity potential for the vortex. Therefore, if we add a vortex of strength $-\Gamma$ at $(x_0, 0)$ to one of strength $+\Gamma$ at $(-x_0, 0)$ the velocity potential for the resulting flow will be given by the right-hand side of Eq. 13 with Γ replacing Λ, and the equipotential lines will be described by the circles in Fig. 3. Furthermore, if $x_0 \to 0$ and $2x_0\Gamma$ remains constant, a vortex doublet is formed whose equipotential lines coincide with the streamlines of the source-sink doublet shown in Fig. 4. The streamlines of this vortex doublet, on the other hand, coincide with the equipotential lines of the source-sink doublet which are circles tangent to the y axis with centers on the x axis.

Similarly, if we add a vortex of strength $+\Gamma$ at $(0, y_0)$ to one of strength $-\Gamma$ at $(0, -y_0)$, and let $\kappa = 2y_0\Gamma$ while $y_0 \to 0$, the same streamlines as shown in Fig. 4

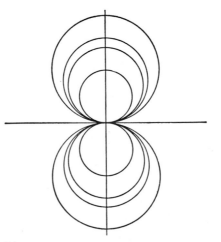

Fig. 4 Streamlines of a doublet.

are produced. It follows that Eqs. 15 will apply to a vortex doublet on the y axis as well as to a source-sink doublet on the x axis, where κ refers also to the strength of the vortex doublet (see Problem 4.5.3).

4.6 Flow About A Circular Cylinder in a Uniform Stream

The stream function for the uniform flow of a fluid with velocity V_∞ in the direction of the positive x axis is given by the expression

$$\psi = +V_\infty y$$

If the uniform flow is added to the doublet, the flow about a circular cylinder in a uniform stream is obtained. The stream function of the combined flow is

$$\psi = V_\infty y - \frac{\kappa y}{2\pi r^2}$$

or, upon letting $\kappa/2\pi V_\infty = a^2$,

$$\psi = V_\infty y \left(1 - \frac{a^2}{r^2}\right) \tag{16}$$

The zero streamline is the x axis and a circle of radius $r = a$. See Fig. 5. This flow is irrotational and satisfies *continuity* at every point outside of the circle $r = a$. Therefore, it may be taken as the true stream function for the uniform flow about a circular cylinder when the velocity at infinity is in the direction of the positive x axis.

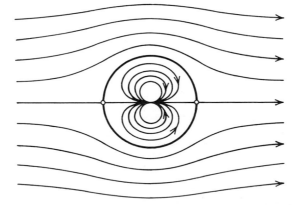

Fig. 5 Streamlines for doublet + uniform flow; synthesis of flow around circular cylinder in uniform flow.

The velocity distribution throughout the flow field is found by differentiation according to Eq. 16, after writing $y = r \sin \theta$,

$$u_r = V_\infty \left(1 - \frac{a^2}{r^2}\right) \cos \theta ; \qquad u_\theta = -V_\infty \left(1 + \frac{a^2}{r^2}\right) \sin \theta$$

At the surface of the cylinder $(r = a)$ $u_r = 0$ and $-u_\theta = 2V_\infty \sin \theta$ (the minus sign occurs because θ is positive counterclockwise). The pressure distribution on the surface, expressed in terms of the pressure coefficient, is by Bernoulli's equation,

$$C_P = \frac{p - p_\infty}{q_\infty} = 1 - 4 \sin^2 \theta \qquad (16a)$$

In the next section these results will be generalized to include circulation around the cylinder.

4.7   Circulatory Flow About a Cylinder in a Uniform Stream

If the stream function for a vortex at the origin from Eqs. 9 is added to Eq. 16, the resulting stream function will satisfy continuity, irrotationality, and the boundary conditions for the circulatory flow about a circular cylinder in a uniform stream.

$$\psi = V_\infty y \left(1 - \frac{a^2}{r^2}\right) + \frac{\Gamma}{2\pi} \ln \left(\frac{r}{a}\right) \qquad (17)$$

The uniform stream is in the direction of the positive x axis, and the circulatory flow is clockwise. The cylinder $r = a$ forms part of the zero streamline. The flow pattern is shown in Fig. 6.

The points where the x axis intersects the circle $(x = \pm a)$ are stagnation points of the flow for $\Gamma = 0$. When circulation in the clockwise sense is added, the stagnation

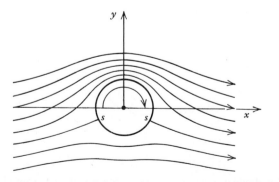

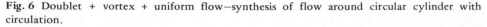
Fig. 6 Doublet + vortex + uniform flow—synthesis of flow around circular cylinder with circulation.

points move down until they coincide at the position $x = 0$, $y = -a$. For further increases in Γ, the stagnation points lie below the cylinder.

To investigate the location of these stagnation points, the velocity components are found by differentiating Eq. 17.

$$u_r = \frac{1}{r}\frac{\partial \psi}{\partial \theta} = V_\infty \left(1 - \frac{a^2}{r^2}\right)\cos\theta$$

$$u_\theta = -\frac{\partial \psi}{\partial r} = -V_\infty \sin\theta \left(1 + \frac{a^2}{r^2}\right) - \frac{\Gamma}{2\pi r} \tag{18}$$

On the surface of the cylinder, $r = a$ and u_r vanishes. The resultant velocity is

$$u_\theta = -2V_\infty \sin\theta - \frac{\Gamma}{2\pi a}$$

For the stagnation value of θ, u_θ must vanish,

$$\sin\theta_s = -\frac{\Gamma}{4\pi a V_\infty}$$

Since $\sin\theta = y/r$, the stagnation position in Cartesian coordinates is

$$x_s = \pm\sqrt{a^2 - y_s^2}\ ; \qquad y_s = -\frac{\Gamma}{4\pi V_\infty} \tag{19}$$

From Eqs. 19, it is apparent that, as Γ becomes large, the stagnation points move downward until $(\Gamma/4\pi V_\infty)^2$ equals a^2; for this condition the stagnation points coincide on the y axis at $(0, -a)$. For $(\Gamma/4\pi V_\infty)^2 > a^2$, Eqs. 19 no longer hold because the stagnation points leave the body. The position of the stagnation points for several values of Γ are shown in Fig. 7.

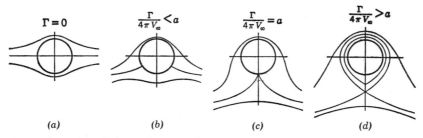

Fig. 7 Stagnation points for several values of Γ.

4.8   Force on a Cylinder with Circulation in a Uniform Steady Flow—the Kutta-Joukowski Theorem

We proceed to apply the momentum theorem to find the force acting on a cylinder in a steady uniform flow. We choose the control surface $\hat{S}$, shown in Fig. 8,

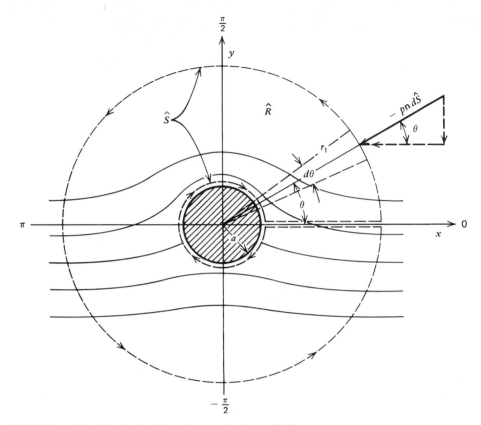

Fig. 8 Control surface in the flow about a circular cylinder.

comprising the outer circle of radius r_1, the inner circle of radius a and the cut connecting them. If we neglect body forces the momentum theorem, expressing the fact that the momentum flux through the surface is equal to the force acting on the boundary, is

$$\iint_{\hat{S}} \mathbf{V}(\rho \mathbf{V} \cdot \mathbf{n}) \, d\hat{S} = - \iint_{\hat{S}} p\mathbf{n} \, d\hat{S} \tag{20}$$

The direction of the integrations, indicated by the arrows, which keeps the region $\hat{R}$ always to the left of the path, is counterclockwise around r_1 and clockwise around a. When the width of the cut approaches zero the two integrals along that part of the path vanish, and if we take into account that at the streamline $r = a$, $\mathbf{V} \cdot \mathbf{n} = u_r = 0$, Eq. 20 may be expressed (as in Eq. 3.29)

$$\int_0^{2\pi} \mathbf{V}\rho u_r r_1 \, d\theta = -\mathbf{F}_e - \int_0^{2\pi} p\mathbf{n}r_1 \, d\theta \tag{21}$$

where $-F_e = -\int_{2\pi}^0 pna \, d\theta = -iF_x - jF_y$ is the force exerted on the fluid by unit length of the inner cylinder. If one considers any segment of width dy parallel to the x axis, it is clear from Fig. 8 that since the streamline configuration and therefore the pressures acting on the boundaries are symmetrical about the y axis, F_x must vanish. Thus the x component of the momentum of an element of fluid entering the contour at the left must equal that when it leaves at the right, and the force acting on the inner cylinder becomes

$$F_e = jF_y = j\left(\underbrace{-2\int_{-\pi/2}^{\pi/2} r_1 p \sin\theta \, d\theta}_{\text{(I)}} - \underbrace{\int_0^{2\pi} \rho r_1 u_r v \, d\theta}_{\text{(II)}}\right) \tag{22}$$

where v is the y component of the velocity.

We evaluate first the pressure integral (I) by the use of Bernoulli's equation

$$p = p_0 - \tfrac{1}{2}\rho V^2$$

When this equation is substituted in (I) above, since the flow is irrotational, p_0 is constant and the term $\int_{-\pi/2}^{\pi/2} p_0 r_1 \sin\theta \, d\theta$ vanishes. Thus

$$\text{(I)} = \rho r_1 \int_{-\pi/2}^{\pi/2} V^2 \sin\theta \, d\theta \tag{23}$$

V^2 can be found from Eqs. 18.

$$V^2 = u_r^2 + u_\theta^2$$

$$V^2 = V_\infty^2 \left(1 - \frac{a^2}{r_1^2}\right)^2 \cos^2\theta + V_\infty^2 \left(1 + \frac{a^2}{r_1^2}\right)^2 \sin^2\theta$$

$$+ \frac{V_\infty \Gamma \sin\theta}{\pi r_1}\left(1 + \frac{a^2}{r_1^2}\right) + \left(\frac{\Gamma}{2\pi r_1}\right)^2$$

Since

$$\int_{-\pi/2}^{\pi/2} \cos^2\theta \sin\theta \, d\theta = \int_{-\pi/2}^{\pi/2} \sin^3\theta \, d\theta = \int_{-\pi/2}^{\pi/2} \sin\theta \, d\theta = 0$$

Eq. 23 can be written

$$\text{(I)} = \rho r_1 \int_{-\pi/2}^{\pi/2} \frac{V_\infty \Gamma}{\pi r_1}\left(1 + \frac{a^2}{r_1^2}\right)\sin^2\theta \, d\theta = \frac{1}{2}\rho V_\infty \Gamma \left(1 + \frac{a^2}{r_1^2}\right) \tag{24}$$

If $r_1 = a$, the calculation gives the force components simply by an integration of the pressure force over the body surface. That is, since $u_r = 0$ at $r_1 = a$, the term

(II) in Eq. 22 vanishes, and by the symmetry conditions above and Eq. 24 the force components are

$$F_x = 0; \qquad F_y = \rho V_\infty \Gamma \tag{25}$$

Equations 25 give the force on the circular cylinder. The remainder of the analysis consists of evaluating term (II) of Eq. 22 over the outer boundary with $r_1 > a$ and reasoning that, since the result is again Eqs. 25 even as $r_1 \to \infty$, the force acting on a cylinder is independent of its cross section.

At $r = r_1$ we find, from Eq. 17,

$$
\begin{aligned}
v &= -\frac{\partial \psi}{\partial x} = -V_\infty \frac{2a^2 xy}{r_1^4} - \frac{\Gamma}{2\pi} \frac{x}{r_1^2} \\
&= -2V_\infty \frac{a^2}{r_1^2} \cos\theta \sin\theta - \frac{\Gamma}{2\pi} \frac{\cos\theta}{r_1}
\end{aligned}
\tag{26}
$$

Then, by the use of Eqs. 18 and 26 term (II) of Eq. 22 becomes

$$\text{(II)} = \rho r_1 V_\infty \left(1 - \frac{a^2}{r_1^2}\right) \int_0^{2\pi} \left(2V_\infty \frac{a^2}{r_1^2} \cos^2\theta \sin\theta + \frac{\Gamma}{2\pi} \frac{\cos^2\theta}{r_1}\right) d\theta$$

The first integral makes no contribution, so that

$$\text{(II)} = \frac{1}{2} \rho V_\infty \Gamma \left(1 - \frac{a^2}{r_1^2}\right)$$

With the above equation and Eq. 24, Eq. 22 becomes

$$F_y = \frac{1}{2} \rho V_\infty \Gamma \left(1 + \frac{a^2}{r_1^2}\right) + \frac{1}{2} \rho V_\infty \Gamma \left(1 - \frac{a^2}{r_1^2}\right) \tag{27}$$

or

$$F_y = \rho V_\infty \Gamma; \qquad F_x = 0 \tag{28}$$

In vector notation,

$$\mathbf{F} = \rho \mathbf{V}_\infty \times \mathbf{\Gamma}$$

where $\mathbf{\Gamma}$ is the circulation vector; its direction is determined by the right hand screw rule. The force acts normal to $\mathbf{\Gamma}$ and $\mathbf{V}_\infty$.

We may conclude from the following reasoning that, since Eqs. 25 and 28 are identical, the force acting on a cylinder is independent of its cross section. The shape of the zero streamline conforming with the surface of a given closed body is matched in a flow field by the superposition of the uniform flow and a unique configuration of sources, sinks and vortices of various strengths within the body. For a closed body continuity demands that the net source strength be zero, and for

$r \gg a$, where a is a characteristic length of the body (e.g., chord of an airfoil, diameter of a cylinder, etc.) the flow field of the resulting source-sink pairs and vortices approaches that of a cluster of doublets and vortices near the origin. As $r \to \infty$ the distances between the doublets and vortices become insignificant and the flow field there approaches that generated by a single doublet with circulation equal to the sum of the vortex strengths within the body. Another example of this limiting process occurs when Eq. 13 is viewed from a great distance so that x_0 becomes negligible and the resulting form for ψ is that of the doublet (Eqs. 15). Thus, in the limit, the forces acting are given by Eqs. 28, independent of the shape of the body; this conclusion is expressed by the *Kutta-Joukowski theorem* which may be stated:

The force acting per unit length on a cylinder of any cross section is equal to $\rho \mathbf{V}_\infty \times \mathbf{\Gamma}$.

If the cross section is an airfoil and $\mathbf{V}_\infty$ is the relative velocity in the x direction, the Kutta-Joukowski theorem yields

$$L' = \rho V_\infty \Gamma; \qquad D' = 0$$

where L' and D' are the lift and drag, respectively, in the y and x directions per unit spanwise length of the airfoil.

The prediction of zero drag may be generalized to apply to a three-dimensional body of any shape in an irrotational, steady, incompressible flow. This result was long known as "D'Alembert's paradox." However, its explanation in terms of the pressure distribution belies its connotation as a "paradox." For an airfoil, for instance, the zero drag results simply from the circumstance that at an angle of attack the high speed and therefore the low pressure in the flow around the leading edge (termed leading edge suction) generates a thrust that just balances a rearward force over the aft regions.

We show in the next chapter that airfoil theory based on the Kutta-Joukowski law and the "Kutta condition" described below predicts with remarkable accuracy the magnitude and distribution of the lift of airfoils up to angles of attack of around $15°$ at flight Reynolds numbers. The prediction of zero drag is of course at total variance with reality, which is not surprising considering that viscosity and compressibility have been neglected; contributions from these effects to the drag are described in later chapters.

4.9 Bound Vortex

It was shown in the last section that the force on a body is determined entirely by the circulation around it and by the free stream velocity. In an identical manner, it can be shown that the force on a vortex that is stationary relative to a uniform flow is given by the Kutta-Joukowski law. The vortex that represents the circulation around the body departs in its characteristics from that of a vortex in the external flow in that it does not remain attached to the same fluid particles but instead it remains *bound* to the body. This vortex that represents the circulatory flow around the wing is called a *bound vortex* in order to distinguish it from a vortex that moves with the flow.

Then, as far as resultant forces are concerned, a bound vortex of proper strength in a uniform stream is completely equivalent to a body with circulation in a uniform stream.

4.10 Kutta Condition

The Kutta-Joukowski theorem states that the force experienced by a body in a uniform stream is equal to the product of the fluid density, stream velocity, and circulation and has a direction perpendicular to the stream velocity. In Section 4.2, it was stated that one and only one irrotational flow can be found that satisfies the boundary conditions at infinity and at the body, provided the circulation is specified. If the circulation is not specified, the conditions at infinity and the geometry of the body do not determine the flow pattern.

In order to find the force on a body that is submerged in a streaming fluid (or, equivalently, on a body moving through a stationary fluid), it is necessary to know the value of the circulation. Yet the theory indicates that the geometry of the body and the stream velocity do not determine the circulation.

The above discussion applies to an inviscid flow, but in a viscous fluid (however small the viscosity), the circulation is fixed by the imposition of an empirical observation. Experiments show that when a body with a sharp trailing edge is set in motion, the action of the fluid viscosity causes the flow over the upper and lower surfaces to merge smoothly at the trailing edge; this circumstance, which fixes the magnitude of the circulation around the body, is termed the "Kutta condition," which may be stated as follows: *A body with a sharp trailing edge in motion through a fluid creates about itself a circulation of sufficient strength to hold the rear stagnation point at the trailing edge.*

The flow around an airfoil at an angle of attack in an inviscid flow develops no circulation and the rear stagnation point occurs on the upper surface; the streamlines are shown schematically in Fig. 9. Figure 10 is a sketch of the streamlines around an airfoil in a viscous flow, indicating the smooth flow past the trailing edge, as observed in practice, and termed the Kutta condition.

The sequence of events for the development of the flow around an airfoil starting impulsively from rest in a viscous fluid is indicated by Fig. 11. The Helmholtz laws (Section 2.13), specify that in an irrotational inviscid flow the circulation around a contour enclosing a number of tagged fluid elements is invariable throughout its

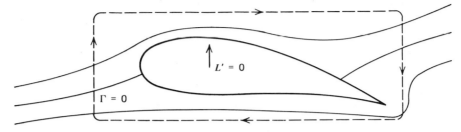

Fig. 9 Flow around an airfoil with zero circulation.

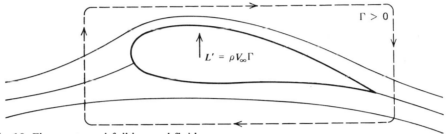

Fig. 10 Flow past an airfoil in a real fluid.

motion; thus the circulation around a contour that includes the airfoil and its "wake," being zero before the motion began, must remain zero. The establishment of the Kutta condition, therefore, requires the formation of the so-called starting vortices (see Fig. 11) with a combined circulation equal and opposite to that around the airfoil. The induced flow caused by the bound vorticity added to that caused by the starting vortices in the wake, will be just sufficient to accomplish the smooth flow at the trailing edge.

The starting vortices are left behind as the airfoil moves farther and farther from its starting point, but during the early stages of the motion, Fig. 11 indicates that their induced velocities assist those induced by the bound vortex, to satisfy the Kutta condition. It follows that the bound vortex will not be as strong in the early stages when it is being helped by the starting vortices as it is after the flow is fully established when the bound vorticity must be strong enough by itself to move the rear stagnation point to the trailing edge. In fact, unsteady flow theory and experi-

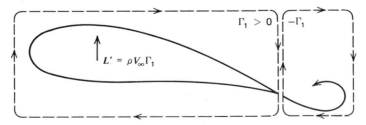

Fig. 11 Starting vortex during early stage of motion.

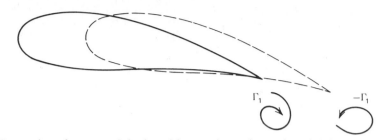

Fig. 12 Generation of vortex pair by impulsive starting and stopping of airfoil.

ment show that the bound vortex at the initial instant of uniform impulsive translation is only about half as intense as it is finally (see Goldstein, 1938, p. 460), when
steady flow is achieved. Thus the lift begins with half its steady flow value; it
reaches 90 percent of its steady value after it has traveled about 3 chord lengths.
One of the many practical examples of this phenomenon occurs during the passage
of an aircraft through a sharp gust; fortunately, the effect is to *attenuate* the unsteady forces one would calculate if the velocities induced by the starting vortices
are neglected.

If an airfoil is started and stopped impulsively the bound vortex is shed into the
fluid and we have, as in Fig. 12, a vortex pair of equal and opposite strengths. This
flow will be discussed further in connection with Fig. 18.

One can see clearly the effect of the bound vorticity on the flow over the upper
and lower surfaces of an airfoil in Fig. 13. This sketch indicates the motion of two
fluid elements A and B that are, at a given instant, near the forward stagnation
point. At subsequent instants they are located at A', A'' and B', B'', respectively,
since the velocity components induced by the bound vorticity cause B to traverse
the upper surfaces more rapidly than does A the lower surface.

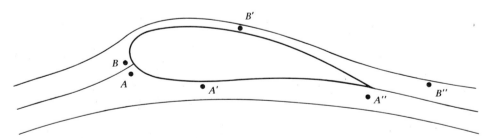

Fig. 13 Successive positions of fluid elements A and B while traversing airfoil surface.

4.11 Numerical Solution of Flow Past Two-Dimensional Symmetric Bodies

It has been shown in Section 4.4 that the superposition of a uniform flow and a
line source results in a two-dimensional body open downstream. When a line sink
of the same strength is added to the source on the downstream side, an oval-shaped

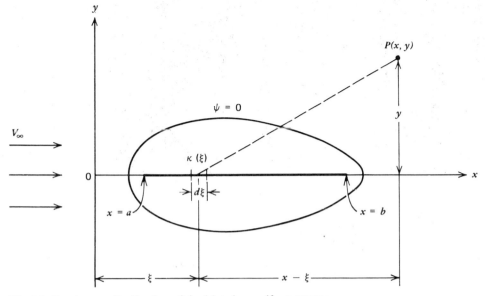

Fig. 14 Continuous distribution of doublets in a uniform stream.

closed body will be formed instead. As the source-sink pair approach each other to form a doublet, the body becomes a circular cylinder as demonstrated in Section 4.6. Thus by adding a uniform flow and a distribution of doublets (or sources and sinks of zero total strength), closed bodies of various shapes can be generated.

Consider a stream of uniform horizontal velocity V_∞ and a continuous distribution of doublets, of strength $2\pi\kappa$ per unit length along the x axis, within the range between $x = a$ and $x = b$. The distribution $\kappa(x)$ will determine the shape of a symmetrical body such as shown in Fig. 14. The quantity $2\pi\kappa$ is termed the **doublet density** so that the total doublet strength, K, within the body is $\int_a^b 2\pi\kappa\, dx$. At a given point $P(x, y)$, the doublets contained within the small interval $d\xi$, located at a distance ξ from the origin, contribute $d\psi$ to the stream function at that point. We find from the first of Eqs. 15 that

$$d\psi = -\frac{\kappa(\xi)y\, d\xi}{(x - \xi)^2 + y^2}$$

The stream function at P, for the superposition of the uniform flow and the doublet distribution, is, therefore,

$$\psi = V_\infty y - \int_a^b \frac{\kappa(\xi)y}{(x - \xi)^2 + y^2}\, d\xi \tag{29}$$

The shape of the body described by $\psi = 0$ is controlled by varying the distribution $\kappa(\xi)$. The distance between the leading edge of the body and the first doublet at $x = a$ and that between the last doublet at $x = b$ and the trailing edge must be nonzero if the radii of curvature at these points are finite. For a prescribed body contour the determination of the function $\kappa(\xi)$ requires the solution of the integral equation, Eq. 29, which is usually difficult. However, such a problem can be solved numerically to any desired degree of accuracy.

In the numerical method we divide the doublet region into n segments of equal width $\Delta\xi$, as shown in Fig. 15. We designate by $\kappa_j\Delta\xi$, the total doublet strength within the j segment, whose center is at a distance ξ_j from the origin; κ_j is taken as constant, equal to the average of the "exact" distribution $\kappa(x)$ within the segment. κ_j will of course vary from one segment to another. The doublets within the j segment will contribute $\Delta\psi_j$ to the stream function at a given point P in the field. This contribution may be written

$$\Delta\psi_j = - \frac{\kappa_j\Delta\xi y_P}{(x_P - \xi_j)^2 + y_P^2}$$

and, for the superimposed uniform flow V_∞ and doublet distribution, the approximate formula corresponding to the exact form of Eq. 29 is

$$\psi_P = V_\infty y_P - \sum_{j=1}^{n} \frac{\kappa_j\Delta\xi y_P}{(x_P - \xi_j)^2 + y_P^2} \tag{29a}$$

We now apply this formula to n points on a body of prescribed shape; we know that $\psi = 0$ everywhere on the body. Thus, we have only to apply Eq. 29a at these points to obtain a set of n simultaneous linear algebraic equations, the solution of which yields the doublet densities $\kappa_1 \ldots \kappa_n$; the velocity and pressure distributions over the body follow immediately.

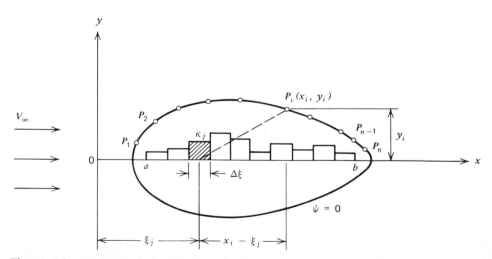

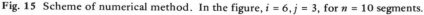

Fig. 15 Scheme of numerical method. In the figure, $i = 6, j = 3$, for $n = 10$ segments.

We designate $P_i(x_i, y_i)$ as any one of the n points on the surface and, with a more convenient notation, write

$$\psi_i = 0 = V_\infty y_i - \sum_{j=1}^{n} c_{ij} \kappa_j \qquad (30)$$

where

$$c_{ij} = \frac{y_i \Delta \xi}{(x_i - \xi_j)^2 + y_i^2}$$

is the contribution of a doublet of unit density ($\kappa_j = 1$) at ξ_j to the stream function at the point P_i on the body.

Equations 30 may be expanded into the n simultaneous equations as follows

$$c_{11} \kappa_1 + c_{12} \kappa_2 + \cdots + c_{1n} \kappa_n = V_\infty y_1$$

$$c_{21} \kappa_1 + c_{22} \kappa_2 + \cdots + c_{2n} \kappa_n = V_\infty y_2$$

$$\cdots \qquad\qquad (31)$$

$$c_{n1} \kappa_1 + c_{n2} \kappa_2 + \cdots + c_{nn} \kappa_n = V_\infty y_n$$

As n approaches infinity the numerical result for the doublet density distribution, κ_j, approaches the exact solution. By utilizing a program for solving simultaneous linear equations, the solution for a reasonably large number of segments can readily be found on a digital computer.

Once the doublet strength distribution is known, the stream function at any point in the flow can be computed by Eq. 29a. Successively the velocity and pressure fields can be obtained by taking derivatives of ψ and then substituting into Bernoulli's equation.

As an example, the dimensionless pressure distribution on the surface of a nonlifting airfoil, based on a numerical computation with 52 segments, is given in Fig. 16. It demonstrates excellent agreement with the exact solution.

For a uniform flow past an asymmetric body, a similar numerical computation can be carried out by distributing doublets or source segments along a curved path. However, because of the asymmetry, the points $P_i(x_i, y_i)$ leading to Eqs. 30 and 31 must be distributed over both the upper and lower surfaces; as a result acceptable accuracy generally requires that n, the number of segments, be increased over that for a symmetrical body.

Instead of the stream function, we may use the velocity potential of the combined flow and require that the flow be tangent, that is, that $\partial \phi / \partial n = 0$ (Eq. 7), at the surface of the body. The two methods are equivalent but the boundary conditions for the velocity potential are generally more involved.

By the Kutta-Joukowski theorem a body will generate lift only if there exists a circulation around it. In the next chapter, a numerical method is described for determining the lift of a planar wing.

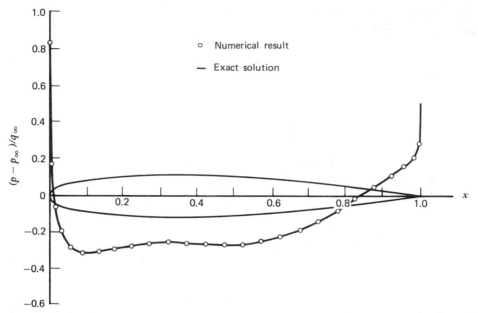

Fig. 16 Distribution of dimensionless pressure on a symmetric airfoil at zero angle of attack. (Courtesy of P. E. Rubbert, Boeing Company.)

4.12 Aerodynamic Interference—Method of Images

To this point we have dealt with features of the flow about single bodies in an infinite fluid. While these are of practical interest by themselves, they often designate the starting point for more practical problems, in which there are many bodies and there is mutual interference among the flows, to an extent depending on the bodies' sizes and their distances from each other. Some examples are the effect on aircraft performance of the proximity of the ground and the flow fields of other aircraft or obstructions; the effect of the wind tunnel boundaries on the measured aerodynamic characteristics of a model under test; flight characteristics as affected by mutual flow interferences among the various component parts of an aircraft. The "method of images" in one form or another is used to determine the magnitude of these interferences.

A simple example shown in Fig. 17 illustrates the distortion of the flow generated by a source at $y = a$ in the presence of a plane wall at $y = 0$. The boundary condition imposed by the wall is $v = 0$ at $y = 0$ and its influence is calculated by observing that the effect of a wall is identical with that of an "image source" of *equal strength* at $y = -a$. The velocity vector diagram shows that the superposition of the two sources results in a flow everywhere tangent to the wall at $y = 0$. The stream function of the combined flow is obtained by the method analogous to that used to obtain Eq. 13 for the source-sink pair. Thus, for the source and its image,

$$\psi = \frac{\Lambda}{2\pi}\left[\tan^{-1}\left(\frac{y-a}{x}\right) + \tan^{-1}\left(\frac{y+a}{x}\right)\right]$$

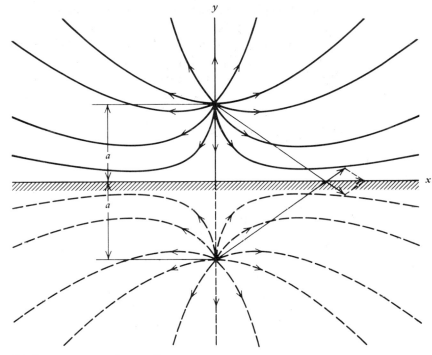

Fig. 17 A source near a plane wall.

where Λ is the strength of cach source. If the flow is established without external constraints on the sources, each of the sources will move away from the median plane at a velocity initially equal to $\Lambda/(2\pi \cdot 2a)$. The origin $(0, 0)$ is a stagnation point of the flow and the flow along the y axis (use the polar coordinate form from Eqs. 9 with Eqs. 3)

$$v_0 \equiv (v)_{x=0} = \frac{\Lambda}{2\pi}\left(-\frac{1}{a-y} + \frac{1}{a+y}\right)$$

where the first term is the velocity in the absence of the image, that is, in the absence of the plane at $y = 0$; the second term is the velocity generated by the image and shows that the flow from the image decreases v at $y < a$ and increases it at $y > a$. The formula shows that $v_0 \to \pm\infty$ as $y \to a$ but if we imagine the fluid issuing from a pipe of small finite radius, having the same density as the fluid, the pipe is in an updraft from the image and, if it is not constrained, its upward velocity is $\Lambda/(2\pi \cdot 2a)$. Bernoulli's equation shows that the velocity distribution from the image contributes a downward pressure gradient and therefore an upward force on a body in the field.

The arrangement is roughly similar to that for an air cushion vehicle in which a downward-blowing air jet is deflected by the ground or by a water surface. The high pressure region so created is confined under a casing whose upper surface is exposed to the atmosphere. The resulting pressure force which can be analyzed quantitatively as an image effect plays an essential role in sustaining the vehicle at a given altitude.

When a line vortex of circulation Γ is placed parallel to and at a distance a above a ground plane at $y = 0$, the stream function is the sum of that associated with the vortex itself and that of its image. After substituting the proper values for r into the first entry in Eqs. 9, the stream function of the flow is

$$\psi = \frac{\Gamma}{2\pi}[\ln \sqrt{x^2 + (y - a)^2} - \ln \sqrt{x^2 + (y + a)^2}]$$

$$= \frac{\Gamma}{4\pi} \ln \left\{ \frac{x^2 + (y - a)^2}{x^2 + (y + a)^2} \right\}$$

Some streamlines are plotted in Fig. 18. In the absence of constraints the vortex and the image will move parallel to the plane to the left at a velocity $\Gamma/(2\pi \cdot 2a)$, the velocity induced at the center of one vortex by the other.

If another vortex of equal and opposite circulation is placed parallel to the first and at the same height above the plane surface the flow configuration is that of the impulsive starting and stopping of an airfoil (see Fig. 12) in the presence of the wall. The equivalent actual and image vortex configuration is shown in Fig. 19,

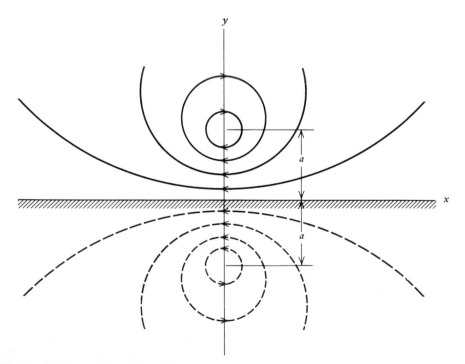

Fig. 18 A vortex near a plane wall.

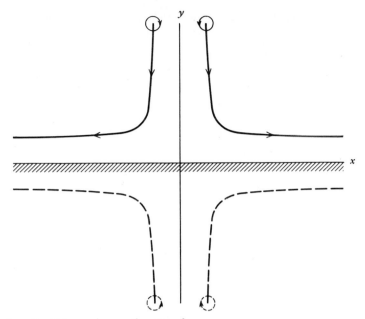

Fig. 19 Motion of a vortex pair near the ground.

where it is shown to approximate a cross section of the trailing vortex system of an airplane wing near the ground. Each vortex moves along the path indicated in the figure because of the velocities induced at its center by the other three.

Another example of the image effect demonstrates that the singularities introduced to achieve a given streamline shape must take into account other singularities in the flow. For instance, while the doublet of Eq. 15 added to a uniform flow generates the flow (Eq. 16) about a circular cylinder moving in an undisturbed fluid, Fig. 20 shows the effect of placing the doublet near a wall parallel to the axis of the

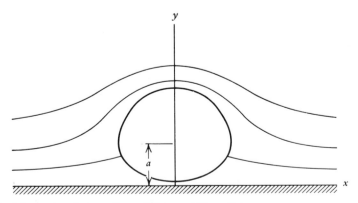

Fig. 20 A doublet near a plane wall parallel to a uniform flow.

doublet. The velocities induced by the image doublet required to simulate the effect of the wall is seen to distort significantly the zero streamline of the flow.

It is necessary to add singularities and, necessarily, their images to preserve the circular shape of the cylinder or, indeed, the shape of any body in the vicinity of the wall or other bodies. The entire system including the images then establishes the pressure distribution and the force acting on the body.

Since the total circulation around the body and images is zero, so is the total force acting on the body and the wall. The force acting on the cylinder alone is therefore equal and opposite to that exerted on the wall.

It must be emphasized that the method of images as presented here is based on the superposition principle described in Section 4.3. It follows that in appreciable areas of rotational flow, such as in the large wakes behind bluff bodies, its use is impaired. Other examples of flow interference and the validity of their treatment by the image method will be described in succeeding chapters.

Example

The configuration of images shown in Fig. 18 can be used to compute the effect of the proximity of the ground on the lift of an airfoil and to demonstrate that the reaction to the lift of an aircraft must ultimately be exerted by the surface of the earth. The flow field about an airfoil at an angle of attack situated at $y = a$ above the plane $y = 0$ in a flow field with uniform velocity V_∞ at infinity is simulated by the superposition of the uniform flow and the vortex and image configuration of Fig. 18. The above expression for ψ is the stream function for the vortex circulation Γ at $y = a$ and its image at $y = -a$ with circulation $-\Gamma$; after adding the stream function $V_\infty y$ to represent the uniform flow and differentiating, we obtain the velocity distribution at the ground plane $y = 0$:

$$u_w = V_\infty - \frac{a\Gamma}{\pi} \frac{1}{x^2 + a^2} ; \qquad v_w = 0$$

The pressure at infinity, p_∞, and p_w at the plane $y = 0$ are related by Bernoulli's equation,

$$p_w + \tfrac{1}{2}\rho u_w^2 = p_\infty + \tfrac{1}{2}\rho V_\infty^2$$

After substituting the above expression for u_w the pressure increment on the ground plane caused by the presence of the airfoil becomes

$$p_w - p_\infty = \frac{\rho a\Gamma}{\pi(x^2 + a^2)} \left[V_\infty - \frac{a\Gamma}{2\pi(x^2 + a^2)} \right]$$

The lift acting on the vortex per unit span is equal and opposite to the force acting on the ground per unit distance along the z axis. Thus, the integrated pressure force on the ground plane becomes, after simplification,

$$L' = \int_{-\infty}^{\infty} (p_w - p_\infty) dx = \rho V_\infty \Gamma \left(1 - \frac{\Gamma}{4\pi a V_\infty} \right)$$

The lift is decreased from that of a single vortex because the flow velocity at the vortex is decreased by the ratio $\Delta V_\infty / V_\infty = \Gamma / 4\pi a V_\infty$, where ΔV_∞ is the velocity induced at $y = a$ by the image at $y = -a$.

The effect of the ground on the lift of a *finite* wing, caused by the presence of the "trailing vortices," is to *increase* the lift. The effect is described in Chapter 6.

4.13 The Method of Source Panels

It has been shown in Section 4.11 that flow about a given body can be generated by distributing singularities (sources, sinks, vortices) within its enclosure at specific locations such that the body surface becomes a streamline of the flow. Alternatively, this flow may be achieved by replacing the surface by a "source sheet," the strength of which varies over the surface in a manner such that at every point the normal velocity generated by the source sheet just balances the normal component of the free stream velocity. If the body generates lift, vortex sheets (Chapter 5) may also be introduced to provide circulation (Hess and Smith, 1967). In this section, we introduce the panel method in its simplest form, that of determination of the flow about a nonlifting body; flow around lifting bodies will be treated in the next two chapters.

The source sheet method may be adapted to computer methods by replacing the body by a finite number of "source panels," each of constant strength (see Fig. 22), instead of by a source sheet of continuously varying strength.

The step-by-step process by which a number of discrete line sources arranged along a line normal to a flow evolve into a source panel forming a streamline of the flow is illustrated in Fig. 21. The complex flow pattern of Fig. 21a is the *upper half* of the streamline pattern resulting from the superposition of a uniform flow V_∞ and five identical line sources, each of strength Λ_a, located at equal intervals along a line normal to V_∞; the method is identical with that used to construct Fig. 1 except that in Fig. 21a four more sources are placed on the y axis, two above and two below the x axis. The (equal) source strengths are chosen great enough so that the zero streamline ($\psi = 0$) encloses all of the sources and, thus, may be visualized as the surface of a rigid body, since the external flow contains no singularities. If the same total source strength ($5 \Lambda_a$) is distributed among 11 equally spaced sources of equal strength, the streamline pattern changes to that shown in Fig. 21b; the waviness exhibited by the zero streamline in Fig. 21a is not evident but further

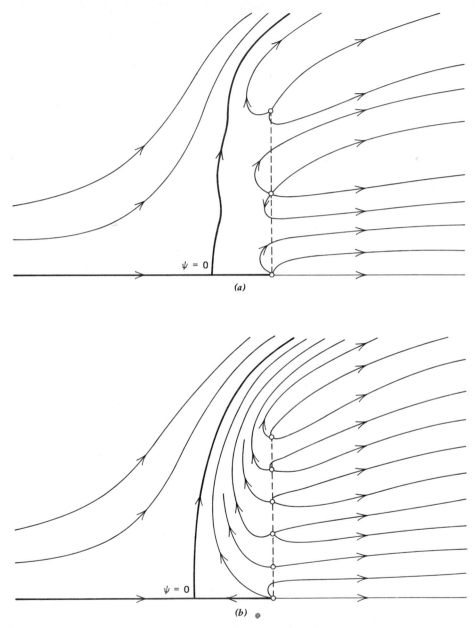

Figs. 21 Flow pattern resulting from distributing m identical line sources along the dashed line in the presence of a uniform flow. Because of symmetry only the upper half of the entire flow is shown. Circles indicate the locations of the line sources. (a) $m = 5$. (b) $m = 11$ while the total source strength is kept the same as in the previous case. (c) $m = 101$ with the same total source strength. (d) $m = 101$ but the source strength is reduced. (e) $m \to \infty$ with $\lambda/2 = -V_\infty$; dashed streamline is schematic. (f) Boundary conditions at inclined panel.

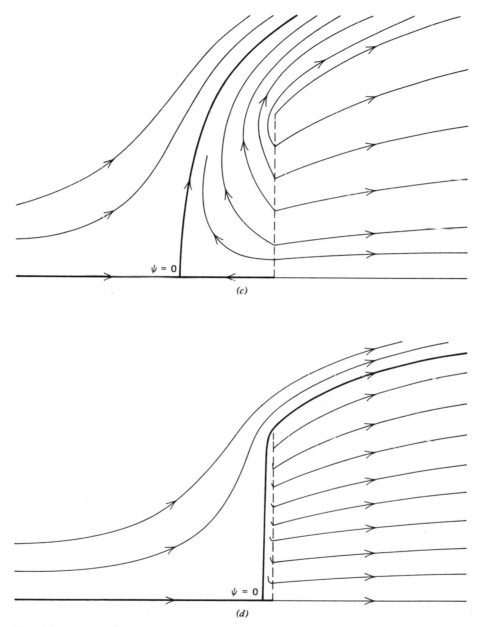

$\psi = 0$

(c)

$\psi = 0$

(d)

Figs. 21 (*Continued*).

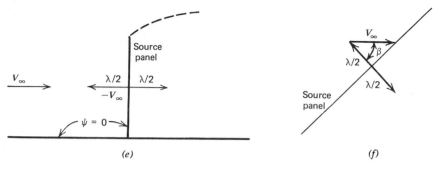

(e)

(f)

Figs. 21 (*Continued*).

downstream the contributions of the individual sources can still be detected. If the total source strength is distributed instead among 101 equally spaced sources of equal strength, all of the streamlines shown in Fig. 21c have lost the waviness identifiable with individual sources. Then, if the strength of each source is reduced uniformly, the zero streamline, that is, the body surface, will approach the line along which the sources are located; one example is shown in Fig. 21d; we see that the zero streamline approaches the line of sources as the total strength decreases.

While the computer does not enable passage to the limit of an infinite number of sources of finite total strength, it is clear that, for the example of Figs. 21a-d, that limit will be the "source panel" shown in Fig. 21e, in which the streamline emanating from the upper edge is depicted schematically. The panel is characterized by a uniform "source density" λ, defined as the volume of fluid discharged per unit area, of the panel. The discharge velocities will be $\pm\lambda/2$ and $\lambda/2 = -V_\infty$ is the boundary condition required in Fig. 21e to make the source panel a streamline of the flow. In Fig. 21f, the source panel is at an angle to the flow and the velocity at the surface for the single source panel shown; for a *closed* body being approximated by *several* source panels the sum of the contributions from each to the normal velocity component must balance $V_\infty \cos \beta$ at each panel.

We consider now a two-dimensional body in a uniform flow of speed V_∞. The body surface is replaced by m source panels of different lengths, s_j, and uniform strengths, λ_j. As shown in Fig. 22 adjacent panels intersect at "boundary points" of the surface; "control points," or the points at which the resultant flow is required to be tangent to the panel surfaces, are chosen to be the midpoints of the panels, as designated in the figure. The mathematical questions that arise because external velocities approach infinity along the lines of intersection of two panels are resolved by Hess and Smith, 1967; this aspect is illustrated in the example in Section 5.9.

The source distribution on the jth panel causes an induced flow field whose velocity potential at any point in the flow (x_i, y_i) is, according to Eqs. 2.44, ex-

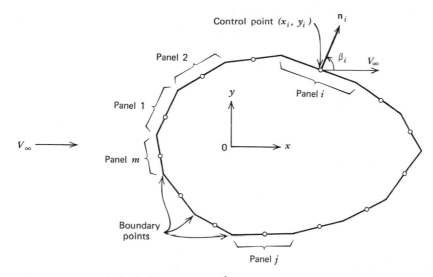

Fig. 22 Replacement of a body by source panels.

pressed by

$$\int_j \ln r_{ij} \cdot \frac{\lambda_j \, ds_j}{2\pi}$$

where $\lambda_j \, ds_j$ is the source strength for the element ds_j at the point (x_j, y_j) on the panel, that is, it is the area flow rate, or the rate of volume flow through ds_j per unit length along the z axis, from the source, and $r_{ij} = \sqrt{(x_i - x_j)^2 + (y_i - y_j)^2}$; the path of integration covers the entire length of the jth panel. Thus the velocity potential for the flow resulting from the superposition of the uniform flow and the m source panels is given by

$$\phi(x_i, y_i) = V_\infty x_i + \sum_{j=1}^{m} \frac{\lambda_j}{2\pi} \int_j \ln r_{ij} \, ds_j \qquad (32)$$

in which the constant factor $\lambda_j/2\pi$ has been taken out of the integral. Since Eq. 32 is valid everywhere in the flow field, we now let (x_i, y_i) be the coordinates of the control point on the ith panel where the outward normal vector is n_i. The angle between n_i and V_∞ is β_i as shown in Fig. 22. The approximate boundary condition at the body surface is that at each control point the resultant normal velocity from all of the superimposed flows vanishes; thus

$$\frac{\partial}{\partial n_i} \phi(x_i, y_i) = 0$$

establishes that the control point of each panel is on a streamline of the flow. Accordingly, we differentiate Eq. 32 to obtain

$$\sum_{i=1}^{m} \frac{\lambda_j}{2\pi} \int_j \frac{\partial}{\partial n_i} (\ln r_{ij}) \, ds_j = -V_\infty \cos \beta_i$$

Each term under the summation represents the integrated contribution of the *j*th panel to the normal velocity component at the *i*th panel. The term representing the contribution of the *i*th panel is simply $\lambda_i/2$, in line with the discussion of Figs. 21*e* and *f* (see Problem 4.13.1), and the above equation becomes

$$\frac{\lambda_i}{2} + \sum_{\substack{j \neq i}}^{m} \frac{\lambda_j}{2\pi} \int_j \frac{\partial}{\partial n_i} (\ln r_{ij}) \, ds_j = -V_\infty \cos \beta_i \qquad (33)$$

where the summation is carried out for all values of m except $j = i$.

For a given panel configuration both n_i and β_i are specified at each control point. After the evaluation of the integrals in Eq. 33 for all values of i, a set of m simultaneous algebraic equations is obtained which enables us to solve for λ_j. With known panel strengths, the velocity and pressure at *any* point in the flow can be computed by taking derivatives of ϕ expressed in Eq. 32 and using Bernoulli's equation.

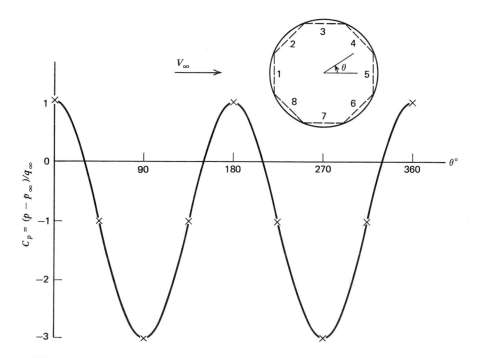

Fig. 23*a* Pressure coefficient on a circular cylinder in a uniform stream obtained by using eight source panels (marked by X) in comparison with the exact solution.

As a specific example the source panel technique is applied to solving the problem of a uniform flow past a circular cylinder, whose exact solution has been found analytically in Section 4.6. As sketched in Fig. 23a the surface of the cylinder is replaced by eight source panels of equal widths, whose orientations are so arranged that the first panel is facing the oncoming flow.

As a sample calculation, let us compute the contribution to the normal velocity at panel 3 by the source distribution on panel 2. This contribution is the product of $\lambda_3/2\pi$ and the integral

$$I_{32} \equiv \int_2 \frac{\partial}{\partial n_3} (\ln r_{32}) \, ds_2$$

We let (x_3, y_3) be the coordinates of the control point on panel 3 (the origin is at the center of the cylinder) and (x_2, y_2) be those of an arbitrary point on panel 2; (x_2, y_2) is at a distance s_2 from the lower end of the panel (see Fig. 23b). Thus $r_{32} = \sqrt{(x_3 - x_2)^2 + (y_3 - y_2)^2}$ and the integral becomes

$$I_{32} = \int_0^{l_2} \frac{(x_3 - x_2)(\partial x_3/\partial n_3) + (y_3 - y_2)(\partial y_3/\partial n_3)}{(x_3 - x_2)^2 + (y_3 - y_2)^2} \, ds_2 \qquad (34)$$

in which l_2 (= 0.7654) is the total length of panel 2, $\partial x_3/\partial n_3 = \cos \beta_3 = 0$, and $\partial y_3/\partial n_3 = \sin \beta_3 = 1$. Panel 2 makes an angle θ_2 (= 45°) with the x axis; therefore, $\sin \theta_2 = \cos \theta_2 = 0.7071$. With the substitutions that

$$x_3 = 0; \quad y_3 = 0.9239$$

$$x_2 = -0.9239 + 0.7071 s_2; \quad y_2 = 0.3827 + 0.7071 s_2$$

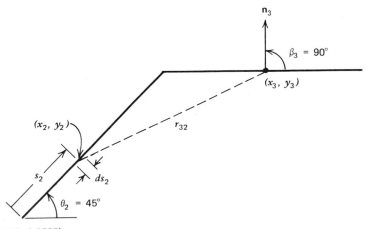

Fig. 23b Designations for sample calculation.

the right-hand side of Eq. 34 is expressed in terms of s_2 only and, after some manipulation, the integral may be written in the form

$$I_{32} = \int_0^a \frac{b - cs_2}{s_2^2 - es_2 + f} \, ds_2$$

where $a = 0.7654$, $b = 0.5412$, $c = 0.7071$, $e = 2.0720$, $f = 1.1464$. The form of the result is determined by the constants in the denominator. For the present case $(4f - e^2) = b^2 > 0$

$$I_{32} = -\frac{c}{2} \, [\ln (s_2^2 - es_2 + f)]_0^a + \frac{2b - ce}{b} \left[\tan^{-1} \left(\frac{2s_2 - e}{b} \right) \right]_0^a$$

$$= 0.3528$$

The remaining integrals contained in Eq. 33 can be evaluated in a similar fashion for every combination of i and j. Thus, by letting $\lambda_j' = \lambda_j / 2\pi V_\infty$ and $i = 1, 2, \ldots,$ 8, Eq. 33 becomes, when expressed in the more convenient matrix form:

$$
\begin{pmatrix}
3.1416 & .3528 & .4018 & .4074 & .4084 & .4074 & .4018 & .3528 \\
.3528 & 3.1416 & .3528 & .4018 & .4074 & .4084 & .4074 & .4018 \\
.4018 & .3528 & 3.1416 & .3528 & .4018 & .4074 & .4084 & .4074 \\
.4074 & .4018 & .3528 & 3.1416 & .3528 & .4018 & .4074 & .4084 \\
.4084 & .4074 & .4018 & .3528 & 3.1416 & .3528 & .4018 & .4074 \\
.4074 & .4084 & .4074 & .4018 & .3528 & 3.1416 & .3528 & .4018 \\
.4018 & .4074 & .4084 & .4074 & .4018 & .3528 & 3.1416 & .3528 \\
.3528 & .4018 & .4074 & .4084 & .4074 & .4018 & .3528 & 3.1416
\end{pmatrix}
\begin{pmatrix}
\lambda_1' \\ \lambda_2' \\ \lambda_3' \\ \lambda_4' \\ \lambda_5' \\ \lambda_6' \\ \lambda_7' \\ \lambda_8'
\end{pmatrix}
=
\begin{pmatrix}
1.0000 \\ .7071 \\ .0000 \\ -.7071 \\ -1.0000 \\ -.7071 \\ .0000 \\ .7071
\end{pmatrix}
$$

The above set of equations has the solution

$$\lambda_1' = 0.3765, \quad \lambda_2' = 0.2662, \quad \lambda_3' = 0, \quad \lambda_4' = -0.2662,$$

$$\lambda_5' = -0.3765, \quad \lambda_6' = -0.2662, \quad \lambda_7' = 0, \quad \lambda_8' = 0.2662,$$

the sum of the λ''s is zero, as it must be to represent a closed body, since all of the panels are of the same length.

With only eight panels, the pressure coefficient computed at the control points are shown in Fig. 23a to be in coincidence with the theoretical distribution (Section 4.6)

$$C_p = \frac{p - p_\infty}{q_\infty}$$

$$= 1 - 4 \sin^2 \theta$$

This example demonstrates the power of the panel method, although the excellent agreement with the exact distribution can be misleading. The panel method is, after all, a method of approximation, and its accuracy depends on the shape of the body and on the panel configuration chosen. For example, when eight source panels of unequal lengths are used to approximate the surface of an elliptical cylinder, the resultant pressure distribution does not coincide with the analytical curve at all control points; the errors can be reduced by increasing the number of panels.

In a similar manner, the panel method can be applied to a two-dimensional nonlifting airfoil. In the case of a lifting body vortex panels are needed in addition to source panels to generate a circulation around the body, and the Kutta condition has to be satisfied. The numerical technique is discussed in detail in Section 5.9.

Three-dimensional nonlifting bodies can be represented by the use of finite source panels as illustrated in Fig. 24. The boundary points, which are the intersections of

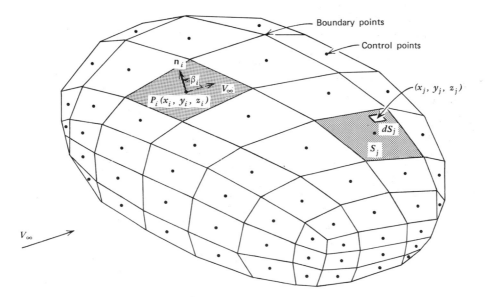

Fig. 24 Source panel representation of a three-dimensional body.

the source panels with the body surface, and control points are indicated. The computation procedure is more complicated in that more panels are needed and the line integrals in Eq. 33 become surface integrals; however, no new concepts are involved. Application to lifting three-dimensional bodies is discussed in Chapter 6.

CHAPTER 5

Aerodynamic Characteristics of Airfoils

5.1 Introduction

The history of the development of airfoil shapes is long and involves many names in philosophy and the sciences (Giacomelli, 1943). By the beginning of the twentieth century, the methods of classical hydrodynamics had been successfully applied to airfoils, and it became possible to predict the lifting characteristics of certain airfoil shapes mathematically. These special shapes, which lent themselves to precise mathematical treatment, did not represent the optimum in airfoil performance, and workers in the field resorted to experimental methods guided by theory to determine the characteristics of arbitrarily shaped airfoils.

In 1929, the National Advisory Committee for Aeronautics (NACA) began studying the characteristics of systematic series of airfoils in an effort to find the shapes that were best suited for specific purposes. Families of airfoils constructed according to a certain plan were tested and their characteristics recorded. The airfoils were composed of a **thickness envelope** wrapped around a **mean camber line** in the manner shown in Fig. 1. The mean line lies halfway between the upper and lower surfaces of the airfoil and intersects the chord line at the leading and trailing edges.

The various families of airfoils are designed to show the effects of varying the important geometrical variables on the important aerodynamic characteristics, such as lift, drag, and moment, as functions of angle of attack. The geometrical variables include the maximum camber z_c of the mean line and its distance x_c behind the leading edge; the maximum thickness t_{max} and its distance x_t behind the leading edge; the radius of curvature of the surface at the leading edge, r_0, and the angle between the upper and lower surfaces at the trailing edge. Theoretical studies and wind tunnel experiments show the effects of these variables in a way to facilitate the choice of shapes for specific applications. The reader is referred to Abbott and

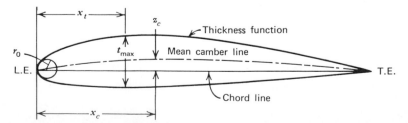

Fig. 1 Important geometrical variables.

Doenhoff (1949) for means of identifying the various families of airfoil shapes; we are concerned here with geometrical features insofar as they have an important effect on aerodynamic characteristics of airfoils.

The lifting characteristics of an airfoil below the stall are negligibly influenced by viscosity. Further, the resultant of the pressure forces on the airfoil (magnitude, direction and line of action) is only slightly influenced by the thickness function, provided that the ratio of maximum thickness to chord t_{max}/c is small, the maximum mean camber z_c is small, and the airfoil is operating at a small angle of attack. These three conditions are usually fulfilled in the normal operation of the usual types of airfoils. As a consequence, the overall lifting characteristics of an airfoil are well predicted by the thin-airfoil theory outlined in the following sections. In this approximate theory, viscosity is neglected and the airfoil is replaced with its mean camber line. The kinematic problem is resolved into one of finding a flow pattern that has one streamline coincident with the mean camber line. The **bound vortex sheet** described in the following section is used to construct the pattern. The distribution of the pressure jump across the camber line is given by the Kutta-Joukowski law and the overall lifting characteristics are determined from the integral of the pressure forces.

Finally, the source panel method described in Section 4.13 for arbitrary nonlifting bodies is extended to determination of the lift of airfoils of arbitrary thickness and camber. The extension involves the superposition of source and vortex panels on the airfoil surface and adjusting their strengths so the surface is a streamline of the flow and the Kutta condition is satisfied.

5.2 The Vortex Sheet

Extension of the vortex flow, analyzed as a single vortex filament in Section 2.12, is here extended to a surface, termed a vortex sheet; this extension provides a means for analysis of the flow around lifting surfaces and bodies. The vortex sheet, shown schematically in Fig. 2, may be thought of as an infinite number of vortex filaments each of infinitesimal strength and each extending to infinity or to a boundary of the flow. The strength of the sheet, γ, termed the circulation density, is defined as a limiting form of Stokes' theorem.

$$\gamma = \lim_{\Delta s \to 0} \left[\frac{1}{\Delta s} \oint \mathbf{v} \cdot d\mathbf{l} \right]$$

where Δs is the width of the sheet enclosed by the contour. If the circulation is clockwise, γ is a positive number (it has the dimensions of a velocity) and $\gamma\, ds$ is the circulation per unit width at the point.

If the elements of the sheet are straight, doubly infinite vortex filaments, then the velocity induced in the surrounding field at any point P by an element ds of

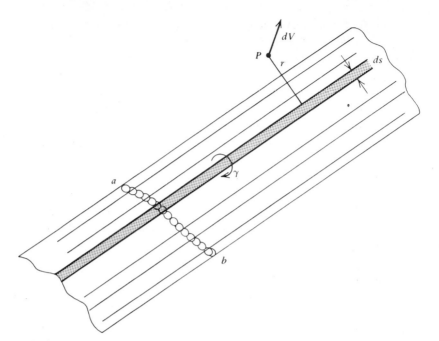

Fig. 2 Velocity induced at point P by element ds of a vortex sheet.

the sheet is in Eq. 2.50,

$$dV = \frac{\gamma\, ds}{2\pi r} \tag{1}$$

where dV is normal to r in Fig. 2.

It is apparent that the velocity field induced by a vortex sheet will satisfy *continuity* at all points in the field because each element individually induces a field that satisfies *continuity* at all points. *Irrotationality* is satisfied at all points in the field by the same argument. The argument fails at the sheet itself where the value of the curl **V** is nonzero.

It can easily be seen that the velocity is finite at all points in the field, excluding the points of the sheet, by realizing that the velocity at point P induced by any finite increment of the sheet is finite, and therefore the velocity at P induced by the entire sheet is finite. For points on the sheet, the demonstration is more involved, but it can be shown that the velocity induced at any point on the sheet by the entire sheet is finite, except for the points a and b, that is, at the edges of the sheet. At these points, the velocity is infinite.

No discontinuities in the velocity occur anywhere in the field except at the sheet itself. We will show that the velocity component parallel to the sheet jumps by an amount equal to the strength of the sheet at the point where the contour crosses the sheet. We consider the cross section of the sheet shown in Fig. 3 and integrate

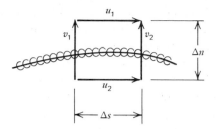

Fig. 3 Contour for evaluating circulation density.

the vorticity over the area $\Delta s \Delta n$, then let these dimensions approach zero. Thus,

$$\bar{\gamma} \Delta s = - \iint_{\hat{S}} \omega \, d\hat{S} = u_1 \Delta s - v_2 \Delta n - u_2 \Delta s + v_1 \Delta n$$

where $\bar{\gamma}$ is the average circulation density over Δs. In the limit, as Δn and $\Delta s \to 0$ so that $(v_1 - v_2) \to 0$ and the expression above reduces to $\gamma \, ds$ where γ is the circulation density at a point x on the sheet. It follows that

$$\gamma(x) = u_1 - u_2 \tag{2}$$

which is the jump in tangential velocity across the sheet.

5.3   The Vortex Sheet in Thin-Airfoil Theory

In thin-airfoil theory, the airfoil is replaced with its mean camber line. The flow pattern is built up by placing a bound vortex sheet on the camber line and so adjusting its strength that the camber line becomes a streamline of the flow. Points on the camber line (and therefore on the vortex sheet) lie outside the field of flow. The velocity pattern, then, is composed of a uniform stream plus the field induced by the vortex sheet.

Continuity and *irrotationality* are satisfied at every point in the field. The velocity at infinity is that of the uniform stream because a vortex sheet can make no contribution at infinity. At the camber line the resultant of the uniform stream and the field induced by the sheet is parallel to the camber line.

According to the discussion of Chapter 4, in order to establish the uniqueness of an irrotational flow, it is necessary to specify not only the velocity at infinity and the direction of the velocity at the body but also the circulation around the body. It is apparent that the circulation around the sheet is simply the strength of the entire sheet. It follows that for a sheet of given total strength, there is only one distribution of vortex strength that will make the sheet a streamline when the field of the sheet is combined with the uniform stream. It is this distribution that is sought.

The circulation around the body is established by the Kutta condition. In Section 4.10, it was shown that the Kutta condition means that there can be no velocity discontinuity at the trailing edge. In terms of the vortex strength distribution along the mean camber line, the Kutta condition must be interpreted as fixing the strength of the vorticity at the trailing edge at zero. Therefore, the Kutta condition removes the difficulty of an infinite velocity at the trailing edge of the vortex sheet. The infinite velocity at the leading edge remains, which means that the flow pattern at the leading edge predicted by the theory cannot be correct.

In summary, it can be said that the resultant of the uniform stream and the field induced by a vortex sheet satisfies *continuity* and *irrotationality* and has a value at infinity equal to that of the uniform stream. One and only one distribution of vortex strength can be found of given total strength, which, when combined with a uniform stream, makes the vortex sheet a streamline. The total strength of the sheet is fixed by the Kutta condition:

$$\gamma(\text{T.E.}) = 0 \tag{3}$$

In order for the resultant of the uniform stream and the velocity induced by the sheet to be parallel to the sheet, the normal components of the uniform stream and induced velocity must sum to zero. The geometry, drawn out of scale for clarity, is shown in Fig. 4. It should be remembered that, for wings in common use, the maximum mean camber is of the order of 2 percent of the chord. An increment of induced velocity is given by Eq. 1. The normal component of the increment is

$$dV_{in} = \frac{\gamma \, ds}{2 \pi r} \cos \delta_3 \tag{4}$$

Using the relations

$$r = \frac{x_0 - x}{\cos \delta_2}$$

$$ds = \frac{dx}{\cos \delta_1}$$

and integrating from the leading edge to the trailing edge,*

$$V_{in} = - \frac{1}{2\pi} \int_{\text{L.E.}}^{\text{T.E.}} \frac{\gamma \, dx}{x_0 - x} \frac{\cos \delta_2 \cos \delta_3}{\cos \delta_1} \tag{5}$$

*The integral of Eq. 5 has an infinite integrand at $x = x_0$. The induced velocity is the *principal value* of this integral. For a discussion of the limiting process involved in taking the principal value of an integral see Sokolnikoff (1939). Integrals of this type occur frequently in the present and the following chapter. In each instance it is the principal value to which reference is being made.

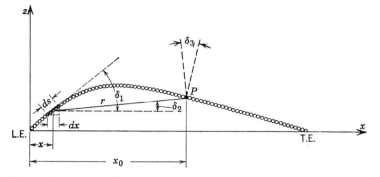

Fig. 4 Field induced by vortex sheet.

The negative sign is uscd because clockwise circulation and outward along the nor-
mal to the upper surface are considered positive. The three angles δ_1, δ_2, and δ_3
are function of x.

The component of the free stream normal to the mean camber line at P is given
by

$$V_{\infty n} = V_\infty \sin\left[\alpha - \tan^{-1}\left(\frac{dz}{dx}\right)_0\right] \tag{6}$$

The subscript $_0$ on the angle indicates a chordwise station corresponding to the
coordinate x_0 in Fig. 4. The angle of attack as shown in Fig. 5 is taken as positive.
$(dz/dx)_0$ is the slope of the mean camber line relative to the chord at the station
x_0.

The sum of Eqs. 5 and 6 must be zero if the mean camber line is to be a streamline of the flow.

$$V_{in} + V_{\infty n} = 0 \tag{7}$$

The central problem of thin airfoil theory is to find a γ distribution that satisfies
Eqs. 3 and 7. In the next section a simplification of Eq. 7 is introduced which
leads to the concept of the planar wing.

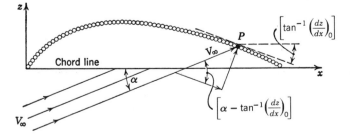

Fig. 5 Velocity components establishing boundary condition.

5.4 Planar Wing

The three angles δ_1, δ_2, and δ_3 in Eq. 5 are small if the maximum mean camber is small. This is the usual situation in practical airfoils, and so to a good approximation the cosines of the three angles can be set equal to unity. Then Eq. 5 becomes

$$V_{in} = -\frac{1}{2\pi} \int_{\text{L.E.}}^{\text{T.E.}} \frac{\gamma \, dx}{x_0 - x} \tag{8}$$

But Eq. 8 represents the velocity induced on the x axis by a vortex sheet lying on the x axis. Therefore, *the simplification introduced above is equivalent to satisfying boundary conditions on the x axis instead of at the mean camber line.*

The same order of approximation is made in Eq. 6, the additional assumption being that the angle of attack is small. Setting the sine and tangent equal to the angle, Eq. 6 becomes

$$V_{\infty n} = V_{\infty} \left[\alpha - \left(\frac{dz}{dx} \right)_0 \right]$$

The boundary condition at the airfoil corresponding to Eq. 7 becomes

$$\frac{1}{2\pi} \int_0^c \frac{\gamma \, dx}{x_0 - x} = V_{\infty} \left[\alpha - \left(\frac{dz}{dx} \right)_0 \right] \tag{9}$$

Equation 9 represents the condition of zero flow normal to the mean camber line. The condition is applied at the x axis, however, instead of at the mean camber line.

This technique, referred to as the *planar wing approximation*, is used throughout thin-wing theory. It appears again in Chapter 6 in connection with the finite wing and in Chapter 12 in the development of supersonic wing theory.

5.5 Properties of the Symmetrical Airfoil

The distribution that will satisfy Eqs. 3 and 9 will be found first for the case $dz/dx = 0$. This corresponds to a symmetrical airfoil or one in which the chord line and mean camber line are coincident. It is convenient to change coordinates, as illustrated in Fig. 6, by letting

$$x = \tfrac{1}{2} c (1 - \cos \theta)$$

where c is the chord of the airfoil. θ becomes the independent variable and θ_0 corresponds to x_0. Then the conditions to be satisfied are, from Eqs. 3 and 9,

$$\gamma(\pi) = 0$$

$$\frac{1}{2\pi} \int_0^{\pi} \frac{\gamma \sin \theta \, d\theta}{\cos \theta - \cos \theta_0} = V_{\infty} \alpha \tag{10}$$

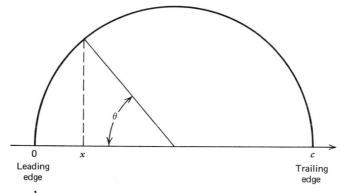

Fig. 6 Plot of $x = \frac{1}{2}c(1 - \cos\theta)$.

It can be readily verified that the γ distribution that satisfies both of Eqs. 10 is*

$$\gamma = 2\alpha V_\infty \frac{1 + \cos\theta}{\sin\theta} \tag{11}$$

To verify that Eq. 11 satisfies the second of Eqs. 10, it is neccessary to show that

$$\int_0^\pi \frac{1 + \cos\theta}{\cos\theta - \cos\theta_0}\, d\theta = \pi$$

This can be done with the help of the definite integral (Glauert, 1937, p 92)

$$\int_0^\pi \frac{\cos n\theta}{\cos\theta - \cos\theta_0}\, d\theta = \pi \frac{\sin n\theta_0}{\sin\theta_0} \tag{12}$$

Then, with $n = 1$, Eq. 11 satisfies the second of Eqs. 10. This integral occurs several times in both thin-airfoil theory and finite-wing theory. An evaluation of it may be found in other textbooks on aerodynamics (Glauert 1937, p. 92; Mises 1945, p. 208; Kármán and Burgers 1943, p. 173).

That Eq. 11 satisfies the first of Eqs. 10 can be shown by evaluating the indeterminate form as $\theta \to \pi$.

In terms of x, Eq. 11 becomes

$$\gamma = 2\alpha V_\infty \sqrt{\frac{c - x}{x}} \tag{11a}$$

The lift per unit area at a given location is given by

$$\Delta p = \rho V_\infty \gamma \tag{13}$$

*A direct method for finding the γ distribution consists of transforming the flow about a circle into the flow about a flat plate by conformal mapping. See Kármán and Burgers (1943).

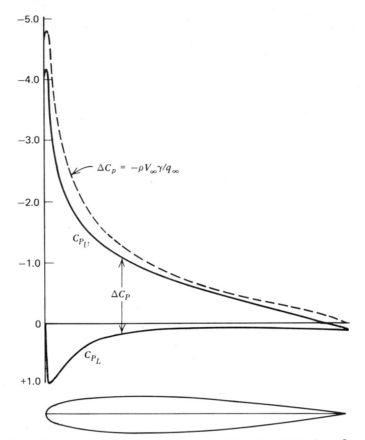

Fig. 7 Distribution of pressure coefficient and ΔC_p on NACA 0012 airfoil at $\alpha = 9°$.

and is numerically equal to the difference in pressure between the upper and lower surfaces at the point. Figure 7 shows chordwise plots of the pressure coefficients (Eq. 3.11), C_{p_L}, C_{p_U} for the lower and upper surfaces, and,

$$\Delta C_p = \frac{p_U - p_L}{q_\infty} = -\frac{\Delta p}{q_\infty}$$

for an NACA 0012 airfoil at an angle of attack of $9°$.

The lift per unit span, L', follows from Eq. 13,

$$L' = \int_0^c \Delta p\, dx = \rho V_\infty \int_0^c \gamma\, dx \tag{14}$$

and for the distribution of Eq. 11,

$$L' = \rho V_\infty \int_0^\pi 2\alpha V_\infty \frac{1 + \cos\theta}{\sin\theta} \frac{c}{2} \sin\theta\, d\theta \tag{15}$$

The appropriate dimensionless parameter is the **sectional lift coefficient** defined by

$$c_l = \frac{L'}{q_\infty c} \tag{16}$$

After evaluating the integral, Eqs. 15 and 16 lead to the simple result:

$$c_l = 2\pi\alpha \equiv m_0\alpha \tag{17}$$

where m_0 is the slope of the c_l versus α curve. Thus, thin-airfoil theory indicates that the sectional lift coefficient for a symmetrical airfoil is directly proportional to the geometric angle of attack, where the geometric angle of attack is defined as the angle between the flight path and the chord line of the airfoil. Further, when the geometric angle of attack is zero, the lift coefficient is zero. The moment of the lift about the leading edge of the airfoil is given by

$$M'_{L.E.} = -\int_0^c \Delta p x \, dx$$

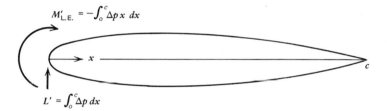

Fig. 8 Load system on an airfoil.

A stalling moment is taken as positive (clockwise in Fig. 8). Again using Eqs. 11 and 13 and defining a sectional moment coefficient $c_{m\,L.E.} = M'_{L.E.}/q_\infty c^2$, the value of the sectional moment coefficient becomes

$$c_{m\,L.E.} = -\frac{\pi\alpha}{2}$$

or, in terms of the lift coefficient,

$$c_{m\,L.E.} = -\frac{c_l}{4} \tag{18}$$

The center of pressure of a force is defined as the point about which the moment vanishes. If we consider the lift concentrated at that point its moment will just balance $M'_{L.E.}$. Thus

$$M'_{L.E.} + L'x_{c.p.} = 0$$

and, by use of Eqs. 17 and 18,

$$x_{c.p.} = \frac{c}{4} \tag{19}$$

at all angles of attack.

The lifting characteristics have now been completely determined in magnitude, direction, and line of action. Summarizing, for a symmetrical airfoil:

1. The sectional lift coefficient is directly proportional to the geometric angle of attack and is equal to zero when the geometric angle of attack is zero.
2. The lift curve slope m_0 equals 2π.
3. The center of pressure is at the quarter chord for all values of the lift coefficient.

5.6 Vorticity Distribution for the Cambered Airfoil

The method of determining the properties of a cambered airfoil is essentially the same as that for the symmetrical airfoil. However, because of the dependence of these properties on the mean-camber-line shape, the actual computations are more involved. The properties of the cambered airfoil must include those of the symmetrical airfoil as a special case. Again, the central problem is finding a γ distribution that satisfies Eqs. 3 and 9. Using the transformation (Fig. 6)

$$x = \tfrac{1}{2} c(1 - \cos \theta)$$

Eqs. 3 and 9 become

$$\gamma(\pi) = 0 \tag{20}$$

$$\frac{1}{2\pi} \int_0^\pi \frac{\gamma \sin \theta \, d\theta}{\cos \theta - \cos \theta_0} = V_\infty \left[\alpha - \left(\frac{dz}{dx} \right)_0 \right] \tag{21}$$

The γ distribution that satisfies Eq. 21 may be represented as the sum of two parts. One part involves the shape of the mean camber line *and* the angle of attack and has the form of the γ distribution for the symmetrical airfoil as given by Eq. 11. This part is written

$$2V_\infty A_0 \frac{1 + \cos \theta}{\sin \theta}$$

It will be shown later that $A_0 = \alpha$ when the airfoil is symmetrical; that is, when $dz/dx = 0$. The other part of the γ distribution depends only on the shape of the mean camber line and is finite everywhere including the point at the leading edge. It is convenient to express this part as a Fourier series and, following the form used for the first part we write

$$2V_\infty \sum_{n=1}^{\infty} A_n \sin n\theta$$

for the contribution of the camber.

The total γ distribution is the sum of the two parts and may be written

$$\gamma = 2V_\infty \left[A_0 \frac{1 + \cos\theta}{\sin\theta} + \sum_{n=1}^{\infty} A_n \sin n\theta \right] \tag{22}$$

When $\theta = \pi$, $\gamma = 0$ for all values of the coefficients; thus Eq. 20 is satisfied.

It remains to find the values of A_0 and A_n that will make Eq. 22 satisfy Eq. 21. To this end, Eq. 22 is substituted in Eq. 21, giving

$$\frac{1}{\pi} \int_0^\pi \frac{A_0(1 + \cos\theta)}{\cos\theta - \cos\theta_0} d\theta + \frac{1}{\pi} \int_0^\pi \sum_{n=1}^{\infty} \frac{A_n \sin n\theta \sin\theta}{\cos\theta - \cos\theta_0} d\theta = \alpha - \left(\frac{dz}{dx} \right)_0 \tag{23}$$

The first integral on the left-hand side of Eq. 23 is of the form of the relation shown in Eq. 12. The second infinite series of integrals may also be evaluated by Eq. 12 if the trigonometric identity $\sin n\theta \cdot \sin\theta = \frac{1}{2} [\cos(n-1)\theta - \cos(n+1)\theta]$ is used. After performing these integrations and rearranging terms, Eq. 23 becomes

$$\frac{dz}{dx} = (\alpha - A_0) + \sum_{n=1}^{\infty} A_n \cos n\theta \tag{24}$$

The station subscript has been dropped for it is understood that Eq. 24 applies to any chordwise station. The coefficients A_0 and A_n must satisfy Eq. 24 if Eq. 22 is to represent the γ distribution that satisfies the condition of parallel flow at the mean camber line.

It will be observed that Eq. 24 has the form of the cosine series expansion of dz/dx. For a given mean camber line, dz/dx is a known function of θ, and, therefore, the values of A_0 and A_n may be written directly as

$$A_0 = \alpha - \frac{1}{\pi} \int_0^\pi \frac{dz}{dx} d\theta \tag{25}$$

$$A_n = \frac{2}{\pi} \int_0^\pi \frac{dz}{dx} \cos n\theta \, d\theta \tag{26}$$

Equations 25, 26, and 22 determine the γ distribution of the cambered airfoil in terms of the geometric angle of attack and the shape of the mean camber line. For zero camber, $A_0 = \alpha$ and $A_n = 0$. Equation 22 becomes

$$\gamma = 2V_\infty\alpha\,\frac{1 + \cos\theta}{\sin\theta}$$

which was shown to be the γ distribution for the symmetrical airfoil in Section 5.5.

5.7 Properties of the Cambered Airfoil

The lift and moment coefficients for the cambered airfoil are found in the same manner as for the symmetrical airfoil.

$$c_l = \frac{1}{q_\infty c}\int_0^c \Delta p\ dx$$

$$c_{m\text{L.E.}} = -\frac{1}{q_\infty c^2}\int_0^c \Delta px\ dx$$

where Δp equals $\rho V_\infty \gamma$ and where γ is given by Eq. 22. After performing the integrations, the lift and moment coefficients become

$$c_l = 2\pi A_0 + \pi A_1 \tag{27}$$

$$c_{m\text{L.E.}} = -\tfrac{1}{2}\pi(A_0 + A_1 - \tfrac{1}{2}A_2) \tag{28}$$

The moment coefficient in terms of the lift coefficient may be written

$$c_{m\text{L.E.}} = -\tfrac{1}{4}c_l + \tfrac{1}{4}\pi(A_2 - A_1) \tag{29}$$

The center of pressure position behind the leading edge is found by dividing the moment about the leading edge by the lift.

$$x_{\text{c.p.}} = \frac{c}{4} - \frac{\pi c}{4}\frac{A_2 - A_1}{c_l} \tag{30}$$

From Eq. 26, it can be seen that A_1 and A_2 are independent of the angle of attack. They depend only on the shape of the mean camber line. Therefore, Eq. 30 shows that the position of the center of pressure will vary as the lift coefficient varies. The line of action of the lift, as well as the magnitude, must be specified for each angle of attack.

It will be observed from Eq. 29, that, if the load system is transferred to a point behind the leading edge by a distance equal to 25 percent of the chord, the moment coefficient about this point will be independent of angle of attack.

$$c_{m_{c/4}} = \tfrac{1}{4}\pi(A_2 - A_1) \tag{31}$$

Equivalent load systems for the three locations, the leading edge, the quarter chord point, and the center of pressure are shown in Fig. 9. Note that (1) $M'_{c/4}$ is

$$M'_{L.E.} = \left[-c_l + \pi(A_2 - A_1) \right] \tfrac{1}{4} q_\infty c^2$$

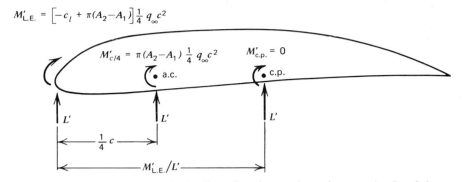

$$M'_{c/4} = \pi(A_2 - A_1) \tfrac{1}{4} q_\infty c^2 \qquad M'_{c.p.} = 0$$

a.c.　　　　　c.p.

L'　　　　　L'　　　　　L'

$\tfrac{1}{4}c$

$M'_{L.E.}/L'$

Fig. 9 Equivalent load systems at the leading edge, the aerodynamic center (a.c.) and the center of pressure (c.p.).

dependent only on the geometry of the section, and (2) $M'_{c.p.} = 0$, but the location of c.p. $= M'_{L.E.}/L'$ can vary between $\pm\infty$ as the lift varies.

The load system is commonly specified as a lift and a constant moment acting at the quarter chord. The point about which the moment coefficient is independent of the angle of attack is called the **aerodynamic center** of the section, and the moment coefficient about the aerodynamic center is given the symbol c_{mac}. Because this moment is a constant for all angles of attack, including the angle of attack that gives zero lift, it is frequently called the **zero-lift moment**. A moment in the absence of a resultant force is a couple. The zero-lift moment, therefore, is a couple. According to thin-airfoil theory, the aerodynamic center is at the quarter chord point, and, therefore, the moment coefficient about the aerodynamic center is given by Eq. 31. After A_2 and A_1 are replaced with their equivalents from Eq. 26, the c_{mac} becomes

$$c_{mac} = \frac{1}{2} \int_0^\pi \frac{dz}{dx} (\cos 2\theta - \cos \theta)\, d\theta \tag{32}$$

The influence of the mean-camber-line shape on the c_{mac} will be shown later. For symmetrical airfoils, c_{mac} is zero.

Replacing the coefficients A_0 and A_1 with their equivalents from Eqs. 25 and 26, the lift coefficient becomes

$$c_l = 2\pi \left[\alpha + \frac{1}{\pi} \int_0^\pi \frac{dz}{dx} (\cos \theta - 1)\, d\theta \right] \tag{33}$$

The lift coefficient varies linearly with the geometric angle of attack, and the slope of the lift curve m_0 is 2π (see Fig. 10). The lift coefficient, however, is not zero when the geometric angle of attack is zero, as it is for the symmetrical airfoil. The value of the geometric angle of attack which makes the lift coefficient zero is called

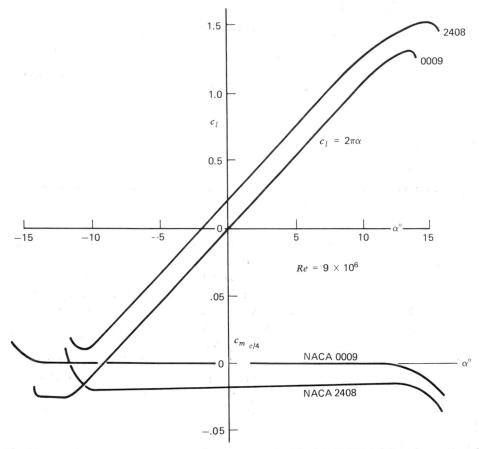

Fig. 10 c_l and $c_{m_{c/4}}$ versus α curves for a symmetric (NACA 0009) airfoil and a cambered airfoil (NACA 2408).

the **angle of zero lift**, α_{L0}. From Eq. 33,

$$\alpha_{L0} = -\frac{1}{\pi} \int_0^{\pi} \frac{dz}{dx} (\cos\theta - 1)\, d\theta \tag{34}$$

Experimental results such as those shown in Fig. 10 indicate remarkable agreement with the forgoing formulas based on thin-wing theory. The experiments were carried out on a symmetrical airfoil (NACA 0009) of 9 percent maximum thickness and on a cambered airfoil (NACA 2408) with maximum mean camber of 2 percent located at $x = 0.4c$ and maximum thickness of 8 percent. We note the following points of comparison between theory and experiment. (1) The c_l versus α curves show that, $dc_l/d\alpha \cong 2\pi$ in good agreement with Eqs. 17 and 27 in the range $-10° < \alpha < +10°$ for both sections, though for the cambered section c_l continues to increase for a short range. (2) When the c_l curves depart markedly from the linear relation, the airfoils "stall" and the theory becomes invalid because, as we show in later chapters, viscous effects cause failure of the Kutta condition. (3) The aerodynamic center and center of pressure coincide for the symmetrical section (in the unstalled range), as follows from $c_{m_{c/4}} = 0$ (Eqs. 18 and 19). (4) For positive camber $c_{m_{c/4}} = $ constant and is negative, as is predicted by Eq. 31. (5) The angle of zero lift α_{L0} is zero for the symmetrical section and negative for positive camber, as is indicated by Eq. 34.

In Fig. 11 an airfoil is shown set at a geometric angle of attack equal to the angle of zero lift. A line on the airfoil parallel to the flight path V_∞ and passing through

Fig. 11 Orientation of zero lift.

the trailing edge when the airfoil is set at the orientation of zero lift is called the **zero-lift line** of the airfoil. For symmetrical airfoils, the zero-lift line coincides with the chord line.

The **absolute angle of attack** is defined as the angle included between the flight path and the zero-lift line and is given the symbol α_a. From Fig. 12

$$\alpha_a = \alpha - \alpha_{L0} \qquad (35)$$

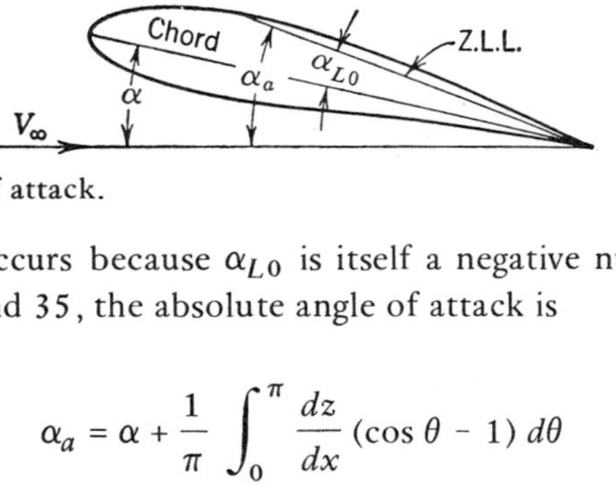

Fig. 12 Absolute angle of attack.

The negative sign occurs because α_{L0} is itself a negative number on normal airfoils. From Eqs. 34 and 35, the absolute angle of attack is

$$\alpha_a = \alpha + \frac{1}{\pi} \int_0^\pi \frac{dz}{dx} (\cos\theta - 1)\, d\theta \qquad (36)$$

Then Eq. 33 may be written

$$c_l = 2\pi\alpha_a \qquad (37)$$

Airfoil characteristics for symmetrical $(dz/dx = 0)$ and cambered sections may be summarized as follows:

(a) $m_0 = 2\pi$.
(b) $\alpha_{L0} = -(1/\pi) \int_0^\pi (dz/dx)(\cos\theta - 1)\, d\theta$.

(c) Aerodynamic center is $c/4$ behind the leading edge, as it is for the symmetrical airfoil.

(d) $c_{mac} = c_{mc/4} = \frac{1}{2} \int_0^\pi (dz/dx) (\cos 2\theta - \cos \theta) \, d\theta$.

(e) Center of pressure (Eq. 30) is at $x = c/4$ for the symmetrical airfoil ($A_2 = A_1 = 0$) and varies with c_l for a cambered section.

Example.

Consider an airfoil whose mean camber line is represented by the parabola

$$z = 4z_m \left[\frac{x}{c} - \left(\frac{x}{c} \right)^2 \right]$$

where z_m is the maximum height of the camber at the mid-chord. The slope of this camber line is

$$\frac{dz}{dx} = 4 \frac{z_m}{c} \left(1 - 2 \frac{x}{c} \right) = 4 \frac{z_m}{c} \cos \theta$$

and from Eqs. 25 and 26

$$A_0 = \alpha, \quad A_1 = 4 \frac{z_m}{c}, \quad A_n = 0 \text{ for } n \geqslant 2$$

Then from Eqs. 27–31,

$$\alpha_{L0} = -2 \frac{z_m}{c}; \quad c_l = 2\pi \left(\alpha + 2 \frac{z_m}{c} \right)$$

$$c_{m L.E.} = -\frac{1}{2} \pi \left(\alpha + 4 \frac{z_m}{c} \right); c_{mac} = -\pi \frac{z_m}{c}$$

$$x_{c.p.} = \frac{c}{4} + z_m \left/ \left(2\alpha + 4 \frac{z_m}{c} \right) \right.$$

The last expression shows that the center of pressure is at the midchord at zero angle of attack, and moves toward the aerodynamic center as α (or c_l) increases.

A quantitative comparison between tests and theory of the dependence of the moment coefficient about the aerodynamic center (c_{mac}) on the maximum mean camber and position of maximum mean camber for the NACA five-digit sections is shown in Fig. 13. The sections with the reflexed mean lines were calculated to give zero values of c_{mac} (see Section 5.8); the agreement with experiment for these sections as well as for those with simple camber is quite good.

Representative experimental values of those section characteristics that are mainly dependent on the potential flow have been tabulated in Fig. 14. It will be observed that the lift curve slope and position of the aerodynamic center are closely

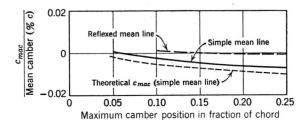

Fig. 13 Theory versus experiment for effect of camber on c_{mac}. (Courtesy NASA.)

Representative Experimental Values of the Section
Characteristics (Courtesy NASA)

Section Designation	$\dfrac{m_0}{2\pi}$	α_{L0} (degrees)	a.c. $\dfrac{x}{c}$ aft of L.E.	c_{mac}
0009	0.995	0	0.25	0
2412	0.985	−1.9	0.243	−0.05
2415	0.97	−1.9	0.246	−0.05
2418	0.935	−1.85	0.242	−0.05
2421	0.925	−1.85	0.239	−0.045
2424	0.895	−1.8	0.228	−0.04
4412	0.985	−3.9	0.246	−0.095
23012	0.985	−1.2	0.241	−0.015
64_3-418	1.06	−2.9	0.271	−0.07
65_3-418	1.03	−2.5	0.266	−0.06
66_3-418	1.00	−2.5	0.264	−0.065

Fig. 14 Representative experimental values of the section characteristics. (Courtesy NASA.)

predicted by the thin-airfoil theory; the variation of the angle of zero lift and the moment coefficient about the aerodynamic center with section shape also follow the predictions of thin-airfoil theory.

The above discussion summarizes the behavior of wing sections in low-speed flow when the wing is in an unstalled position, that is, when the angle of attack is sufficiently low so that the boundary layer remains attached to the body and thus the Kutta condition can be satisfied. A discussion of the maximum lift coefficient and the character of the stall is reserved for later chapters.

5.8 The Flapped Airfoil

The term $(\cos\theta - 1)$ in Eq. 34 vanishes at the leading edge where $\theta = 0$; and its absolute value reaches maximum at the trailing edge where $\theta = \pi$. Thus the portion of the mean camber line in the vicinity of the trailing edge powerfully influences

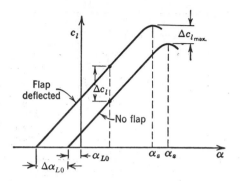

Fig. 15 Effect of flap deflection on lift curve.

the value of α_{L0}. It is on this fact that the aileron as a lateral-control device and the flap as a high-lift device are based. A deflection downward of a portion of the chord at the trailing edge effectively makes the ordinates of the mean camber line more positive in this region. As a consequence, α_{L0} becomes more negative and the lift at a given geometric angle of attack is increased. These results are shown in Fig. 15. The lift curve is displaced to the left as a result of an increase in α_{L0} negatively. The gain in lift at the given geometric angle of attack is shown as Δc_l. If the rear portion of the trailing edge is deflected upward, an opposite displacement of the lift curve results and the lift at a given geometric angle of attack is decreased.

The success of the flap as a high-lift device is based on the fact that, though the stalling angle α_s is reduced by the deflection of a flap, the reduction is not great enough to remove the gain arising from the shift of the curve as a whole. The increase in maximum lift coefficient, $\Delta c_{l_{max}}$, is shown in Fig. 15.

The influence of small flap deflections on the section properties can be predicted by thin-airfoil theory (Glauert, 1924, 1927). Because all angles are small, it is sufficient to find the properties of a symmetrical airfoil at zero angle of attack with flap deflected. These may be added directly to the properties of the cambered airfoil at any angle of attack. In Fig. 16, the mean camber line of a symmetrical airfoil at zero angle of attack is shown with a trailing-edge flap deflected through an angle η. If the leading and trailing edges are connected by a straight line and if this is treated as a fictitious chord line, the problem reduces to that of a cambered airfoil at an angle of attack α' (see Fig. 17). Let E be the ratio of the flap chord to the total chord. hc is the length shown in Fig. 17. From the leading edge to the hinge line, the slope of the mean camber line is $h/(1 - E)$. From the hinge line to

Fig. 16 Airfoil with flap deflected.

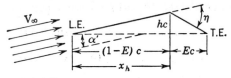

Fig. 17 Geometry of flapped airfoil.

the trailing edge, the slope is $-h/E$. The formulas of Sections 5.6 and 5.7 apply directly. From Eqs. 25 and 26,

$$A_0 = \alpha' - \frac{1}{\pi} \int_0^\pi \frac{dz}{dx} \, d\theta$$

$$A_n = \frac{2}{\pi} \int_0^\pi \frac{dz}{dx} \cos n\theta \, d\theta$$

The integrals must be evaluated in two parts: from the leading edge to the hinge line θ_h, and from the hinge line to the trailing edge. A_0 becomes

$$A_0 = \alpha' - \frac{1}{\pi} \left[\frac{h}{1-E} \theta \right]_0^{\theta_h} - \frac{1}{\pi} \left[-\frac{h}{E} \theta \right]_{\theta_h}^\pi$$

After substituting in the limits and using the realtions

$$\frac{h}{1-E} + \frac{h}{E} = \eta$$

$$\alpha' + \frac{h}{E} = \eta$$

the value of A_0 becomes

$$A_0 = \frac{\eta(\pi - \theta_h)}{\pi} \tag{38}$$

In a similar manner, the values of A_n are found to be

$$A_n = \frac{2\eta \sin n\theta_h}{n\pi} \tag{39}$$

These equations, when substituted into Eqs. 27 and 31, yield *incremental* aerodynamic characteristics Δc_l and Δc_{mac} due to the flap deflections. Thus Eqs. 27 and 31, respectively, become

$$\Delta c_l = 2\pi A_0 + \pi A_1$$

$$\Delta c_{mac} = \tfrac{1}{4} \pi (A_2 - A_1)$$

Using the relations shown in Eqs. 38 and 39, these equations become

$$\Delta c_l = [2(\pi - \theta_h) + 2 \sin \theta_h] \eta \qquad (40)$$

$$\Delta c_{mac} = [\tfrac{1}{2} \sin \theta_h (\cos \theta_h - 1)] \eta \qquad (41)$$

The incremental lift coefficient and moment coefficient about the aerodynamic center vary linearly with the flap deflection. The magnitudes depend on the parameter θ_h, which is related to the distance x_h of the hinge line behind the leading edge by the expression

$$x_h = \tfrac{1}{2} c(1 - \cos \theta_h)$$

By reference to Fig. 15, $\Delta \alpha_{L0}$ is given by the formula

$$\Delta \alpha_{L0} = (\Delta c_l / 2\pi) = [(\pi - \theta_h) + \sin \theta_h] \eta / \pi \qquad (42)$$

These equations show that the incremental values of c_l, c_{mac}, and α_{L0} vary linearly with the flap deflection. The magnitudes are shown to be strong functions of θ_h, which is related to the hinge location by the formula following Eq. 41, so that, as $0 < \theta_h < \pi$, $0 < x_h/c < 1$. The flap-chord ratio follows from $E = 1 - x_h/c$.

A comparison between theory and experiment for $-\Delta \alpha_{L0}/\eta$ versus E is shown in Fig. 18. The cross-hatched band includes all of the experimental results collected from various sources by Abbott and Doenhoff (1949) for 21 airfoil sections with maximum thicknesses up to $0.21c$. The gaps at the hinge line were sealed and the flaps were contoured to the shape of the section for zero flap deflection. Near $E = 0$ the agreement between theory and experiment is poor but improves steadily

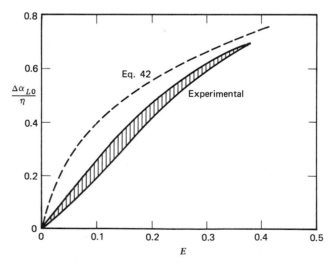

Fig. 18 Experiment versus theory for effect of flap-chord ratio, E, on $\Delta \alpha_{L0}/\eta$ based on flap deflection range from $0°$ to $20°$. (Courtesy of NASA.)

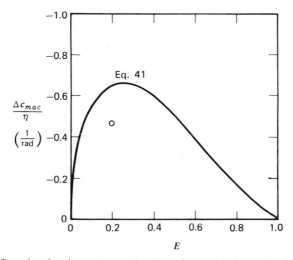

Fig. 19 Effect of flap-chord ratio on $\Delta c_{mac}/\eta$. Experimental point at $E = 0.2$.

until at $E = 0.4$ where the theoretical curve is only 7 percent too high. The poor agreement for low E is ascribed to the boundary layer, which attains a thickness of a few percent of the airfoil chord near the trailing edge; for small flap-chord ratios the entire flap chord is immersed in this relatively thick layer of retarded fluid, so that the response to flap deflection will be significantly less than that predicted by theory.

Equation 41, plotted in Fig. 19, shows that the effect of flap deflection is also to increase c_{mac} negatively. It follows from an examination of Eqs. 33 and 34 that positive flap deflection has qualitatively the same effect on α_{L0} and c_{mac} as an increase in the camber or a rearward movement of the position of maximum camber of the mean line; negative flap deflection of course has the opposite effect. The experimental point shown at $E = 0.2$, taken with $0° < \eta < 20°$, indicates that the theory overestimates $-\Delta c_{mac}$ by about 40 percent; for greater flap angles the flap stalls and $-\Delta c_{mac}$ tends to level off or even decrease. However, as we show later with a slot at the flap leading edge, stall is delayed and $-\Delta c_{mac}$ continues its increase to higher flap angles.

Perring (see Allen, 1938), by representing a given mean line as the superposition of many flaps, showed that the aerodynamic characteristics of a given mean line can be determined. Also, by deflecting the flaps near the trailing edge negatively, c_{mac} and α_{L0} can be made to vanish or become positive. The concept is illustrated in Fig. 20. Figure 13 shows measurements of c_{mac} for an airfoil with such a "reflexed mean line."

As is indicated in Fig. 15, the effect of the flap is to increase the maximum lift and decrease the angle of stall. Since the stall is a viscous effect detailed treatment will be delayed until Chapter 19 on "Boundary Layer Control."

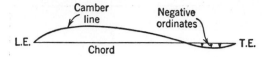

Fig. 20 Reflexed mean line representing a multiple flap system with $\eta > 0$ forward and $\eta < 0$ near the trailing edge.

5.9 Numerical Solution of the Thin-Airfoil Problem

The analytical method of Sections 5.5 and 5.6 requires the use of Eq. 11 as a starting point for determining the circulation density for a thin airfoil. In addition to the mathematical analyses, we describe here, without recourse to Eq. 11, an approximate numerical method the results of which show excellent agreement with the analytical solution; the method is easily adapted to machine calculation.

As a first step in the method, the vortex sheet situated on the mean camber line $z(x)$ is replaced by n discrete vortices with strengths γ_j, located at x_j, where $j = 1, 2, \ldots, n$, as indicated in Fig. 21. At points on the mean line but midway

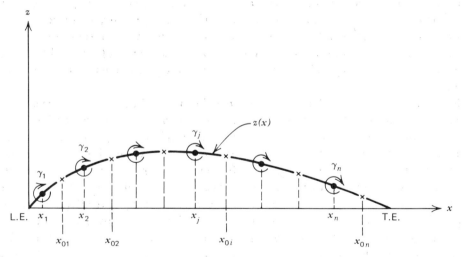

Fig. 21 Vortex configuration for numerical method of solution of planar-wing problem.

between the line vortices, n control points are chosen having abscissas x_{0i}, $i = 1$, $2, \ldots, n$. After evaluating Eq. 9 at these control points and replacing the integral by a summation, we obtain n simultaneous algebraic equations.

If the intervals between vortices near the trailing edge are small enough, application of Eq. 9 at the last control point will satisfy approximately the Kutta condition. Figure 22 gives a comparison of the increment of lift coefficient $\Delta c_l = \Delta p / q_\infty$ for a symmetrical airfoil approximated by 40 equally spaced vortices along the chord line.

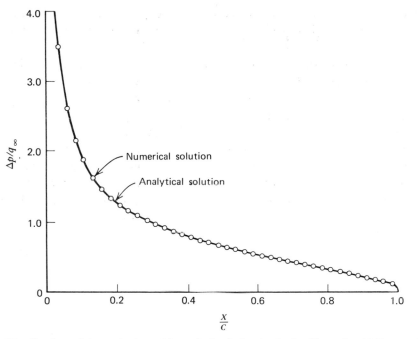

Fig. 22 Distribution of $\Delta c_l = \Delta p/q_\infty$. Numerical solution is obtained by using 40 line vortices. (Courtesy of P. E. Rubbert, Boeing Company.)

5.10 The Airfoil of Arbitrary Thickness and Camber

The analytical method of Sections 5.4 and 5.6 and the numerical method of Section 5.9 give remarkably accurate results for the thin slightly cambered airfoils of conventional aircraft. On the other hand, determination of the aerodynamic characteristics of thick, highly cambered, slotted surfaces, with single or multiple flaps and mutual interference effects among wings, fuselages, nacelles, and so forth require in general the use of numerical methods such as the *source panel* representation described in Section 4.13. Since the method as described there applies only to nonlifting bodies, to treat lifting bodies it is necessary to introduce circulation, the strength of which is fixed by the Kutta condition. The accuracy of the method in practical flow problems is limited only by the skill of the designer in representing adequately the surface by source and vortex panels, by the number of simultaneous linear algebraic equations the computer can handle expeditiously, and by the accuracy to which the effects of viscosity and, at high flow speeds, compressibility can be included in the computation.

The following method is only one variation of the use of the panel method; it involves representation of the airfoil by *both* source and vortex panels. The airfoil and wing problems can be solved by means of a vortex panel distribution alone, but calculation of fuselage and nacelle characteristics and their interference flows dictate the use of source and, possibly, doublet as well as vortex panels. Results of an analysis of a multiple flap configuration using vortex panels only are given in Section 19.4.

The numerical method is based on an approximation of the airfoil shape made up of a number of panels that serve as both *source* and *vortex* panels; their number, shape, and placement are dictated by experience and knowledge of the aerodynamic theory. We assume here that the panels, m in number, are planar, each with constant source strength and circulation density. The computational procedure is directed to finding the source densities, λ_i, that establish the circulation-free (zero lift) flow, and the circulation densities, γ_i, that establish the Kutta condition for the flow with lift, for each of the panels for a given angle of attack of the airfoil. Knowing λ_i and γ_i for each of the m panels and V_∞ we can calculate the velocity and pressure distributions and the lift and moment coefficients.

Some results of the method are shown in Fig. 23, reproduced by courtesy of A.M.O. Smith and the Douglas Aircraft Company. The results of three calculations,

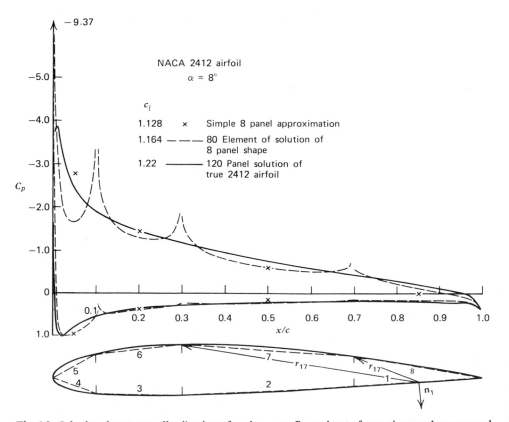

Fig. 23 Calculated pressure distributions for three configurations of superimposed source and vortex panels. (Courtesy of A.M.O. Smith, Douglas Aircraft Company.)

varying from rough to almost exact, of the pressure distribution over the NACA 2412 airfoil at $8°$ angle of attack are shown.

The first calculation was made on a polygon ($m = 8$) roughly approximating the airfoil shape, as shown in the figure. The "boundary points" are the intersections of contiguous source panels; the condition that the polygon be a streamline is met approximately by applying the condition of zero normal velocity component at "control points," specified as the midpoints of the panels. Then for the ith panel the boundary condition is, according to Eq. 4.33,

$$\frac{\lambda_i}{2} + \sum_{j \neq i}^{m} \frac{\lambda_j}{2\pi} \int_j \frac{\partial}{\partial n_i} (\ln r_{ij}) \, ds_j = -V_\infty \cos \beta_i \tag{43}$$

where the integral is taken over the jth panel, the summation extends over all but the ith panel (the contribution from that panel is represented by the term $\lambda_i/2$) and r_{ij} varies between the distances from the control point of the ith panel to the two boundary points of the jth panel; the limiting values of r_{17} are indicated in Fig. 23. Equation 43 represents m simultaneous linear algebraic equations for the m unknowns λ_i, for $i = 1, 2, \ldots, m$.

The general method of solution for the velocity distribution for the no-lift condition at any given angle of attack α comprises the following steps: (1) Eq. 43 is formulated for $\alpha = 0°$ and for $\alpha = 90°$ under the conditions that $|\mathbf{V}_\infty| = 1$, and zero circulation (see Fig. 24a, b); the two sets of equations are solved to yield two sets of source panel densities $\overline{\lambda}_{ai}$ and $\overline{\lambda}_{bi}$, where the bar indicates that the quantities are dimensionless since they refer to unit V_∞; (2) the tangential velocities at the control point of each panel, under the condition of zero normal velocity at the surface with unit V_∞ at, respectively, $\alpha = 0$ and $90°$ are

$$\overline{V}_{ai} = 1 \cdot \sin \beta_{ai} + \sum_{j \neq i}^{m} \frac{\overline{\lambda}_{aj}}{2\pi} \int_j \frac{\partial}{\partial s_i} (\ln r_{ij}) \, ds_j$$

$$\overline{V}_{bi} = 1 \cdot \sin \beta_{bi} + \sum_{j \neq i}^{m} \frac{\overline{\lambda}_{bj}}{2\pi} \int_j \frac{\partial}{\partial s_i} (\ln r_{ij}) \, ds_j$$

where $\beta_{bi} = \beta_{ai} + 90°$, and β_{ai} is the angle between V_∞ and the normal to the ith panel for $\alpha = 0°$. (3) At an angle of attack α the free-stream velocity $\mathbf{V}_\infty$ is decomposed into a horizontal component $V_\infty \cos \alpha$ and a vertical component $V_\infty \sin \alpha$, as shown in Fig. 24c. Then by linear combination, the tangential velocity at each panel for zero lift at this angle of attack is

$$V_{oi} = V_\infty (\overline{V}_{ai} \cos \alpha + \overline{V}_{bi} \sin \alpha) \tag{44}$$

A crucial point in applying the method to a *lifting* body is the superposition of vortex panels of such strengths as to satisfy the Kutta condition at the trailing edge

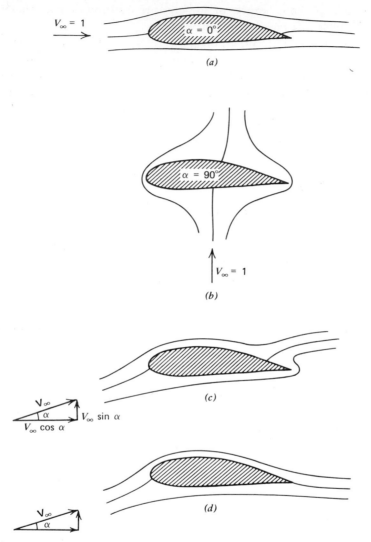

Fig. 24 Four stages in determination of flow about an airfoil: (*a*) airfoil at $\alpha = 0°$, $\Gamma = 0$; (*b*) $\alpha = 90°$, $\Gamma = 0$; (*c*) superposition of (*a*) and (*b*) to determine flow at arbitrary α, $\Gamma = 0$; (*d*) flow (*c*) with Γ determined by Kutta condition.

while preserving the airfoil shape as a streamline; in the present application the Kutta condition is satisfied approximately by requiring that the flow at the control points on the rearmost panels (1 and 8 in Fig. 23) be not only parallel to the surface but that the velocities there be of equal magnitude (Hess, 1971). It is therefore important for the meeting of the Kutta condition that the rearmost panels be very short so that the flow at their midpoints will effectively represent that at the trailing edge.

Next the effects of the vortex panels coincident with the source panels are calculated. The circulation densities γ_i of the panels are adjusted to preserve the panels as streamlines and to satisfy the Kutta condition, as illustrated for the exact shape in Fig. 24d. The velocity potential at the ith control point (x_i, y_i) associated with the vorticity distribution on the jth panel is, from Eqs. 2.44,

$$\int_j \tan^{-1}\left(\frac{y_i - y_j}{x_i - x_j}\right) \cdot \frac{\gamma_j \, ds_j}{2\pi}$$

and the condition of zero outward normal velocity at the ith panel is achieved by summing the normal derivatives of this expression. Thus

$$\sum_{j=1}^{m} \frac{\gamma_j}{2\pi} \int_j \frac{\partial}{\partial n_i} \left(\tan^{-1} \frac{y_i - y_j}{x_i - x_j}\right) ds_j = 0 \tag{45}$$

where the integral extends over the jth panel and the summation extends over all panels, noting that γ_i contributes no normal velocity component at the ith panel. Equation 45 is homogeneous, and therefore if it were applied at every one of the m control points, a trivial solution of vanishing circulation densities would result. Thus, an additional relation is required to connect the γ_i's with the angle of attack. This relation derives from the approximate Kutta condition that the velocities be of equal magnitude at control points 1 and m, that is, on the two rearmost panels. The additional relation is therefore

$$V_{01} + \frac{1}{2}\gamma_1 + \sum_{j=2}^{m} \frac{\gamma_j}{2\pi} \int_j \frac{\partial}{\partial s_1} \left(\tan^{-1} \frac{y_1 - y_j}{x_1 - x_j}\right) ds_j$$

$$= V_{0m} + \frac{1}{2}\gamma_m + \sum_{j=1}^{m-1} \frac{\gamma_j}{2\pi} \int_j \frac{\partial}{\partial s_m} \left(\tan^{-1} \frac{y_m - y_j}{x_m - x_j}\right) ds_j \tag{46}$$

where $m = 8$, 80, and 120, respectively, for the three calculations represented in Fig. 23. The factors $1/2$ arise from the definition of the circulation density which states that γ is equal to the total jump in velocity across the sheet, that is, twice the velocity on one side of the sheet. The values of V_{01} and V_{0m} are computed from Eq. 44.

Equation 46 and the $(m - 1)$ equations, obtained by applying Eq. 45 at all control points except one around the airfoil, form a complete set for the determination of the γ_i's at a particular angle of attack α. Then the tangential velocity distribution at the surface at this angle of attack is

$$V_i = V_{0i} + \frac{1}{2}\gamma_i + \sum_{j \neq i}^{m} \frac{\gamma_j}{2\pi} \frac{\partial}{\partial s_i} \left(\tan^{-1} \frac{y_i - y_j}{x_i - x_j}\right) ds_j \tag{47}$$

and, by Bernoulli's equation,

$$p_i - p_\infty = \frac{1}{2}\rho(V_\infty^2 - V_i^2) = q_\infty\left(1 - \frac{V_i^2}{V_\infty^2}\right)$$

$$c_{pi} = \frac{p_i - p_\infty}{q_\infty} = 1 - \left(\frac{V_i}{V_\infty}\right)^2 \tag{48}$$

The application of the method leads to the results shown in Fig. 23. The "eight panel, eight control point" solution yields eight values for c_p; the "eight panel, eighty control point" solution yields the broken curve with peaks at the points where adjacent panels intersect (boundary points); the "120 panel 120 control point" solution yields, for all practical purposes, the exact solution. Surprisingly, the values of c_l, found by integrating the three solutions, are not far different; that for the roughest approximation deviates by only 7 percent from the exact value.

The panel method outlined here is considerably more cumbersome than the exact method for a single airfoil. However, the great power of the method emerges for flow calculations on multiple surfaces, such as airfoils with flaps and slots or cascades representing axial compressors or turbines and many other problems, for which exact methods are in general not available (see Hess, 1971). Corrections for compressibility and for viscous effects at high speeds or high angles of attack are taken up in the following chapters.

5.11 Summary

The aerodynamic characteristics of airfoils of moderate camber and thickness have been derived on the hypothesis, verified by experimental results, that the shape of the "mean line" and the Kutta condition determine, to a good approximation, the aerodynamic characteristics. We show that for thin airfoils, $dc_l/d\alpha = 2\pi$ and $x_{ac} = \frac{1}{4}c$. However, x_{ac} and c_{mac} are shown to be strongly influenced by the maximum mean camber and its location along the mean line; for an airfoil with flaps these variables were interpreted, respectively, in terms of magnitude of flap deflection and flap-chord ratio (that is, location of the hinge line). Experimental results are shown to indicate that the airfoil thickness may be as high as $0.2c$ and the maximum mean camber as high as $.04c$ without affecting significantly the accuracy of the characteristics predicted.

These limitations do not exist however for the panel method, taken up in the foregoing section. The method is effective for airfoils of any thickness or camber. Compressibility and viscous effects can be introduced, as is pointed out in following chapters.

It must be kept in mind that the correctness of the panel method for determining the flow properties around arbitrary bodies *requires* that the flows induced by the several panels be *superposable*, that is, that their kinematic properties must satisfy *linear* differential equations.   This condition is satisfied for incompressible irrotational flows that satisfy continuity (Section 4.2).

$$\nabla^2 \phi = \frac{\partial^2 \phi}{\partial x^2} + \frac{\partial^2 \phi}{\partial y^2} + \frac{\partial^2 \phi}{\partial z^2} = 0$$

Two-dimensional flows of constant vorticity also satisfy this superposition criterion since they satisfy the linear equation

$$\nabla^2 \psi = \text{constant}$$

CHAPTER
6

The Finite Wing

6.1 Introduction

It was shown in Section 4.8, from momentum considerations, that a vortex that is stationary with respect to a uniform flow experiences a force of magnitude $\rho V_\infty \Gamma$ in a direction perpendicular to V_∞. The Kutta-Joukowski theorem, which was derived in Chapter 4, states that the force experienced by unit span of a right cylinder of any cross section whatever is $\rho V_\infty \Gamma$ and it is directed perpendicular to V_∞. It follows that a stationary line vortex normal to a moving stream is the equivalent of a two-dimensional wing as far as the resultant force is concerned.

The airfoil-vortex analogy also forms the basis for calculating the properties of the *finite* wing. However, since the lift, and, therefore, the circulation, are zero at the wing tips and vary throughout the span, flow components appear that were not present in the airfoil theory of Chapter 5; in this chapter we treat these components and show their effect on the airfoil section properties along the span.

We show further how the section properties and the Helmholtz vortex theorems of Section 2.13 enable us to describe and calculate the flow field and aerodynamic characteristics of finite wings.

6.2 Flow Fields Around Finite Wings

Consider a wing of span b in a uniform flow of velocity V_∞ represented by a *bound vortex* (AB of Fig. 1) of constant circulation Γ; by the Kutta-Joukowski law a force $\rho V_\infty \Gamma$ normal to V_∞ is exerted on the vortex. The Helmholtz laws require however that the bound vortex cannot end at the wing tips; it must form a complete circuit, or, it must extend to infinity or to a boundary of the flow. As was described in Section 4.10, these laws require further that at the beginning of the motion a *starting vortex* of strength equal and opposite to that of the bound vortex, indicated by CD in Fig. 1, be formed. The vortex laws are satisfied by including the *trailing vortices* BC and AC of strength Γ.

The resulting velocity field of Fig. 1 is comprised of the uniform flow V_∞ with a superimposed downward flow within the rectangle $ABCD$ and an upward flow outside. The flow however is unsteady, since the starting vortex moves downstream with the flow and the trailing vortices AC and BD are therefore increasing in length at the rate V_∞.

The objective of a study of the finite wing is to modify the airfoil characteristics

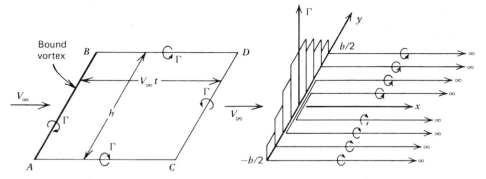

Fig. 1 Vortex configuration soon after start of flow past bound vortex AB.

Fig. 2 Superposition of horseshoe vortices in steady flow.

derived in Chapter 5 to take account of the velocities induced at the wing by the starting and trailing vortices.

We note first (Section 2.12) that the velocity induced by a given vortex varies with the reciprocal of the distance from the vortex. Therefore, as time goes on, the starting vortex recedes from the wing position and relatively soon after the start the velocities it induces at the wing are negligible compared with those induced by the portions of the trailing vortices near the wing. Quantitatively, in practice $b \ll V_\infty t$ for steady flight and the configuration becomes essentially a *horseshoe vortex* fixed to the wing and extending to infinity.

Actual finite wings are made up of a superposition of horseshoe vortex elements of various strengths, as shown schematically in Fig. 2. An infinite number of these lead to a continuous distribution of circulation and therefore of the lift as a function of y extending over $-b/2 < y < b/2$. In steady flight the vortices will in general be symmetrically placed, as in Fig. 2; the deflection of ailerons, for instance, would be represented by the addition of a horseshoe of one sign near $+b/2$ and one of opposite sign near $-b/2$.

From a physical standpoint we can visualize the formation of the trailing vortices with the help of the head-on view of the wing in Fig. 3. The flow field that develops as a consequence of the circulation around the wing is initiated by an underpressure (−) over the upper surface and an overpressure (+) over the lower surface. The indicated flow from high to low pressure at the wing tips signifies the formation of the trailing vortices.

We can describe the formation of the trailing vortices in the terminology of the vortex laws of Section 2.13 and the Kutta condition of Section 4.10. The circula-

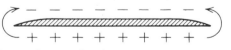

Fig. 3 Formation of trailing vortices at wing tips.

tion about the wing is generated as a consequence of the action of viscosity in establishing the Kutta condition at the trailing edge. The boundary layer which forms adjacent to the surface is a rotational flow resulting from the viscous shearing action; the rotating fluid elements spill over the wing tips as indicated in Fig. 3 at the rate required to form trailing vortices with circulation equal to that around the wing. After leaving the wing tips the trailing vortices follow the streamlines of the flow and, in conformity with the vortex laws, the circulation around them remains constant.

The superposition of several horseshoe vortices then leads to the trailing vorticity field of Fig. 2; we see that at each point on the wing where the lift changes, a trailing vortex is generated, equal in strength to the change in circulation about the wing at that point. In the limit when the circulation distribution becomes continuous, as in Fig. 4, the change in circulation at any point is an infinitesimal and the strength $d\Gamma$ of the trailing vortex starting at a given point on the wing is given by

$$d\Gamma = \frac{d\Gamma}{dy}\, dy \tag{1}$$

The trailing vortex system then becomes a vortex sheet of zero total strength since it is comprised of the superimposed flow fields of elementary horseshoe vortices the trailing branches of which are vortex pairs of equal and opposite strengths.

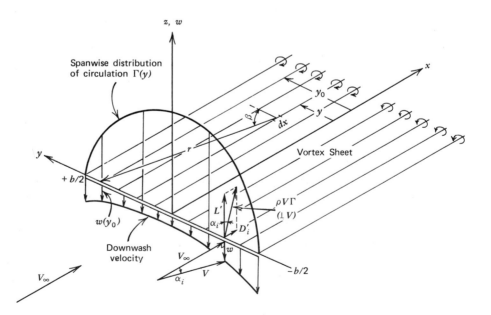

Fig. 4 Downwash velocity w induced by the trailing vortices. Resultant velocity V and forces L' and D' per unit span.

6.3 Downwash and Induced Drag

The main problem of finite wing theory is the determination of the distribution of airloads on a wing of given geometry flying at a given speed and orientation in space. The analysis is based on the assumption that at every point along the span the flow is essentially two-dimensional; thus the resultant force per unit span at any point is that calculated for the airfoil by the methods of Chapter 5, but at an angle of attack corrected for the influence of the vortex configurations of the previous section. The effect is illustrated in Fig. 4, in which the bound vortex with circulation varying along the span represents a wing for which the center of pressure at each spanwise point lies on the y axis. The lift distribution is continuous and the trailing vortices therefore form a vortex sheet of total circulation zero, since the flow field is that of an infinite number of infinitesimally weak horseshoe vortices, the cross section of each being a vortex pair of zero total circulation. The vortices are assumed to lie in the $z = 0$ plane and to be parallel to the x axis; the effect on the flow at a given point on the bound vortex is therefore a *downwash*, w (positive in $+z$ direction), whose magnitude at each point is given by the integrated effect of the continuous distribution of semi-infinite vorticity over the range $-b/2 < y < b/2$.

Before proceeding with the detailed calculation of the downwash we may easily see its qualitative effect on the forces, as shown in the lower portion of Fig. 4. The resultant velocity at the wing has two components V_∞ and $w(y)$ at each point; these define an *induced angle of attack*,

$$\alpha_i(y) = \tan^{-1} \frac{w}{V_\infty}$$

(We note from Fig. 4 that $w < 0$ for $L' > 0$.) By the Kutta-Joukowski law the force on the bound vortex per unit span has the magnitude $\rho V \Gamma$ and is normal to V, that is, it is inclined to the z axis at the angle α_i. This force has a lift component normal to V_∞ given by

$$L' = \rho V \Gamma \cos \alpha_i$$

and a drag component, termed the *induced drag*,

$$D_i' = -\rho V \Gamma \sin \alpha_i$$

The primes in these formulas designate the force per unit span. In most practical applications the downwash is small, that is $|w| \ll V_\infty$, and it follows that

$$\tan \alpha_i \simeq \sin \alpha_i \simeq \alpha_i \tag{2}$$

and the above formulas become

$$\alpha_i(y) = \frac{w}{V_\infty} \tag{3}$$

$$L'(y) = \rho V_\infty \Gamma \tag{4}$$

$$D_i'(y) = -L'\alpha_i = -\rho w \Gamma \tag{5}$$

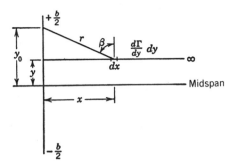

Fig. 5 Diagram for calculation of downwash contribution from a trailing vortex filament.

We note particularly that the induced drag D_i' is a component of the Kutta-Joukowski force in the direction of V_∞, that is, in the plane of flight; in later chapters we calculate drag contributions due to viscosity and compressibility.

We now calculate the downwash and induced angle of attack, according to the diagram shown in the upper part of Fig. 4, the essential features of which are shown in the top view of the $z = 0$ plane shown in Fig. 5; the downwash w is positive outward. By means of the Biot-Savart law (Section 2.12), we express the increment of downwash at the point $(0, y_0)$ induced by the element dx of the vortex filament of strength $d\Gamma$ extending from $(0, y)$ to infinity in the $+x$ direction; this increment is designated $dw_{y_0 y}$. The Biot-Savart law expressed in these symbols is

$$dw_{y_0 y} = -\frac{d\Gamma}{4\pi} \frac{\cos \beta \, dx}{r^2} \tag{6}$$

We arrive at the sign in this equation by observing from Eq. 1 that $d\Gamma$ takes the sign of $(d\Gamma/dy)$ and since, from Fig. 5, all other terms in the Eq. 6 are positive, $dw_{y_0 y} > 0$ for $d\Gamma < 0$. The entire vortex filament at y contributes the downwash

$$w_{y_0 y} = -\frac{d\Gamma}{4\pi} \int_0^\infty \frac{\cos \beta \, dx}{r^2} = -\frac{d\Gamma}{4\pi} \frac{1}{y_0 - y} \tag{7}$$

This integration was carried out in detail in Section 2.12. Here however $w_{y_0 y}$ is an infinitesimal contribution from the filament at y, which by Eq. 1 is of strength $(d\Gamma/dy)dy$. The *total* downwash at y_0 is w_{y_0} so we write

$$w_{y_0 y} = dw_{y_0}$$

After integrating Eq. 7 and dividing by V_∞ we obtain

$$\alpha_i(y_0) = \frac{w_{y_0}}{V_\infty} = -\frac{1}{4\pi V_\infty} \int_{-b/2}^{b/2} \frac{d\Gamma/dy}{y_0 - y} \, dy \tag{8}$$

This equation gives the amount by which the downwash alters the angle of attack of the wing as a function of the coordinate y_0 along the span. It will be discussed in greater detail in the next section.

6.4 The Fundamental Equations of Finite Wing Theory

The fundamental equation that must be solved to find the circulation distribution for a finite wing is expressed as the equation connecting the three angles: α_a, the *absolute angle of attack* (see Fig. 6), that is, the angle between the direction of

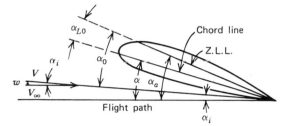

Fig. 6 Fundamental diagram of finite-wing theory.

the flow for zero lift (Z.L.L.) at a given y_0 and the flight velocity vector $\mathbf{V}_\infty$; the induced angle of attack, α_i; and the effective angle of attack, α_0. This relation is

$$\alpha_a = \alpha_0 - \alpha_i = \alpha - \alpha_{L0} \tag{9}$$

The latter relation connects the "aerodynamic angles of attack" α_a, α_0, α_i with the "geometric," α and α_{L0}, that is, those measured relative to the chord line at a given spanwise station. The effective angle of attack α_0 is a section property and thus must satisfy the formula derived in Chapter 5:

$$c_l = m_0 \alpha_0 \tag{10}$$

where $m_0 = 2\pi$ according to thin wing theory; actually, as was pointed out in Chapter 5, m_0 varies slightly according to the airfoil shape. We define another lift curve slope by writing

$$c_l = m \alpha_a \tag{11}$$

where by Eq. 9, m is a function of α_i. The relation between m_0 and m follows from substitutions of Eqs. 10 and 11 in Eq. 9. Thus

$$\frac{c_l}{m} = \frac{c_l}{m_0} - \alpha_i$$

$$\tag{12}$$

$$m = \frac{m_0}{1 - m_0 \alpha_i / c_l} = \frac{m_0}{1 - \alpha_i / \alpha_0}$$

We note from these equations that $m \leqslant m_0$ for $c_l > 0$.

Equation 9 may be put in a form for solution of a given problem by first writing

$$L' = \rho V_\infty \Gamma = m_0 \alpha_0 q_\infty c = m_0 \alpha_0 \tfrac{1}{2} \rho V_\infty^2 c$$

from which,

$$\alpha_0 = \frac{2\Gamma}{m_0 V_\infty c} \tag{13}$$

Then, the fundamental equation in its final form is obtained by substituting Eqs. 8 and 13 into the first of Eqs. 9:

$$\alpha_a(y_0) = \left(\frac{2\Gamma}{m_0 V_\infty c}\right)_{y_0} + \frac{1}{4\pi V_\infty} \int_{-b/2}^{b/2} \frac{d\Gamma/dy}{y_0 - y}\, dy \tag{14}$$

The only unknown in this integro-differential equation is the circulation Γ, and its solution for all spanwise stations y_0 solves the airload distribution problem for a given wing. Unfortunately, its solution can be obtained in a straightforward manner for only a few special cases; the most important of these, the elliptical lift distribution, will be taken up in the next section.

6.5   The Elliptical Lift Distribution

Equation 14 is readily solved if the Γ distribution is assumed to be known and the chord distribution $c(y)$ is taken as the unknown. This problem of finding a chord distribution that corresponds to a given circulation distribution simply involves the solution of an algebraic equation. A very important special case is the elliptical circulation distribution, for, as will be shown, this distribution represents the wing of minimum induced drag.   Fortunately, the properties of wings of arbitrary plan forms that do not differ radically from the usual shapes are close to those of the elliptical wing. It is therefore customary to write the properties of wings of arbitrary plan form in terms of the properties of the elliptical wing and a correction factor. In this section, the properties of a wing with an elliptical circulation distribution are analyzed.

If Γ_s represents the circulation in the plane of symmetry, the elliptical variation of circulation with span is written

$$\Gamma = \Gamma_s \sqrt{1 - \left(\frac{y}{b/2}\right)^2} \tag{15}$$

The induced angle of attack for an elliptical Γ distribution is found by substituting Eq. 15 in Eq. 8

$$\alpha_i(y_0) = -\frac{\Gamma_s}{4\pi V_\infty} \int_{-b/2}^{+b/2} \frac{\dfrac{d}{dy}\sqrt{1 - \left(\dfrac{y}{b/2}\right)^2}}{y_0 - y}\, dy$$

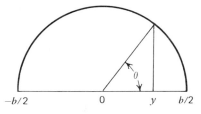

Fig. 7 Plot of $y = \frac{1}{2} b \cos \theta$.

The integral is evaluated easily if the trigonometric substitution (Fig. 7) $y = (b/2) \cdot \cos \theta$ is made. The equation becomes

$$\alpha_i(\theta_0) = \frac{-\Gamma_s}{2\pi b V_\infty} \int_\pi^0 \frac{\frac{d}{d\theta} \sin \theta}{\cos \theta_0 - \cos \theta} \, d\theta$$

$$= \frac{\Gamma_s}{2\pi b V_\infty} \int_0^\pi \frac{\cos \theta \, d\theta}{\cos \theta_0 - \cos \theta}$$

The integral will be recognized as that which occurred in Eq. 5.12. The value of the integral was given as

$$\int_0^\pi \frac{\cos n\theta \, d\theta}{\cos \theta - \cos \theta_0} = \frac{\pi \sin n\theta_0}{\sin \theta_0} \tag{16}$$

The case at hand corresponds to $n = 1$. The value of the induced angle of attack then becomes

$$\alpha_i = -\Gamma_s/2b V_\infty \tag{17}$$

Equation 17 indicates that α_i at any point along the lifting line is constant if the Γ distribution is elliptical. Therefore, if the absolute angle of attack α_a of every spanwise station is the same, Eq. 9 indicates a constant effective angle of attack. Thus from Eqs. 9, 10, and 5, respectively,

$$\alpha_0 = \alpha_a + \alpha_i; \qquad c_l = m_0 \alpha_0; \qquad c_{d_i} = \frac{D_i'}{q_\infty c} = -c_l \alpha_i \tag{18}$$

where c_{d_i} is the sectional-induced drag coefficient. Thus, if the sectional-lift curve slopes are independent of y, the sectional-lift coefficients and induced drag coefficients will be independent of y. Summarizing, for wings with an elliptical Γ distribution and constant lift curve slope and absolute angle of attack, the nondimensional section properties will not vary along the span.

Under these conditions, the product $m_0 \alpha_0 c$ must vary elliptically, for

$$L' = \rho V_\infty \Gamma_s \sqrt{1 - \left(\frac{y}{b/2}\right)^2} = m_0 \alpha_0 \frac{1}{2} \rho V_\infty^2 c$$

$$m_0 \alpha_0 c = \frac{2\Gamma_s}{V_\infty} \sqrt{1 - \left(\frac{y}{b/2}\right)^2}$$

It should be observed that this equation indicates an elliptical planform *only* if the product $m_0 \alpha_0$ is independent of y. On the other hand, for a nonelliptical planform, since m_0 is nearly constant, α_0 must be a specific function of y; that is, the wing must be twisted, if the equation is to be satisfied; this condition could occur only at a specific attitude of the wing. Thus, only an untwisted elliptical planform will yield an elliptical lift distribution at all angles of attack up to the stall.

The wing properties are found by integrating the section properties across the span. The wing-lift coefficient C_L is defined as the total wing lift L divided by the product of the dynamic pressure q_∞ and the wing area S. Then, since the lift per unit span for a wing with an elliptical Γ distribution is

$$L' = \rho V_\infty \Gamma_s \sqrt{1 - \left(\frac{y}{b/2}\right)^2} \qquad (19)$$

we can write the wing-lift coefficient and integrate to find

$$C_L = \frac{1}{\frac{1}{2}\rho V_\infty^2 S} \int_{-b/2}^{+b/2} L' \, dy = \frac{\Gamma_s \pi b}{2 V_\infty S} \qquad (20)$$

The wing-lift coefficient and sectional-lift coefficient are equal when the sectional-lift coefficients are constant along the span, that is, with constant c_l and $S = \int_{-b/2}^{b/2} c \, dy$, we have

$$C_L = \frac{1}{q_\infty S} \int_{-b/2}^{b/2} c_l q_\infty c \, dy = c_l \qquad (21)$$

If Eq. 20 is solved for Γ_s and this value is used in Eq. 17, the expression for the induced angle of attack becomes, for an elliptical Γ distribution under the condition described by Eq. 21,

$$\alpha_i = \frac{w}{V_\infty} = -\frac{C_L}{\pi \mathcal{R}} = -\frac{c_l}{\pi \mathcal{R}} \qquad (22)$$

where $\mathcal{R}$ is the aspect ratio of the wing and is defined

$$\mathcal{R} = \frac{b^2}{S}$$

The wing-induced drag coefficient (Eqs. 18) is given by

$$C_{D_i} = c_{d_i} = -C_L \alpha_i = C_L^2 / \pi \mathcal{R} \tag{23}$$

The expression for the lift curve slope for a section of a finite wing of constant sectional-lift coefficient may now be completed.   The value of α_i/c_l is $-1/\pi \mathcal{R}$.   Therefore, Eq. 12 becomes

$$m = \frac{m_0}{1 + m_0/\pi \mathcal{R}} \tag{24}$$

Thus for the elliptical Γ distribution, the c_l versus α_a curve has the same slope at every spanwise section; its magnitude is equal to that for the C_L versus α curve.

Example

A wing with an elliptical planform and an elliptical lift distribution has an aspect ratio of 6 and a span of 12 m. The wing loading is 900 N/m² when flying at a speed of 150 km/hr at sea level. We shall compute the induced drag for this wing.

The projected area and the total lift of the wing are, respectively.

$$S = \frac{b^2}{\mathcal{R}} = 24 \text{ m}^2 \qquad L = 21{,}600 \text{ N}$$

Since both the planform and the lift distribution are elliptical, the sectional-lift coefficients are constant along the span (Eq. 21) and are equal to the wing-lift coefficient. Thus

$$c_l = C_L = L/q_\infty S = 0.845$$

The induced drag is, by using Eq. 23,

$$D_i = \tfrac{1}{2}\rho V_\infty^2 C_{D_i} S = \frac{L C_{D_i}}{C_L} = \frac{L C_L}{\pi \mathcal{R}} = 969 \text{ N}$$

which is 4.49 percent of the total lift.

The computation shows that for a given lift coefficient and wing area the induced drag varies inversely with the aspect ratio.

6.6   The Arbitrary Circulation Distribution

It is convenient to represent an arbitrary circulation distribution in terms of an infinite series, the first term of which describes the elliptical distribution. We first observe that Eq. 15 may be written

$$\Gamma = \Gamma_s \sin \theta \tag{25}$$

Glauert (1937) considered a circulation distribution expressed, instead, by a Fourier series of which the above was the first term.

A dimensionally correct Fourier representation of an arbitrary symmetrical circulation distribution that includes all of the significant variables is

$$\Gamma = \tfrac{1}{2} m_{0s} c_s V_\infty \sum_{n=1}^\infty A_n \sin n\theta \tag{26}$$

where the subscript s refers as before to values at the plane of symmetry and the A_n's are dimensionless coefficients that vary with spanwise station and angle of attack. With the substitution $y = (b/2) \cos \theta$ (see Fig. 7) and Eq. 26, Eq. 14 becomes

$$\alpha_a(\theta_0) = \frac{(m_0 c)_s}{(m_0 c)_{\theta_0}} \sum_{n=1}^\infty A_n \sin n\theta_0 + \frac{m_{0s} c_s}{4\pi b} \int_0^\pi \frac{\dfrac{d}{d\theta}\left(\displaystyle\sum_{n=1}^\infty A_n \sin n\theta \right) d\theta}{\cos \theta - \cos \theta_0} \tag{27}$$

In this equation θ_0 refers to a specific spanwise station, while in the integration θ varies from 0 to π. After performing the differentiation and integration, with the aid of Eq. 16, we may drop the subscript on θ and Eq. 27 reduces to

$$\alpha_a(\theta) = \underbrace{\frac{m_{0s} c_s}{m_0 c} \sum_{n=1}^\infty A_n \sin n\theta}_{\alpha_0} + \underbrace{\frac{m_{0s} c_s}{4b} \sum_{n=1}^\infty n A_n \frac{\sin n\theta}{\sin \theta}}_{-\alpha_i} \tag{28}$$

As is indicated in the derivation of Eq. 14, at a given station θ, the first group on the right of Eq. 28 is the effective section angle of attack, α_0, and the second is the induced section angle of attack α_i. In Section 6.8 we outline the solution of Eq. 28 for practical cases.

The sectional lift and induced drag coefficients are readily found by substituting from Eqs. 26 and 28, that is

$$c_l = \frac{\rho V_\infty \Gamma}{q_\infty c} = \frac{m_{0s} c_s}{c} \sum_{n=1}^\infty A_n \sin n\theta \tag{29}$$

$$c_{d_i} = -c_l \alpha_i = \frac{m_{0s}^2 c_s^2}{4bc} \left(\sum_{n=1}^\infty A_n \sin n\theta \right)\left(\sum_{k=1}^\infty k A_k \frac{\sin k\theta}{\sin \theta} \right) \tag{30}$$

where in the last series in Eq. 30 the subscript is changed to k to avoid confusion when multiplying the two series.

The coefficient of total lift of the wing is given by the weighted integral of Eq.

29, that is,

$$C_L = \int_{-b/2}^{b/2} \frac{c_l q_\infty c \, dy}{q_\infty S} = \frac{m_{0s} c_s}{S} \int_0^\pi \sum_{n=1}^\infty A_n \sin n\theta \cdot \frac{b}{2} \sin \theta \, d\theta \qquad (31)$$

The integration is carried out easily by interchanging the orders of the integration and the summation and by use of the definite integrals (Peirce Tables No. 485)

$$\int_0^\pi \sin n\theta \sin k\theta \, d\theta = \begin{cases} 0 \text{ for } n \neq k \\ \dfrac{\pi}{2} \text{ for } n = k \end{cases} \qquad (32)$$

Thus, since $k = 1$ in Eq. 31, all of the integrals except that for $n = 1$ vanish, and the wing-lift coefficient becomes

$$C_L = \frac{m_{0s} c_s \pi b}{4S} A_1 \qquad (33)$$

We note therefore that the wing-lift coefficient for an *arbitrary symmetrical* circulation distribution is proportional to A_1 and thus is independent of all other Fourier coefficients. We note from Eq. 20 that for an *elliptical* lift distribution C_L is proportional to Γ_s which, as is shown by Eq. 25, is also a first Fourier coefficient. In Section 6.9 an empirical rule is described, showing that for nonelliptical planforms the spanwise distribution of c_l can also be related to the elliptical distribution.

The coefficient of total induced drag for the wing is calculated by using Eq. 30 and expressing dy in terms of $d\theta$. Thus

$$C_{D_i} = \int_{-b/2}^{b/2} \frac{c_{d_i} q_\infty c}{q_\infty S} \, dy = \frac{m_{0s}^2 c_s^2}{8S} \int_0^\pi \sum_{n=1}^\infty \sum_{k=1}^\infty k A_n A_k \sin n\theta \sin k\theta \, d\theta \qquad (34)$$

As we did with Eq. 31, we interchange the orders of the integration and the summations, and, since the factor $kA_n A_k$ is independent of θ, the multiplication of the terms in the two series leaves us with an infinite number of integrals of the exact form of Eq. 32. Since only the squared terms $(k = n)$ survive the integration, Eq. 34 simplifies to

$$C_{D_i} = \frac{m_{0s}^2 c_s^2 \pi}{16S} \sum_{n=1}^\infty n A_n^2 \qquad (35)$$

We can now prove that, for a given lift coefficient and aspect ratio, the induced drag coefficient is a minimum for the elliptical lift distribution. For the elliptical lift distribution, designated by the subscript *el*, Eq. 23 is

$$(C_{D_i})_{el} = \frac{C_L^2}{\pi R}$$

If this equation, with the expression for C_L from Eq. 33, is substituted into Eq. 35 we get for an arbitrary symmetrical lift distribution,

$$C_{D_i} = (C_{D_i})_{el} \sum_{n=1}^{\infty} \frac{nA_n^2}{A_1^2} = (C_{D_i})_{el}(1 + \sigma) \tag{36}$$

where

$$\sigma = \sum_{n=2}^{\infty} (nA_n^2/A_1^2) \tag{37}$$

Since the correction factor, σ, comprising only squared terms, *is always positive* it follows that the induced drag coefficient is a minimum for the elliptical lift distribution at a given lift coefficient and aspect ratio.

Similarly, we can use the general relation Eq. 12 to show that the slope of the section lift curve slope m for an arbitrary lift distribution can be related to the elliptical distribution by a correction factor τ, that is,

$$m = \frac{m_0}{1 - \dfrac{m_0 \alpha_i}{c_l}} = \frac{m_0}{1 + \dfrac{m_0}{\pi R}(1 + \tau)} \tag{38}$$

The expression for τ in terms of the Fourier coefficients is found easily by the use of Eq. 29 and α_i designated in Eq. 28.

Glauert (1937) tabulated values for the correction factors σ and τ. He finds that the corrections are small even for appreciable departures from the elliptical distribution, as is demonstrated below for a rectangular planform. General methods of solution of Eq. 28 are outlined in Section 6.9.

It is important to keep in mind that the above analyses are based on replacing the wing by a single bound vortex. For low aspect ratio or highly swept wings, as pointed out in Section 6.12, the lift distribution is represented by a number of bound vortices at different chordwise locations. For this configuration Eqs. 30 and 38 do not describe the spanwise variations of c_{d_i} and m.

Example

The analysis of this section is now applied to compute the characteristics of an untwisted rectangular wing of aspect ratio 6 flying at an angle of attack α. Assume that the airfoil is uncambered so that the absolute angle of attack α_a is equal to α everywhere along the span (see Fig. 6). Furthermore, since the wing sections do not vary with the spanwise location, we have $c = c_s$ and $m_0 = m_{0s} = 2\pi$. Thus Eq. 28 becomes

$$\sum_{n=1}^{\infty} A_n \sin n\theta \left(1 + \frac{n\pi}{2 R \sin \theta}\right) = \alpha$$

For a symmetrically loaded wing, the coefficients A_n vanishes for even values of n (see Problem 6.6.1). If four values of n are taken and we choose $R = 6$, every station $\theta = \cos^{-1} (2y/b)$ satisfies the equation

$$A_1 \sin \theta \left(1 + \frac{\pi}{12 \sin \theta}\right) + A_3 \sin 3\theta \left(1 + \frac{\pi}{4 \sin \theta}\right) + A_5 \sin 5\theta \left(1 + \frac{5\pi}{12 \sin \theta}\right)$$

$$+ A_7 \sin 7\theta \left(1 + \frac{7\pi}{12 \sin \theta}\right) = \alpha$$

It is sufficient to choose four stations along one-half of the span because of the symmetry of the rectangular wing. For $\theta = \pi/8, \pi/4, 3\pi/8,$ and $\pi/2$, we obtain a set of four simultaneous linear equations for the coefficients A_n

$$0.6445A_1 + 2.8200A_3 + 4.0840A_5 + 2.2153A_7 = \alpha$$

$$0.9689A_1 + 1.4925A_3 - 2.0161A_5 - 2.5397A_7 = \alpha$$

$$1.1857A_1 - 0.7080A_3 - 0.9249A_5 + 2.7565A_7 = \alpha$$

$$1.2618A_1 - 1.7854A_3 + 2.3090A_5 - 2.8326A_7 = \alpha$$

The solutions of the set of equations is

$$A_1 = 0.9174\alpha, \quad A_3 = 0.1104\alpha, \quad A_5 = 0.0218\alpha, \quad A_7 = 0.0038\alpha$$

The wing-lift coefficient is, from Eq. 33,

$$C_L = \pi^2 \frac{A_1}{2} = 4.5273\alpha$$

Based on the values

$$(C_{D_i})_{el} = C_L^2/\pi R = 1.0874\alpha^2$$

and

$$\sigma = \frac{3A_3^2 + 5A_5^2 + 7A_7^2}{A_1^2} = 0.0464$$

the induced drag coefficient for the wing is computed from Eq. 36:

$$C_{D_i} = (C_{D_i})_{el}(1 + \sigma) = 1.1379\alpha^2$$

which is approximately 5 percent higher than $(C_{D_i})_{el}$. On the other hand, in comparison with the lift coefficient of a wing of the same aspect ratio but with an elliptical lift distribution, which is calculated using Eqs. 11, 21, and 24.

$$(C_L)_{el} = \frac{m_0 \alpha}{1 + m_0/\pi R} = 1.5\pi\alpha = 4.7124\alpha$$

160 *The Finite Wing*

so that the slope of the lift coefficient curves, $dC_L/d\alpha$, of the rectangular wing is approximately 4 percent lower.

The result verifies the statement at the beginning of Section 6.5, that the properties of wings of arbitrary plan forms are close to those of the elliptical wing.

6.7 Comparison with Experiment

Classical test results on rectangular planforms reported by Prandtl (1921) are shown in Fig. 8. Seven wings of aspect ratios ($b^2/S = b/c$) varying from 1 to 7 were tested in a wind tunnel. In Fig. 8a, C_L is plotted versus the geometrical angle of

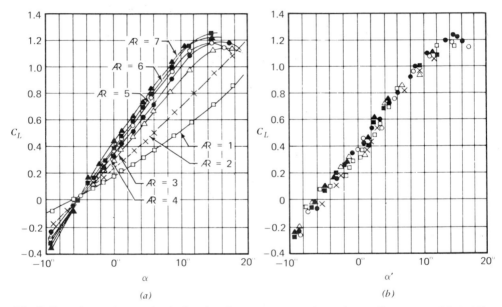

Fig. 8 Experimental test of calculated α_i for rectangular wings of $1 < R < 7$ (Prandtl, 1921).

attack α; in Fig. 8b, these experimental points are corrected to $R = 5$ by use of Eq. 22. Then these corrected results are plotted against the new geometrical angle of attack α' for $R = 5$, given by

$$\alpha' = \alpha + \frac{C_L}{\pi}\left(\frac{1}{5} - \frac{1}{R}\right)$$

The plots in Fig. 9 are termed "polar plots" for the wing. In Fig. 9a, C_L is plotted against C_D, defined by

$$C_D = C_{D_0} + C_{D_i}$$

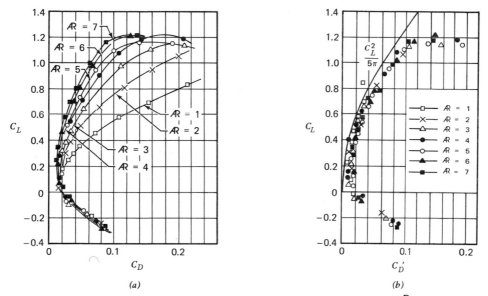

(a)   (b)

Fig. 9 Experimental tests of calculated C_{D_i} for rectangular wings of $1 < \mathcal{R} < 7$ (Prandtl, 1921).

where C_{D_0} is the "profile drag," that is, the viscous drag; C_{D_0} is a minimum at small C_L, increases slowly until the stall, after which it increases very rapidly. In Fig. 9*b*, the points are corrected to $\mathcal{R} = 5$ by use of Eq. 23; these corrected results are plotted against the new drag coefficient C_D' for $\mathcal{R} = 5$ given by

$$C_D' = C_D + \frac{C_L^2}{\pi}\left(\frac{1}{5} - \frac{1}{\mathcal{R}}\right)$$

The C_{D_i} curve for an elliptical lift distribution at aspect ratio 5 is shown for comparison.

Figures 8*b* and 9*b* verify therefore that significant departures from the elliptical distribution can be tolerated without causing appreciable corrections to the induced drag and angle of attack in the range $1 < \mathcal{R} < 7$. We note however that in Fig. 8*a* the C_L versus α plot for $\mathcal{R} = 1$ has a noticeable curvature; the curvature becomes more pronounced as the aspect ratio decreases further so that the assumption that the slope, m, is independent of angle of attack becomes more hazardous.

6.8 The Twisted Wing—Basic and Additional Lift

In order to obtain desirable aerodynamic force and moment distributions along the span the wing is often twisted, geometrically or aerodynamically, or both. These terms are defined as follows: (1) *Geometric twist* is achieved by twisting the axis of the wing so that the geometric angle of attack α varies spanwise; *geometric washout* signifies that α decreases from root to tip. (2) *Aerodynamic twist* is achieved by changing the airfoil section from root to tip effecting a spanwise variation of camber and position of maximum camber; these changes affect the spanwise variation of absolute angle of attack and the center of pressure (see Section 5.8). For a fixed chordwise position of maximum mean camber, *aerodynamic washout* could be achieved by decreasing the camber from root to tip.

For a wing of given twist and planform in steady motion along a given rectilinear path the spanwise distributions of absolute angle of attack, α_a, and of chord, c, are known. To determine numerically the spanwise distributions of sectional lift and drag we select k spanwise stations $\theta_1 \cdots \theta_k$, at which wing chords are respectively $c_1 \cdots c_k$. Then, after changing the upper limits of the summations in Eq. 28 from ∞ to k, we write for the jth station,

$$\alpha_{aj} = \frac{m_{0s}c_s}{m_{0j}c_j} \sum_{n=1}^{k} A_n \sin n\theta_j + \frac{m_{0s}c_s}{4b} \sum_{n=1}^{k} nA_n \frac{\sin n\theta_j}{\sin \theta_j} \tag{39}$$

In general, the accuracy of a given calculation is increased by increasing k or by decreasing the relative spanwise interval

$$y_{j+1} - y_j = \tfrac{1}{2} b (\cos \theta_{j+1} - \cos \theta_j)$$

in those regions where the sectional properties are expected to change rapidly, such as near the wing tips and in the vicinity of nacelles or engine pods.

Application of Eq. 39 to the k spanwise stations yields a set of k simultaneous equations, which is solved for the coefficients $A_1 \cdots A_k$.

The local values of the effective and induced angles of attack at the jth station are, by reference to Eq. 28, given by the first and second summations of Eq. 39:

$$\alpha_{0j} = \frac{m_{0s}c_s}{m_{0j}c_j} \sum_{n=1}^{k} A_n \sin n\theta_j \tag{40}$$

$$\alpha_{ij} = \alpha_{0j} - \alpha_{aj} = -\frac{m_{0s}c_s}{4b} \sum_{n=1}^{k} nA_n \frac{\sin n\theta_j}{\sin \theta_j} \tag{41}$$

With reference to Eq. 12, the local slope of the c_l versus α_a dependence is given by

$$m_j = \frac{m_{0j}}{1 - \alpha_{ij}/\alpha_{0j}} \tag{42}$$

Substitution of Eqs. 40 and 41 into Eq. 42 gives the slope at each of the k stations for the α_a distribution represented by Eq. 39, that is, for one given attitude of the wing relative to the flight path.

The above calculation outlines the procedure for finding the distributions of aerodynamic characteristics for a given $\alpha_a(y)$. Calculations covering the entire range of

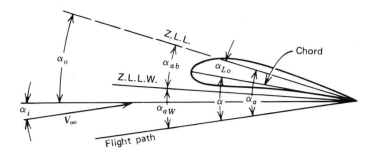

Fig. 10 Zero-lift line of wing.

α_a (assuming the wing is unstalled at every section) are facilitated with the help of Fig. 10. In this figure, which supplements Fig. 6, angles are expressed relative to the direction designated Z.L.L.W., the "zero lift line of the wing," that is, the direction of V_∞ for which the total lift of the wing is zero. The angle α_a can by this terminology be expressed as

$$\alpha_a = \alpha_{ab} + \alpha_{aW} \qquad (43)$$

where α_{ab}, termed the basic angle of attack, is the local absolute angle of attack for zero total lift of the wing and thus depends at a given station only on the wing twist; α_{aW}, termed the additional absolute angle of attack, is measured from Z.L.L.W. to the flight path.

The corresponding spanwise circulation contributions designated respectively Γ_b and Γ_a are shown schematically along with their sum

$$\Gamma = \Gamma_a + \Gamma_b$$

in Fig. 11. They satisfy the relations

$$\rho V_\infty \int_{-b/2}^{b/2} \Gamma_b \, dy = \int_{-b/2}^{b/2} L_b' \, dy = 0$$

$$L' = L_a' + L_b' = \rho V_\infty (\Gamma_a + \Gamma_b)$$

The corresponding lift coefficients are defined by

$$c_{l_a} = \frac{L_a'}{q_\infty c}, \quad c_{l_b} = \frac{L_b'}{q_\infty c}$$

The local lift coefficient c_l is expressed by

$$c_l = c_{l_b} + c_{l_a} \qquad (44)$$

where c_{l_b} depends on α_{ab}, that is, on the twist of the wing, and is thus independent of α_{aW}; c_{l_a}, the additional lift coefficient, is dependent on α_{aW} and is thus inde-

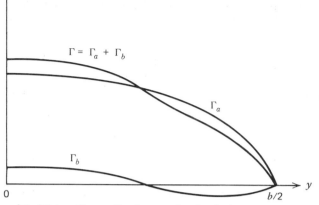

Fig. 11 "Basic" and "additional" contributions to the circulation.

pendent of wing twist. Equation 44 may be written

$$c_l = c_{l_b} + c'_{l_a} C_L \tag{45}$$

where c'_{l_a} is introduced for ease of presentation of data. Determination of the aerodynamic characteristics for a given wing at any angle of attack within the range of completely unstalled flow resolves itself into finding c_{l_b} and c'_{l_a} as functions of y. The procedure is given below.

Equations 39 are solved for two angles of attack, that is, for two sets of α_{aj}, designated α_{aj_1} and α_{aj_2}, to obtain two sets of coefficients $A_{11} \cdots A_{k1}$ and $A_{12} \cdots A_{k2}$. Within the limitations of the lifting line representation (Eq. 42)

$$m_j = \frac{dc_{lj}}{d\alpha_{aj}} \tag{46}$$

is a constant at a given station, independent of α_a but varies from station to station. Thus, to a good approximation, $m_{j1} = m_{j2} = m_j$ and

$$c_{lj_1} = m_j \alpha_{aj_1} ; \qquad c_{lj_2} = m_j \alpha_{aj_2} \tag{47}$$

If these coefficients are integrated spanwise with respect to y, we obtain

$$C_{L_1} = \frac{1}{S} \int_{-b/2}^{b/2} c_{l_1} c \, dy; \qquad C_{L_2} = \frac{1}{S} \int_{-b/2}^{b/2} c_{l_2} c \, dy \tag{48}$$

Then we write Eq. 45 for the two angles of attack

$$c_{lj_1} = c_{lbj} + c'_{laj} C_{L_1}$$
$$c_{lj_2} = c_{lbj} + c'_{laj} C_{L_2} \tag{49}$$

After substituting for c_{lj_1}, c_{lj_2}, C_{L_1}, and C_{L_2} from Eqs. 47 and 48, solution of the simultaneous equations 49 yields c_{l_b} and c'_{l_a} as functions of y. Thus the equations can be solved for these coefficients and c_l versus α_a determined throughout the range of angles of attack for which the wing is unstalled at all sections.

The spanwise distribution of induced drag can be found by means of Eqs. 18,

$$c_{d_i} = -c_l \alpha_i \tag{50}$$

and α_i can be obtained easily by the formula

$$\alpha_i = \alpha_0 - \alpha_a = \frac{c_l}{m_0} - \frac{c_l}{m}$$

so that

$$c_{d_i} = c_l^2 \left(\frac{m_0 - m}{m_0 m} \right) \tag{51}$$

We may also define a "weighted mean slope" for the entire wing as

$$\overline{m} = \frac{1}{S} \int_{-b/2}^{b/2} mc \, dy$$

so that

$$C_L = \overline{m} \alpha_a W \tag{52}$$

It must be emphasized that these results were derived for wings of large enough aspect ratio and small enough sweepback so that they can be represented by a lifting line normal to the flight path. In particular Eqs. 50 and 51 are restricted to such planforms; thus wings with appreciable sweepback or small aspect ratio must be treated by other methods, such as the source-doublet-vorticity panel representation treated briefly in Sections 4.13, 5.9, and 6.13.

6.9 Approximate Calculations of Additional Lift

We showed in Section 6.6 that the total lift coefficient of an arbitrary circulation distribution is proportional to the coefficient of $\sin \theta$ in the Fourier series describing the distribution. With this clue, Schrenk (1940) examined experimental results for many untwisted planforms and devised the approximate rule that the distribution of the *additional lift*, that is, the lift associated with the chord distribution without twist, is nearly proportional at every point to the ordinate that lies halfway between the elliptical and actual chord distributions for the same total area and span. Thus

$$L_a' = \frac{1}{2} \left[c + c_{sE} \sqrt{1 - \left(\frac{y}{b/2} \right)^2} \right] \frac{L}{S}$$

where c is the *actual* chord, and c_{sE} is the chord at the plane of symmetry for the elliptical planform of the same area and span. Thus

$$S = \int_{-b/2}^{b/2} c \, dy = \frac{\pi}{4} b c_{sE}$$

By use of the relations

$$c'_{l_a} = \frac{L'_a}{q_\infty c C_L} \qquad \text{and} \qquad \bar{c} = \frac{S}{b}$$

we find Schrenk's approximate relation

$$c'_{l_a} = \frac{1}{2} \left[1 + \frac{\bar{c}}{c} \sqrt{1 - \left(\frac{y}{b/2}\right)^2} \right] \tag{53}$$

This relation shows clearly the effect of taper on the spanwise lift distribution. The dotted line in Fig. 12 are drawn halfway between the actual chord distributions and that for an ellipse of the same area and semi-span.

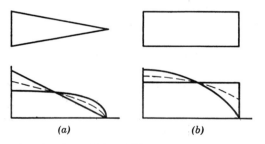

(a)   (b)

Fig. 12 Schrenk approximation for additional lift.

It can be seen that the effect of taper is to increase the load in the outboard portion above that which would occur if the additional lift were proportional to the chord. For stress analysis purposes the assumption of a lift distribution proportional to the chord is unconservative if the ratio of tip chord to root chord is less than 1/2.

6.10 Other Characteristics of a Finite Wing

The center of pressure, aerodynamic center and moments about finite wings are calculated as the weighted mean averages of the section characteristics.

The fore and aft location of the center of pressure (see Fig. 13), an important design parameter, will then be the weighted average of x_{cp}, the locations for the sections. The distance from a reference line to the center of pressure line for the entire wing, X_{CP}, for a given angle of attack is

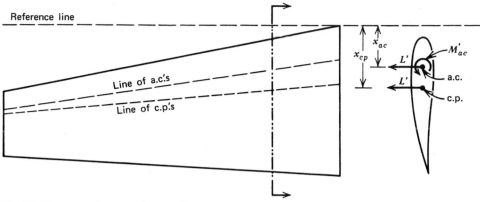

Fig. 13 Moments about a reference line.

$$X_{CP} = -\frac{M_{RL}}{L} \tag{54}$$

where M_{RL}, the moment about the reference line, is given by

$$M_{RL} = -\int_{-b/2}^{b/2} L' x_{cp}\, dy = -\int_{-b/2}^{b/2} L' x_{ac}\, dy + \int_{-b/2}^{b/2} M'_{ac}\, dy$$

In this equation x_{ac} and M'_{ac} are functions only of y, since they are independent of angle of attack; however L' is a function of y, therefore. of the sectional absolute angle of attack, α_a. It follows that M_{RL} will be (Fig. 10) a function of α_{aW}, the absolute angle of attack of the wing (see Eq. 52), and Eq. 54 shows that X_{CP}, will also be a function of α_{aW}. If we express L' in terms of the basic and additional contributions Eq. 54 becomes

$$X_{CP} = \frac{\displaystyle\int_{-b/2}^{b/2} [-(c_{l_b} + c'_{l_a} C_L) x_{ac} c + c_{mac} c^2]\, dy}{\displaystyle\int_{-b/2}^{b/2} c'_{l_a} C_L c\, dy} \tag{55}$$

We see from Fig. 13 that we may describe the loading at a section in terms of the sectional lift L', acting at the a.c., and M'_{ac}, or, in terms of L' acting at the c.p. (since $M'_{cp} \equiv 0$). Then X_{AC}, the distance from the reference line to the *wing* aerodynamic center is given by

$$X_{AC} = \frac{\displaystyle\int_{-b/2}^{b/2} c'_{l_a} c x_{ac}\, dy}{\displaystyle\int_{-b/2}^{b/2} c'_{l_a} c\, dy} \tag{56}$$

This equation states that the aerodynamic center of the wing (A.C.) is located at the centroid of the additional lift.

It follows that the c_{l_b} and c_{mac} terms in Eq. 55 must determine M_{AC}, the moment about the A.C.; this follows from the definition of M_{AC} as independent of angle of attack and, therefore of the additional lift. Thus, if we define Δx_{ac} as the distance from the A.C. line to the *section* aerodynamic center (a.c.) and we define $C_{MAC} = M_{AC}/q_{\infty}\bar{c}S$, where $\bar{c} = S/b$ is the mean chord, we have

$$C_{MAC} = \int_{-1/2}^{1/2} \left[-c_{l_b} \frac{\Delta x_{ac}c}{\bar{c}^2} + c_{mac} \left(\frac{c}{\bar{c}}\right)^2 \right] d\left(\frac{y}{b}\right) \tag{57}$$

The integrand of this equation can also be interpreted as the result of the transfer of moments, from the a.c. at a given y to the A.C., the centroid of the additional lift of the wing. Thus

$$M_{AC} = \int_{-1/2}^{1/2} (M'_{ac} - L'_b \Delta x_{ac}) d\left(\frac{y}{b}\right)$$

6.11   Stability and Trim of Wings

A wing is termed **statically stable** if, as a result of a small angular disturbance from equilibrium in steady flight, an aerodynamic moment is generated tending to return the wing to equilibrium. In Fig. 14, a wing cross section is shown with the load system acting at the aerodynamic center. Considering the wing as a rigid body, any unbalanced moments will cause it to rotate about its center of gravity (C.G.). If, as in Fig. 14, the C.G. is behind the aerodynamic center and M_{CG} is zero or balanced externally, the increment of lift, $+\Delta L$, that results from an increment in angle of attack will cause a moment $+\Delta M_{CG}$ (in the direction of stall). Conversely, if the angle of attack decreases, the resulting ΔL and ΔM_{CG} will both be negative. In either case $dM_{CG}/dL > 0$; thus, since the moment generated is in the direction to increase the deviation from equilibrium, this inequality identifies the configuration of Fig. 14 as *unstable*. Therefore, the wing is *stable* if $X_{AC} > X_{CG}$ so that $dM_{CG}/dL < 0$.

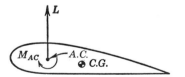

Fig. 14 Load system on rigid wing.

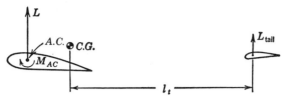

Fig. 15 Tail stabilizer.

For most wing configurations, the C.G. lies behind the aerodynamic center as in Fig. 14, and stability is generally achieved by placing a horizontal stabilizer behind the wing. From Fig. 15 it can be seen that the tail contributes a stabilizing moment when the wing-tail configuration is disturbed from equilibrium, and, by a proper adjustment of the tail area and tail length l_t, this stabilizing moment can easily be made to outweigh the destabilizing effect of the wing.

Stability is not the entire consideration. For steady equilibrium flight the airplane must be *trimmed*, meaning that the net moment acting must vanish. In view of the above discussion the airplane is stable if $dM_{CG}/dL < 0$; if trim is also to be achieved in steady flight $(L > 0)$ it is necessary that $M_{CG} = 0$ at $L = L_{trim}$. Thus to satisfy both conditions it is required that $M_{CG} > 0$ at $L < L_{trim}$. In terms of the coefficients the two conditions, trim and stability, for steady equilibrium flight are, respectively,

$$C_{MCG} > 0 \quad \text{at} \quad C_L = 0 \tag{58}$$

$$\frac{dC_{MCG}}{dC_L} < 0 \tag{59}$$

These conditions are illustrated in Fig. 16 (note that at $C_L = 0$, $C_{MCG} = C_{MAC}$ is designated C_{M_0}) which shows schematically the dependence of C_{MCG} on C_L for the wings with C.G. ahead of and behind the A.C., for the tail (behind A.C.), and for the wing-tail combination. We see that the wing A alone trims but is unstable if the C.G. is behind the A.C., and it is stable but does *not* trim if the C.G. is ahead of the A.C.; both stability and trim are achieved by adding the tail. For a conventional wing (wing B) with positive camber $C_{M_0} < 0$ so that the tail is required for steady equilibrium flight.

The contribution of the basic lift to C_{M_0} may be either positive or negative, depending upon the twist and sweep of the wing. Consider the sweptback wing shown in Fig. 17. The direction of sweep of a wing is determined by the inclination of the line of aerodynamic centers. Presume that the wing is set at $C_L = 0$ so that the lift acting at any section is the basic lift. Then, if the wing is washed out at the tips, the lift acting on the outboard sections will be down, whereas the lift acting on the inboard sections will be up. This is a consequence of the fact that the integral of the basic lift must be zero. The moment about the aerodynamic center of basic lift so distributed will be positive.

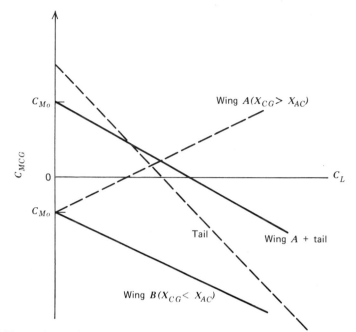

Fig. 16 Stability and trim for conventional wing normal to flight path ($C_{M_0} < 0$), for tail, and for wing + tail.

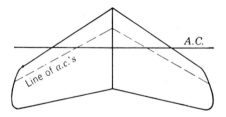

Fig. 17 Sweptback wing.

By the same argument, it can be shown that a combination of sweep-forward and washin at the tips will also result in a positive contribution to the C_{MAC} by the basic lift.

The reflex airfoil shown in Fig. 5.20 can be designed for $C_{M_0} > 0$ so with the C.G. ahead of the A.C. it would be stable and could be trimmed without a tail surface.

In summary, the C_{M_0} of a wing may be made positive by reflexing the trailing edges of the wing sections or by providing the proper amount of twist and sweep.

The combination of sweepback and washout is useful in flying wing design. The sweepback moves the aerodynamic center of the wing rearward, thereby facilitating the stability condition (Eq. 59) which requires that the C.G. lie in front of the aerodynamic center. The combination of sweepback and washout obtains a positive C_{M_0}, which is a necessary condition for trim (Eq. 58).

6.12 Higher Approximations

Refinements of Prandtl's lifting line theory are necessary for highly swept and for low aspect ratio wings. The lifting line may for instance be swept backward at the angle of the wing in which case, instead of calculating the downwash solely by integrating over the trailing vortex sheet as in Eq. 8, one must include a contribution from the bound vortex as well.

For low aspect ratio wings the extension of the methods of Sections 6.2 and 3 to the lifting surface is effected by arranging elementary bound vortices over the surface. These, with their trailing vortices, as shown in Fig. 18, form a configuration

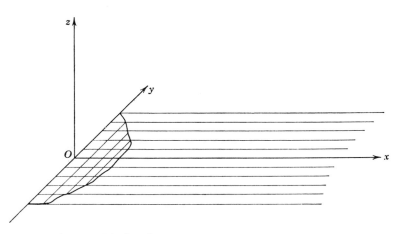

Fig. 18 Lifting surface model of a wing.

of horseshoe vortices appropriate to the spanwise and chordwise lift distribution. The portion of the $z = 0$ plane representing the wing is covered by a lattice of vortex filaments extending in the spanwise and chordwise directions; the wake portion of the plane contains only chordwise (streamwise) filaments. The problem is to find the vorticity distribution in the $z = 0$ plane such that the flow induced by both bound and trailing vortices will cancel the component of the free stream velocity normal to the surface *and* will satisfy the Kutta condition at the trailing edge (see Ashley and Landahl, 1965, Chap. 7).

6.13 The Complete Airplane

Theoretical and experimental studies along with advances in computer technology in recent years have culminated in the capability of determining with remarkable accuracy the aerodynamic characteristics of complete aircraft. The three significant developments that have led to this capability are (1) the concepts of source and doublet planels and their utilization, along with vorticity panels, to determine

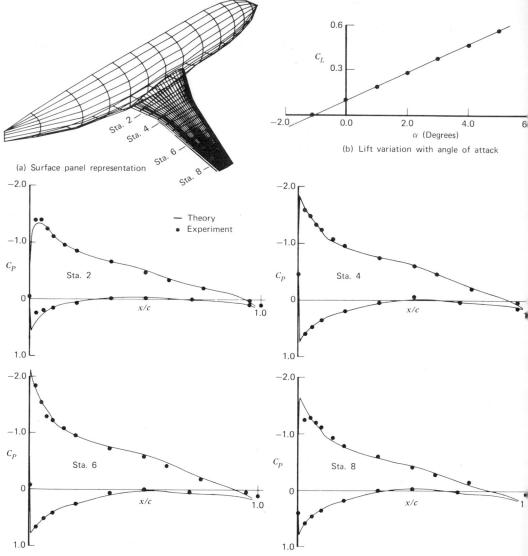

(a) Surface panel representation

(b) Lift variation with angle of attack

— Theory
• Experiment

(c) Wing surface pressure distributions

Fig. 19 Comparison of numerical and experimental results for a Boeing 737 wing-body model. (Courtesy of P. E. Rubbert and G. R. Saaris, The Boeing Company.)

the irrotational velocity and pressure distributions over the aircraft; (2) advances in semiempirical methods for introducing viscous boundary layer and wake effects; (3) development of computers that can handle expeditiously literally thousands of simultaneous linear algebraic equations. The general method for the use of source panels to establish an arbitrary nonlifting shape as a stream surface in a flow was described in Section 4.13; the superposition of vorticity panels to satisfy the Kutta condition for two-dimensional lifting airfoils was described in Section 5.9.

Determination of the effects of the trailing vorticity field of the lifting surfaces and mutual aerodynamic interference effects among fuselage, wing, tail, engines, and so forth, introduce considerable complication into the calculations but, to the extent that the flows are incompressible and inviscid so that superposition is justified, no new concepts are involved. Effects of compressibility and viscosity, which are described in the following chapters, must still be introduced in the computations. For subsonic aircraft these effects are introduced as corrections that do not alter significantly the general scheme of the calculations.

Details of the method are described elsewhere (Hess and Smith, 1967; Hess, 1971, 1972; Rubbert and Saaris, 1969, 1972). Figure 19 shows a perspective view of the surface panel representation of the Boeing 737, and the comparison between theory and experiment for C_L versus α and for pressure distributions at four spanwise stations.

6.14 Interference Effects

A few of the significant interference effects involving finite wings, so far not mentioned, are given below.

1. *Ground Effect.*

In an example in Section 4.12 the image effect was used to calculate the lift force per unit length of a vortex parallel to a plane surface (Fig. 4.18) and normal to a flow along the surface. The configuration is similar to that used to calculate the ground effect on an infinite wing. As was pointed out there, the ground effect on a finite wing must take into account the effect of not only the bound vortex but also the trailing vortices. To visualize this effect one may consider Fig. 4.19 as the cross section of the trailing vortices and their images near the ground. It is clear that the images induce an upwash between the above-ground vortices; its effect will be to decrease the downwash at the wing and therefore the lift at the same angle of attack and $dC_L/d\alpha$ will accordingly be increased by an amount which decreases with height above the ground. The effect is shown in Fig. 20 where the subscript h refers to values at heights $2h/b$ above the ground (Goranson, 1944).

2. *Wind Tunnel Boundary Effect.*

When a model of an aircraft is palced in a wind tunnel the measured characteristics must be corrected for the velocities induced by the images required to establish the boundary condition of no flow through the solid boundaries.

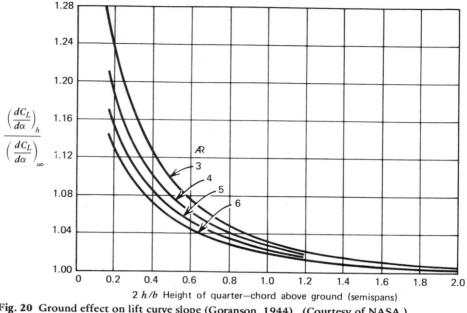

Fig. 20 Ground effect on lift curve slope (Goranson, 1944). (Courtesy of NASA.)

Example.

Consider a wing of span b in an airstream of radius R with impervious walls as shown in cross section in Fig. 21. A and B are the trailing vortices of the wing; A' and B' are images of A and B calculated to establish the circle R as a streamline of the flow. As far as the flow components in the plane of the paper are concerned the vortices originate at the points shown and extend to infinity into the paper (the Helmholtz laws are satisfied by the bound vortices extending from A to B, from A' to the right to infinity, and from B' to the left to infinity). The reader may verify that, if the circulations of A' and B' are of equal and opposite magnitudes respectively to A and B, and, with $\overline{OA} = \overline{OB} = b/2$, $\overline{OA'} = \overline{OB'} = R^2/\frac{1}{2}b$, the circle $r = R$ is a streamline of the

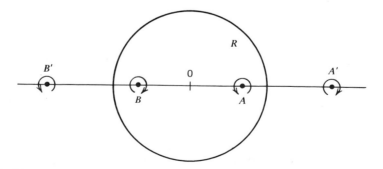

Fig. 21 Trailing vortices and images of wing AB in circular wind tunnel (Glauert, 1937).

flow. This result is accomplished by applying the formula for the stream function from Eqs. 2.44. At the center of the wing the upwash induced by the (semi-infinite) image vortices is

$$\Delta w = 2 \cdot \frac{\Gamma}{4\pi} \frac{b/2}{R^2} = \frac{\Gamma b}{4\pi R^2}$$

In terms of the equation for the lift coefficient, with

$$C_L S_W \cdot \tfrac{1}{2} \rho V_\infty^2 = \rho V_\infty \Gamma b$$

where S_W is the area of the wing, the upwash becomes

$$\Delta w = \frac{C_L S_W V_\infty}{2 S_T}$$

where $S_T = \pi R^2$, the cross-sectional area of the tunnel. The correction to the angle of attack of the wing may be written

$$\Delta \alpha = \frac{\Delta w}{V_\infty} = \frac{C_L S_W}{2 S_T}$$

It follows that the correction to the drag coefficient is

$$\Delta C_D = \frac{1}{2} \frac{C_L^2 S_W}{S_T}$$

The calculation assumes, in addition to a circular cross section and concentrated trailing vortices separated by the wing span and that $b/2R$, S_W/S_T, and C_L be small enough so that new variables, such as blockage of the tunnel airstream, interference with the wall boundary layers and excessive deflection of the airstream do not exert significant effects.

Many other interference effects, for instance, those between the blades of rotors, between appendages and lifting surfaces, between viscous wakes and nearby surfaces, etc. All of these affect the aerodynamic characteristics and each can exert a predominant effect for some configurations.

6.15 Concluding Remarks

The preceding sections are intended to show the basis of finite-wing theory and to show some of the methods used for determining the spanwise lift distribution; this distribution, along with the section properties calculated by the methods of Chapter 5, determine the complete aerodynamic characteristics of the wing.

The source, doublet, and vorticity panel method is the most accurate, though it is remarkable how effectively and how easily the elliptical lift distribution can be used for *preliminary* design purposes. The real power of the panel method lies in its capability to evaluate aerodynamic interference effects and in the framework that has been developed by means of which thickness, compressibility, and viscous effects can be introduced. Also, this method has been applied effectively to flow through channels and complicated passages, evaluation of boundary layer control methods, characteristics of flow past multiple surfaces, jet interference effects, and other flow problems over a range of engineering fields.

The effects of compressibility and viscosity were passed over lightly in the above paragraph. Their introduction into the panel method or into any of the analyses of this chapter can be effective only if carried out in the light of a thorough understanding of compressible and viscous fluid flows. In terms of the pertinent dimensionless parameters, the calculated results become highly suspect over the range of Mach numbers within which shock waves occur and over the range of Reynolds numbers and angles of attack within which viscosity effects are significant. The following chapters are designed to provide a background for the evaluation of compressibility and viscous effects and their influence on design of aircraft and fluid flow devices.

Another effect that is not mentioned is the determination of the lift of three dimensional bodies, for example, the fuselage. The difficulty lies in the circumstance that the Kutta condition of smooth flow at the trailing edge is a two-dimensional concept, and we still have nothing to repalce it for a three-dimensional body. Fortunately, the lift of the fuselage is small and the lift of the wing may be calculated as if it extends well into the fuselage body.

CHAPTER 7

Introduction to Compressible Fluids

7.1 Scope

Chapters 7 through 13 describe the changes that occur in the aerodynamic characteristics of bodies in high-speed flight. In the preceding chapters, the basic concepts of the flow of a perfect fluid past solid bodies were studied. The agreement with theory and the discrepancies between the theory and experiments were pointed out. Some of these discrepancies result from treating the fluid as nonviscous. Others arise as a consequence of assuming air to be incompressible.

The similarity parameter* characterizing compressible flows is the Mach number, which is the ratio of the airspeed to the speed of sound. Whereas incompressible fluid considerations are a good approximation to low Mach number flows, as the Mach number increases density variations throughout the flow become greater and greater and the approximation involved in taking the density constant becomes poorer and poorer.

Since air is both viscous and compressible, the flow will depend on Mach number, Reynolds number, and other similarity parameters to be defined in Chapter 14. It is permissible, however, to postulate that the effect of viscosity is confined to a thin boundary layer (except where flow separation occurs) so that the main flow is approximately dependent only on Mach number. Accordingly, the analysis in these chapters will generally be limited to compressible nonviscous flows; experimental tests demonstrate the extent to which viscosity alters these results.

The physical quantities that were considered constant in previous chapters and now must be treated as variables are density and temperature. The two additional unknowns require additional equations for the solution of compressible flow problems and these are obtained from empirical information associated with energy concepts and the behavior of an ideal gas. The empirical principles are described briefly in Chapter 8 and the necessary mathematical equations are developed from them. In Chapters 9 and 10, the equations are manipulated to expose the nature of subsonic and supersonic flow. The section on compressibility is concluded with an introduction to wing theory in Chapter 12.

In Chapters 2 and 3, the principles of conservation of mass and momentum, which apply to elements of fixed identity, were formulated in terms of field properties. The resulting equations are perfectly general and apply whether the fluid is compressible or not. From these conservation principles, the equations of continuity, equilibrium, irrotationality, and the associated concepts of stream function and

---

*Similarity parameters are discussed in Chapter 1 and Appendix A.

velocity potential were derived. The remainder of this chapter is devoted to a brief discussion of each of the derived relationships from the viewpoint of compressible flow application.

7.2 Equation of Continuity—Stream Function

The equation of continuity for the general case of unsteady, compressible flow is given by Eq. 2.13.

$$\text{div } \rho \mathbf{V} = - \frac{\partial \rho}{\partial t} \tag{1}$$

It may be readily verified, by expansion into Cartesian form, that Eq. 1 may be written

$$\rho \text{ div } \mathbf{V} = - \frac{\mathcal{D} \rho}{\mathcal{D} t} \tag{2}$$

For ρ constant, the substantial derivative is zero, and Eq. 2 takes the familiar form of incompressible-flow theory.

A stream function for two-dimensional steady compressible flows may be defined, provided that the density ρ is included in the definition. In accordance with the argument of Section 2.6 a stream function ψ' can be found such that the speed at any point in the flow is given by

$$|\mathbf{V}| = \frac{1}{\rho} |\text{grad } \psi'| \tag{3}$$

and the direction of the velocity is given by the family of curves

$$\psi' = \text{constant} \tag{4}$$

The component of velocity in any direction s is given by differentiating the stream function at right angles to the left.

$$V_s = \frac{1}{\rho} \cdot \frac{\partial \psi'}{\partial n} \tag{5}$$

where the direction of n is normal to that of s.

7.3 Irrotationality—Velocity Potential

The vorticity vector $\boldsymbol{\omega}$ defined in Section 2.7 does not involve density, and, therefore, the three components of vorticity of a flow, compressible *or* incompressible, are

$$\omega_x = \frac{\partial w}{\partial y} - \frac{\partial v}{\partial z}$$

$$\omega_y = \frac{\partial u}{\partial z} - \frac{\partial w}{\partial x} \tag{6}$$

$$\omega_z = \frac{\partial v}{\partial x} - \frac{\partial u}{\partial y}$$

In the absence of viscosity, the vorticity of a flow that was originally at rest is always zero, that is

$$\boldsymbol{\omega} = \text{curl } \mathbf{V} = 0 \tag{7}$$

for irrotational velocity fields; it follows that a velocity potential ϕ exists, such that

$$\mathbf{V} = \text{grad } \phi \tag{8}$$

or, in Cartesian form,

$$u = \frac{\partial \phi}{\partial x}; \qquad v = \frac{\partial \phi}{\partial y}; \qquad w = \frac{\partial \phi}{\partial z} \tag{9}$$

The existence and implications of the existence of a velocity potential ϕ were shown in Section 2.11. The compressibility of the fluid in no way alters that discussion.

7.4 Equation of Equilibrium—Bernoulli's Equation

Euler's equation expressing dynamic equilibrium of an element of mass of an inviscid, compressible *or* incompressible, fluid was derived in Section 3.4. Compressible fluids are invariably gases, for which gravity forces may be neglected, so that the equation may be written

$$\frac{\mathcal{D}\mathbf{V}}{\mathcal{D}t} + \frac{\text{grad } p}{\rho} = 0 \tag{10}$$

In Section 3.5, the dot product of terms in Euler's equation with an increment of streamline ds led to the result, for steady flow,

$$d\frac{V^2}{2} + \frac{dp}{\rho} = 0 \tag{11}$$

Equation 11 is the equilibrium equation in differential form. If ρ is constant, an integration of Eq. 11 leads to the familiar form of Bernoulli's equation. If ρ is variable, a relation between ρ and p must be found before the middle term can be integrated.

7.5   Criterion for Superposition of Compressible Flows

In Section 4.2, it was shown that the irrotationality condition in terms of the stream function for an incompressible flow leads to the Laplace equation

$$\nabla^2 \psi = 0 \tag{12}$$

and, similarly, the equation of continuity in terms of the velocity potential leads to the Laplace equation

$$\nabla^2 \phi = 0 \tag{13}$$

Thus the velocity field for an incompressible flow is completely determined by a linear differential equation with one dependent variable, the scalar ϕ or ψ. Therefore, solutions can be added (superimposed) to form new solutions, and, in this way, complicated flows can be analyzed as the sum of a number of simple flows. For instance, in Section 4.6, the flow around a circular cylinder was obtained by superimposing a uniform flow and a doublet, which is itself a superposition of source and sink flow.

An analogous procedure is not justified for compressible flows unless, as is pointed out below, the density variation is small enough to be neglected. For a compressible flow, the irrotationality condition in terms of the stream function leads to (z component)

$$\text{curl}_z \, \mathbf{V} = -\frac{\partial}{\partial x}\left(\frac{1}{\rho}\frac{\partial \psi'}{\partial x}\right) - \frac{\partial}{\partial y}\left(\frac{1}{\rho}\frac{\partial \psi'}{\partial y}\right) = 0 \tag{14}$$

Similarly, the equation of continuity for steady flow in terms of the velocity potential leads to

$$\text{div} \, \rho \mathbf{V} = \frac{\partial}{\partial x}\left(\rho\frac{\partial \phi}{\partial x}\right) + \frac{\partial}{\partial y}\left(\rho\frac{\partial \phi}{\partial y}\right) = 0 \tag{15}$$

Since the density in a compressible flow is a function of the velocity, the terms in parentheses in Eqs. 14 and 15 are nonlinear. Therefore separate solutions $\phi(x, y)$ or $\psi'(x, y)$ cannot be added to obtain new solutions, except for flows in which the density variation can be neglected.

Expansion of Eq. 15 indicates in more detail the circumstances under which the superposition is justified. This expansion becomes, after substituting for u and v, from Eqs. 9,

$$\rho \nabla^2 \phi + u\frac{\partial \rho}{\partial x} + v\frac{\partial \rho}{\partial y} = 0 \tag{16}$$

The last two terms in this equation, expressed vectorially by $(\mathbf{V} \cdot \nabla)\rho$, are non-linear; they represent the convective derivative of the density, that is, the rate of change of the density of a fluid element as it is convected along a streamline in a steady flow. Accordingly, if

$$(\mathbf{V} \cdot \nabla)\rho \ll \rho \nabla^2 \phi \qquad (17)$$

Eq. 16 is nearly linear and superposition is justified.

For flows that are *everywhere* subsonic, condition 17 is shown to be satisfied to a good enough approximation to enable linearization of Eq. 16; as a result we show that compressible flow problems in this flow range can be solved by applying a simple "stretching" factor, dependent on the "local" Mach number, to an equivalent incompressible flow. For flows *everywhere* supersonic past slender bodies condition 17 is satisfied to the extent that superposition is again justified if the Mach number is not too high or the shock waves are not too intense.

CHAPTER 8

The Energy Relations

8.1 Introduction

In Section 3.2 it is pointed out that Bernoulli's equation for an incompressible inviscid flow expresses the conservation of mechanical energy. If we neglect the effect of gravity (as we do for gas flows) the equation (Eq. 3.7) becomes

$$p + \tfrac{1}{2}\rho V^2 = p_0 = \text{constant} \tag{1}$$

The terms on the left may be regarded respectively as the potential and kinetic energies per unit volume and the equation states that their sum, p_0, is the total mechanical energy per unit volume. A decrease in kinetic energy per unit volume, according to Eq. 1, is accompanied by an equal increase in pressure energy so that the sum of the two energies is conserved. Thus, mechanical energy is conserved. Equation 1 represents simply one more implication of the principle of conservation of momentum expressed by Euler's equation.

When the gas is compressible, mechanical energies are *not* conserved, and it is no longer possible to deduce an energy relation from the momentum principle. An entirely separate empirical fact, the first law of thermodynamics, must be introduced. *Intrinsic energy*, a concept derived from the first law, enters the energy balance. Furthermore, the first law encompasses other energies in addition to mechanical work, thereby generalizing the energy conservation principle. The energy forms of interest in aerodynamics are thermal and mechanical. The laws governing the transfer from one form to the other are derived and applied in this and the following chapters to problems of practical importance.

Intrinsic energy involves temperature, a gas characteristic that has not been considered so far. To relate temperature to density and pressure requires another empirical fact, *the equation of state*, which is discussed in Section 8.2, together with other characteristics of an ideal gas.

In Sections 8.3 and 8.4, the first law of thermodynamics is formulated and used to develop an equation that expresses conservation of energy for a gas in motion. The use of the energy equation with and without viscous dissipation and heat addition is illustrated by examples. In Section 8.5, the concept of *reversibility* is introduced, and special relations among temperature, pressure, and density are derived for adiabatic reversible flows.

The direction in which an energy exchange may proceed is governed by the second law of thermodynamics, also an empirical observation. The second law is formulated in terms of *entropy* in Section 8.6.

Finally, for the special case of adiabatic reversible flows, the relation between density and pressure developed from the first law is introduced into the momentum equation and the integration of the momentum equation is carried out. The resulting energy equation is called Bernoulli's equation for compressible flow. However, unlike its counterpart for incompressible flow, the compressible Bernoulli equation stands as an independent energy relation that cannot be deduced from purely momentum considerations.

The thermodynamics concepts are stated in summary form in the present chapter. For details, the reader is referred to texts on thermodynamics (see, e.g., Zemansky, 1943).

8.2   Characteristics of an Ideal Gas—Equation of State

The following table lists various quantities used in this chapter, together with the units in which they are measured. In aerodynamics, it is convenient to measure thermal energy in Newton-meters (joules), thereby avoiding the confusion of converting between mechanical and thermal units.

In the table, the first 12 quantities represent characteristics of the gas. In contrast, the last 2, heat and work, represent energies in transit and are not characteristics of the gas. The gas characteristics may be directly observable, they may be quantities defined in terms of observable characteristics, or they may be quantities deduced from experiment. The *state* of a gas is fixed when all its characteristics have definite values. It will be seen later that not all of the characteristics are independent, and therefore the state of the gas may be fixed by specifying a limited number of characteristics.

In addition to the thermodynamic characteristics listed in the table, there are mechanical characteristics such as displacement and velocity that determine the potential and kinetic energies of the gas as a whole. In the following material, the term *state* is used to mean the thermodynamic state. The thermodynamic characteristics are described briefly in the following paragraphs.

1. *Equation of State and the Gas Constant R*

The **ideal*** or thermally perfect gas in equilibrium obeys the equation of state

$$p = \rho R T \tag{2}$$

where the gas constant R depends on the molecular weight of the gas being considered.[†] For a gas of fixed composition, R is constant. For air, based on 21 percent oxygen and 79 percent nitrogen by volume,

$$R = 287 \text{ Nm/kg}^\circ\text{K (joules/kg}^\circ\text{K)}$$

---

*The molecular model of an ideal gas has been described in Section 1.3.
[†] In terms of a *universal* gas constant $R' = nR$ where n is the number of kilograms of gas equal to the molecular weight of the gas (n = 28.97 for air). Then R' = 8314 in SI units.

Table of Symbols[a]

Symbol	Name	Units
p	Pressure	N/m^2
ρ	Density	kg/m^3
$v = 1/\rho$	Specific volume	m^3/kg
T	Absolute temperature	degrees Kelvin
u	Specific intrinsic energy	Nm/kg (joule/kg)
e	Specific internal energy	Nm/kg (joule/kg)
h	Specific enthalpy	Nm/kg (joule/kg)
S	Specific entropy	$Nm/kg\,^\circ K$ (joule/kg$^\circ$K)
c_p	Constant pressure specific heat	$Nm/kg\,^\circ K$ (joule/kg$^\circ$K)
c_v	Constant volume specific heat	$Nm/kg\,^\circ K$ (joule/kg$^\circ$K)
R	Gas constant	$Nm/kg\,^\circ K$ (joule/kg$^\circ$K)
γ	Specific heat ratio c_p/c_v	
$^\circ K$	Degrees Kelvin	
$^\circ C$	Degrees Centigrade	
$\hat{R}$	Control volume fixed in the field	
$\hat{S}$	Control surface fixed in the field	
$\hat{R}_1$	Control volume moving with the fluid	
$\hat{S}_1$	Control surface moving with the fluid	
q	Heat transfer per unit mass	Nm/kg (joule/kg)
w	Work transfer per unit mass	Nm/kg (joule/kg)

[a]In previous chapters, u, v, and w have been used to denote the Cartesian components of velocity. In this chapter, u, v, and w have the meanings given in the table and the symbol $\mathbf{V}$ is used to indicate velocity.

At the extremely high temperatures encountered in very high Mach number flight, dissociation, ionization, and the formation of new compounds cause the average molecular weight of air to decrease with an attending rise in the gas constant R. In this text, R will be considered to have the constant value given above unless stated otherwise.

The state of a gas of given molecular weight (and therefore all its thermodynamic characteristics) is, according to Eq. 2, fixed when any two of the three variables p, ρ, T are known.

2. *Intrinsic Energy and Constant Volume Specific Heat*

The intrinsic energy is a characteristic deduced from experiment and is discussed in Section 8.3 in connection with the first law of thermodynamics. Because intrinsic energy is determined by the state of the gas, it may be written*

$$u = u(v, T)$$

$$du = \left(\frac{\partial u}{\partial v}\right)_T dv + \left(\frac{\partial u}{\partial T}\right)_v dT$$

The Joule-Thomson experiments show that for an ideal gas

$$\left(\frac{\partial u}{\partial v}\right)_T = 0$$

and therefore the intrinsic energy depends only on temperature. By definition, the constant volume specific heat is

$$c_v \equiv \left(\frac{\partial u}{\partial T}\right)_v = \left(\frac{\partial u}{\partial T}\right)_p \tag{3}$$

so that a change in intrinsic energy is given by:

$$\Delta u = \int_{T_1}^{T_2} c_v \, dT \tag{4}$$

Because c_v is defined in terms of characteristics u and T, it is itself a characteristic of the gas. For temperatures less than $600°\mathrm{K}$, c_v for air is practically constant and has the value

$$c_v = 717 \ \mathrm{Nm/kg°K} \ (\mathrm{joules/kg°K})$$

For the very high temperatures encountered in hypersonic flight, c_v can rise to several times the value indicated above. Unless stated otherwise, c_v in this text will be treated as a constant.

3. *Enthalpy and Constant Pressure Specific Heat*

The enthalpy h is a characteristic defined as

$$h = pv + u \tag{5}$$

*The notation $(\partial u/\partial v)_T$ indicates that u has been differentiated with respect to v while holding T constant. Note that if the subscript T had been omitted, the notation would not specifically indicate that the quantity being differentiated is an explicit function of both v and T.

Also by definition, the constant pressure specific heat is

$$c_p \equiv \left(\frac{\partial h}{\partial T}\right)_p = \left(\frac{\partial h}{\partial T}\right)_v \tag{6}$$

The last equality follows from the Joule-Thomson law above. Equations 2, 4, and 5 show the enthalpy to be a function only of temperature. Therefore

$$dh = \left(\frac{\partial h}{\partial T}\right)_p dT = \left(\frac{\partial h}{\partial T}\right)_v dT$$

so that a change in enthalpy is given by

$$\Delta h = \int_{T_1}^{T_2} c_p \, dT \tag{7}$$

c_p is a characteristic of the gas because it is defined in terms of the characteristics h and T. For temperature less than $600°K$, c_p for air is practically constant and has the value

$$c_p = 1004 \text{ Nm/kg}°\text{K (joules/kg}°\text{K)}$$

The comments made about c_v concerning its variation with temperature also apply to c_p.

It should be noted that an absolute value of intrinsic energy u is not defined, and therefore it is appropriate to speak only of changes in intrinsic energy as indicated in Eq. 4. Because h is defined in terms of u, an absolute value of h does not exist either.

4. *Relation Between the Gas Constant and the Specific Heats*

From the definition of enthalpy, Eq. 5,

$$\frac{\partial h}{\partial T} = p \frac{\partial v}{\partial T} + v \frac{\partial p}{\partial T} + \frac{\partial u}{\partial T}$$

$$\left(\frac{\partial h}{\partial T}\right)_p = p \left(\frac{\partial v}{\partial T}\right)_p + \left(\frac{\partial u}{\partial T}\right)_p$$

Using Eqs. 6, 2, and 3 in the above,

$$c_p = R + c_v \tag{8}$$

5. *Ratio of Specific Heats*

The specific heat ratio

$$\gamma = \frac{c_p}{c_v} \tag{9}$$

occurs frequently in compressible flow theory. For air, at $288°K$

$$\gamma = 1.4$$

γ is close to this value for temperatures under $600°K$. As the temperature rises γ decreases towards unity. In this text γ will be assumed equal to 1.4 unless stated otherwise.

8.3 First Law of Thermodynamics

The principle of conservation of energy, like that of momentum stated in Section 3.7, applies to a group of particles of fixed identity. In thermodynamic terminology, the group of particles of fixed identity is called a system and everything outside the group of particles is called the surroundings of the system. The first law of thermodynamics stems from the fundamental experiments of Joule which demonstrated that heat and work are entities of the same kind. If a system goes through a succession of states such that the initial and final states are the same, then the net heat transferred across the boundary is equal to the net work transferred across the boundary. Adopting the standard convention that work transferred *out* of the system and heat transferred *into* the system are positive, the first law may be formulated:

$$\oint (\delta q - \delta w) = 0 \qquad (10)$$

where the line integral represents a succession of state changes in which the initial and final states are identical, and δq and δw represent increments in the quantities.* As indicated in Section 8.2, the state of an ideal gas is completely determined by the two variables p and v. Therefore, Eq. 10 applies to a succession of states represented by the locus of points from A to B to A in the $p - v$ diagram of Fig. 1.

Because by the first law the line integral around a closed path is zero, the line integral of $\delta q - \delta w$ between any two states A and B depends only on A and B and therefore a new gas characteristic e is defined.[†]

$$\Delta e = e_B - e_A = \int_A^B (\delta q - \delta w) \qquad (11)$$

*The increments are expressed as δq and δw for the reason that q and w are not unique functions of p and v, and they therefore cannot be expressed as differentials dq and dw in the p, v plane. On the other hand e, the specific intrinsic energy (Eq. 11) is for an ideal gas a function of T only.

[†]The mathematical argument has been elaborated in connection with the derivation of the velocity potential in Section 2.11.

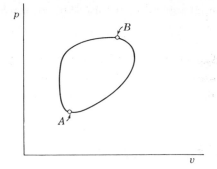

Fig. 1 $p - v$ Diagram.

e is called the specific internal energy of the gas. The absolute value of e is not defined but the difference in internal energy Δe between any two states A and B can
be found by measuring the net heat and work transfers across the boundaries of the
system as the system changes from state A to state B by any path whatever.

The rate of increase of energy within the elementary volume ΔR enclosed by the
surface ΔS may be written*

$$\frac{d}{dt}(\rho e)\Delta R = \left[\frac{\delta}{\delta t}(\rho q) - \frac{\delta}{\delta t}(\rho w)\right]\Delta R = \left(\frac{\delta q'}{\delta t} - \frac{\delta w'}{\delta t}\right)\Delta S \tag{12}$$

where q' and w' are defined, respectively, as the heat and work transfer per unit
area in Nm/m^2. Applied to the volume $\hat{R}_1$ enclosed by the surface $\hat{S}_1$, Eq. 12
becomes

$$\frac{d}{dt}\iiint_{\hat{R}_1} \rho e \, d\hat{R}_1 = \frac{\delta}{\delta t}\iint_{\hat{S}_1} (q' - w')d\hat{S}_1 \tag{13}$$

where $\hat{R}_1$ comprises particles of fixed identity, independent of the time.

The conversion to field representation of the conservation of energy principle follows the procedure used in Section 3.7 in which the same conversion was made of
the conservation of momentum principle.

Consider the control surface $\hat{S}$ in Fig. 2. At time t, the control surface $\hat{S}$ which is
fixed in space contains a tagged set of fluid particles. At time t_1, these particles
will have moved to the region enclosed by the dotted curve $\hat{S}_1$.

Let E_A, E_B, and E_C be the internal energy of the fluid in regions A, B, and C,
respectively, at time t, the particles have internal energy $E_A(t) + E_B(t)$ and, at time
t_1, internal energy $E_B(t_1) + E_C(t_1)$. The internal energy change during the interval
$t_1 - t$ is

$$[E_B(t_1) - E_B(t)] + [E_C(t_1) - E_A(t)] \tag{14}$$

*The notation d/dt is to be interpreted as a derivative in the ordinary sense.

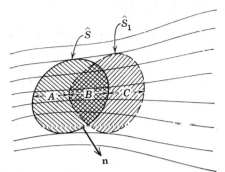

Fig. 2 Conversion to field representation.

$E_C(t_1)$ is the internal energy of the fluid that has passed through $\hat{S}$ during the interval, and $E_A(t)$ is the internal energy of the fluid that has entered $\hat{S}_1$ during the interval. The time rate of change of internal energy is given by the limit of expression 14 as $t_1 \to t$

$$\lim_{t_1 \to t} \left[\frac{E_B(t_1) - E_B(t)}{t_1 - t} + \frac{E_C(t_1) - E_A(t)}{t_1 - t} \right] \tag{15}$$

In the limit as $t_1 \to t$, $\hat{S}_1$ coincides with $\hat{S}$, and the first term in (15) becomes the time rate of change of internal energy of the fluid in region $\hat{R}$ enclosed by $\hat{S}$. This is written as the integral

$$\frac{d}{dt} \iiint_{\hat{R}} e\rho \, d\hat{R}$$

The second term in (15) is the energy flux through $\hat{S}$, outward being counted positive. In integral form, the second term is written

$$\iint_{\hat{S}} e\rho \mathbf{V} \cdot \mathbf{n} \, d\hat{S}$$

The conservation of energy principle Eq. 13 is finally written

$$\frac{d}{dt} \iiint_{\hat{R}} e\rho \, d\hat{R} = - \iint_{\hat{S}} e\rho \mathbf{V} \cdot \mathbf{n} \, d\hat{S} + \frac{\delta}{\delta t} \iint_{\hat{S}} (q' - w') \, d\hat{S} \tag{16}$$

All terms in the above equation are field properties. The heat transfer term can be written in terms of the temperature gradient at the surface and the work transfer term in terms of general surface stresses. Equation 16 can then be reduced to a differential equation relating field properties at each point in the fluid. This analysis has been carried out in Appendix B, Section 6.

In the next section, a useful form of Eq. 16 is derived for a special case.

8.4 Steady Flow Energy Equation

For steady flow, the conservation of energy principle, Eq. 16, applied to a fixed control volume is

$$\iint_{\hat{S}} e\rho \mathbf{V} \cdot \mathbf{n} \, d\hat{S} = \frac{\delta}{\delta t} \iint_{\hat{S}} q' \, d\hat{S} - \frac{\delta}{\delta t} \iint_{\hat{S}} w' \, d\hat{S} \tag{17}$$

The net energy flux out of the fixed control volume $\hat{R}$ shown in Fig. 3 is equal to the rate of heat transfer into the control volume minus the rate of work transfer out of the control volume. The terms of Eq. 17 are interpreted below.

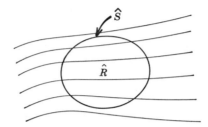

Fig. 3 Control volume.

1. *Internal Energy Flux*

In the flow processes treated in this text internal energy is stored in the gas by thermodynamic and mechanical means. Associated with the thermodynamic state of the gas is an intrinsic energy u as described in Section 8.2. For our purposes we may consider the intrinsic energy to be stored in the random molecular motion and other microscopic properties of the gas. From Eq. 4, assuming c_v constant

$$u = c_v T + u_0 \tag{18}$$

The constant u_0 is added because the zero level of intrinsic energy is undefined.

The mechanical storage consists of kinetic energy of the ordered motion $V^2/2$ and potential energy $-gz$ arising from the position of the system within the gravitational field. The total internal energy is

$$e = c_v T + u_0 + \frac{V^2}{2} - gz \tag{19}$$

The value of e given by Eq. 19 is substituted into the left side of Eq. 17 and integrated around the closed contour $\hat{S}$. The constant u_0 can make no contribution to the integral, and in the problems we are concerned with gz changes so slightly over the contour $\hat{S}$ that its contribution is essentially zero. The left-hand side of Eq. 17 becomes

$$\iint_{\hat{S}} \left(\frac{V^2}{2} + c_v T \right) \rho \mathbf{V} \cdot \mathbf{n} \, d\hat{S} \tag{20}$$

2. *Rate of Heat Transfer*

If a temperature gradient exists across the control surface there will be a rate of heat transfer that is easily expressed in integral form. This has been done in Appendix B, Section 6. Heat may also be transferred to the control volume at a specific rate by a combustion process. In this chapter the total rate of heat addition to the fluid inside the control surface will be designated $M\dot{q}$ where M is the mass within the control volume and q is the heat transferred per unit mass.

The thermal conductivity of air is small and unless temperature gradients are large, the heat transfer can frequently be considered negligible. This circumstance arises when the control surface lies outside regions of viscous dissipation like the boundary layer of bodies or shock waves. If the rate of heat transfer across the control surface is zero, the flow is *adiabatic*.

3. *Work Rate*

Pressure and shearing stresses at the control surface boundary lead to a work rate that has been developed in detail in Appendix B. There may also be a work rate through the control surface if a propeller, turbine, or other device is present. Work of this nature will be called machine work w_m and the work rate is given the symbol $M\dot{w}_m$.

In many applications involving viscous dissipation it is possible to choose the control surface in such a way that the surface shearing stresses can be neglected. This leads to a great simplification in formulating the work rate due to surface stresses. Under these conditions, the surface stress is entirely pressure and the work rate for an increment of surface $d\hat{S}$ is $p\mathbf{V} \cdot \mathbf{n}\, d\hat{S}$, or for the entire control surface

$$\iint_{\hat{S}} \frac{p}{\rho} \rho \mathbf{V} \cdot \mathbf{n}\, d\hat{S} \tag{21}$$

The steady flow energy equation applied to a fixed control volume is written by substituting expressions 20 and 21 in Eq. 17.

$$\iint_{\hat{S}} \left(\frac{V^2}{2} + c_v T + \frac{p}{\rho} \right) \rho \mathbf{V} \cdot \mathbf{n}\, d\hat{S} = \dot{Q} - \dot{W}_m \tag{22}$$

$\dot{Q}$ and $\dot{W}_m$ represent rates of heat and work transfer respectively.

It is understood that the control surface must be chosen in such a way that the surface shearing stresses may be neglected. Using the equation of state $p/\rho = RT$ and the relation $c_v + R = c_p$, the energy equation can be simplified further to read

$$\iint_{\hat{S}} \left(\frac{V^2}{2} + c_p T \right) \rho \mathbf{V} \cdot \mathbf{n}\, d\hat{S} = \dot{Q} - \dot{W}_m \tag{23}$$

Three applications of Eq. 23 follow.

Example 1

A turbo machine is indicated schematically in Fig. 4. Air enters the machine at 288°K with a speed of 150 m/sec. The specific enthalpy drop across the machine is 6000 Nm/kg and the exhaust velocity is 300 m/sec. Heat is added at the rate of 50,000 Nm/kg. How much work per kilogram of air passing through the machine is delivered to the compressor?

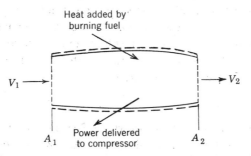

Fig. 4 Application of energy equation to a flow with heat and work transfer.

Solution

Choose a control surface indicated by the dotted line in Fig. 4 and assume the exit and entrance conditions are steady and uniform over the areas A_1 and A_2, respectively. Equation 23 for this case may be written:

$$\left(\frac{V^2}{2} + c_p T\right)_2 (\rho VA)_2 - \left(\frac{V^2}{2} + c_p T\right)_1 (\rho VA)_1 = \dot{Q} - \dot{W}_m$$

From conservation of mass the flux into the machine $(\rho VA)_1$ is equal to the mass flux out $(\rho VA)_2$. Dividing through by the mass flux and rearranging:

$$\frac{\dot{W}_m}{\rho VA} = \frac{\dot{Q}}{\rho VA} + \left(\frac{V_1^2}{2} - \frac{V_2^2}{2}\right) + c_p(T_1 - T_2)$$

The left-hand side of the equation is the work per unit mass.

$$\text{Work per unit mass} = 50,000 + \frac{150^2}{2} - \frac{300^2}{2} + 6000$$

$$= 22,250 \text{ Nm/kg}$$

Example 2

The shock wave at the nose of the wedge shown in Fig. 5 is a region in which viscous dissipation results in a change of flow properties. If the wedge is traveling through sea level air at a speed of 600 m/sec, and the temperature

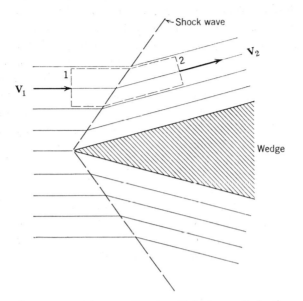

Fig. 5 Application of energy equation to a flow in which viscous dissipation occurs.

of the air behind the shock is $390°K$, what is the speed of the air behind the shock measured relative to the shock?

Solution

Let the axis of reference be attached to the wedge. Then sea level air appears to approach the shock at a uniform steady velocity of 600 m/sec. Choose a control surface whose lateral sides coincide with streamlines, as shown in Fig. 5. Assuming viscosity effects to be small everywhere except at the shock itself, there will be no heat conducted across the control surface and Eq. 23 may be written

$$\left(\frac{V^2}{2} + c_p T\right)_1 (\rho V A)_1 = \left(\frac{V^2}{2} + c_p T\right)_2 (\rho V A)_2$$

Using the conservation of mass principle and rearranging:

$$\frac{V_2^2}{2} = c_p(T_1 - T_2) + \frac{V_1^2}{2}$$

or, after substituting in the numbers

$$V_2 = 394 \text{ m/sec}$$

where V_2 is an average value across A_2. Because of the two-dimensional nature of the problem, V_2 has the same value no matter where the control surface is chosen. Therefore, it is correct to identify V_1 as the velocity upstream of the shock and V_2 the velocity downstream of the shock.

An essential feature of the application of Eq. 23 illustrated above is the fact that viscous dissipation within the control volume is permissible. It is only required that shearing stresses and heat transfer across the control surface be zero. Within the boundary layer, where viscous dissipation is general, a control surface that satisfies these requirements cannot be drawn. On the other hand, in flows containing isolated regions of viscous dissipation, a control surface approximately satisfying the above requirements is easily drawn.

Example 3

As a final example of the application of Eq. 23, consider a flow in which viscous dissipation and heat conduction are everywhere zero. Such a flow is approximated by the region outside the boundary layer of a body moving through air at constant speed, as shown in Fig. 6. The control surface in the

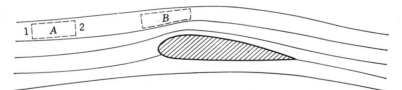

Fig. 6 Application of energy equation to reversible flow.

figure has equal values of $c_p T + V^2/2$ over faces 1 and 2. Because the control surface could have been drawn anywhere on the stream tube, for example at B, it follows that $c_p T + V^2/2$ *has the same value at every point on the stream tube, or in the limit, on a streamline.* This result could have been obtained in a mathematical fashion by applying the divergence theorem to Eq. 23.

Next, the question of variation of $c_p T + V^2/2$ from streamline to streamline can be settled by observing that the flow at infinity has uniform temperature and speed. Therefore, the sum must have the same value for each streamline at infinity. Since there can be no variation along streamlines, it follows that

$$c_p T + \frac{V^2}{2} = \text{constant} \tag{24}$$

is satisfied *at every point in the flow.*

Flows in which the viscous dissipation and heat conduction are everywhere zero are in the category of reversible processes, described in the next section.

8.5 Reversibility

Let a system of particles of fixed identity exchange energy with its surroundings. If the process can be reversed so that the system of particles and its surroundings are restored in all respects to their initial conditions, the exchange process is said to be reversible.

For example, consider the process shown schematically in Fig. 7. Let heat flow from surroundings A to system B. As a consequence, system B does work on surroundings C. If the work can be recovered from C to pump the heat back from B to A in such a manner that A, B, and C all return to their original conditions, the process is reversible.

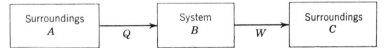

Fig. 7 Schematic diagram of thermodynamic system and surroundings.

The possibility of devising a process of the type outlined above cannot be settled by a theoretical proof. It is a fact of experience that a reversible process *has never been devised*, and this empirical observation is embodied in the second law of thermodynamics which is stated in Section 8.6.

All fluid flow processes are irreversible as a result of viscosity effects and heat conduction. For example, consider the system of particles of fixed identity contained in the control volume $\hat{R}_1$ of Fig. 8.

Viscous dissipation resulting from shearing stresses at the surface and heat conduction through the surface $\hat{S}_1$ are responsible for the fact that the system and surrounding flow can never, after an elapsed interval, return to its original condition. This is true no matter how small the volume $\hat{R}_1$. Therefore, it is true for every vanishingly small subsystem within $\hat{R}_1$. Then it may be said, if there are any viscosity or heat transfer effects *within or on* the surface $\hat{S}_1$, the system will undergo an irreversible change of state.

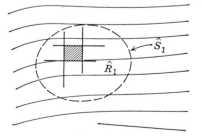

Fig. 8 Control volume.

If the control surface of Fig. 8 lies outside the boundary layer and contains no shock wave the requirement for reversibility is very nearly met. Under these conditions, special relations exist among the thermodynamic variables p, ρ, and T that are useful in the following chapters. They may be derived from the momentum and energy equations in the following manner:

The energy equation (24) is satisfied at every point in an adiabatic reversible flow. Therefore, its differential form may be written

$$c_p \, dT + V \, dV = 0 \tag{25}$$

The first term is the differential of the enthalpy which may be written from Eqs. 5 and 4 as

$$c_p \, dT = p \, dv + v \, dp + c_v \, dT \tag{26}$$

Equations 25 and 26 give the result

$$p \, dv + c_v \, dT + \frac{dp}{\rho} + d\,\frac{V^2}{2} = 0 \tag{27}$$

The differential form of the equilibrium equation for a nonviscous fluid (gravity neglected) is given by Eq. 11 of Chapter 7.

$$d\,\frac{V^2}{2} + \frac{dp}{\rho} = 0$$

Therefore, the sum of the first two terms of Eq. 27 are independently zero. Replacing p with its equivalent from the equation of state and dividing through by T leads to

$$R\,\frac{dv}{v} + c_v\,\frac{dT}{T} = 0$$

$$d \ln (v^R T^{c_v}) = 0 \tag{28}$$

$$\frac{T^{c_v}}{\rho^R} = \text{constant}$$

Using Eqs. 8 and 9, which relate the gas constant to the specific heats, the following relations may be deduced from Eq. 28.

$$\frac{T}{\rho^{\gamma-1}} = C_1$$

$$\frac{p}{\rho^\gamma} = C_2 \tag{29}$$

$$\frac{T}{p^{(\gamma-1)/\gamma}} = C_3$$

Equations 29 hold for *adiabatic reversible* flows.

8.6   Second Law of Thermodynamics

The irreversibility of real fluid flow processes suggests that there are limitations on the direction in which energy exchanges can take place. The second law of thermodynamics defines the direction in which a state change can occur, and when formulated in terms of *entropy* it provides a quantitative measure of the degree of irreversibility.

Consider a system of particles of fixed identity enclosed by the surface $\hat{S}_1$. According to the first law (Section 8.3)

$$\delta q = de + \delta w \qquad (30)$$

Internal energy depends on the state of the gas, but the heat and work transfers are not characteristics of the gas. If the equation is divided by the temperature we shall show that, under specialized circumstances, the ratio $\delta q/T$ is the differential of a quantity that depends only on the state of the gas. This quantity is called the *entropy*. The statement may be demonstrated by considering the system of fluid particles within the surface $\hat{S}_1$ along the stream tube shown in Fig. 9. The work

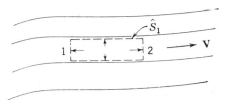

Fig. 9 Control surface.

transfer across the surface arises in general from machine work, w_m, and the work done by the surface pressure and shearing stresses. We assume w_m to be zero and the fluid to be nonviscous, so that only normal stresses exist at the surface. The work of the normal stresses can be written in simple form. The pressures on the lateral faces are normal to the flow direction and therefore do no work. At face 2, the pressure stress is in the direction of the displacement and therefore positive work is done at the rate $p_2 A_2 V_2$ (work/sec). The flow rate at face 2 is $\rho_2 V_2 A_2$ (mass/sec). Therefore the work per unit mass at face 2 is $+p_2/\rho_2$. Similarly, the work per unit mass at face 1 of the boundary is $-p_1/\rho_1$. For the entire boundary, the work transfer at the surface is the sum of these two terms, which in the limit for small amounts of work is written

$$\delta w = \frac{p_2}{\rho_2} - \frac{p_1}{\rho_1} = d\frac{p}{\rho} \qquad (31)$$

The differential of internal energy is written from Eq. 19 (gravity neglected) for an ideal gas,

$$de = c_v \, dT + d\frac{V^2}{2} \qquad (32)$$

Substituting Eqs. 31 and 32 into 30 and dividing through by T yields

$$\frac{\delta q}{T} = \frac{1}{T}\left(d\,\frac{V^2}{2} + \frac{dp}{\rho}\right) + c_v\,\frac{dT}{T} - \frac{p}{\rho T}\frac{d\rho}{\rho}$$

The term in the parentheses must vanish in a steady nonviscous flow and the remainder can be written

$$\frac{\delta q}{T} = d(\ln T^{c_v} - \ln \rho^R) \tag{33}$$

Thus, for an ideal gas, $\delta q/T$ becomes an exact differential of the thermodynamic variables T and ρ.

The assumption that allowed the work term to be written in differential form was precisely the assumption of reversibility. Therefore $\delta q/T$ is an exact differential of state variables *only* if the flow is reversible. Under these conditions, $\delta q/T$ is defined as the differential of the specific entropy.

$$dS = \left(\frac{\delta q}{T}\right)_{\text{rev.}}$$

$$\tag{34}$$

$$S_2 - S_1 = \int_1^2 \left(\frac{\delta q}{T}\right)_{\text{rev.}}$$

Like internal energy, the entropy has no absolute base, and therefore we can give values only to entropy changes unless an arbitrary base is assigned.

From Eqs. 33 and 34, the change in entropy ΔS between states 1 and 2 is

$$\Delta S = \int_1^2 \left(\frac{\delta q}{T}\right)_{\text{rev.}} = \ln\left[\left(\frac{T_2}{T_1}\right)^{c_v}\left(\frac{\rho_1}{\rho_2}\right)^R\right] \tag{35}$$

The second law of thermodynamics is a completely independent physical principle and is one of the empirical laws on which aerodynamics is based. It may be stated in terms of entropy change in the following fashion:

$$\Delta S - \int_1^2 \frac{\delta q}{T} = \int_1^2 \left(\frac{\delta q}{T}\right)_{\text{rev.}} - \int_1^2 \frac{\delta q}{T} \geqslant 0 \tag{36}$$

Let a system of particles at state 1 exchange energy with its surroundings and pass into state 2. From the initial and final states, ΔS can be computed using Eq. 35. Its value will always be equal to or greater than $\int_1^2 \delta q/T$. The equal sign corresponds to a reversible process.

For adiabatic processes, the entropy rule reduces to

$$\Delta S \geqslant 0 \tag{37}$$

If the process is adiabatic and reversible

$$\Delta S = 0 \tag{38}$$

Such processes are called *isentropic*.

8.7 Bernoulli's Equation for Isentropic Compressible Flow

At each point in an isentropic flow, the energy relation given by Eq. 24 must be satisfied.

$$c_p T + \frac{V^2}{2} = c_p T_1 + \frac{V_1^2}{2} \tag{39}$$

Station 1 is a position at which all flow properties are assumed known. From the last of Eqs. 29

$$\frac{T}{p^{(\gamma-1)/\gamma}} = \frac{T_1}{p_1^{(\gamma-1)/\gamma}} \tag{40}$$

Substituting Eq. 40 in Eq. 39 and replacing T_1 with $p_1/\rho_1 R$ from the equation of state:

$$\left(\frac{c_p}{R} \frac{p_1^{1/\gamma}}{\rho_1}\right) p^{(\gamma-1)/\gamma} + \frac{V^2}{2} = \frac{c_p}{R} \frac{p_1}{\rho_1} + \frac{V_1^2}{2}$$

Finally, from Eqs. 8 and 9,

$$\frac{c_p}{R} = \frac{c_p}{c_p - c_v} = \frac{\gamma}{\gamma - 1} \tag{41}$$

and the energy equation in terms of p and V becomes

$$\left(\frac{\gamma}{\gamma - 1} \frac{p_1^{1/\gamma}}{\rho_1}\right) p^{(\gamma-1)/\gamma} + \frac{V^2}{2} = \frac{\gamma}{\gamma - 1} \frac{p_1}{\rho_1} + \frac{V_1^2}{2} \tag{42}$$

Equation 42 is called Bernoulli's equation for compressible flow, and, unlike its incompressible counterpart, it cannot be deduced from the conservation of momentum principle alone. It will be recalled that *Eq. 39 is valid for adiabatic flows whereas Eq. 42 has the additional restriction of isentropy, imposed by the use of Eq. 40.* For this reason, Eq. 39 is frequently referred to as the *strong* form of the energy equation and Eq. 42 as the *weak* form.

Equation 42 can also be obtained by integrating the differential form of the momentum equation *provided* we use the relation between p and ρ for isentropic flow (Eqs. 29),

$$\frac{p}{\rho^\gamma} = \frac{p_1}{\rho_1^\gamma} \tag{43}$$

But Eq. 43 was derived from the first law of thermodynamics. Therefore, its use in integrating the momentum equation makes the result a consequence of the first law. The derivation is carried out starting with Eq. 7.11, gravity omitted.

$$d\,\frac{V^2}{2} + \frac{dp}{\rho} = 0 \tag{44}$$

From Eq. 43,

$$\int \frac{dp}{\rho} = \left(\frac{\gamma}{\gamma - 1} \frac{p_1^{1/\gamma}}{\rho_1} \right) p^{(\gamma-1)/\gamma} \tag{45}$$

Substituting Eq. 45 in Eq. 44 and integrating

$$\frac{V^2}{2} + \left(\frac{\gamma}{\gamma - 1} \frac{p_1^{1/\gamma}}{\rho_1} \right) p^{(\gamma-1)/\gamma} = \frac{V_1^2}{2} + \frac{\gamma}{\gamma - 1} \frac{p_1}{\rho_1} \tag{46}$$

Equations 46 and 42 are identical.

8.8   Static and Stagnation Values

The pressure, density, and temperature of a gas at rest were discussed in Section 1.3. The same interpretation of these properties applies to a gas in motion providing the measuring instrument moves with the gas. For example, in Eq. 39, in order to measure T, the thermometer must move with the speed V. Relative to the thermometer, the gas is at rest and the temperature recorded will be the stream or *static* value. The same argument applies to the pressure in Eq. 42. Whenever the terms temperature, pressure, or density are used without a modifying adjective, the *static* value is implied.

To arrive at the meaning of *stagnation* value, we must visualize a hypothetical flow in which the ordered velocity is reduced to zero isentropically. The position of zero ordered velocity is called a stagnation point and the values of the gas characteristics at a stagnation point are called *stagnation* values. Stagnation values will be given a subscript or superscript (0). If station 1 in Eqs. 39 and 42 is a stagnation point, then the two equations would be written

$$c_p T + \frac{V^2}{2} = c_p T_0 \tag{47}$$

$$\left(\frac{\gamma}{\gamma - 1} \frac{p_0}{\rho_0} \right) \left(\frac{p}{p_0} \right)^{(\gamma-1)/\gamma} + \frac{V^2}{2} = \frac{\gamma}{\gamma - 1} \frac{p_0}{\rho_0} \tag{48}$$

The stagnation pressure, density, and temperature in an isentropic flow have the same values at every point in the flow. In a nonisentropic flow, the stagnation values vary from point to point.

The restriction of isentropy for constant *stagnation temperature* T_0 may be lightened from the following consideration. Equation 39 applies to adiabatic flows with viscous dissipation provided T, V, T_1, and V_1 are interpreted as average values over a properly constructed control surface, as explained in Section 8.4. Therefore, it follows that Eq. 47 applies to adiabatic irreversible flows with the terms interpreted as averages.

Example 4

Find the stagnation temperature, pressure, and density at every point in the flow of Example 2, Section 8.4. Assume an entropy rise across the shock of 26 Nm/kg°K.

Solution

Only two points need to be considered because the entropy, and therefore the stagnation values, change only across the shock. Upstream of the shock:

$$V_1 = 600 \text{ m/sec}$$

$$p_1 = 1.013 \times 10^5 \text{ N/m}^2$$

$$\rho_1 = 1.226 \text{ kg/m}^3$$

$$T_1 = 288° \text{ K}$$

From Eq. 47,* $\qquad T_1^0 = T_1 + \dfrac{V_1^2}{2c_p} = 468° \text{ K}$

From Eqs. 29, $\qquad \dfrac{T_1}{\rho_1^{\gamma-1}} = \dfrac{T_1^0}{(\rho_1^0)^{\gamma-1}}$

$$\rho_1^0 = \rho_1 \left(\frac{T_1^0}{T_1}\right)^{1/(\gamma-1)} = 4.126 \text{ kg/m}^3$$

$$p_1^0 = \rho_1^0 R T_1^0 = 5.54 \times 10^5 \text{ N/m}^2$$

Downstream of the shock:

$$T_2 = 390° \text{ K}$$

$$T_2^0 = T_1^0 = 468° \text{ K}$$

$$\Delta S = 26 \text{ Nm/kg°K}$$

*When subscripts are used to denote positions in the fluid, the symbol 0 indicating stagnation value is appended as a superscript.

From Eq. 35, $\rho_2 = \rho_1 \left(\dfrac{T_2}{T_1}\right)^{c_v/R} \dfrac{1}{\exp \dfrac{\Delta S}{R}} = 2.39 \text{ kg/m}^3$

From Eqs. 29, $\rho_2^0 = \rho_2 \left(\dfrac{T_2^0}{T_2}\right)^{1/(\gamma-1)} = 3.77 \text{ kg/m}^3$

From Eq. 2, $p_2^0 = \rho_2^0 R T_2^0 = 5.063 \times 10^5 \text{ N/m}^2$

CHAPTER 9

Some Applications of One-Dimensional Compressible Flow

9.1 Introduction

In one-dimensional flow theory, the ordered velocity is assumed to be unidirectional, and the equations that describe the flow may be written as though the flow consists of one component only. As a result, the theory is simplified greatly, and it is possible to deduce relatively simple relations that expose the fundamental nature of compressible flow.

Many important flow phenomena approximate one-dimensional flow closely enough to make the theory developed here applicable. In the first two sections, the speed of sound is defined and the isentropic flow parameters are expressed in terms of Mach number. Application is made to the basic one-dimensional flow represented by the Laval nozzle in Section 9.6. The last two sections deal with the influence of friction and heat addition in one-dimensional flows.

9.2 Speed of Sound

Consider the propagation of a disturbance of infinitesimal proportions through a fluid that is at rest. The geometry of the disturbance, that is, whether it is plane or spherical, does not influence the following argument. It will be shown in Chapter 10, that, because the disturbance is infinitesimal, the compression or expansion of the fluid as the disturbance passes through it is accomplished reversibly. The speed with which such a disturbance travels through a fluid is established by the properties of the fluid. Because a sustained sound is simply a succession of small disturbance of this nature, the speed with which a small disturbance travels is called the speed of sound. In contrast, a large disturbance such as an explosion wave does not compress the fluid reversibly as it passes through the fluid, and the speed of propagation is very different from that of sound. The speed of propagation of large disturbances is treated in the next chapter.

The speed with which a spherical sound wave is propagated through a fluid is the same as that with which a plane sound wave is propagated. For simplicity, the plane wave is treated. Consider a portion of a plane wave contained within the cylinder shown in Fig. 1. Let the wave be traveling to the left with a velocity u, and let it be required to find the relation between the speed of propagation u and the properties of the fluid. In order to make the flow steady, attach the axis of reference to the wave so that the wave is stationary and the fluid passes through it from left to right. In the steady isentropic flow through a tube of constant cross section, a continuous change in the properties of the fluid does not occur because nothing about the system will promote such a change. However, a discontinuity in the flow properties may be premised; the mechanism by which a discontinuity develops is discussed in the next chapter. The point of view is taken that the wave is a region at which discontinuities in the flow properties occur. Upstream of the wave, the fluid velocity, pressure, and density have the constant values u, p, and ρ, respectively. Downstream of the wave the properties are also constant and equal to $u + \Delta u$, $p + \Delta p$, and $\rho + \Delta \rho$. The properties upstream and downstream of the wave

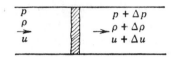

Fig. 1 Plane wave in a tube.

are related by the equations of *continuity* and *momentum* which, when applied to the example illustrated in Fig. 1, become

$$\rho u = (\rho + \Delta \rho)(u + \Delta u)$$

$$p - (p + \Delta p) = (\rho + \Delta \rho)(u + \Delta u)^2 - \rho u^2 \tag{1}$$

Upon expanding, and using the *continuity* relation in the *momentum* equation, Eqs. 1 become

$$\frac{\Delta u}{u} = -\frac{\Delta \rho}{\rho} - \frac{\Delta \rho \Delta u}{\rho u}$$

$$u \Delta u = -\frac{\Delta p}{\rho}$$

If the second of the above equations is divided by the first, there results

$$u^2 = \frac{-\Delta p}{-\Delta \rho - \Delta \rho (\Delta u/u)} \tag{2}$$

For very small discontinuities, u^2 approaches the value of the ratio of the change in pressure across the wave to the change in density. In the limit, u becomes the velocity of sound a and we have from Eq. 2

$$a = \sqrt{\frac{dp}{d\rho}} \tag{3}$$

Because the process is isentropic, the density and pressure are related by

$$\frac{p}{\rho^\gamma} = c$$

from which

$$a^2 = \frac{dp}{d\rho} = \gamma \frac{p}{\rho} \tag{4}$$

From Eqs. 3 and 4,

$$a = \sqrt{\gamma p / \rho} \tag{5}$$

and, by using the equation of state,

$$a = \sqrt{\gamma R T} \tag{6}$$

Equation 6 shows that the speed of sound is a function only of the temperature, that is, of the intrinsic energy of the gas. For air, Eq. 6 becomes

$$a = 20.04 \sqrt{T} \ \text{m/sec} \tag{7}$$

where T is in degrees Kelvin.

The above formulas give the speed of sound at a given point in a steady flow, even though the flow may be nonisentropic; the restriction is however that *the speed is that of a small-amplitude wave, as distinct from that of a shock wave*, to be analyzed in Section 10.5.

In the atmosphere, since the temperature decreases with height (up to 11 km altitude) so does the speed of sound. (See Fig. 1.2 and Table 3 at the end of the book.)

The variation of the speed of sound with velocity in an adiabatic flow may be seen by using the energy equation developed in the last chapter:

$$c_p T + \tfrac{1}{2} V^2 = c_p T_0 \tag{8}$$

After replacing T and T_0 with $a^2/\gamma R$ and $a_0^2/\gamma R$, respectively, and using the relation

$$\frac{c_p}{\gamma R} = \frac{c_p}{\gamma(c_p - c_v)} = \frac{1}{\gamma - 1}$$

Eq. 8 becomes

$$\frac{V^2}{2} + \frac{a^2}{\gamma - 1} = \frac{a_0^2}{\gamma - 1} \tag{9}$$

The local speed of sound decreases from a_0 at $V = 0$ to zero at the speed

$$V_m = \sqrt{\frac{2}{\gamma - 1}} \, a_0 \tag{10}$$

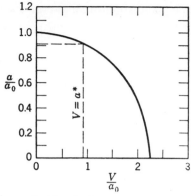

Fig. 2 Variation of speed of sound with velocity.

V_m is, therefore, the speed that is approached as the temperature approaches absolute zero. A plot of a/a_0 versus V/a_0 for air, where $\gamma = 1.4$, is shown in Fig. 2. Setting $V = a^*$ in Eq. 9, we see that for a speed equal to the speed of sound

$$a^* = \sqrt{\frac{2}{\gamma + 1}} \, a_0 = 0.913 a_0 \tag{11}$$

for air; that is, a^*, which is the speed of sound corresponding to a Mach number of unity, bears a definite relation to a_0. It is often convenient to express the right side of Eq. 9 in terms of a^* instead of a_0. Thus

$$\frac{V^2}{2} + \frac{a^2}{\gamma - 1} = \frac{a_0^2}{\gamma - 1} = \frac{\gamma + 1}{2(\gamma - 1)} a^{*2} = \frac{V_m^2}{2} \tag{12}$$

It is noteworthy, for analyses that follow, that Eq. 12 is not restricted to isentropic flows. It applies as well to steady adiabatic irreversible flow.

9.3 Flow Relations in Terms of Mach Number

It was stated in Chapter 1 that if pressure changes within a flow are great enough the elasticity of the fluid must be considered, together with the viscosity and the density, as physical properties entering into the derivation of similarity parameters governing the flow. The elasticity is defined as the change in pressure per unit change in specific volume, that is, $\rho\,dp/d\rho$. Hence by the analysis of Section 9.2 the elasticity is given by ρa^2, and so we may take a the speed of sound as a physical property, which along with the density determines the elasticity. The dimensional analysis of Appendix A led to the definition of the Mach number.

$$M = \frac{V}{a} \tag{13}$$

The importance of Mach number and γ in determining the physical properties of steady compressible flows can be demonstrated by means of Eq. 12 for adiabatic flow and by Eq. 8.48 for isentropic flow; the latter may be expressed in the form

$$\left(\frac{p}{p_0}\right)^{(\gamma-1)/\gamma} = 1 - \frac{\gamma-1}{2} V^2 \frac{\rho_0}{\gamma p_0} = 1 - \frac{\gamma-1}{2} \frac{V^2}{a_0^2} \tag{14}$$

where $a_0^2 = \gamma p_0/\rho_0$ is the speed of sound at a stagnation point. To put this equation in terms of the Mach number we multiply and divide the last term by a^2, and since the term a^2/a_0^2 can be written

$$\frac{a^2}{a_0^2} = \frac{T}{T_0} = \frac{p}{p_0}\frac{\rho_0}{\rho} = \left(\frac{p}{p_0}\right)^{(\gamma-1)/\gamma} \tag{15}$$

In these equations, since a can be expressed as a function only of the temperature the first dependence in Eqs. 15 is not restricted to isentropic flow; however, the last dependence, in terms of p/p_0, requires the use of the isentropic relation between p and ρ, and it is therefore restricted to isentropic flow.

Then the *isentropic* relation, Eq. 14, becomes

$$\frac{p}{p_0} = \left(1 + \frac{\gamma-1}{2} M^2\right)^{-\gamma/(\gamma-1)} \tag{16}$$

A relation, applicable to any *adiabatic* flow, follows after dividing the first of Eqs. 12 by a^2:

$$\frac{\gamma-1}{2} M^2 + 1 = \frac{a_0^2}{a^2} \tag{17}$$

or in terms of the temperature ratio.

$$\frac{T}{T_0} = \left(1 + \frac{\gamma-1}{2} M^2\right)^{-1} \tag{18}$$

An *isentropic* relation follows after dividing Eq. 16 by Eq. 18:

$$\frac{\rho}{\rho_0} = \frac{p/RT}{p_0/RT_0} = \frac{pT_0}{p_0 T} = \left(1 + \frac{\gamma-1}{2} M^2\right)^{-1/(\gamma-1)} \tag{19}$$

Another useful *adiabatic* relation is

$$\frac{p_0 - p}{\frac{1}{2}\rho V^2} = 2 \frac{p}{\rho V^2}\left(\frac{p_0}{p} - 1\right) = \frac{2}{\gamma M^2}\left(\frac{p_0}{p} - 1\right) \tag{20}$$

By use of Eq. 16 the *isentropic* relation is

$$\frac{p_0 - p}{\frac{1}{2}\rho V^2} = \frac{2}{\gamma M^2}\left[\left(1 + \frac{\gamma-1}{2} M^2\right)^{\gamma/(\gamma-1)} - 1\right] \tag{21}$$

This relation will be discussed in the next section.

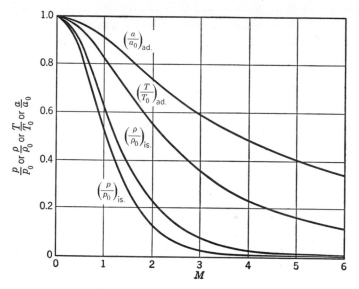

Fig. 3 Flow parameters versus Mach number for adiabatic (ad) and isentropic (is) flows.

Plots of Eqs. 16 through 19 are shown in Fig. 3 versus Mach number. Subscripts *is* and *ad* indicate the degree of generality of the relations.

The variables in the equations are M and γ. In aerodynamics, except at high temperatures, γ has a very nearly constant value of 1.40, so for the analyses that follow the only variable is the Mach number.

9.4 Measurement of Flight Speed (Subsonic)

Equations 21, the right side of which is unity for an incompressible fluid flow, provides a ready means for determining the error involved in neglecting compressibility in determining speed of flight or of gas flows in general. At Mach 0.5, for instance, the right side of the equation is equal to 1.064; thus, for a given density, neglect of the compressibility leads to an underestimation of the pressure by 6.4 percent or the velocity by 3.2 percent.

As the speed of flight increases, two main difficulties arise: (1) measurement of p and p_0 (as given by the pitot-static tube) now determines, according to Eq. 16, the Mach number instead of the airspeed; (2) the flow interference due to adjacent parts of the aircraft grows rapidly with the Mach number so that the *position error* also increases. If the Mach number is obtained by means of pitot tube measurements of p and p_0, the speed can be computed, if we can obtain the speed of sound. The easiest way to do this is to measure T_0 by a suitable stagnation temperature thermometer. Then, knowing the Mach number, compute the ambient temperature T by Eq. 18. The speed of flight then follows from the formula $V = Ma$.

Since compressibility effects on the aerodynamic characteristics are a function of Mach number, an indispensable instrument for high-speed aircraft is the *machmeter*. Equation 16 is the fundamental equation on which this device operates; p and p_0 are measured by means of a pitot tube, and through complicated linkages or a computer network the Mach number is displayed.

The above method is applicable only to the measurement of subsonic speeds. At supersonic speeds, a shock wave forms ahead of the pitot tube, and the isentropic formulas derived in the present chapter are not applicable.

9.5 Isentropic One-Dimensional Flow

In a channel with a small rate of change of cross section or between nearly parallel streamlines, the velocity components normal to the mean flow direction are small compared to the total velocity. As a consequence, the flow may be analyzed as though it were one-dimensional. This means that the flow parameters are constant in planes that are normal to the mean flow direction. In many practical problems, the conditions for one-dimensional flow are very nearly fulfilled.

Simple considerations of the equilibrium and continuity equations for isentropic flow lead to important conclusions regarding the nature of isentropic subsonic and

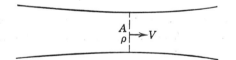

Fig. 4 Stream tube.

supersonic flows. The flow through the streamtube of Fig. 4 is such as to obey the continuity equation

$$\rho V A = \text{constant} \tag{22}$$

as long as the angle between the streamlines is small enough so that we may consider the flow substantially one-dimensional.

Taking the logarithmic derivative, Eq. 22 becomes

$$\frac{dV}{V} + \frac{d\rho}{\rho} + \frac{dA}{A} = 0 \tag{23}$$

The differential form of the equilibrium equation, gravity neglected, is given by Eq. 7.11

$$d\frac{V^2}{2} + \frac{dp}{\rho} = 0$$

and may be written

$$VdV + \frac{d\rho}{\rho}\frac{dp}{d\rho} = VdV + a^2\frac{d\rho}{\rho} = 0$$

By substituting for $d\rho/\rho$ from Eq. 23 we get

$$VdV - a^2\left(\frac{dV}{V} + \frac{dA}{A}\right) = 0$$

and, after dividing through by a^2 and collecting terms,

$$\frac{dV}{V}(M^2 - 1) = \frac{dA}{A}$$

After introducing ds, the element of length along a streamline, the above equation becomes

$$(M^2 - 1)\frac{1}{V}\frac{dV}{ds} = \frac{1}{A}\frac{dA}{ds} \qquad (24)$$

Three cases must be considered.

1. $M < 1$.

dV/ds and dA/ds are then opposite in sign. Hence a subsonic flow decelerates ($dV/ds < 0$) in an expanding channel ($dA/ds > 0$) and accelerates in a converging channel.

2. $M > 1$.

In this case, dA/ds and dV/ds have the same sign. Hence a supersonic flow accelerates in an expanding channel and decelerates in a converging channel. Thus the behavior of the supersonic airstream is opposite to that of a subsonic.

3. $M = 1$.

In this case, $dA/ds = 0$ or $dV/ds = \infty$. The latter condition is ruled out on physical grounds. Then we see that sonic flow is not possible in a converging or expanding channel. We can rule out the large section of a channel (see Fig. 5, station B) by remembering that a supersonic flow accelerates and a subsonic flow decelerates in an expanding channel; hence it could never reach sonic speed at the end of a divergence. We conclude, therefore, that *the sonic flow can occur only at a constriction* (throat), such as station A in Fig. 5.

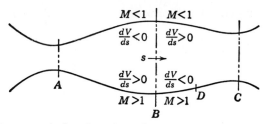

Fig. 5 Subsonic and supersonic flow in a channel.

9.6 The Laval Nozzle

We shall here develop the one-dimensional *Laval nozzle* equations for isentropic flow through channels of varying cross section. They follow simply from the continuity relation Eq. 22 and the relation between V and p/p_0 given by Eq. 14. Let m be the rate of mass flow. Then, from continuity,

$$m = \rho VA = \rho_0 VA \, \frac{\rho}{\rho_0} = \rho_0 VA \left(\frac{p}{p_0}\right)^{1/\gamma}$$

after substituting for V from Eq. 14

$$\frac{m}{A} = \sqrt{\frac{2\gamma}{\gamma - 1} p_0 \rho_0 \left(\frac{p}{p_0}\right)^{2/\gamma} \left[1 - \left(\frac{p}{p_0}\right)^{(\gamma-1)/\gamma}\right]} \tag{25}$$

This is known as the St. Venant equation.

It is conveninet to write Eq. 25 in terms of the ratio of the area at a point to the critical area, A^*, that is, the throat area corresponding to sonic speed for the given mass flow. Introducing $M = 1$ in Eq. 16, the *critical value* of the pressure ratio is

$$\left(\frac{p}{p_0}\right)^* = \left(\frac{\gamma + 1}{2}\right)^{-\gamma/(\gamma-1)}$$

Substituting this value in Eq. 25

$$\frac{m}{A^*} = \sqrt{\gamma \left(\frac{2}{\gamma + 1}\right)^{(\gamma+1)/(\gamma-1)} p_0 \rho_0}$$

whence

$$\left(\frac{A}{A^*}\right)^2 = \frac{\gamma - 1}{2} \frac{\left(\dfrac{2}{\gamma + 1}\right)^{(\gamma+1)/(\gamma-1)}}{\left[1 - \left(\dfrac{p}{p_0}\right)^{(\gamma-1)/\gamma}\right]\left(\dfrac{p}{p_0}\right)^{2/\gamma}} \tag{26}$$

Substituting for p/p_0 in terms of M from Eq. 16,

$$\left(\frac{A}{A^*}\right)^2 = \frac{1}{M^2}\left[\frac{2}{\gamma+1}\left(1 + \frac{\gamma-1}{2}M^2\right)\right]^{(\gamma+1)/(\gamma-1)} \tag{27}$$

These are the fundamental equations connecting pressure ratio and Mach number with area ratio. They are plotted in Fig. 6 and the numerical data are tabulated in

Fig. 6 Area ratio versus pressure ratio and Mach number.

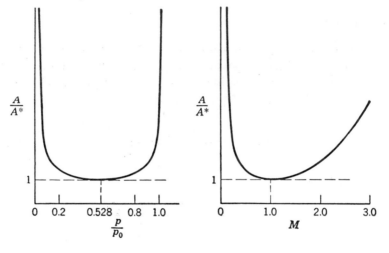

Tables 4 and 5 at the end of the book. Several features of these curves will be noted. The curves consist of a subsonic and a supersonic branch, so that A/A^* is double-valued except at $M = 1$ and at

$$\frac{p}{p_0} = \left(\frac{2}{\gamma+1}\right)^{\gamma/(\gamma-1)} = 0.528$$

(for air). Since isentropic flow must follow these curves, in order to obtain subsonic flow in one part of the channel and supersonic in another, a throat of area A^* must intervene. If a throat of area A^* does not occur in the channel the flow remains supersonic or subsonic throughout.

Downstream of a throat of area A^*, the flow can either accelerate supersonically or decelerate subsonically, depending on the pressure ratio at the exit of the channel. This fact is demonstrated by the curves of Fig. 7. Curves 1 and 4 show completely subsonic and completely supersonic flows, respectively; the area does not reach A^* for the mass flows involved in these cases. For the remaining curve, which branches at the throat into a subsonic and a supersonic branch ending at 2 and 3, respectively, the throat area has the critical value A^*. Curves 1, 2, and 3 are possible flows from a reservoir for the end pressure ratios indicated on the right. For 2

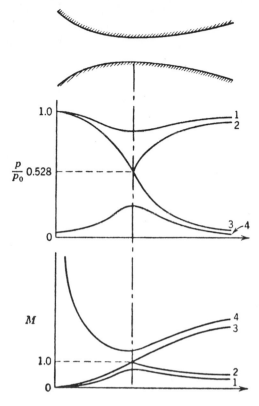

Fig. 7 Laval nozzle.

and 3, the maximum speed at the throat is sonic.   Flow 4 must have a throat of area A^* somewhere upstream, since otherwise supersonic flow would never have been reached.   It is clear that exit pressure ratios in the range between points 2 and 3 cannot be obtained isentropically.   It will be shown in the next chapter that these flows involve shock waves.

9.7   One-Dimensional Flow with Friction and Heat Addition

Extensions of the analysis of Section 9.5 lead to an indication of the effects of friction and heat addition on the one-dimensional flow of a compressible fluid.   In Fig. 8 is indicated a portion of stream tube in a one-dimensional steady flow.   The dotted line represents a fixed control surface on which are acting pressure and shearing stresses as indicated.   The shearing stress is assumed to act only on the lateral faces and has an average value τ.   The area over which the shear stress acts is f.   The pressure stress on the lateral faces is taken to be the average of the pressure stresses on the normal faces.   The heat per unit mass added between stations 1 and 2 is q.   The equations that determine the flow are given below.

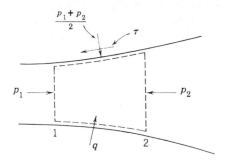

Fig. 8 Stream tube in one-dimensional steady flow.

1. *Conservation of Mass*

$$\rho_2 V_2 A_2 - \rho_1 V_1 A_1 = 0$$

$$\Delta(\rho V A) = 0$$

2. *Conservation of Momentum*

$$\rho_2 V_2^2 A_2 - \rho_1 V_1^2 A_1 = p_1 A_1 - p_2 A_2 - \tau f + \left(\frac{p_1 + p_2}{2}\right)(A_2 - A_1)$$

$$\rho V A \Delta V = -\Delta(pA) - \tau f + \frac{p_1 + p_2}{2} \Delta A$$

3. *Conservation of Energy*

$$\left(c_p T_2 + \frac{V_2^2}{2}\right) - \left(c_p T_1 + \frac{V_1^2}{2}\right) = q$$

$$\Delta\left(c_p T + \frac{V^2}{2}\right) = q$$

4. *Equation of State*

$$\frac{p_2}{\rho_2 T_2} = \frac{p_1}{\rho_1 T_1}$$

$$\Delta\left(\frac{p}{\rho T}\right) = 0$$

Ignoring higher order effects, these four equations may be expanded to read

$$\frac{\Delta \rho}{\rho} = -\frac{\Delta V}{V} - \frac{\Delta A}{A} \qquad (28)$$

$$\frac{\Delta p}{\rho} = - V\Delta V - \frac{\tau f}{\rho A} \qquad (29)$$

$$\frac{\Delta T}{T} = \frac{q}{c_p T} - \frac{V\Delta V}{c_p T} \qquad (30)$$

$$\frac{\Delta \rho}{\rho} = \frac{\Delta p}{p} - \frac{\Delta T}{T} \qquad (31)$$

Equation 29 can be written

$$\frac{p}{\rho}\frac{\Delta p}{p} = - V\Delta V - \frac{\tau f}{\rho A}$$

and using the relation $a^2 = \gamma p/\rho$,

$$\frac{\Delta p}{p} = - V\Delta V\frac{\gamma}{a^2} - \frac{\gamma}{a^2}\frac{\tau f}{\rho A} \qquad (32)$$

Equating expressions 28 and 31 and eliminating $\Delta T/T$ and $\Delta p/p$ with Eqs. 30 and 32, respectively, leads to

$$-\frac{\gamma}{a^2}V\Delta V - \frac{\gamma}{a^2}\frac{\tau f}{\rho A} - \frac{q}{c_p T} + \frac{V\Delta V}{c_p T} = -\frac{\Delta V}{V} - \frac{\Delta A}{A}$$

$$\frac{\Delta V}{V}\left(-\gamma M^2 + 1 + \frac{V^2}{c_p T}\right) = \frac{\gamma}{a^2}\frac{\tau f}{\rho A} + \frac{q}{c_p T} - \frac{\Delta A}{A} \qquad (33)$$

Using the relation $T = a^2/\gamma R$ in the left-hand side of Eq. 33 leads to the final result

$$\frac{\Delta V}{V} = \frac{1}{M^2 - 1}\left(\frac{\Delta A}{A} - \frac{q}{c_p T} - \frac{\gamma\tau f}{\rho A a^2}\right) \qquad (34)$$

This equation shows that the convergence of the stream tube ($\Delta A < 0$), friction, and heat addition cause acceleration of a subsonic flow and deceleration of a supersonic flow. Further, in the presence of heat addition or friction, sonic flow is reached in a slightly divergent channel, rather than at the throat (see Bailey, 1944).

The postulation of one-dimensional viscous flow is to a certain extent anomalous, since in channel flow friction acts at the walls and destroys the one-dimensional nature of the flow. No difficulty arises, however, if we assume fully developed flow in the sense that velocity profiles remain similar so that cross components of the velocity do not exist; then each elementary stream tube is subject to the above analysis. The measurements of Keenan and Neumann (1945) and of Froessel (1938) indicate that departures from one-dimensionality are not of great importance in fully developed high-speed flow through a tube.

9.8   Heat Addition to a Constant Area Duct

Quantitative relations for the flow changes across a frictionless constant-area duct resulting from the introduction of heat are of interest in the study of aero-thermodynamic powerplants.   The subject is treated here because it represents a simple application of the conservation principles applied to one-dimensional diabatic flow.*   The conservation equations set down at the beginning of the last section when specialized to constant area frictionless flow may be written:

$$\rho_1 V_1 = \rho_2 V_2 \tag{35}$$

$$p_1 - p_2 = \rho_2 V_2^2 - \rho_1 V_1^2 \tag{36}$$

$$c_p T_1 + \tfrac{1}{2} V_1^2 + q = c_p T_2 + \tfrac{1}{2} V_2^2 \tag{37}$$

$$\frac{p_1}{\rho_1 T_1} = \frac{p_2}{\rho_2 T_2} \tag{38}$$

Stations 1 and 2 in Eqs. 35 through 38 refer to the entrance and exit of the constant area duct shown in Fig. 9.

Eliminating density from Eqs. 35 and 38

$$V_1 \frac{p_1}{T_1} = V_2 \frac{p_2}{T_1}$$

$$p_1 \frac{M_1}{\sqrt{T_1}} = p_2 \frac{M_2}{\sqrt{T_2}}$$

$$\frac{p_2}{p_1} = \frac{M_1}{M_2} \sqrt{\frac{T_2}{T_1}} \tag{39}$$

From Eq. 36,

$$p_1 - p_2 = \frac{p_2}{R T_2} V_2^2 - \frac{p_1}{R T_1} V_1^2$$

$$p_1 (1 + \gamma M_1^2) = p_2 (1 + \gamma M_2^2)$$

$$\frac{p_2}{p_1} = \frac{1 + \gamma M_1^2}{1 + \gamma M_2^2} \tag{40}$$

*In an actual combustion chamber the cross-sectional areas will vary, and frictional effects will not be entirely absent. Also, the introduction of fuel changes the mass flow slightly and the products of combustion will have some effect on the gas constant R. However, these effects generally are small enough to make the results of this analysis applicable as a guide to the behavior of a combustion chamber.

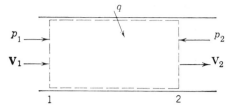

Fig. 9 Constant area duct.

Equating the expressions for p_2/p_1 given by Eqs. 39 and 40 and solving for the temperature ratio

$$\frac{T_2}{T_1} = \frac{M_2^2}{M_1^2} \left(\frac{1 + \gamma M_1^2}{1 + \gamma M_2^2} \right)^2 \qquad (41)$$

From Eq. 37,

$$T_1 \left(c_p + \frac{V_1^2}{2T_1} + \frac{q}{T_1} \right) = T_2 \left(c_p + \frac{V_2^2}{2T_2} \right)$$

$$T_1 \left(1 + \frac{\gamma R}{2c_p} M_1^2 + \frac{q}{c_p T_1} \right) = T_2 \left(1 + \frac{\gamma R}{2c_p} M_2^2 \right)$$

$$\frac{T_2}{T_1} = \frac{1 + [(\gamma - 1)/2] M_1^2 + (q/c_p T_1)}{1 + [(\gamma - 1)/2] M_2^2} \qquad (42)$$

Equating the expressions for T_2/T_1 given by Eqs. 41 and 42,

$$\frac{M_1^2 \left(1 + \dfrac{\gamma - 1}{2} M_1^2 \right)}{(1 + \gamma M_1^2)^2} + \frac{M_1^2}{(1 + \gamma M_1^2)^2} \frac{q}{c_p T_1} = \frac{M_2^2 \left(1 + \dfrac{\gamma - 1}{2} M_2^2 \right)}{(1 + \gamma M_2^2)^2} \qquad (43)$$

By defining a function $\phi(M)$ as

$$\phi(M) = \frac{M^2 \left(1 + \dfrac{\gamma - 1}{2} M^2 \right)}{(1 + \gamma M^2)^2}$$

Equation 43 can be written

$$\phi(M_2) = \phi(M_1) + \frac{M_1^2}{(1 + \gamma M_1^2)^2} \frac{q}{c_p T_1} \qquad (44)$$

A knowledge of the entrance Mach number and temperature and the heat per unit mass added between entrance and exit is sufficient to compute the exit Mach number from Eq. 44. The pressure and temperature ratios can then be computed from

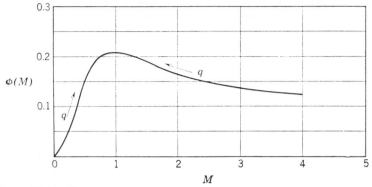

Fig. 10 Heat addition diagram.

Eqs. 40 and 41, respectively. The Mach number and static characteristics at the exit can be used to calculate the stagnation characteristics at the exit. In this manner, all exit characteristics can be found.

A plot of $\phi(M)$ is shown in Fig. 10. The figure indicates that heat added subsonically will raise the Mach number and heat added supersonically will lower it. This conclusion was found in the qualitative analysis of Section 9.7. At any Mach number, there is a critical amount of heat that will drive the stream Mach number to unity. This is the condition for *thermal choking*. Heat added in excess of the critical amount will cause a decrease in mass flux through the duct.

CHAPTER
10
Waves

10.1 Establishment of a Flow Field

The disturbance caused by a body that is in motion through a fluid is propagated throughout the fluid with the velocity of sound. The speed of sound in an incompressible flow (ρ = constant) must be infinite according to the expression $a^2 = dp/d\rho$, and therefore the pressure variations in the fluid caused by the motion of the body are felt instantaneously at all points in the fluid. It follows that when the speed of motion of a body is much less than the speed of sound, the flow will closely resemble in all details that of an incompressible fluid, in which the flow field extends to infinity in all directions.

The pressure field generated by a pressure source such as occurs at the nose of a pointed body moving at various speeds relative to the speed of sound is illustrated in Fig. 1; the curves are the circular (or spherical) wave fronts of the pressure disturbances generated at five equally spaced instants, at intervals Δt previous to the instant of observation. The pressure source is stationary in (a), moves at $M = 0.5$ in (b), at $M = 1$ in (c), and at $M = 2.0$ in (d). In (a) the field, designated by the concentric circles, extends uniformly in all directions; (b), (c), and (d) indicate the distorted pressure fields as seen by an observer moving with the source at the Mach numbers 0.5, 1.0, and 2.0. At $M = 0.5$, there is some distortion of the field. At $M = 1$, the wave fronts propagate at the same speed of the source so that they do not extend beyond the vertical plane; thus a "zone of silence" results, which is the limiting "Mach cone" shown in (d) for $M = 2$. In (d) we define the "Mach angle," μ, which we can describe quantitatively as the envelope of the wave fronts generated by the moving source; during the previous interval $5\Delta t$ the source moved from -5 to 0 at a speed of V, while the wave front, moving at speed a arrived at the point $-5'$. After drawing the radius perpendicular to the wave front, we see that

$$\mu = \sin^{-1}\frac{a\Delta t}{V\Delta t} = \sin^{-1}\frac{1}{M} \tag{1}$$

Figure 2 is a shadowgraph of a sphere moving (in a firing range) at Mach 3 over a plate with equally spaced holes along the line of flight. We see that the waves generated by the passage of the shock over the holes form an envelope inclined at the Mach angle to the flight direction. We shall refer to this photograph in various later sections to point out other of the many flow phenomena it illustrates.

From Eq. 1, it can be seen that as the Mach number decreases towards unity, the Mach angle becomes larger and larger, finally becoming $90°$ as the Mach number

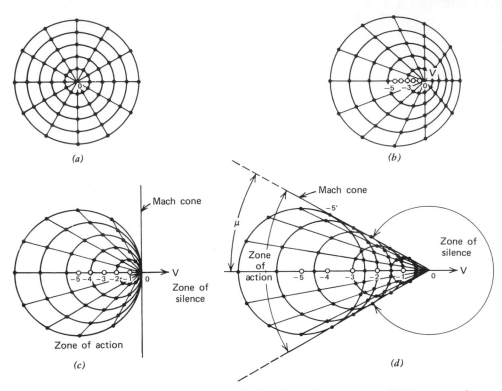

Fig. 1 Point source moving in compressible fluid. (*a*) Stationary source. (*b*) Source moving at half the speed of sound. (*c*) Source moving at the speed of sound. (*d*) Source moving at twice the speed of sound. (Adapted from von Kármán, 1947.)

reaches unity. For Mach numbers less than unity, the waves travel ahead of the body and Eq. 1 has no meaning.

The above description is independent of whether the flow is two- or three-dimensional; that is, the body may be a razor blade or needle parallel to the flow. The Mach wave will be conical in the latter case and wedge-shaped in the former.

10.2   Mach Waves

It was shown in the last section that the fluid surrounding a cone or wedge that is moving at a supersonic velocity can be divided into two parts. The part ahead of the Mach wave can receive no signals from the body. Consequently, any deflection of the stream due to the presence of the body must begin at the Mach wave. In the preceding treatment, it was specified that the cone or wedge be extremely thin; that is, the deflection of the stream due to the body is infinitesimal. If the deflection is finite the line of wave fronts is a *shock wave* rather than a Mach wave. The subject

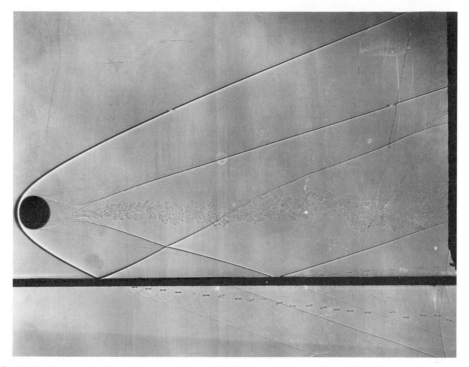

Fig. 2 Shadowgraph of a sphere moving at Mach 3 parallel to a plate with holes equally spaced parallel to the flight direction. (Courtesy U.S. Army Ballistics Research Laboratory, Maryland.)

of shocks is discussed in Section 10.5. It is the purpose of this section to derive some quantitative relations between the infinitesimal flow deflection through a Mach wave and the flow parameters. It is understood that the changes in the flow parameters through a Mach wave will also be infinitesimal.

Consider the wall shown in Fig. 3 which changes its direction by an amount $d\theta$ at the point A. If the fluid is to follow the wall, then its direction will also change by $d\theta$ at the point A. The disturbance to the flow will not be felt ahead of the Mach line originating at A. This case is comparable to that of a wedge moving through a fluid at rest if the axis of reference is attached to the wedge. It is different from the wedge case in that the flow deflection is in the opposite sense. The wall is assumed to extend to infinity in a direction normal to the plane of the paper. If the only bend in the wall is at the point A, then all the turning will occur at the Mach line springing from A. The two-dimensional nature of the problem dictates that the turning must be identical at all points on the wave from A to infinity. Therefore, the change in flow properties across the wave must be the same at all points along the wave.

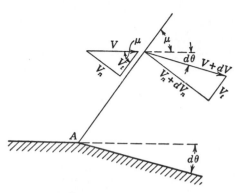

Fig. 3 Mach wave generated by an infinitesimal flow deflection.

Because there is no pressure differential along the wave, the tangential component of the velocity cannot change in crossing the wave. This follows from considerations of continuity and equilibrium and is proved in the next section. Thus the change in the velocity of the fluid in crossing the wave results entirely from the change in the component of the velocity normal to the wave. The changes in the magnitude and direction of the velocity are functions of the change in the normal component, and, therefore, they are related to each other. The rate of change of magnitude with direction is found in the following manner.

The magnitude of the velocity or speed is given by

$$V = \sqrt{V_n^2 + V_t^2} \tag{2}$$

Because V_t is constant, the differential of the speed is

$$dV = \frac{V_n}{V} \, dV_n \tag{3}$$

The direction of the velocity referred to the direction of the wave may be observed from Fig. 3 to be

$$\mu = \tan^{-1}\left(\frac{V_n}{V_t}\right) \tag{4}$$

The change in flow direction across the wave is the differential of Eq. 4. Calling this $d\theta$,

$$d\theta = \frac{V_t}{V^2} \, dV_n \tag{5}$$

The rate of change of speed with deflection angle is obtained by dividing Eq. 3 by Eq. 5:

$$\frac{dV}{d\theta} = V \frac{V_n}{V_t} \tag{6}$$

From Eq. 4, $(V_n/V_t) = \tan \mu$, and since $\sin \mu = 1/M$, Eq. 6 may be written

$$\frac{dV}{d\theta} = \frac{V}{\sqrt{M^2 - 1}} \qquad (7)$$

The above is a scalar equation that relates change in speed to change in the velocity direction. From Fig. 3, it can be seen that an increase in speed is associated with a clockwise change in direction, and therefore the positive sense must be counted clockwise. The flow about a wedge corresponds to a deflection opposite to that shown in Fig. 3. For such deflections the speed of the stream is decreased.

Pressure variation dp across the Mach wave is found with the aid of the differential form of the equilibrium equation (Eq. 7.11)

$$V \, dV + \frac{dp}{\rho} = 0$$

$$\frac{dV}{V} = -\frac{dp}{\rho V^2} \qquad (8)$$

The relation $a^2 = \gamma p/\rho$ is used to remove ρ in Eq. 8. Thus

$$\frac{dV}{V} = -\frac{1}{\gamma M^2}\frac{dp}{p} \qquad (9)$$

Substituting Eq. 9 in Eq. 7,

$$\frac{dp}{d\theta} = -\frac{\gamma M^2}{\sqrt{M^2 - 1}} p \qquad (10)$$

Because $dp/d\theta$ is a negative number, flow deflection as shown in Fig. 3 is accompanied by a decrease in pressure; that is, the fluid has been expanded. The Mach wave in this case is an expansion wave. For the flow about a wedge, the deflection causes a compression of the fluid. For this situation the Mach wave is a compression wave. An expansion of the fluid is accompanied by an increase in velocity and a decrease in pressure; a compression of the fluid is accompanied by a decrease in velocity and an increase in pressure.

It must be remembered that the relations derived in this section are valid only for infinitesimal flow deflections. The disturbance caused by the deflection is infinitesimal and the condition of isentropy here assumed is valid.

10.3 Large Amplitude Waves

Important concepts can be gained by a qualitative consideration of the sequence of events for waves of large amplitude. Consider a wave traveling to the right and having a distribution of pressure as shown in Fig. 4. As the wave passes through the fluid, the fluid is compressed from d to e, expanded from e to f, and compressed back to ambient pressure again from f to g. Regions de and fg are the compressive portions of the wave and the region ef is the expansion portion.

The steepening of the compression regions de and fg and the associated flattening of the expansion region ef occur as a result of two effects. (1) The pressure wave moving to the right exerts a pressure force $-\partial p/\partial x$ per unit volume in the $+x$ direction on the fluid elements in the regions de and fg, and in the $-x$ direction on the elements in ef. In response to these forces, as indicated in Fig. 4, the gas, which has velocity $u = 0$ at d, accelerates to a maximum at e, then decelerates to a maximum negative value at f, then accelerates again to zero at g. The pressure wave propagates at a velocity $u + a$; thus, the crest e overtakes the point d, and g overtakes the point f, causing the pressure distribution to distort initially toward the shape shown in Fig. 5 and, finally to that shown in Fig. 6. (2) An effect in the same direction is indicated by the equation $a = a_0(p/p_0)^{(\gamma-1)/2\gamma}$ (Eq. 9.15) which shows that a is a maximum at e and a minimum at f.

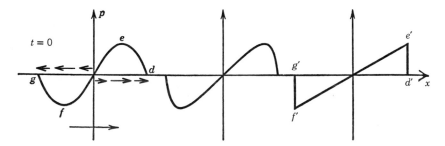

Figs. 4, 5, and 6 Development of a shock wave.

The two effects above, the second of which is the smaller, combine to form the two pressure discontinuities, termed shock waves, at the points where g' and f' and where d' and e' coincide in Fig. 6, connected by an expansion region; we show in Section 10.5 that the speed of propagation of a shock wave depends on the pressure jump across it and is always greater than that of a small wave. If the wave of Fig. 4 moved to the left instead of the right, the segment fe would steepen into a shock and fg and ed would flatten. The above behavior was predicted by Riemann in 1860, the quantitative analysis is given by Liepmann and Roshko (1957).

The different behavior of expansion and compression regions is illustrated in Fig. 7 for a double wedge section moving to the left at a supersonic speed. In the terminology of Fig. 3, the angle of turning of the flow is negative at A and C, where the flow turns into itself. Therefore, by Eq. 10, $\overline{AA_1}$ and $\overline{CC_1}$ are compression waves; they form at the tangents to the wave fronts generated at previous positions of the points A' and A'', C' and C'', respectively. By the reasoning above, these waves, being finite compressions, remain sharp as they propagate outward along

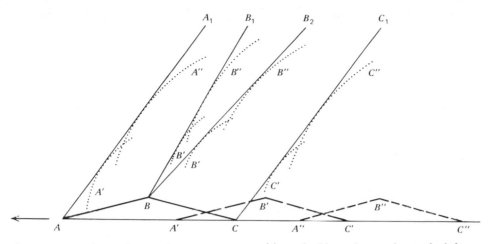

Fig. 7 Compression and expansion waves generated by a double wedge moving to the left at a supersonic speed. The letter designations indicate circular wave fronts and their centers.

$\overline{AA_1}$ and $\overline{CC_1}$. On the other hand, at B the angle of turn is positive and, therefore, the expansion waves formed at B'' and B' become less steep as they propagate outward, and by the time the body reaches the position ABC the expansion is bounded by the lines $\overline{BB_1}$ and $\overline{BB_2}$, representing respectively the Mach directions for the flows along the surfaces $\overline{AB}$ and $\overline{BC}$.

10.4 Prandtl-Meyer Flow

The flow deflection at point B in Fig. 7 is such that the fluid expands as it turns. It was shown in the last section that an expansion wave (a disturbance through which the fluid is expanded) of finite amplitude becomes less steep as it is propagated outward. As the fluid passes through the wave bounded by $\overline{BB_1}$ and $\overline{BB_2}$, the flow parameters change gradually from their values upstream of $\overline{BB_1}$ to their values downstream $\overline{BB_2}$. The finite change in properties can be considered as being accomplished by an infinite number of infinitesimal changes across a series of Mach waves lying within the fan B_1BB_2. The relation between speed change and flow deflection across any one of these infinitesimal waves has been given by Eq. 7, which may be written in the form

$$d\theta = \frac{\sqrt{M^2 - 1}}{V}\,dV \qquad (11)$$

The total deflection through the fan is the integral of Eq. 11. To perform the integration, $\sqrt{M^2 - 1}/V$ must be expressed in terms of V. This expression can be written

$$\frac{\sqrt{M^2 - 1}}{V} = \sqrt{\frac{1}{a^2} - \frac{1}{V^2}} \qquad (12)$$

The speed of sound a can be written in terms of speed of the stream by using the energy relation as given by Eq. 9.9.

$$\frac{V^2}{2} + \frac{a^2}{\gamma - 1} = \frac{a_0^2}{\gamma - 1} \tag{13}$$

We solve for a in Eq. 13 and substitute this value in Eq. 12. Then

$$\frac{\sqrt{M^2 - 1}}{V} = \frac{\dfrac{\gamma + 1}{2} V - \dfrac{a_0^2}{V}}{\sqrt{\dfrac{1 - \gamma^2}{4} V^4 + \gamma a_0^2 V^2 - a_0^4}} \tag{14}$$

By means of Eq. 14, $d\theta/dV$ in Eq. 11 can be expressed entirely in terms of V and the integration can be performed. Let $\theta = 0$ be the flow direction that corresponds to a Mach number of unity; that is, $V = a^*$. Then the deflection θ from this flow direction is given by

$$\theta = \int_{a^{*2}}^{V^2} \frac{\dfrac{\gamma + 1}{4} d(V^2)}{\sqrt{\dfrac{1 - \gamma^2}{4} V^4 + \gamma a_0^2 V^2 - a_0^4}} - \int_{a^{*2}}^{V^2} \frac{\dfrac{a_0^2}{2V^2} d(V^2)}{\sqrt{\dfrac{1 - \gamma^2}{4} V^4 + \gamma a_0^2 V^2 - a_0^4}} \tag{15}$$

Both integrals in Eq. 15 are in standard form (Peirce, 1929, Nos. 161, 183). The equation becomes[†]

$$\theta = -\frac{1}{2} \sqrt{\frac{\gamma + 1}{\gamma - 1}} \left[\sin^{-1} \left(\frac{1 - \gamma^2}{2} \frac{V^2}{a_0^2} + \gamma \right) - \sin^{-1} \left(\frac{1 - \gamma^2}{2} \frac{a^{*2}}{a_0^2} + \gamma \right) \right]$$

$$- \frac{1}{2} \left[\sin^{-1} \left(\gamma - 2 \frac{a_0^2}{V^2} \right) - \sin^{-1} \left(\gamma - 2 \frac{a_0^2}{a^{*2}} \right) \right] \tag{16}$$

a^*/a_0 and V/a_0 are easily found in terms of Mach number. For $V = a = a^*$, Eq. 13 becomes

$$\left(\frac{a^*}{a_0} \right)^2 = \frac{2}{\gamma + 1} \tag{17}$$

Also from Eq. 13

$$\left(\frac{V}{a_0} \right)^2 = \frac{2}{\gamma - 1} \left[1 - \left(\frac{a}{a_0} \right)^2 \right] \tag{18}$$

[†]The relation between speed and direction as given by Eq. 16 is valid for isentropic compressive deflections as well as for expansive deflections; it forms the basis for the method of characteristics for analyzing two- and three-dimensional supersonic flows, (e.g., Liepmann and Roshko, 1957).

$(a_0/a)^2$ has been given by Eq. 9.17. Substituting this value in Eq. 18 yields

$$\left(\frac{V}{a_0}\right)^2 = \frac{M^2}{1 + \frac{\gamma - 1}{2} M^2} \tag{19}$$

It is convenient to replace $(a^*/a_0)^2$ and $(a_0/V)^2$ in Eq. 16 by their equivalents from Eqs. 17 and 19. The flow deflection may then be expressed in terms of Mach number

$$\theta = -\frac{1}{2} \sqrt{\frac{\gamma + 1}{\gamma - 1}} \left[\sin^{-1} \left(\frac{\frac{1 - \gamma^2}{2} M^2}{1 + \frac{\gamma - 1}{2} M^2} + \gamma \right) - \frac{\pi}{2} \right]$$

$$- \frac{1}{2} \left[\sin^{-1} \left(\gamma - 2 \frac{1 + \frac{\gamma - 1}{2} M^2}{M^2} \right) + \frac{\pi}{2} \right] \tag{20}$$

For $M = 1$, $\theta = 0$ as it should because θ is measured from the flow direction corresponding to $M = 1$. As M increases, θ increases. Finally, when M becomes infinite, θ reaches the value $130.5°$. A plot of θ versus M appears in Fig. 8 and is tabulated in Table 3 at the end of the book. The change in Mach number associated with a

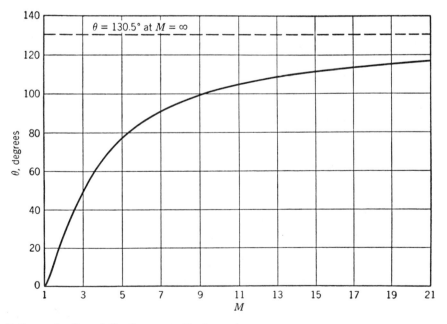

Fig. 8 Expansive flow deflection versus Mach number.

given flow deflection may be computed from this curve in the following manner: Read the flow deflection θ_i corresponding to the initial Mach number. This represents the amount that the stream has been deflected from the $M = 1$ direction. Add to θ_i the desired deflection $\Delta\theta$. The sum $\theta_f = \theta_i + \Delta\theta$ is the amount that the stream must be turned to bring it from the $M = 1$ direction to the direction corresponding to the Mach number after the deflection $\Delta\theta$. This final Mach number is the abscissa of the point on the curve for which θ_f is the ordinate. For example, if a stream with a Mach number of 2 is deflected through $20°$, the Mach number after deflection will be 2.83. It should be remembered that expansive flow is isentropic. Therefore, the Mach number at any point during the turning determines all the other properties of the flow. The *static-stagnation* ratios versus Mach number are given by Eqs. 16, 18, and 19 of Chapter 9.

10.5 Finite Compression Waves

The flow deflection at point A in Fig. 7 is in a direction that results in a compression of the fluid. The line AA_1 is the line of disturbance fronts. Because the disturbances created at the nose of the body are finite compressive disturbances, they will steepen as they are propagated outward, and, as explained in Section 10.3, in a very short period of time a discontinuity or *shock wave* develops. Unlike an expansion wave, the change in flow properties across a shock wave is abrupt. The device of integrating infinitesimal changes is therefore not available.

In order to find the changes in flow properties when the fluid passes through a shock wave, it is necessary to consider continuity, equilibrium, and the conservation of energy across the shock. These conditions will show that the speed of propagation of the finite disturbance is greater than the speed of sound. They also show that the compression of the fluid in passing through a shock is not an isentropic process.

Figure 9 is an enlarged view of the leading edge of the body shown in Fig. 7. The axis of reference is now attached to the body so that it appears that the body is stationary in a moving fluid. The flow parameters upstream p_1, ρ_1, and V_1 are

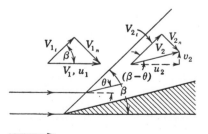

Fig. 9 Flow through a shock wave.

known. Let it be required to find the flow parameters downstream, p_2, ρ_2, and V_2.[†] Two more parameters enter the problem: the deflection angle θ and the wave angle β. It will be shown presently that four equations are available for the solution of the problem and therefore four unknowns may be determined. In addition to p_2, ρ_2, and V_2, either β or θ can be taken as an unknown. In the following, it is assumed that the wave angle β is known and the deflection angle that corresponds to this wave angle and the upstream conditions will be determined.

The four equations available for the solution of the unknowns follow. Reference should be made to Fig. 9 for the notation used.

1. *Equation of Continuity*

Conservation of mass requires that, for any unit area of the wave, the mass flux entering must equal the mass flux leaving.

$$\rho_1 V_{1n} = \rho_2 V_{2n} \tag{21}$$

2. *Equations of Equilibrium*

The pressure and velocity changes across the wave are related by the momentum theorem, which is given by Eq. 3.29. Consider the region $\hat{R}$ shown in Fig. 10. Let

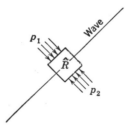

Fig. 10 Equilibrium in shock flow.

the faces parallel to the wave be of unit area. According to the momentum theorem, equilibrium normal to the wave is expressed by

$$p_1 - p_2 = \rho_2 V_{2n}^2 - \rho_1 V_{1n}^2 \tag{22}$$

Parallel to the wave, there is no change in pressure, and equilibrium is expressed by

$$0 = \rho_2 V_{2n} V_{2t} - \rho_1 V_{1n} V_{1t} \tag{23}$$

3. *Energy Equation*

The energy equation for a gas in motion as given by Eq. 9.12 is applicable to any adiabatic flow whether it is reversible or not, and therefore the value of a^* is con-

[†] Equally important applications occur where the downstream parameters are known and the upstream parameters are the unknowns. For example, in wind-tunnel work, the downstream parameters are frequently measured and the upstream values are computed.

stant throughout the flow. Then the following energy equations are applicable where $a*$ has the same value in both equations.

$$\frac{V_{1n}^2 + V_{1t}^2}{2} + \frac{\gamma}{\gamma - 1}\frac{p_1}{\rho_1} = \frac{\gamma + 1}{2(\gamma - 1)}a^{*2}$$

$$\frac{V_{2n}^2 + V_{2t}^2}{2} + \frac{\gamma}{\gamma - 1}\frac{p_2}{\rho_2} = \frac{\gamma + 1}{2(\gamma - 1)}a^{*2}$$

(24)

From Eqs. 21 and 23 it follows that

$$V_{1t} = V_{2t}$$

(25)

The tangential component of velocity is the same on both sides of a shock. This fact was utilized in the treatment of infinitesimal waves in Section 10.2.

Since the tangential components of the velocity do not change when the fluid passes through the wave, the velocity downstream is established once V_{2n} is determined. From Eqs. 21 and 22

$$p_2 - p_1 = \rho_1 V_{1n}(V_{1n} - V_{2n})$$

(26)

Again using continuity,

$$\frac{p_2}{\rho_2 V_{2n}} - \frac{p_1}{\rho_1 V_{1n}} = V_{1n} - V_{2n}$$

p_2/ρ_2 and p_1/ρ_1 may be eliminated by employing the energy relations Eqs. 24. The station subscript is dropped from V_t because $V_{1t} = V_{2t}$.

$$\frac{\gamma + 1}{2\gamma}a^{*2}\left(\frac{1}{V_{2n}} - \frac{1}{V_{1n}}\right) + \frac{\gamma - 1}{2\gamma}\left[\left(V_{1n} + \frac{V_t^2}{V_{1n}}\right) - \left(V_{2n} + \frac{V_t^2}{V_{2n}}\right)\right] = V_{1n} - V_{2n}$$

which rearranges to the form

$$\left(\frac{\gamma + 1}{2\gamma}\frac{a^{*2}}{V_{1n}V_{2n}} - \frac{\gamma - 1}{2\gamma}\frac{V_t^2}{V_{1n}V_{2n}} - \frac{\gamma + 1}{2\gamma}\right)(V_{1n} - V_{2n}) = 0$$

The above equation is satisfied when either factor is zero. The solution for which $V_{1n} - V_{2n} = 0$ corresponds to a shock wave of zero intensity or a Mach wave. The first factor set equal to zero gives the expression

$$V_{1n}V_{2n} = a^{*2} - \frac{\gamma - 1}{\gamma + 1}V_t^2$$

(27)

From Fig. 9, $V_{1n} = V_1 \sin \beta$ and $V_t = V_1 \cos \beta$. These expressions are substituted into Eq. 27 with the result

$$V_{2n} = \frac{a^{*2}}{V_1 \sin \beta} - \frac{\gamma - 1}{\gamma + 1}V_1\frac{\cos^2 \beta}{\sin \beta}$$

which may be written

$$V_{2n} = \frac{V_1}{\sin \beta} \left[\left(\frac{a^*}{a_0} \frac{a_0}{V_1} \right)^2 - \frac{\gamma - 1}{\gamma + 1} \cos^2 \beta \right]$$

or, after replacing a^*/a_0 and a_0/V_1 with their equivalents from Eqs. 17 and 19

$$V_{2n} = \frac{V_1}{\sin \beta} \left(\frac{\gamma - 1}{\gamma + 1} \sin^2 \beta + \frac{2}{\gamma + 1} \frac{1}{M_1^2} \right) \qquad (28)$$

From Fig. 9, V_{2n} is related to the wave angle β and deflection angle θ by

$$V_{2n} = V_t \tan (\beta - \theta) = V_1 \cos \beta \tan (\beta - \theta)$$

which may be equated to the value of V_{2n} given by Eq. 28 with the result

$$\tan (\beta - \theta) = \frac{1}{\sin \beta \cos \beta} \left(\frac{\gamma - 1}{\gamma + 1} \sin^2 \beta + \frac{2}{\gamma + 1} \frac{1}{M_1^{\,2}} \right)$$

From the above, it can be seen that only two of the three variables β, θ, and M_1 are independent. An explicit solution for the deflection angle in terms of the upstream Mach number and wave angle is

$$\theta = \beta - \tan^{-1} \left[\frac{1}{\sin \beta \cos \beta} \left(\frac{\gamma - 1}{\gamma + 1} \sin^2 \beta + \frac{2}{\gamma + 1} \frac{1}{M_1^{\,2}} \right) \right] \qquad (29)$$

From Eq. 29, it may be readily verified that when the wave angle is equal to the Mach angle

$$\beta = \mu = \sin^{-1} \frac{1}{M_1}$$

the deflection goes to zero as is to be expected, because under these conditions the wave is of infinitesimal strength. When $\beta = \pi/2$ in Eq. 29, corresponding to a shock wave normal to the flow, the deflection again goes to zero.

A plot of Eq. 29 for the complete range of Mach numbers is shown in Fig. 11. Two points should be observed.

(1) For any value of M_1, there are two wave angles that produce the same deflection θ. The wave represented by the larger value of β (dashed curve) is termed a *strong shock*. The smaller value of β (solid curve) corresponds to the *weak shock*. The waves are so named because the strong shock produces the greater entropy rise.

(2) For any value of M_1, there is a maximum deflection angle. The maximum occurs along the solid curve that separates the strong and weak shocks. Flow deflections in excess of the maximum are not consistent with the basic relations, Eqs. 21 through 24, used in the derivation of Eq. 29.

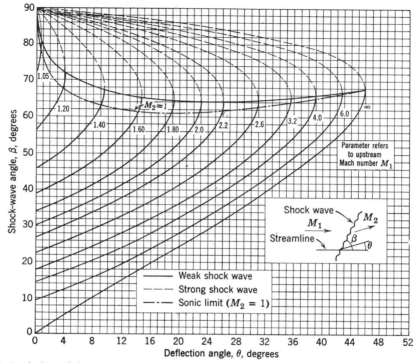

Fig. 11 Variation of shock angle β with flow deflection angle θ. Perfect gas, $\gamma = 1.4$. (Courtesy of the National Advisory Committee for Aeronautics.) A more versatile chart for rough calculations is included with the tables at the end of the book.

There is included with the tables in the back of the book a shock chart that gives more information on oblique shocks than Fig. 11 though it is restricted to $M_1 < 5.5$ and the resolution is not as great.

Equation 29 is applicable to conical shocks as well as plane shocks.` In the case of the plane shock illustrated in Fig. 9 the deflection angle θ is the same as the angle of the wedge that produces the shock. In the three-dimensional case, the deflection angle θ is *not* the same as the cone angle that generates the conical shock. This point will be discussed in greater detail in Section 10.9.

10.6   The Characteristic Ratios as Functions of Mach Number

In this section it is shown that the pressure, density, temperature, and stagnation pressure ratios across a shock wave are functions of the normal component of the upstream Mach number only. From Fig. 9, it can be seen that the normal component of the upstream Mach number is given by

$$M_{1n} = M_1 \sin \beta \qquad (30)$$

M_{1n} in the following is referred to as the *normal Mach number*.

The pressure ratio can be derived from Eq. 26, which after substitution from Eq. 27 for $V_{1n}V_{2n}$ becomes

$$\frac{p_2}{p_1} = 1 + \frac{\rho_1}{p_1} \left(V_{1n}^2 - a^{*2} + \frac{\gamma - 1}{\gamma + 1} V_t^2 \right) \tag{31}$$

ρ_1/p_1 is replaced by its equivalent γ/a_1^2 and V_t^2 by $V_1^2 - V_{1n}^2$. Then Eq. 31 becomes

$$\frac{p_2}{p_1} = 1 + \frac{2\gamma}{\gamma + 1} \left(\frac{V_{1n}}{a_1} \right)^2 - \gamma \left(\frac{a^*}{a_1} \right)^2 + \frac{\gamma(\gamma - 1)}{\gamma + 1} M_1^2 \tag{32}$$

$(a^*/a_1)^2$ can be obtained from the first of Eqs. 24. Then, with $(V_{1n}/V_1) = \sin \beta$, Eq. 32 becomes

$$\frac{p_2}{p_1} = \frac{2\gamma}{\gamma + 1} M_{1n}^2 - \frac{\gamma - 1}{\gamma + 1} \tag{33}$$

The density ratio across the shock may be found by multiplying the right-hand side of Eq. 26 by $(V_{1n} + V_{2n})$. Equation 26 becomes

$$p_2 - p_1 = (V_{1n}^2 - V_{2n}^2) \frac{\rho_1 V_{1n}}{V_{1n} + V_{2n}} \tag{34}$$

By using the equation of continuity, Eq. 34 becomes

$$V_{1n}^2 - V_{2n}^2 = (p_2 - p_1) \left(\frac{1}{\rho_1} + \frac{1}{\rho_2} \right) \tag{35}$$

But, from Eqs. 24,

$$V_{1n}^2 - V_{2n}^2 = \frac{2\gamma}{\gamma - 1} \left(\frac{p_2}{\rho_2} - \frac{p_1}{\rho_1} \right) \tag{36}$$

Equations 35 and 36 are equated and after some rearrangement we have

$$\frac{\rho_2}{\rho_1} = \frac{\dfrac{\gamma + 1}{\gamma - 1} + \dfrac{p_1}{p_2}}{1 + \dfrac{\gamma + 1}{\gamma - 1} \dfrac{p_1}{p_2}} \tag{37}$$

Equation 37 is known as the Rankine-Hugoniot relation. From Eqs. 37 and 33, it can be seen that ρ_2/ρ_1 is a function only of M_{1n}. The expression for temperature ratio follows directly from the equation of state.

$$\frac{T_2}{T_1} = \frac{p_2}{p_1} \frac{\rho_1}{\rho_2} \tag{38}$$

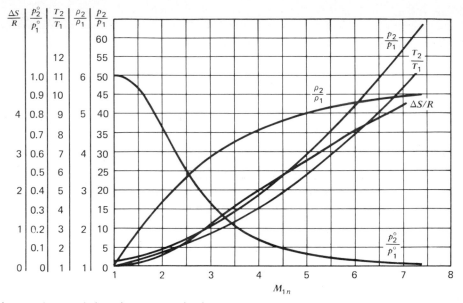

Fig. 12 Characteristic ratios across a shock wave versus normal Mach number.

Pressure, density, and temperature ratios across a shock wave in air ($\gamma = 1.4$) are plotted against the normal Mach number in Fig. 12. The numerical data are included in Table 6 at the end of the book. From Eqs. 33, 37, and 38, it can be seen that the pressure and temperature ratios become infinite and that the density ratio approaches 6 as M_{1n} becomes infinite.

The change in entropy between the thermodynamic states on the two sides of a shock wave, ΔS, is given by Eq. 8.35,

$$\Delta S = \ln \left[\left(\frac{T_2}{T_1} \right)^{c_v} \left(\frac{\rho_1}{\rho_2} \right)^R \right] = R \ln \left[\left(\frac{T_2}{T_1} \right)^{1/(\gamma-1)} \left(\frac{\rho_1}{\rho_2} \right) \right] \tag{39}$$

With the aid of Eqs. 33, 37, and 38, Eq. 39 may be expressed in terms of the normal Mach number. It can be readily verified that for normal Mach numbers less than unity, Eq. 39 indicates an entropy loss and Eq. 33 indicates a pressure decrease in passing through the shock. An entropy loss is ruled out by the second law of thermodynamics, and a pressure decrease means that the wave is an expansion discontinuity. Equations 33 and 40, therefore, prove that expansion discontinuities cannot exist. This is the same conclusion reached in the discussion of Section 10.3.

Since when we compare the stagnation with the static (or ambient) properties of the gas we assume that the gas was brought to rest isentropically, we may substitute in Eq. 39 the stagnation properties on the two sides of the shock. Thus, Eq. 39 may be written

$$\Delta S = R \ln \left[\left(\frac{T_2^0}{T_1^0} \right)^{1/(\gamma-1)} \left(\frac{\rho_1^0}{\rho_2^0} \right) \right]$$

This equation is simplified by noting that, since the flow through the shock is adiabatic, $T_2^0 = T_1^0$, and, as a consequence, the equation of state shows that $\rho_1^0/\rho_2^0 = p_1^0/p_2^0$. Then the last equation becomes

$$\Delta S = R \ln (p_1^0/p_2^0) \tag{40}$$

The ratio of the stagnation pressures across the shock follows immediately from the comparison of Eq. 39 and 40:

$$\frac{p_1^0}{p_2^0} = \left(\frac{T_2}{T_1}\right)^{1/(\gamma-1)} \left(\frac{\rho_1}{\rho_2}\right) = \left(\frac{p_2}{p_1}\right)^{1/(\gamma-1)} \left(\frac{\rho_1}{\rho_2}\right)^{\gamma/(\gamma-1)} \tag{41}$$

A crucial point in the analysis of supersonic flow over thin bodies is the circumstance that the entropy increase through weak shock waves is small. To demonstrate this result, Eq. 39 is expressed in terms of M_{1n} by means of Eqs. 32, 37, and 38. The resulting expression can be expanded in a series in powers of $(M_{1n}^2 - 1)$ (see Liepmann and Roshko, 1957, p. 60). The result is

$$\frac{\Delta S}{R} \cong \frac{2\gamma}{(\gamma + 1)^2} \frac{(M_{1n}^2 - 1)^3}{3} \tag{42}$$

Equation 40, plotted in Fig. 12 as a function of M_{1n} for $\gamma = 1.4$, shows that for M_{1n} less than about 1.5, the entropy increase through a shock wave is very small. The increase, by Eq. 42, is proportional to the cube of the deviation of M_{1n}^2 from unity. This relation signifies that the flow is isentropic through Mach waves and is very nearly isentropic through weak shocks. As the value of M_{1n} increases, the loss in stagnation pressure increases. It is evident, therefore, that energy is dissipated within a shock wave just as it is dissipated in the flow through a tube with friction or in a boundary layer or wake.

Finally, it remains to interpret the normal Mach number. By changing the axes of reference so that the wave is traveling through a stationary fluid with a speed corresponding to a Mach number M_1, it can be seen that M_{1n} is simply the ratio of the velocity of propagation of the wave to the speed of sound. When the normal Mach number is unity, the shock is a Mach wave and there are no discontinuities in the flow parameters across it. As the speed of propagation becomes greater, the discontinuity in the flow parameters becomes greater. Thus, the intensity of the discontinuity or shock is a function of the speed of propagation of the wave.

10.7 Normal Shock Wave

If the shock is normal to the flow, $\beta = \pi/2$, and because $V_t = 0$, Eq. 27 becomes

$$V_1 V_2 = a^{*2} \tag{43}$$

Equation 43 indicates that if V_1 is greater than a^*, V_2 must be less than a^*. Entropy considerations have shown that the Mach number upstream of a shock must be supersonic. Therefore, for normal shock waves, the flow downstream of the shock is subsonic.

Equation 43 may be written

$$\frac{V_1}{a_1} \frac{a_1}{a_0} \frac{V_2}{a_2} \frac{a_2}{a_0} = \left(\frac{a^*}{a_0}\right)^2 \tag{44}$$

From Eq. 12 of Chapter 9, $(a^*/a_0)^2 = 2/(\gamma + 1)$. From Eq. 17 of the same chapter,

$$\frac{a}{a_0} = \left(\frac{\gamma - 1}{2} M^2 + 1\right)^{-1/2}$$

These values are substituted in Eq. 44. Then

$$\frac{M_2}{\sqrt{\frac{1}{2}(\gamma - 1)M_2^2 + 1}} = \frac{2}{\gamma + 1} \frac{\sqrt{\frac{1}{2}(\gamma - 1)M_1^2 + 1}}{M_1}$$

This is readily solved for M_2^2.

$$M_2^2 = \frac{(\gamma - 1)M_1^2 + 2}{2\gamma M_1^2 - (\gamma - 1)} \tag{45}$$

As M_1^2 increases, M_2^2 decreases, finally approaching the value $(\gamma - 1)/2\gamma = 1/7$ (for air) as M_1 approaches infinity. A tabulation of M_2 for a range of values of M_1 is included in Table 4 at the end of the book.

10.8   Plane Oblique Shock Waves

The wedge in Fig. 13 forces the stream to turn through an angle w. Providing w is not too great for the upstream Mach number M_1 considered, the turning can be accomplished by a plane shock wave as shown by the dotted line. In this circumstance, w corresponds to the angle θ of Section 10.5. With the values of M_1 and θ

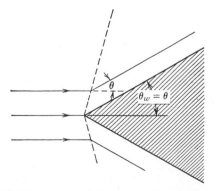

Fig. 13 Wedge with attached shock.

known, the wave angle β can be read from Fig. 11. Of the two choices for β, it is an experimental fact that the one corresponding to the *weak* shock usually occurs.

β and M_1 fix the value of the normal Mach number M_{1n}, which in turn determines the change in flow properties across the shock. For the two-dimensional case, the oblique shock wave divides the flow field into two uniform regions of different entropy. For example, if the upstream Mach number is 2 and the wedge angle is 14°, the wave angle is 44° from Fig. 11. Then the normal Mach number from Eq. 30 is 1.39. Opposite $M_{1n} = 1.39$ in Table 6 may be read the ratios of pressure, density, and so forth across the shock. To deduce the downstream Mach number, remember that even though there is an entropy rise across the shock the flow is adiabatic and there is no change in stagnation temperature; that is, $T_1^0 = T_2^0$. We can write

$$\frac{T_2}{T_2^0} = \frac{T_2}{T_1}\frac{T_1}{T_1^0}\frac{T_1^0}{T_2^0} = (1.248)\,(0.5556)\,(1) = 0.694$$

T_1/T_1^0 corresponds to $M_1 = 2$ and was read from Table 5 at the end of the book. From the same table, corresponding to $T_2/T_2^0 = 0.694$, the value of M_2 is found to be 1.48.

The Mach number downstream of a *weak* oblique shock is supersonic for all turning angles up to a degree of the maximum turning angle. In the example above, for $M_1 = 2$ the maximum turning angle is 23°, and M_2 is sonic for $\theta \simeq 22.7°$ and supersonic for θ less than this value. The boundary between subsonic and supersonic downstream Mach numbers is indicated by the dashed line marked $M_2 = 1$ in Fig. 11. The Mach number downstream of *strong* oblique shocks is subsonic.

The above discussion assumes the wedge angle does not exceed the maximum turning angle for the M_1 considered. If the wedge angle exceeds the maximum, then a *detached* curved shock appears ahead of the wedge as shown in Fig. 14. At

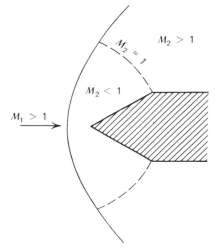

Fig. 14 Wedge with detached shock.

the center of the wave $\beta = \pi/2$, which is the limiting value for the strong shock. Proceeding away from the center along the shock, the wave angle decreases. This corresponds to a movement downward and towards the right on a strong shock line of Fig. 11. Until a wave angle corresponding to the $M_2 = 1$ line is reached, the flow immediately downstream of the shock will be subsonic. Thus, pressure pulses from the wedge can be communicated to the fluid behind the central portion of the shock and, as a consequence, the flow curves about the wedge in typical subsonic fashion.

The entropy rise across the detached shock depends on the wave angle, which is changing continuously. Therefore, the entropy of the flow behind the curved shock is not uniform.

10.9   Conical Oblique Shock Waves

As pointed out in Section 10.5, the relation between upstream Mach number M_1, wave angle β, and deflection angle θ is independent of the geometry of the shock. Therefore, if M_1 and θ are given, the change in flow properties across a shock of any configuration may be computed by the methods of the last section.

There is, however, an essential difference between the flow behind a plane shock and that behind an axially symmetric shock as is formed, for instance, in front of a body with a conical nose. The flow behind a conical shock is shown diagrammatically in Fig. 15. Only part of the turning takes place at the shock; the increasing radius of the conical cross section with distance downstream makes it necessary, from continuity considerations, that the streamlines have qualitatively the configuration shown in Fig. 15. Therefore, the cone angle θ_c is greater than the flow deflection θ through the shock. The flow field downstream of the shock, though isentropic, is not uniform, and therefore the flow parameters immediately downstream of the shock are not the same as they are at the surface of the cone.

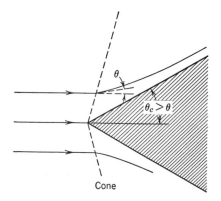

Cone

Fig. 15 Cone with attached shock.

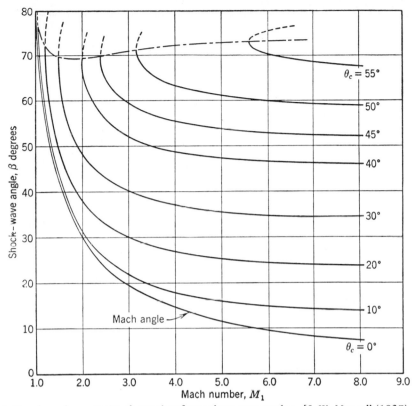

Fig. 16 Shock angle versus Mach number for various cone angles. [J. W. Maccoll (1937). Courtesy of Royal Society of London.]

Taylor and Maccoll (see Taylor and Maccoll, 1933; Maccoll, 1937) gave an exact solution for the supersonic flow past a cone with an attached shock wave. They were able to *patch* an isentropic flow about a cone with a shock flow such that the flow parameters immediately downstream of the shock had the same values from the two solutions. The results of their work are summarized in Figs. 16 and 17, which give the shock wave angle and the pressure coefficient at the surface of the cone as functions of Mach number and cone angle. As for the wedge, there is a maximum cone angle, which is a function of Mach number, beyond which a conical shock will not form. Here again, a detached shock will exist. In keeping with previous remarks, it can be seen that, for a given M_1, the maximum cone angle is greater than the maximum wedge angle. Figure 18 shows an attached and a detached shock on a body with a conical nose at a Mach number of 1.9.[†]

[†]Interferometer, schlieren, and shadowgraph are three devices for detecting shock waves optically. They detect density, density gradient, and rate of change of density gradient, respectively. The photographs of Fig. 18 were taken with a schlieren system. For a description of these techniques see Bradfield (1964).

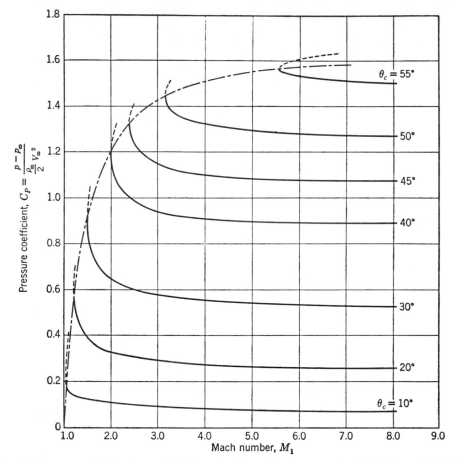

Fig. 17 Pressure coefficient versus Mach number for various cone angles. [J. W. Maccoll (1937). Courtesy of Royal Society of London.]

10.10   Shocks in Tubes

In Section 9.6, the one-dimensional isentropic flow in channels is treated. Referring to Fig. 7 of Chapter 9, if the end pressure is below point 2, only one isentropic flow is possible, namely, point 3. This indicates that only by a change of entropy is it possible to reach end pressures between points 2 and 3. We can fill in this space by plotting isentropic curves for different values of p_0 according to Eq. 9.26. First, it must be shown that the throat area for sonic flow varies with p_0, that is, with the entropy. Consider successive throats in a tube, and let $M = 1$ at each throat. Then

$$\rho_1^* a_1^* A_1^* = \rho_2^* a_2^* A_2^*$$

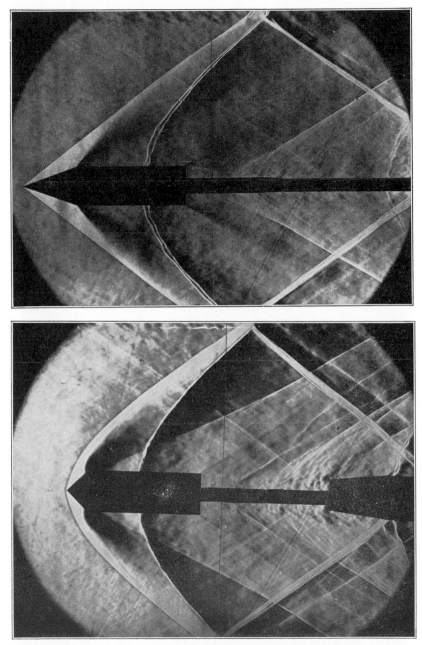

Fig. 18 Schlieren photographs of attached and detached shock waves for flow past cones of different vertex angles. $M = 1.90$. (Courtesy of University of Michigan Aerospace Laboratories.)

where the subscripts refer to the separate throats. Because $a_1^* = a_2^*$ as long as the flow is adiabatic, regardless of changes in entropy, and because ρ_1^* and ρ_2^* are constant multiples of their respective stagnation values, the above equation becomes

$$\rho_1^0 A_1^* = \rho_2^0 A_2^*$$

Because $T_1^0 = T_2^0$, the equation of state enables us to write

$$p_1^0 A_1^* = p_2^0 A_2^*$$

or

$$\frac{p_1^0}{p_2^0} = \frac{A_2^*}{A_1^*} \tag{46}$$

Then, for flow through a tube, if a shock occurs, Eq. 9.26 also represents the flow downstream of a shock in which the new p_0 and A^* are related to the old values through Eq. 46. The result is a series of isentropic curves as shown in Fig. 19. The vertex of each curve occurs at that point in the channel where $A = A_2^*$; the upper portion represents a subsonic, the lower represents a supersonic flow.

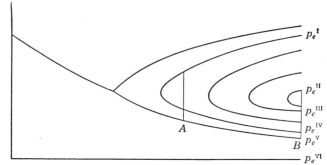

Fig. 19 Nonisentropic flow through a channel.

Since the flow changes from supersonic to subsonic through a normal shock, an end pressure p_e^I can be reached if a shock occurs at A and the flow downstream of the shock is isentropic. As p_e decreases, the shock will move downstream until for $p_e = p_e^{II}$ it is at the exit. The adjustment to still lower exit pressures may take place outside the channel, as shown in Fig. 20. As long as $p_e > p_e^V$, oblique shocks can provide the necessary increase in pressure from the exit to the surroundings. For $p_e = p_e^V$ the isentropic flow without shocks will exist. For $p_e < p_e^V$ the adjustment takes place through expansion waves emanating from the lip.

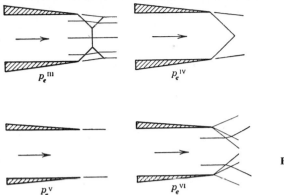

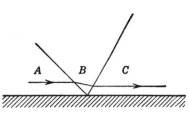

Fig. 21 Shock reflection.

Fig. 20 Adjustments to ambient pressure near lip
of a jet.

10.11 Reflection of Waves

When a wave strikes a plane surface as shown in Fig. 21, the angle of reflection is determined by the condition that the net flow deflection through the two shocks must be zero, that is, the flow must follow the surface in region C. For Mach waves, all effects are linear and, therefore, the angle of reflection will equal the angle of incidence. If the incident wave is a shock, however, the Mach number in region B will be less than that in region A, and the reflected shock required to turn the flow back to parallelism with the wall will be at a smaller angle to the wall than was the incident shock.

If, as in Fig. 22, the wall is deflected at the point of incidence to a direction parallel to the streamlines downstream of the incident shock, no reflected shock occurs. This principle of *absorption* of waves is employed in the design of nozzles to achieve *supersonic* shock-free flow.

If the Mach number in region B is so low that the turning required is greater than the maximum deflection angle θ_m for that Mach number, a so-called *Mach reflection* as shown in Fig. 23 may be formed. In this event, the reflected shock as well

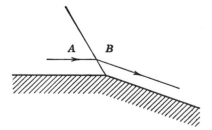

Fig. 22 Shock absorption.

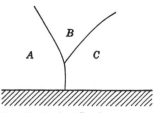

Fig. 23 Mach reflection.

as that which connects the intersection with the wall may be curved; however, at the point where the shock strikes the wall it must be normal so that no turning will result. The flow in at least a portion of region C is subsonic and the streamlines are not parallel to each other.

When a shock wave reflects from a free surface, such as that at the boundary of a jet (see Fig. 20), the boundary condition requires that the pressure be constant and equal to its ambient value, that is, that in the surrounding space. Figure 24 shows the wave configuration for the case of the surrounding pressure less than that in the jet. Expansion waves originate at the lip, and the jet expands. When these waves

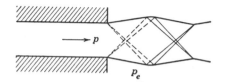

Fig. 24 Shock reflection from a free surface.

reach the opposite boundary of the jet, they must reflect as compression waves if the pressure at the boundary is to be constant. As a result, the "rocket plumes" is formed comprised of the expanding and contracting jet flow; the wavelength of the expansions and contractions depend on the Mach number and pressure of the issuing jet and on the ambient pressure. However, viscosity causes the jet boundaries to become more and more blurred with distance from the jet lip. (See review by Adamson, 1964).

A schlieren photograph of an "under-expanded" axisymmetric jet with sonic flow at the lip is shown in Fig. 25. The curved shock that forms at the boundary is

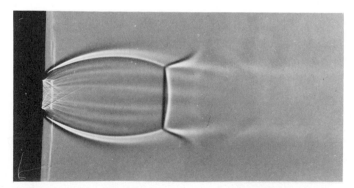

Fig. 25 The expansion of an axisymmetric flow showing the expansion fan, the viscous spreading of the free boundary, and the "Mach disk." The Mach number of the flow at the lip is unity; the ratio of static pressure at the lip to the external pressure, P/Pe = 18.5. (Courtesy of Gas Dynamics Laboratory, Aerospace Engineering Dept., University of Michigan.)

continuously re-enforced by the compression waves generated at the free boundary and the "Mach disk" provides the downstream adjustment to the ambient pressure.

10.12 Flow Boundary Interference in Transonic and Supersonic Wind Tunnels

The wave reflection phenomena described in Section 10.11 are utilized in many applications, among them the design of transonic and low supersonic wind tunnels to minimize tunnel boundary interference. In these tunnels, as well as in low speed tunnels, the object is to simulate the flow phenomena past aircraft in flight in the atmosphere by passing a flow, uniform in direction and velocity, past a stationary model in the wind tunnel. It becomes necessary to correct force measurements on the model for the presence of the tunnel boundaries. For incompressible flow, this calculation is made by the method of images described in Section 4.12.

In supersonic wind tunnels, the attainment of a uniform supersonic flow in a particular region necessarily means that the region is free of expansion or compression waves, either of which would signify nonuniformities in both direction and Mach number. The uniformity is achieved by utilizing the wave reflection phenomena of the previous section in two ways.

1. The two-dimensional nozzle is designed as illustrated schematically in Fig. 26. The flow is sonic at the throat, after which the walls expand to an area ratio (given in Table 5) for the design Mach number M_∞ in the test section. At A in Fig. 26 the flow turns away from itself by an amount $+\delta\theta$, *thus generating* an expansion wave; by the principles described above this wave is *absorbed* when it encounters an equal

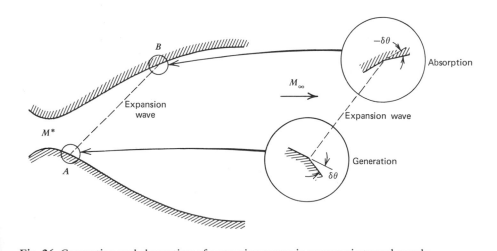

Fig. 26 Generation and absorption of expansion waves in supersonic tunnel nozzle.

and opposite deflection, $-\delta\theta$, at the point B where the wave intersects the opposite wall. If the upper and lower walls are designed throughout to absorb all of the waves generated by the turning of the flow at the opposite wall, the resulting flow issuing from the nozzle at M_∞ will be wave free and therefore uniform in Mach number and direction. The design of the nozzle is carried out by the "method of characteristics" (see Liepmann and Roshko, 1957) by which the nozzle is represented by a multitude of straight line segments at 1° or 2° to each other. The method also takes into account the deflection of each of the expansion waves as it intersects one from the opposite wall.

2. The second effect of reflections involves the waves generated by the model, as indicated in Fig. 27. The flow in the "test rhombus" shown is uniform as long as the waves generated by the body and reflected from the tunnel wall do not inter-

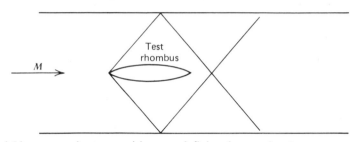

Fig. 27 Model in supersonic stream with waves defining the test rhombus.

sect the body. The size of the test rhombus and, therefore, of the body, the flow about which is to be independent of boundary interference, will depend on the intensity of the shock waves it generates and on the Mach number.

At Mach numbers near unity the shock angles will be near 90° and the rhombus therefore will be so short (or the tunnel must be so large) that other methods of eliminating boundary effects must be found. "Slotted wall" tunnels have been developed which obviate the reflection shown in Figs. 21 and 24. Figure 21 shows that when a wave intersects a solid plane boundary it reflects without change in sign, that is, a compression reflects as a compression and an expansion as an expan-

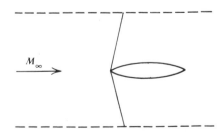

Fig. 28 Model in supersonic tunnel indicating absorption of waves by slotted boundaries.

sion.   Figure 24 shows the opposite, that a wave intersecting a constant pressure boundary reflects with change in sign.   A slotted or porous boundary as shown in Fig. 28 provides a compromise between solid and constant pressure boundaries that effectively absorb both compressions and expansion generated by the body surface. This description is, of course, idealized, but use of the method has yielded reasonably satisfactory test results near Mach unity.

"Blockage" is still a serious problem in this range because however small the model is, at Mach unity the test section becomes a second throat (see Section 9.6).

As was mentioned in the previous section viscous effects on the model and along the tunnel boundaries must be taken into account.

CHAPTER
11

Linearized Compressible Flow

11.1 Introduction

The analyses of the preceding chapters have shown the essential differences between compressible and incompressible flows; these stem from the circumstances that as gas speeds increase the accompanying density decrease becomes significant well before sonic speed is reached locally; then, when the flow becomes completely supersonic, the entire flow configuration is altered. Thus, in subsonic flow any modification of the surface (flap or elevator deflection, etc.) influences the pressure distribution over the entire wing; disturbances in a supersonic flow, on the other hand, have a limited "zone of action." For instance, in Section 5.8 we showed that in incompressible flow, deflection of a flap alters the entire pressure distribution over the wing, while in supersonic flow the analyses of Sections 10.4 and 10.5 indicate that only the pressure in the immediate vicinity of the flap would be altered.

In some cases of interest pressure distributions in supersonic flow may be calculated by means of the shock and Prandtl-Meyer theory of Chapter 10. Generally, however, exact solutions of the finite wing and body problems are not feasible and recourse is made to approximations. By assuming a wing geometry that disturbs the flow only slightly, it is possible to approximate the flow field by the solution of a linear differential equation over much of the Mach number range.

In this chapter, the equation governing the steady flow of a nonviscous compressible fluid is derived and then approximated in accordance with small perturbation theory. In the subsonic range the equation may be linearized and solved by means of a transformation of variables that converts the problem to an equivalent one in incompressible flow. In the supersonic range, the linearized equation is also a good approximation and solutions may be found by superimposing elementary solutions in a manner that will satisfy the boundary conditions of the problem. The procedures used follow closely those introduced earlier in the text in connection with incompressible wing theory. In transonic and hypersonic flows, the linearization is invalid but special methods permit approximate solutions to design problems.

The material presented in the following sections is meant as an introduction to some of the methods of wing and body theory. The underlying principles are described as well as their relation to material presented in previous chapters. For complete mathematical developments and summaries of results, the reader is referred to original papers and to texts on wing theory.

11.2 The Flow Equation

For an incompressible potential flow, the velocity pattern is given by a solution of Laplace's equation $\nabla^2 \phi = 0$, where ϕ is the velocity potential satisfying the boundary conditions of the problem considered. In high-speed flight, the fluid density may not be considered constant and the flow equation is much more complicated. In this section the equation governing the steady flow of a nonviscous compressible fluid is derived by combining the equations of continuity, motion, and energy. Isentropy and irrotationality are assumed. For simplicity, the derivation is carried out in two dimensions. The three-dimensional equation may be derived by following an identical procedure with the z component of velocity considered.

The equations of motion and continuity may be written

$$u \frac{\partial u}{\partial x} + v \frac{\partial u}{\partial y} = -\frac{1}{\rho} \frac{\partial p}{\partial x}$$

$$u \frac{\partial v}{\partial x} + v \frac{\partial v}{\partial y} = -\frac{1}{\rho} \frac{\partial p}{\partial y} \tag{1}$$

$$\frac{\partial \rho u}{\partial x} + \frac{\partial \rho v}{\partial y} = 0$$

Using the relation $a^2 = dp/d\rho$, the first two equations may be written

$$u \frac{\partial u}{\partial x} + v \frac{\partial u}{\partial y} = -\frac{a^2}{\rho} \frac{\partial \rho}{\partial x}$$

$$u \frac{\partial v}{\partial x} + v \frac{\partial v}{\partial y} = -\frac{a^2}{\rho} \frac{\partial \rho}{\partial y}$$

The first of these equations is multiplied by u and the second by v, and the two are added, giving

$$u^2 \frac{\partial u}{\partial x} + uv \frac{\partial u}{\partial y} + uv \frac{\partial v}{\partial x} + v^2 \frac{\partial v}{\partial y} = -\frac{a^2}{\rho} \left(u \frac{\partial \rho}{\partial x} + v \frac{\partial \rho}{\partial y} \right) \tag{2}$$

The third of Eqs. 1 may be expanded and becomes

$$\frac{\partial u}{\partial x} + \frac{\partial v}{\partial y} = -\frac{1}{\rho} \left(u \frac{\partial \rho}{\partial x} + v \frac{\partial \rho}{\partial y} \right) \tag{3}$$

Substituting Eq. 3 in the right side of Eq. 2 gives

$$\left(\frac{u^2}{a^2} - 1 \right) \frac{\partial u}{\partial x} + \frac{uv}{a^2} \left(\frac{\partial u}{\partial y} + \frac{\partial v}{\partial x} \right) + \left(\frac{v^2}{a^2} - 1 \right) \frac{\partial v}{\partial y} = 0 \tag{4}$$

If we introduce the irrotationality condition $(\partial v/\partial x - \partial u/\partial y) = 0$ and the velocity potential, Eq. 4 becomes

$$\left(\frac{u^2}{a^2} - 1\right)\frac{\partial^2 \phi}{\partial x^2} + \left(\frac{2uv}{a^2}\right)\frac{\partial^2 \phi}{\partial x \partial y} + \left(\frac{v^2}{a^2} - 1\right)\frac{\partial^2 \phi}{\partial y^2} = 0 \tag{5}$$

Equation 5 is the steady flow equation for a nonviscous compressible fluid. Its three-dimensional counterpart is

$$\left(\frac{u^2}{a^2} - 1\right)\frac{\partial^2 \phi}{\partial x^2} + \left(\frac{v^2}{a^2} - 1\right)\frac{\partial^2 \phi}{\partial y^2} + \left(\frac{w^2}{a^2} - 1\right)\frac{\partial^2 \phi}{\partial z^2} + \frac{2}{a^2}\left(uv\frac{\partial^2 \phi}{\partial x \partial y}\right.$$

$$\left. + vw\frac{\partial^2 \phi}{\partial y \partial z} + wu\frac{\partial^2 \phi}{\partial z \partial x}\right) = 0 \tag{6}$$

The sonic speed a is a variable. It may be put in terms of the flow speed by using the energy relation, Eq. 9.9, in the form

$$a^2 + \tfrac{1}{2}(\gamma - 1)V^2 = a_\infty^2 + \tfrac{1}{2}(\gamma - 1)V_\infty^2 \tag{7}$$

All velocities may, of course, be written in terms of the velocity potential. Equations 5 and 6 may therefore be written with ϕ as the only dependent variable. Because of the great complexity of the resulting potential equation, an analytical solution for boundary conditions corresponding to shapes of interest in aerodynamics is not possible. Fortunately, many practical flow problems can be approximately represented by a linearized form of the flow equation. This form is developed in the next section.

It should be observed that for an incompressible fluid, a is infinite and Eqs. 5 and 6 reduce to Laplace's equation, solutions of which are treated in earlier chapters.

11.3   Flow Equation for Small Perturbations

A thin body at a small angle of attack in motion through a fluid, in general, disturbs the fluid only slightly. That is, the perturbation caused by the body is small compared to the velocity of the body.

Consider a thin body moving with speed $-V_\infty$ through a fluid. An observer stationed on the body will see a uniform stream V_∞ on which are superimposed perturbation components u', v', and w'. The perturbation components, except at stagnation points, will be small compared to V_∞. Assuming V_∞ to be in the direction of the x axis, the total velocity at any point will be given by the three components.

$$V_\infty + u' = V_\infty + \phi_x$$

$$v' = \phi_y \qquad (8)$$

$$w' = \phi_z$$

The partial derivatives of ϕ are indicated by the appropriate subscripts, thus Eqs. 8 are substituted in Eq. 5 and for simplicity the analysis is restricted to two dimensions. Equation 5 then becomes

$$\left(\frac{V_\infty^2 + 2u'V_\infty + u'^2}{a^2} - 1\right)\phi_{xx} + 2\left(\frac{V_\infty v' + u'v'}{a^2}\right)\phi_{xy} + \left(\frac{v'^2}{a^2} - 1\right)\phi_{yy} = 0 \qquad (9)$$

Since V_∞ is a constant, the second derivative of its potential is zero, and therefore ϕ in Eq. 5 may be regarded as either the perturbation potential or the total potential of the flow. In this chapter, ϕ is taken as the perturbation potential. It will be shown that under certain conditions the bracketed coefficients are constants and Eq. 9 becomes a linear equation in the perturbation potential.

If the velocity is written as the sum of free-stream and perturbation parts in Eq. 7, the local sonic ratio becomes

$$\left(\frac{a}{a_\infty}\right)^2 = 1 - \frac{\gamma - 1}{2}M_\infty^2\left(2\frac{u'}{V_\infty} + \frac{u'^2}{V_\infty^2} + \frac{v'^2}{V_\infty^2}\right)$$

or, neglecting higher order terms in the perturbation to free stream ratio,

$$\left(\frac{a}{a_\infty}\right)^2 \simeq 1 - (\gamma - 1)M_\infty^2\frac{u'}{V_\infty} \qquad (10)$$

Now, if we multiply Eq. 9 by $(a/a_\infty)^2$, neglect in the parentheses all terms of second order in the perturbations, and substitute for $(a/a_\infty)^2$ from Eq. 10, we obtain

$$\left[(M_\infty^2 - 1) + (\gamma + 1)M_\infty^2\frac{\phi_x}{V_\infty}\right]\phi_{xx} + \left(2M_\infty^2\frac{\phi_y}{V_\infty}\right)\phi_{xy} - \phi_{yy} = 0$$

This equation is correct to the second order in the perturbations. The terms involving ϕ_x and ϕ_y are nonlinear but for slender bodies lateral components are small compared with the longitudinal so the middle term is neglected. The equation thus reduces to

$$\left[(M_\infty^2 - 1) + (\gamma + 1)M_\infty^2\frac{\phi_x}{V_\infty}\right]\phi_{xx} - \phi_{yy} = 0 \qquad (11)$$

This equation is termed the *transonic flow equation* because in near sonic flows the second term in the brackets cannot be neglected compared with $(M_\infty^2 - 1)$. Thus transonic flow is governed by a nonlinear differential equation in the velocity perturbations; the equation becomes effectively linear in Mach number ranges for which

$$|M_\infty^2 - 1| \gg (\gamma + 1) M_\infty^2 \frac{\phi_x}{V_\infty} \tag{12}$$

When the Mach number of the flow for a given body is such that this condition is satisfied, Eq. 11 reduces to

$$(1 - M_\infty^2)\, \phi_{xx} + \phi_{yy} = 0 \tag{13}$$

and, if the reduction is carried out in three dimensions, the equation is

$$(1 - M_\infty^2)\, \phi_{xx} + \phi_{yy} + \phi_{zz} = 0 \tag{14}$$

As $M_\infty \to 0$, the equation approaches Laplace's equation, as it must since the flow becomes incompressible. At the higher Mach numbers, angles of attack, or thickness-chord ratios, the resulting increase in $M_\infty^2 \phi_x$ in the inequality (12) tends to invalidate Eq. 14, so that Eq. 11 must be used as the governing equation. Quantitatively, for flows in which $M_\infty \alpha$ or $M_\infty \tau$, where τ is the maximum thickness-chord ratio, reach a value of 0.5 to 0.7, significant errors occur in the solutions of Eq. 14; thus, for example, in a $M_\infty = 5$ flow, if $\alpha = 0.1$ rad or $\tau = 10$ percent, the errors involved in the use of linear theory will be appreciable.

Since M_∞ is constant the coefficients of ϕ_{xx} in Eqs. 13 and 14 are constants and, as we show in the next chapter, an appropriate stretching of the coordinates reduces a compressible subsonic flow problem to that about an "equivalent" body in an incompressible flow, to which the methods of chapters 4 and 5 are again applicable. The results can then be converted to those for the actual body in a compressible flow.

11.4   Steady Supersonic Flows

In incompressible flows (Chapters 1 to 6), we discussed flows about arbitrary bodies in terms of the superposition of source, doublet, and vortex panels and uniform flows; in *completely supersonic linearized flow*, the linearity of Eq. 14 validates superposition of flows about *slender* bodies as well. However, since solutions of Eq. 14 are expressed in terms of $\beta = \sqrt{1 - M_\infty^2}$, which becomes imaginary for $M_\infty > 1$, these solutions are valid for subsonic flow *only*. Therefore, for supersonic applications, Eq. 14 is written in the form

$$(M_\infty^2 - 1)\, \phi_{xx} - \phi_{yy} - \phi_{zz} = 0 \tag{15}$$

Instead of β, the quantity $\lambda = \sqrt{M_\infty^2 - 1}$ occurs importantly in these solutions. Physically the change from Eq. 14 to Eq. 15 with $M_\infty > 1$ signifies very significant changes between supersonic and subsonic compressible flow fields. The differences can be exemplified by considering the supersonic counterparts of the above elementary solutions for incompressible flows all of which must satisfy Laplace's equation $\nabla^2 \phi = 0$, instead of Eq. 15.

These elementary supersonic solutions with *centers at the origin* are:

$$
\left.
\begin{array}{ll}
\text{Source} & \phi = -\dfrac{C}{2\pi h} \\[2ex]
\begin{array}{l}\text{Doublet} \\ \text{(with vertical axis)}\end{array} & \phi = \dfrac{C\lambda^2 z}{2\pi h^3} \\[2ex]
\text{Vortex} & \phi = \dfrac{Cxz}{2\pi\,(y^2 + z^2)\,h}
\end{array}
\right\}
\tag{16}
$$

where

$$
h = \sqrt{x^2 - \lambda^2\,(y^2 + z^2)} \tag{17}
$$

and C is a constant. The reader may verify that these are solutions of Eq. 15.

These solutions and their superposition to solve specific problems are discussed by Heaslet and Lomax (1954). We shall discuss here only the first of the above solutions, that for the supersonic source; the others are referred to in Chapter 13.

At $M_\infty = 0$, Eq. 15 reduces to Laplace's equation, the exact equation for incompressible flow, $\nabla^2 \phi = 0$, and the source solution of Eqs. 16

$$
\phi = \frac{C}{2\pi\sqrt{x^2 + y^2 + z^2}} = \frac{C}{2\pi r}
$$

describes a point source in incompressible flow (see Problem 2.4.3).

The elementary supersonic solution Eqs. 16 will be referred to in Chapter 13 in connection with aircraft configurations for transonic and supersonic flow. They are characterized particularly by the so-called hyperbolic radius h of Eq. 17, illustrated in Fig. 1; we note that $h = 0$ on the surface of the cone with axis along x,

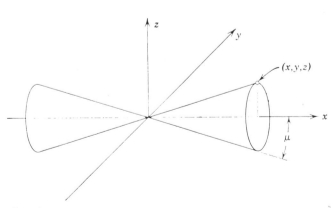

Fig. 1 Supersonic point source.

and with semiapex angle

$$\mu = \sin^{-1}\left(\frac{1}{M_\infty}\right); \qquad \cot\mu = \sqrt{M_\infty^2 - 1} \equiv \lambda$$

Inside the cone h is real; outside it is imaginary.

The term *source* is carried over to the supersonic case, although the previous physical interpretation (that of an incompressible fluid issuing from a point) must be revised. The properties of this linearized source can be described by reference to Fig. 1, which, for a given M_∞, represents the Mach cones with apices at the source. However, the influence of a supersonic source cannot extend upstream in a supersonic flow, and, since,

$$\lambda\sqrt{y^2 + z^2} \qquad \begin{array}{l} > x \text{ outside of the Mach cone} \\[4pt] < x \text{ within the Mach cone} \end{array}$$

h is real within and imaginary outside the Mach cones. For these reasons C, the strength of the source, doublet, or vortex in Eqs. 16 is replaced by zero everywhere outside the *downstream* Mach cone, and within that cone the velocity components of its flow field are ϕ_x, ϕ_y, ϕ_z.

It follows that the supersonic flow about a slender, nonlifting body of revolution can be analyzed by superposing on the main flow the perturbation velocities of a line of supersonic sources, of strength $C(x)$, whose Mach cones intersect the body surface upstream of any given surface point. This problem was solved by von Kármán and Moore in 1934 (see Liepmann and Roshko, 1957, Chap. 9).

Another noteworthy feature of the supersonic flow past a body of revolution emerges when we compare the perturbation velocities, given by superposition of the supersonic source fields with those for two-dimensional linearized flows. For two-dimensional flows, ϕ and its derivatives, that is, the perturbation velocities (Eqs. 8) are constant at every point on a given Mach line; on the other hand the perturbation velocities given by Eqs. 16 decrease with distance from the x axis within the Mach cone.

11.5   The Pressure Coefficient for Small Perturbations

For small perturbations the difference between the pressure p at any point in the flow and the free stream pressure p_∞ is given to first order of approximation by Eq. 7.11 (gravity neglected).

$$p - p_\infty = dp = -\rho\, d\left(\frac{V^2}{2}\right) \tag{18}$$

ρ may be expressed as the sum of a free-stream part ρ_∞ and a perturbation part ρ'. Also, $d(V^2/2)$ is interpreted as the difference $(V^2 - V_\infty^2)/2$.

$$\tfrac{1}{2}(V^2 - V_\infty^2) = V_\infty u' + \tfrac{1}{2}(u'^2 + v'^2 + w'^2)$$

Making these substitutions in Eq. 15 leads to

$$p - p_\infty = -(\rho_\infty + \rho')[V_\infty u' + \tfrac{1}{2}(u'^2 + v'^2 + w'^2)]$$

After neglecting all terms greater than first degree in the perturbations, the above equation becomes

$$p - p_\infty = -\rho_\infty V_\infty u'$$

which leads to the pressure coefficient

$$C_p \equiv \frac{p - p_\infty}{\tfrac{1}{2}\rho_\infty V_\infty^2} = -\frac{2u'}{V_\infty} = -\frac{2\phi_x}{V_\infty} \tag{19}$$

For the flow past slender "fuselage-like" bodies, Lighthill pointed out that the term $\tfrac{1}{2}(v'^2 + w'^2)$ is no longer negligible (see Liepmann-Roshko, 1957) compared with $V_\infty u'$ and as a consequence, Eq. 19 must be replaced by

$$C_p = -2\frac{u'}{V_\infty} - \frac{v'^2 + w'^2}{V_\infty^2} \tag{19a}$$

If the flow is supersonic this result can be expressed in terms of θ, the local angle of inclination of the surface to V_∞. For small inclinations, we may write Eq. 10.10 as

$$\frac{dp}{d\theta} \simeq \frac{p - p_\infty}{\theta} = \frac{\gamma M_\infty^2 p_\infty}{\sqrt{M_\infty^2 - 1}}$$

where the sign has been changed because θ, as used here, is positive in the counter-clockwise direction. But $\gamma p_\infty M_\infty^2 = \rho_\infty V_\infty^2$, so that the pressure coefficient becomes, for supersonic flow,

$$C_p = \frac{2\theta}{\sqrt{M_\infty^2 - 1}} \tag{20}$$

11.6 Summary

The differential equation governing compressible flows is linearized so that its application is limited to slender bodies. The various neglected terms are discussed and reasons for the nonapplication of the equation to transonic and hypersonic flows are pointed out. Some elementary solutions for sources, doublets, and vortices are given. The formula for the pressure coefficient in linearized flow is derived.

CHAPTER
12

Airfoils in Compressible Flows

12.1 Introduction

As a prelude to the consideration of finite wings with sweepback, as well as wing-body combinations in Chapter 13, this chapter is concerned with the application of the linear equation, Eq. 11.13, to the two-dimensional compressible flow past airfoil shapes and their agreement with experiment. Consider first an airfoil shape of zero camber at zero angle of attack in a compressible flow. The pressure coefficient may be expressed by

$$C_p = C_p \left[\frac{t}{c} (x), \tau, M_\infty \right]$$

where t/c is the distribution of relative thickness as a function of position along the chord, and τ designates $(t/c)_{max}$, a member of a given "family" of airfoil shapes (a single proportionality factor relates the thickness distribution of one member to any other member of a family). Our first analyses will be confined to subsonic flow and will be concerned with (1) C_p at a given x/c for a given t/c distribution, and (2) C_p at a given x/c and M_∞ as a function of τ. Then we will define, mainly on the basis of experiment, the limits of validity of the result.

Transonic flow, that is, a subsonic main flow with an embedded supersonic flow region or a supersonic main flow with an embedded subsonic flow region, is treated mainly on the basis of experiment and the physical principles involved.

Supersonic flow past airfoil shapes is treated completely on the basis of linearized flow theory. Comparison with experiment will be presented and discussed.

12.2 Boundary Conditions

As pointed out in previous chapters on airfoil theory, the wing is a barrier to the flow in the sense that there can be no resultant velocity component normal to the surface. In other words, *the sum of the components of the free-stream and pertur-bation velocities normal to the surface must be zero*. With reference to Fig. 1, which represents an airfoil section with exaggerated thickness, camber, and angle of attack, the condition of vanishing normal component at the boundary is expressed by

$$V_{\infty n} + \frac{\partial \phi}{\partial n} = 0 \tag{1}$$

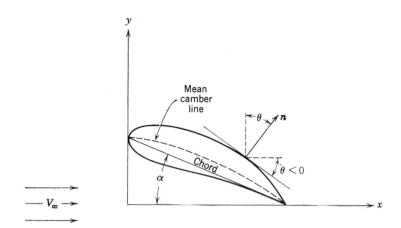

Fig. 1 Section geometry.

θ is the angle between the tangent to the surface and the free-stream direction. Then

$$V_{\infty n} = -V_\infty \sin \theta \simeq -V_\infty \theta \tag{2}$$

For situations in which small perturbation theory is applicable, θ is small and to close approximation it may be taken as the sum of the following three parts (Fig. 2): (a) angle between the free stream and the chord line, that is, the angle of attack, α; (b) angle between the chord line and the tangent to the mean camber line; (c) angle between the tangent to the mean camber line and the tangent to the surface.

As a consequence of the additive character of the three contributions, the flow field may be treated as the superposition of a flat plate at an angle of attack, the mean camber line at *zero* angle of attack, and a symmetrical thickness envelope at

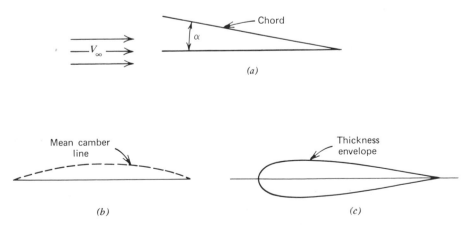

Fig. 2 Resolution of slope θ into parts due to angle of attack, camber, and thickness.

zero angle of attack. (The mean camber line and thickness envelope were defined in Chapter 5.)

Items (a) and (b) contribute to the lift. Item (c) is nonlifting. To find the flow pattern or pressure distribution for the airfoil of Fig. 1 the lifting and nonlifting problems are solved separately and the results added.

Finally, following the procedure of Section 5.4, the *planar wing* approximation is made and the boundary condition expressed by Eq. 1 is satisfied in the $y = 0$ plane. Equation 1 becomes

$$(\phi_y)_{y=0} = V_\infty \theta \tag{3}$$

For the lifting problem, θ has the same sign on the upper and lower surfaces of the wing, and therefore ϕ_y has the same sign at $y = 0+$ and $y = 0-$. For the nonlifting problem ϕ_y has opposite signs at $y = 0+$ and $y = 0-$.

12.3   Airfoils in Subsonic Flow; Prandtl-Glauert Transformation

Our objective here is to transform the two-dimensional *compressible* flow in the x, y plane into an "equivalent" *incompressible* flow in the ξ, η plane, defined in such a way that the aerodynamic characteristics of the compressible flow can be calculated from their incompressible counterparts. To do so, we write Eq. 11.13 in terms of the velocity potential $\Phi(x, y)$ instead of the perturbation potential ϕ. Thus

$$\Phi_{xx} + \frac{\Phi_{yy}}{\beta^2} = 0 \tag{4}$$

where

$$\beta = \sqrt{1 - M_\infty^2}$$

Equation 4 is identical with Eq. 11.13 because $\Phi_{xx} = \phi_{xx}$. We consider an airfoil shape given by $y = f(x)$ in a compressible flow of velocity V_∞. We now show that Eq. 4 also describes an *incompressible* flow with velocity potential

$$\Phi^0(\xi, \eta) = m\Phi(x, y)$$

where m = constant, and

$$\xi = x, \quad \eta = \beta y \tag{5}$$

Then

$$\Phi_{xx} = \frac{\Phi^0_{\xi\xi}}{m}$$

The transformation of the y derivative will be carried out in steps:

$$\frac{\partial \Phi}{\partial y} = \frac{1}{m} \frac{\partial \Phi^0}{\partial y} = \frac{1}{m} \frac{\partial \Phi^0}{\partial \eta} \frac{d\eta}{dy} = \frac{\beta}{m} \frac{\partial \Phi^0}{\partial \eta}$$

$$\frac{\partial^2 \Phi}{\partial y^2} = \frac{\beta}{m} \frac{\partial}{\partial \eta} \left(\frac{\partial \Phi^0}{\partial \eta} \right) \frac{d\eta}{dy} = \frac{\beta^2}{m} \frac{\partial^2 \Phi^0}{\partial \eta^2}$$

Substituting in Eq. 4 we obtain

$$\Phi_{xx} + \frac{\Phi_{yy}}{\beta^2} = \frac{1}{m} (\Phi^0_{\xi\xi} + \Phi^0_{\eta\eta}) = 0$$

which shows that Φ^0 satisfies Laplace equation in the ξ, η plane. Therefore, the incompressible flow represented by $\Phi^0(\xi, \eta)$ is a solution of Eq. 4 if $\Phi^0 = m\Phi$, and the coordinates are related as shown in Eqs. 5. The boundary condition Eq. 3, however, describes a new airfoil shape in the incompressible flow. For the compressible flow, $\theta = df(x)/dx$ is the inclination of the surface at a given point on the surface; for the incompressible flow, we set $\theta_0 = dg(\xi)/d\xi$, and use thin airfoil approximations to evaluate the relation between θ and θ_0 as follows:

$$V_\infty \theta = V_\infty \frac{df}{dx} = \left(\frac{\partial \Phi}{\partial y} \right)_{y=0} = \frac{1}{m} \left(\frac{\partial \Phi^0}{\partial y} \right)_{y=0} = \frac{\beta}{m} \left(\frac{\partial \Phi^0}{\partial \eta} \right)_{\eta=0} = \frac{\beta V_\infty}{m} \frac{dg}{d\xi} = \frac{\beta V_\infty}{m} \theta_0 \quad (6)$$

so that, from the definitions of f and g,

$$\theta = \frac{\theta_0 \beta}{m} \quad (7)$$

Eqs. 5, 6, and 7 represent the essential features of the "Prandtl-Glauert" transformation relating subsonic compressible and incompressible two-dimensional flows. So far the constant m is free. We now evaluate it for two practical cases.

(1) Consider identical airfoils in compressible and incompressible flows. Then, by Eq. 7,

$$m = \beta$$

so that

$$\Phi(x, y) = \frac{\Phi^0(\xi, \eta)}{\beta}$$

Then the x perturbation velocities u' and u'_0 are

$$u' = \Phi_x = \frac{\Phi^0_\xi}{\beta} = \frac{u'_0}{\beta}$$

and by Eq. 11.19 $C_p = -2u'/V_\infty$, so the pressure coefficients for the identical shapes are related by

$$C_p = \frac{C_{p_0}}{\beta} \qquad (8)$$

where C_{p_0} refers to the pressure coefficient for incompressible flow.

(2) For two geometrically similar shapes such that

$$\Phi(x, y) = \Phi^0(\xi, \eta)$$

so that $m = 1$ and $u' = u'_0$. The pressure coefficients are then also equal, provided the corresponding inclinations of the surfaces are related by (Eq. 7)

$$\theta = \theta_0 \beta \qquad (9)$$

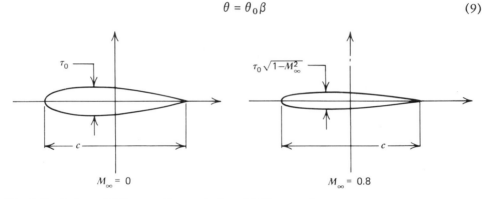

$M_\infty = 0$ $M_\infty = 0.8$

Fig. 3 Equivalent airfoils according to the Prandtl-Glauert rule.

Figure 3 shows two similar symmetrical airfoil shapes with identical C_p distributions, one in an $M_\infty = 0$ flow and the other in an $M_\infty = 0.8$ flow. Their maximum thicknesses are related by

$$\tau = \tau_0 \beta \qquad (10)$$

Equation 9 applies to all deviations of the surface from $0°$, that is, within the validity of the linear theory, the C_p's will be identical only if the thickness, the camber, and the angle of attack (see Fig. 2) are all reduced by the factor $\sqrt{1 - M_\infty^2}$ from their values at $M_\infty = 0$.

Most practical applications, however, utilize the results of case (1) above, expressed locally by Eq. 8. Since the lift coefficient

$$c_l = \frac{1}{q_\infty c} \int_0^c (p_L - p_U)\, dx = \int_0^1 (C_{p_L} - C_{p_U})\, d\frac{x}{c}$$

where the subscripts U and L refer, respectively, to the upper and lower surfaces, Eq. 8 applies as well to the integrated values. Therefore,

$$c_l = \frac{c_{l_0}}{\beta} \tag{11}$$

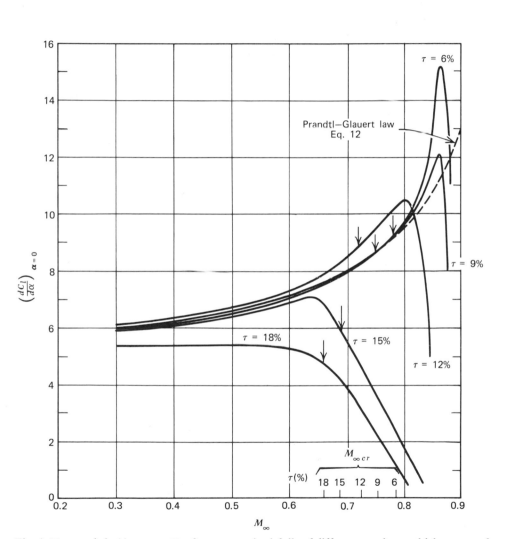

Fig. 4 Measured $dc_l/d\alpha$ versus M_∞ for symmetric airfoils of different maximum thickness τ and comparison with Prandtl-Glauert law for $dc_l/d\alpha = 6.0$ per radian at $M_\infty = 0.3$; $M_{\infty cr}$ for each curve is designated by an arrow and by a point on the abscissa. (Mair and Beavan, see p. 658; Howarth, 1953. Courtesy of Oxford University Press.)

and

$$\frac{dc_l}{d\alpha} = \frac{1}{\beta}\left(\frac{dc_l}{d\alpha}\right)_0 \tag{12}$$

Figure 4 compares the Prandtl-Glauert law, expressed by Eq. 12, with experimentally determined curves of $dc_l/d\alpha$ versus M_∞ for geometrically similar symmetrical airfoil shapes of different maximum thicknesses, τ, at $\alpha = 0°$. We note that the agreement between theory and experiment is good at $M_\infty < 0.63$ for $\tau \leqslant 0.15$ and at $M_\infty < 0.86$ for $\tau = 0.06$.

Three features of the curves require some comment.

(1) The points indicated by arrows designate for each airfoil the M_∞, designated $M_{\infty cr}$, for which the local M becomes unity at some point on the surface; since the linearization that led to Eq. 11.13 in effect substitutes M_∞ for M these arrows designate the upper limit of validity of the Prandtl-Glauert law, respectively, for each of the thicknesses.

(2) The experimental curves for the thinner airfoils show higher values of C_p than predicted by the Prandtl-Glauert law.

(3) The more or less sudden drop of $dc_l/d\alpha$ after reaching a peak is associated with the formation of shock waves and flow separation; this feature will be described in connection with the discussion of transonic flow in Section 12.5 and in later chapters on viscous flow.

As will also be indicated in Section 12.5 (Fig. 7) the lift and moment coefficients of an airfoil are practically unaffected for $M_\infty < M_{\infty cr}$.

12.4   Critical Mach Number

The subsonic critical Mach number, $M_{\infty cr}$, was defined in the previous section as that M_∞ of the external flow for which M, the local Mach number, at some point on the surface reaches unity; the arrows on Fig. 4 indicate $M_{\infty cr}$ for the various airfoils tested.   It was also pointed out that $M_\infty = M_{\infty cr}$ marks the upper limit for the validity of the Prandtl-Glauert law.   The value of $M_{\infty cr}$, if C_p is known at a lower M_∞, may be calculated as follows:

The pressure p and Mach number M at a point on the airfoil are related by Eq. 9.16, that is

$$\frac{p}{p_0} = \left(1 + \frac{\gamma - 1}{2} M^2\right)^{\gamma/(1-\gamma)}$$

Let p_∞ and M_∞ be the pressure and Mach number of the free stream.   Then,

$$\frac{p}{p_\infty} = \left[\frac{1 + \frac{1}{2}(\gamma - 1)M^2}{1 + \frac{1}{2}(\gamma - 1)M_\infty^2}\right]^{\gamma/(1-\gamma)} \tag{13}$$

The pressure coefficient at the point of minimum pressure is

$$\frac{p - p_\infty}{\frac{1}{2}\rho_\infty V_\infty^2} = C_p = \frac{2}{\gamma M_\infty^2}\left(\frac{p}{p_\infty} - 1\right) \tag{14}$$

Substituting Eq. 13 in Eq. 14, the pressure coefficient becomes

$$C_p = \frac{2}{\gamma M_\infty^2}\left\{\left[\frac{1 + \frac{1}{2}(\gamma - 1)M^2}{1 + \frac{1}{2}(\gamma - 1)M_\infty^2}\right]^{\gamma/(1-\gamma)} - 1\right\}$$

If $M = 1$, then by definition $M_\infty = M_{\infty cr}$ and $C_p = C_{p_{cr}}$.

$$C_{p_{cr}} = \frac{2}{\gamma M_{\infty cr}^2}\left\{\left[\frac{\frac{1}{2}(\gamma + 1)}{1 + \frac{1}{2}(\gamma - 1)M_{\infty cr}^2}\right]^{\gamma/(1-\gamma)} - 1\right\} \tag{15}$$

From Eq. 15 may be found the critical Mach number of the airfoil when the pressure coefficient at the point of minimum pressure is known. A plot of $C_{p_{cr}}$ versus $M_{\infty cr}$ is represented by the solid line in Fig. 5.

From the test data at low speeds, the pressure coefficient at the point of minimum pressure may be found. Then, by applying the Prandtl-Glauert rule, C_p at a

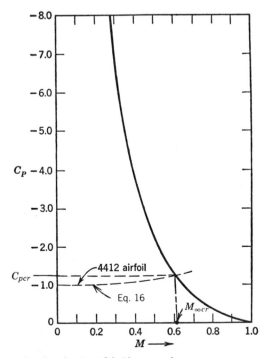

Fig. 5 Critical Mach number by the Prandtl-Glauert rule.

given M_∞ is related to C_{p1} at $M_{\infty 1}$ by

$$C_p = C_{p1} \sqrt{\frac{1 - M_{\infty 1}^2}{1 - M_\infty^2}} \qquad (16)$$

For $M_{\infty 1} = 0$, $C_{p_1} = C_{p_0}$ and the equation reduces to Eq. 8. The pressure coefficient C_p for various values of M_∞ may be computed; the broken line in Fig. 5 represents such a computation for the NACA 4412 airfoil at an angle of attack of $2°$. C_{p_1} used in the computation is the pressure coefficient at the point of minimum pressure, which has the value of -1 at $M_{\infty 1} = 0$ for this airfoil. The intersection with the solid curve occurs at $M_\infty = M_{\infty cr}$ for that shape and that angle of attack.

The critical Mach number $M_{\infty cr}$ decreases with maximum thickness as shown in Fig. 4, as well as with increasing angle of attack and maximum camber for a given family of airfoil profiles.

12.5 Airfoils in Transonic Flows

The flow is transonic, that is, mixed subsonic and supersonic, for $M_\infty > M_{\infty cr}$. Figure 6 shows schematic wave configurations for the range $M_\infty = 0.7$, which is slightly greater than $M_{\infty cr}$, to $M_\infty = 1.3$ for an airfoil at a low angle of attack. Since the "recovery shock" from supersonic to subsonic flow is weak at $M_\infty = 0.70$

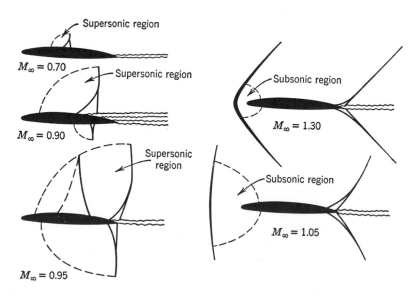

Fig. 6 Flow patterns around an airfoil in transonic flow. (Courtesy of NACA.) Dotted line in $M_\infty = 0.95$ diagram represents a typical expansion wave generated by turning of the flow at the airfoil surface; the solid line represents its reflection as a compression wave from the constant pressure surface representing the "sonic surface."

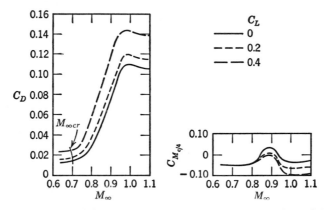

Fig. 7 NACA measurements of forces and moments on typical low-drag airfoil in transonic flow. Aspect ratio 6.4. (Weaver, 1948. Courtesy of NACA.)

and becomes increasingly stronger as M_∞ increases, the resulting alteration of the pressure distribution, lift, drag, and moment characteristics vary drastically through the transonic range. For the airfoils shown, $M_{\infty cr}$ is slightly less than 0.70 and the diagrams at the left show the increase in the area of the region of supersonic flow up to $M_\infty = 0.95$. The change from subsonic to supersonic flow at the "sonic line" occurs smoothly but the change back to subsonic invariably occurs abruptly through a "recovery shock." Near the surface, because of the presence of the boundary layer the shock is bifurcated and is termed a λ shock.* Experimental data for a typical wing are shown in Fig. 7. They show the increase in drag coefficient and moment coefficient associated with passage through the transonic range. Near $M_{\infty cr}$ the supersonic region is small and the flow within it is near-sonic; therefore the recovery shock will be weak and the effect on aerodynamic characteristics will be small. This result is indicated by the results of Figs. 4 and 7 which show that M_∞ can increase significantly beyond $M_{\infty cr}$ before $dc_l/d\alpha$, c_d, and $c_{mc/4}$ are appreciably affected. However, when the effects appear, they are major; $dc_l/d\alpha$ decreases more or less abruptly (Fig. 4) and c_d experiences very large changes (Fig. 7). These changes are associated with the inception of the "shock stall," that is, the flow separation associated with the pressure jump through the shock (see Fig. 6, $M_\infty = 0.9$); this phenomenon is described in some detail in Chapters 18 and 19.

At $M_\infty = 1.05$ and 1.30, the character of the mixed flow changes to a subsonic region near the leading edge embedded in the supersonic external flow. For the particular airfoil shown in Fig. 6, since it has a blunt leading edge, the flow remains mixed and therefore remains strictly a transonic flow as $M_\infty \to \infty$. Sweepback, described in Section 13.3 counteracts this effect.

*As is mentioned in the caption of Fig. 6 the sonic surface is a constant pressure surface and, therefore, expansion waves generated by the turning of the flow along the airfoil surface propagate in the supersonic flow as far as the sonic surface where they reflect as compression waves to maintain the constant pressure surface.

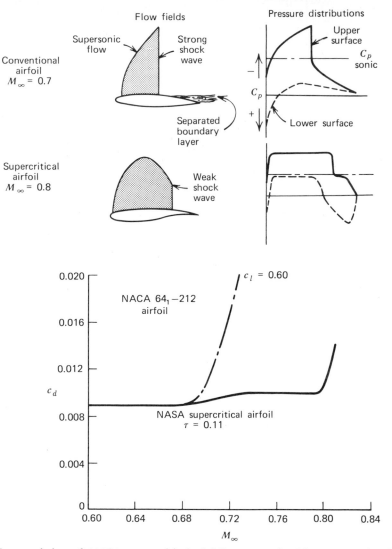

Fig. 8 Characteristics of NASA supercritical airfoil compared with a conventional section (Whitcomb, 1956. Courtesy of NASA.)

Improved airfoil shapes for transonic flows have been reported by Whitcomb in 1952 (see Whitcomb, 1956). Changes in shape from conventional airfoils as indicated in Fig. 8 show remarkable improvements in that the airfoil can be thicker, for greater rigidity, and at the same time the "recovery shock" is decreased in strength and moved rearward; the drag rise associated with the passage through the critical Mach number is delayed by more than 0.1 under cruise conditions. The shape and aerodynamic characteristics are shown in Fig. 8. A review of research on the structure of shocks in transonic flow is given by Sichel (1971).

Shock-free airfoil shapes for given transonic Mach numbers have been designated and tested (see Niewland and Spee, 1973). The shock-free feature is, however, limited to the immediate vicinity of the design Mach number.

In terms of the equations and their approximations discussed in Chapter 11, the linear differential equation, Eq. 11.13, is not valid in transonic flow because the inequality 11.12, with M_∞ designating the local Mach number, is not valid near the sonic line. The theoretical methods available for treating transonic flow are beyond the scope of this book. The von Kármán similarity rules for transonic flows are derived by Liepmann and Roshko (1957, pp. 256ff); the parameters derived provide a remarkable framework for systematizing experiments on two-dimensional bodies.

12.6   Airfoils in Supersonic Flow

The diagrams for $M_\infty > 1$ in Fig. 6 and the accompanying discussion indicate that only with a perfectly sharp leading edge could we eliminate the subsonic flow near the leading edge; also, only then will the linear equation Eq. 11.13 be approximately valid throughout the flow field of an airfoil in a supersonic flow. Actually, however, Eq. 11.13 is valid to a good approximation for the analysis of supersonic flow past airfoils with very small leading edge radius away from the immediate vicinity of the leading edge; the integrated aerodynamic characteristics are particularly well represented.

The analysis of supersonic flow cannot utilize Eq. 4 because, for $M_\infty > 1$, $\beta = \sqrt{1 - M_\infty^2}$ becomes imaginary. Therefore, the equation is written in the form

$$\phi_{xx} - \frac{\phi_{yy}}{\lambda^2} = 0 \qquad (17)$$

where $\lambda = \sqrt{M_\infty^2 - 1}$ and ϕ is the perturbation potential. Mathematically this equation is termed "hyperbolic," while for $M_\infty < 1$, Eq. 4 is "elliptic." The properties of the two equations are quite different, as will be shown by the characters of their solutions. In fact, while Eq. 4 is solved for $M_\infty < 1$ in terms of an "equivalent" incompressible flow, Eq. 17 has a direct simple solution which is expressed in the functional form

$$\phi = f(x - \lambda y) + g(x + \lambda y) \qquad (18)$$

where f and g are arbitrary functions of their arguments. We see immediately that, for $\phi = f(x - \lambda y)$,

$$\phi_{xx} = f''(x - \lambda y); \qquad \phi_{yy} = \lambda^2 f''(x - \lambda y)$$

where f'' signifies the second derivative with respect to its argument, and Eq. 17 is satisfied identically; the same result holds for $g(x + \lambda y)$.

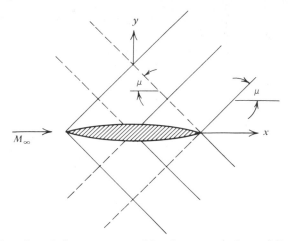

Fig. 9 Two families of Mach lines represented by the two solutions of Eq. 17 given in Eq. 18. The dotted portions are extraneous for the flow shown.

As shown in Fig. 9, $(x - \lambda y)$ and $(x + \lambda y)$ are constants respectively along the straight lines of slopes $dy/dx = \pm 1/\lambda$, which means that f and g in Eq. 18 represent two families of solutions of Eq. 17. From our analyses of waves in Chapter 10, we see that the lines are "Mach lines" of the external flow, that is,

$$\tan \mu = \frac{\pm 1}{\sqrt{M_\infty^2 - 1}} \equiv \frac{\pm 1}{\lambda} \tag{19}$$

where μ is the "Mach angle." In Fig. 9, the solid lines represent the envelopes of pressure perturbations generated by the deflections of the flow around the thin body moving at M_∞. The solid lines emanating from the upper surface are members of the $f(x - \lambda y) =$ const. family, those from the lower are members of the $g(x + \lambda y)$ family. Within the approximations of the linear theory the boundary conditions may be imposed at the chord line (Section 12.2) and the waves generated then intersect the chord line at the Mach angle $\tan^{-1} (1/\lambda)$. Then, if θ is the inclination of the surface to the free stream, the boundary conditions become

$$v' = (\phi_y)_{y=0} = -\lambda (f')_{y=0}$$
$$u' = (\phi_x)_{y=0} = (f')_{y=0} \tag{20}$$

At the surface of the body

$$\theta = \tan^{-1} \left(\frac{dy}{dx} \right)_{y=0} = \frac{v'}{(V_\infty + u')} \simeq \frac{v'}{V_\infty} \tag{21}$$

so that,

$$u' = (f')_{y=0} = -\frac{V_\infty \theta}{\lambda} \tag{22}$$

and the pressure coefficient, C_p (Eq. 11.19) becomes

$$C_p = -\frac{2u'}{V_\infty} = \frac{2\theta}{\lambda} = \frac{2\theta}{\sqrt{M_\infty^2 - 1}} \tag{23}$$

This equation is identical with the previously derived expression given in Eq. 11.20.

Two-dimensional section properties corresponding to those derived in Chapter 5 for incompressible flow can be derived from the integrations of the pressure coefficients as they determine the lifting and nonlifting contributions. The pressure coefficients for the upper and lower surfaces are, from Eq. 23,

$$C_{PU} \equiv C_p(x, 0+) = \frac{2\theta_U}{\lambda}$$

$$C_{PL} \equiv C_p(x, 0-) = -\frac{2\theta_L}{\lambda} \tag{24}$$

The angle θ is positive measured in the counterclockwise direction from the positive x axis.

As is indicated in Fig. 10, the linear theory delineates three separate contributions to pressure distribution: (a) a flat plate along the chord line at an angle of attack α,

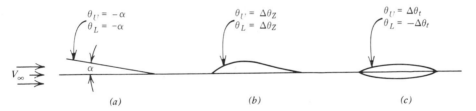

Fig. 10 Resolution of pressure distribution into parts due to angle of attack, camber, and thickness.

(b) the mean camber line at zero angle of attack, and (c) a symmetrical thickness envelope. This division was justified in Section 12.2.

For the lift contribution (a) the slope of the upper surface is the negative of the angle of attack.

$$\theta_L = \theta_U = -\alpha$$

and for the contribution (b), due to camber, the notation $\Delta\theta = \Delta\theta_z$, which is a function of x, is used. In keeping with small perturbation theory, the cosines of all angles are set equal to unity and the sines are set equal to the angle. We shall see that the lift is determined by (a), the moment by (a) and (b), and the drag by (a), (b), and (c).

1. Lift

The lift coefficient is given by

$$c_l = \frac{L'}{q_\infty c} = \int_0^c \frac{p_L - p_U}{q_\infty c} \, dx = \int_0^1 (C_{PL} - C_{PU}) d\frac{x}{c} \tag{25}$$

For the flat plate at an angle of attack, using Eq. 24,

$$c_l = \frac{4\alpha}{\sqrt{M_\infty^2 - 1}} \int_0^1 d\frac{x}{c} = \frac{4}{\sqrt{M_\infty^2 - 1}} \alpha = \frac{4\alpha}{\lambda} \tag{26}$$

From the mean camber line there is no contribution because the integral of the slope is zero. The thickness envelope is nonlifting. Therefore, Eq. 26 represents the entire lift coefficient. Accordingly, the lift curve slope m_0 and the angle of zero lift, α_{L0}, are given by

$$m_0 = 4/\lambda; \qquad \alpha_{L0} = 0 \tag{27}$$

2. Moment

The moment coefficient about the leading edge is

$$c_{mL.E.} = \frac{M_{L.E.}}{q_\infty c^2} = -\int_0^c \frac{p_L - p_U}{q_\infty c^2} x \, dx = -\int_0^1 (C_{PL} - C_{PU})\frac{x}{c} d\frac{x}{c} \tag{28}$$

Both the flat plate at an angle of attack and the mean camber line contribute to $c_{mL.E.}$. From the flat plate, using Eq. 24

$$(c_{mL.E.})_\alpha = \frac{-4\alpha}{\lambda} \int_0^1 \frac{x}{c} d\frac{x}{c} = -\frac{2\alpha}{\lambda} = -\frac{c_l}{2} \tag{29}$$

which indicates that *the center of pressure due to angle of attack is at the midchord*. From the mean camber line, again using Eqs. 24,

$$(c_{mL.E.})_z = +\frac{4}{\lambda} \int_0^1 \Delta\theta_z \frac{x}{c} d\frac{x}{c} = m_0 K_1 \tag{30}$$

the integral K_1 depends only on the shape of the mean line; it is zero for symmetrical airfoils. Because the mean camber line contributes no lift, the moment given by Eq. 30 is a couple. The midchord is the point about which the moment coefficient is independent of angle of attack, and is therefore the aerodynamic center. The sectional moment characteristics are

$$\text{a.c. at midchord;} \qquad c_{mac} = m_0 K_1 \tag{31}$$

3. Drag

All three contributions indicated in Fig. 10 contribute to the drag. The drag due to the flat plate at an angle of attack is the force normal to the plate times $\sin \alpha$. This may be written approximately in coefficient form as*

$$(c_d)_\alpha = c_l \alpha = \frac{c_l^2}{m_0} \tag{32}$$

From the mean camber line, using Eqs. 24,

$$(c_d)_z = -\int_0^c \frac{p_L - p_U}{q_\infty c}\, \Delta\theta_z\, dx = \frac{4}{\lambda}\int_0^1 (\Delta\theta_z)^2\, d\frac{x}{c} = m_0 K_2 \tag{33}$$

the symmetrical thickness envelope contributes to the drag coefficient†

$$(c_d)_t = \frac{4}{\lambda}\int_0^1 (\Delta\theta_t)^2\, d\frac{x}{c} = m_0 K_3 \tag{34}$$

K_2 and K_3 are integrals that depend only on the shape of the section. The total drag coefficient for the infinite wing neglecting viscous effects is

$$c_d = m_0(K_2 + K_3) + \frac{c_l^2}{m_0} \tag{35}$$

The wave drag as given by Eq. 35 has no counterpart in incompressible flow, where the drag of an infinite wing, viscosity neglected, is zero. To the drag indicated by Eq. 35 must be added the skin friction and form drag generated by the fluid viscosity. The part of the wave drag dependent upon *profile* shape can be reduced by proper design. For an airfoil of given thickness it can be shown that the shape having the least profile drag is the symmetrical wedge shown in Fig. 11. The integral K_2 vanishes and K_3 is easily evaluated. The result is that the wave drag at zero lift is

$$c_d = m_0 \left(\frac{t_m}{c}\right)^2$$

Example

The above analysis is applied to a bi-convex airfoil flying at $M_\infty = 2.13$. The upper and lower surfaces of the symmetrical airfoil are circular arcs of the same radius R, and the maximum thickness is 10 percent of the chord.

*The infinite wing in incompressible flow does not show a comparable drag because of the presence of leading edge suction (see Sec. 4.8).
†For these calculations, the cross-product terms ($\alpha\Delta\theta_z$, etc.), which occur when the complete angles $\theta_U = -\alpha + \Delta\theta_z + \Delta\theta_t$ and $\theta_L = -\alpha + \Delta\theta_z - \Delta\theta_t$ are squared, vanish in the integrations because $\alpha = $ constant and $\Delta\theta_z = \Delta\theta_t = 0$ at $x/c = 0$ and 1.

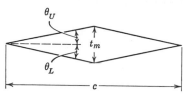

Fig. 11 Section for minimum profile drag.

The airfoil configuration given in Fig. 12 requires that

$$(R - 0.05c)^2 + (0.5c)^2 = R^2$$

which results in $R = 2.525c$. Thus the angle Θ, subtended by half of the chord about a center of curvature, is found to be 0.1993 radians. For the specified Mach number, $m_0 = 4/\sqrt{M_\infty^2 - 1} = 2.13$.

The lift coefficient of this airfoil is, according to Eq. 26,

$$c_l = 2.13\alpha \tag{36}$$

without camber, $K_1 = 0$ and the moment coefficient about the leading edge is described by Eq. 29 alone:

$$c_{m_{\text{L.E.}}} = -\tfrac{1}{2} c_l \tag{37}$$

There is no contribution to the drag from the mean camber line, therefore $K_2 = 0$. For the thin airfoil $x = c/2 + R \sin\theta$ so that $dx \doteq R \, d\theta$. Thus, from Eq. 34,

$$K_3 = \int_0^1 (\Delta\theta_t)^2 \, d\frac{x}{c} = \frac{R}{c} \int_{-\Theta}^{\Theta} \theta^2 d\theta = \frac{2}{3}\frac{R}{c} \Theta^3$$

$K_3 = 0.0133$ for the values already obtained for R and Θ. The drag coefficient is then computed from Eq. 35:

$$c_d = m_0(\alpha^2 + K_2 + K_3) = 2.13\alpha^2 + 0.0284 \tag{38}$$

The angle of attack α in Eqs. 36 and 38 is expressed in radians.

Figure 12 compares measurements (Ferri, 1939) of the aerodynamic characteristics of a bi-convex airfoil at $M_\infty = 2.13$ with the above predictions of linear theory given in Eqs. 36, 37, and 38. The slope of the lift curve, m_0, is seen to be about 2.5 percent below theory. The theoretical α_{L0} is zero for all shapes and, consequently, agrees perfectly with that measured for the symmetrical airfoil used for these measurements but departs from experiment for unsymmetrical shapes. The theoretical drag coefficient is smaller than that measured partly because as mentioned above the theory predicts wave drag only, neglecting that caused by viscosity. The disagreement between theory and experiment for the moment coefficient

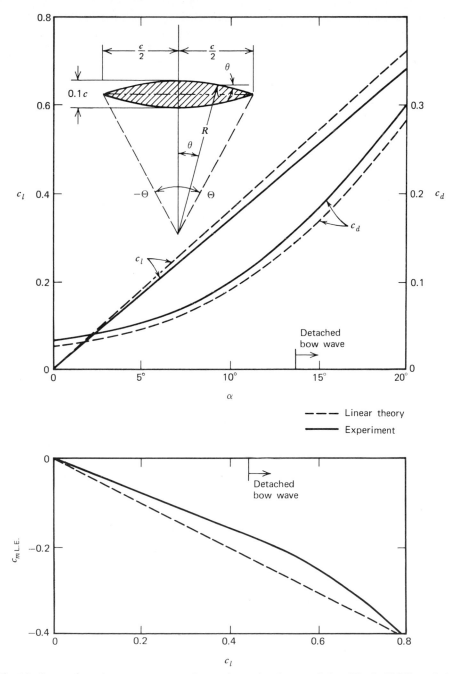

Fig. 12 Comparison between measured aerodynamic characteristics (Ferri, 1939) and those predicted by linear theory at $M_\infty = 2.13$. Airfoil is biconvex, maximum thickness $= 0.1c$.

in Fig. 12 is marked, partly because alteration of the pressure distribution caused by viscous effects are large near the trailing edge where the moment arm is large.

Note that the points designating the angle of attack ($14.7°$) and the lift coefficient (0.42) beyond which the shock is detached do not have an immediate effect on the aerodynamic characteristics; as was pointed out in connection with Fig. 4, the effects are delayed because the detached shock is weak in near-sonic flow. Second-order (inviscid) theories for these flows are available, though considerably more involved than the linear analysis (see Lighthill, 1954). The resulting corrections to the linear results are hardly detectable for lift and drag but for small C_L they nearly close the gap between theory and experiment for $c_{m_{\text{L.E.}}}$ in Fig. 12; that is, until the effects of the bow wave become significant.

CHAPTER 13

Wings and Wing-Body Combinations in Compressible Flow

13.1   Introduction

Knowledge of the aerodynamic characteristics of airfoils forms the starting point for the design of the finite wings and finally the complete aircraft. The final design must take into account mutual flow interferences among the various components as affected by compressibility and (in following chapters) viscosity.

In this chapter we are concerned with the compressibility effects and with the physical concepts governing the flow phenomena. When dealing with the drag we consider the wave drag in two parts—that for the nonlifting configuration and that due to lift (vortex drag).

The Prandtl-Glauert transformation is applied to finite wings in subsonic and supersonic flows and their interpretation for three-dimensional bodies is outlined.

The application of the Area Rule to transonic and supersonic aircraft is described.

13.2   Wings and Bodies in Compressible Flows—The Prandtl–Glauert–Goethert Transformation

The Prandtl-Glauert transformation, applied in Section 12.3 to airfoils in subsonic flow may also be applied to finite wings in subsonic flow. For this analysis we use Eqs. 12.4 and 12.3 in the form:

$$\beta^2 \varphi_{xx} + \varphi_{yy} + \varphi_{zz} = 0 \tag{1}$$

$$\theta = \frac{(\varphi_z)_{z=0}}{V_\infty} \tag{2}$$

If x in Eq. 1 is replaced with

$$x = x_0 \beta \equiv x_0 \sqrt{1 - M_\infty^2} \tag{3}$$

and all other variables are left the same, the equation reduces to

$$\varphi_{x_0 x_0} + \varphi_{yy} + \varphi_{zz} = 0 \tag{4}$$

which is Laplace's equation in the variables x_0, y, z. The transformation has stretched the x coordinate by the factor $1/\sqrt{1 - M_\infty^2}$. Corresponding points in the two fields shown in Fig. 1 have the same y and z coordinates and x_0 and x are re-

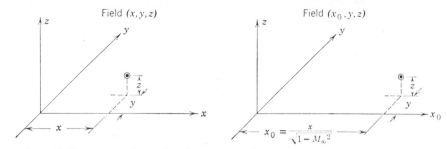

Fig. 1 Prandtl-Glauert transformation for planar surfaces.

lated by Eq. 3. The flow in the x_0, y, z field system is incompressible; in the x, y, z field $M_\infty < 1$. The values of φ *at corresponding points are identical.* Because the z coordinate in the two fields is the same, it must also be true that the values of φ_z *at corresponding points are identical.*

To apply this information, consider a wing in field (x, y, z) that is in the presence of a stream for which $0 < M_\infty < 1$. This case will be referred to in the following as the subsonic wing. In accordance with the boundary condition, Eq. 2, specific values of φ_z will be forced at a group of points in the $z = 0$ plane as represented by the shaded area in Fig. 2*a*. The corresponding points in the $z = 0$ plane of field (x_0, y, z), drawn shaded in Fig. 2*b* must have the same value of φ_z. Consequently, according to Eq. 2 the Mach zero wing of Fig. 2*b* must have the same slope at corresponding points as the subsonic wing of Fig. 2*a*.

Because the flow about the Mach zero wing is governed by Eq. 4, the following conclusion can be drawn:

Equations 1 and 2 may be solved by finding the solution to Eq. 4 for a wing of greater sweep, smaller aspect ratio, and the same section shape.

The manner of defining aspect ratio and sweep will determine the way these quantities are related for the two wings. For the triangular wing in Fig. 2 the leading edge sweep angle σ_0 of the Mach zero wing is related to the leading edge sweep

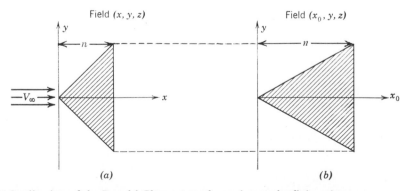

Fig. 2 Application of the Prandtl-Glauert transformation to the finite wing.

angle σ of the subsonic wing by the formula

$$\frac{\tan \sigma_0}{\tan \sigma} = \frac{n/\frac{1}{2}b}{m/\frac{1}{2}b} = \frac{n}{m} = \frac{1}{\beta}$$

$$\sigma = \tan^{-1} (\beta \tan \sigma_0) \tag{5}$$

The geometry is shown in Fig. 3. The aspect ratio R_0 of the Mach zero wing is related to the aspect ratio R of the subsonic wing by the formula

$$\frac{R_0}{R} = \frac{b^2/S_0}{b^2/S} = \frac{S}{S_0} = \frac{c}{c_0} = \beta \tag{6}$$

$$R = R_0/\beta \tag{7}$$

The pressure coefficient, Eq. 11.19, may be written for the subsonic and Mach zero wings as

$$C_p = \frac{-2\varphi_x}{V_\infty}; \qquad C_{p_0} = \frac{-2\varphi_{x_0}}{V_\infty} \tag{8}$$

Using Eq. 3, we can write at corresponding points

$$\varphi_x = \frac{\varphi_{x_0}}{\beta} \tag{9}$$

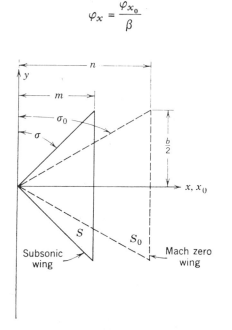

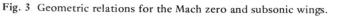

Fig. 3 Geometric relations for the Mach zero and subsonic wings.

Therefore, at corresponding points we have the Prandtl-Glauert rule,

$$C_p = \frac{C_{p_0}}{\beta} \tag{10}$$

that is, the pressure coefficient on the subsonic wing is greater than it is at a corresponding point on the Mach zero wing by the factor $1/\sqrt{1 - M_\infty^2}$. However, for this example the Mach zero wing to which C_{p_0} refers has a different planform for each M_∞.

We will show that the sectional lifts at corresponding spanwise stations on the wings at different Mach numbers are equal. As we did for the two-dimensional wing of the last chapter, we write

$$L' = \int_{L.E.}^{T.E.} (C_{p_L} - C_{p_U}) q_\infty dx = \int_{L.E.}^{T.E.} \frac{(C_{p_L} - C_{p_U})_0}{\beta} q_\infty \beta \, dx_0 = L_0' \tag{11}$$

It follows that, since the spans of the wings are equal and their chord lengths follow the Prandtl-Glauert rule, their total lifts at a given angle of attack will be identical.

On the other hand, for a *given* wing the chord will be fixed at a given station and by the reasoning of the previous chapter we obtain the formula of Eq. 12.12

$$\frac{dc_l}{d\alpha} = \frac{1}{\beta} \left(\frac{dc_l}{d\alpha}\right)_0 \tag{12}$$

The ranges of validity, defined by the critical Mach number, are also discussed in Chapter 12 and illustrated by the experimental results of Figs. 12.4 and 12.7.

For three-dimensional slender bodies, Goethert's extension of the Prandtl Glauert rule (see Sears, 1954) states that: to find the flow about a given body of length l in a *compressible flow*, we consider first an *incompressible* flow about a body of length $l\beta^{-1}$ (other dimensions remaining unchanged) for which the perturbation velocity potential is ϕ; then the perturbation velocity components in the compressible flow about the *original* body are, respectively, $\beta^{-2}\phi_x$, $\beta^{-1}\phi_y$, $\beta^{-1}\phi_z$ at *corresponding points* in the two flows. This rule is applicable to planar flows as well as to slender bodies so it provides a single rule for subsonic aircraft.

For planar flows *everywhere supersonic*, $\beta = \sqrt{1 - M_\infty^2}$ becomes imaginary and, as was pointed out in Chapter 12, Eq. 1 must be replaced by the following differential equation, which is, by analogy with Eq. 12.17, of the "hyperbolic" type:

$$\lambda^2 \varphi_{xx} - \varphi_{yy} - \varphi_{zz} = 0 \tag{13}$$

where $\lambda = \sqrt{M_\infty^2 - 1}$. As for subsonic flow the local inclination of the surface of a body in the flow is given by

$$\theta = \frac{(\varphi_z)_{z=0}}{V_\infty}$$

and the coordinate transformation analogous to that used for subsonic flow (Eq. 3) is

$$x = x_0 \lambda \equiv x_0 \sqrt{M_\infty^2 - 1} \tag{14}$$

But, it is no longer possible to express the aerodynamic characteristics in the *supersonic* flow in terms of those in an incompressible flow. The reason for this change in analysis from that of subsonic flows is that Eq. 13 *cannot* by a stretching (affine) transformation, such as that of Eq. 14, be transformed into Laplace's Equation, $\nabla^2 \varphi = 0$, and thus *there is no equivalent incompressible flow*.

It *is* possible however to express the flow properties in the supersonic field in terms of those for $M_\infty = \sqrt{2}$, for which $\lambda = 1$. Thus the equation analogous to Eq. 5 for the equivalent sweepback angle σ is

$$\tan \sigma = \lambda \tan \sigma_{\sqrt{2}} \tag{15}$$

where σ is the sweepback angle at $M_\infty > 1$ and $\sigma_{\sqrt{2}}$ is that for $M_\infty = \sqrt{2}$. Similarly the equation analogous to Eq. 7 is

$$\mathcal{R} = \frac{\mathcal{R}_{\sqrt{2}}}{\lambda} \tag{16}$$

Thus, since $\lambda \gtrless 1$, respectively, for $M_\infty \gtrless \sqrt{2}$, the equivalent planform for a given M_∞ is stretched if $M_\infty < \sqrt{2}$ and shrunk for $M_\infty > \sqrt{2}$.

The corresponding correction for C_p, which, for subsonic flow was given by Eq. 10, becomes

$$C_p = \frac{C_{p\sqrt{2}}}{\lambda} \tag{17}$$

The section lift at a given location of the transformed planform, is given by

$$L' = \int_{L.E.}^{T.E.} (C_{pL} - C_{pU}) q_\infty dx = \int_{L.E.}^{T.E.} \left[\frac{(C_{pL} - C_{pU})\sqrt{2}}{\lambda} \right] q_\infty \lambda \, dx_{\sqrt{2}} = L'_{\sqrt{2}} \tag{18}$$

that is, the lift at any section is fixed at its value for $M_\infty = \sqrt{2}$. It follows that the total lift of the transformed planform also has a fixed value.

On the other hand, for a given planform the chord will be *fixed* at a given station and we obtain the supersonic counterpart of Eq. 12.

$$\frac{dc_l}{d\alpha} = \frac{1}{\lambda} \left(\frac{dc_l}{d\alpha} \right)_{\sqrt{2}} \tag{19}$$

Since λ increases as M_∞ increases $dc_l/d\alpha$ decreases as well.

13.3   Influence of Sweepback

The utilization of sweepback, the beneficial effects of which are applicable to a broad range of flows, are most widely applied to high speed aircraft. In this application its main effect is to increase the critical Mach number of flight and thus the cruising speeds by delaying the drag increase and other compressibility effects on the wings.

The sweepback effect illustrates an "independence principle" in that if an infinite cylindrical body of uniform cross-section is at an angle of yaw to the flow the chordwise (normal to the axis) and spanwise flows can be calculated separately. Consider an infinite wing of uniform cross section inclined to the flow direction of a M_∞ stream as shown in Fig. 4. The x coordinate is taken chordwise, the y spanwise.

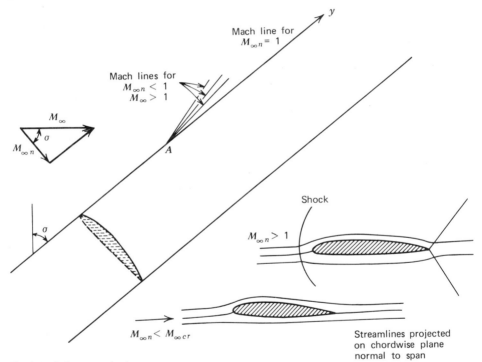

Fig. 4 Infinite yawed wing.

Since the section is uniform along its span the velocity and pressure distributions are identical at every section, that is, there is no spanwise pressure gradient (i.e., $\partial p/\partial y = 0$). Therefore, the flow configuration over the yawed wing will resemble closely that over a two-dimensional wing in a free stream Mach number $M_{\infty n} = M_\infty \cos \sigma$, where σ is defined as the sweepback angle. In other words, the shock wave configurations resemble those of Fig. 12.6 when $M_{\infty n}$ reaches the values given in the figure.

At supersonic M_∞, as long as the Mach lines of the flow originating at an arbitrary point A project forward, they warn the upstream flow of the presence of the wing so that the approaching chordwise flow component will be subsonic. Any shock waves that form on the surface will be oblique and the *local* flow component at their upstream face must be supersonic.

According to the above reasoning the distribution of pressure coefficient

$$C_{pn} = \frac{p - p_\infty}{\frac{1}{2}\rho_\infty V_\infty^2 \cos^2 \sigma} = \frac{(p/p_\infty) - 1}{\frac{1}{2}\gamma M_{\infty n}^2} \tag{20}$$

is dependent only on $M_{\infty n}$. Measurements are shown in Fig. 5 for Mach numbers 0.65, 0.69, and 0.85 at sweepback angles respectively of $0°$, $20°$ and $40°$ so that $M_{\infty n} = M_\infty \cos \sigma = 0.65$ for each set.

These measurements show remarkable agreement with the Prandtl-Glauert theory; accordingly in line with the discussion of Sections 12.4 and 12.5 a recovery shock will be delayed until M_n, the local Mach number of the flow normal to the leading edge at a point on the surface, is greater than unity. This condition then defines

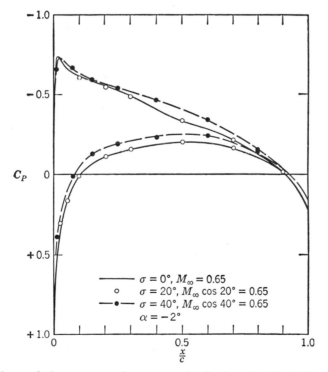

Fig. 5 Comparison of three measured pressure distributions by Lippisch and Beushausen (1946) at $\sigma = 0°$, $20°$, $40°$ and $M_\infty = 0.65$, 0.69, 0.85, respectively; $M_{\infty n} = M_\infty \cos \sigma = 0.65$ for the three tests.

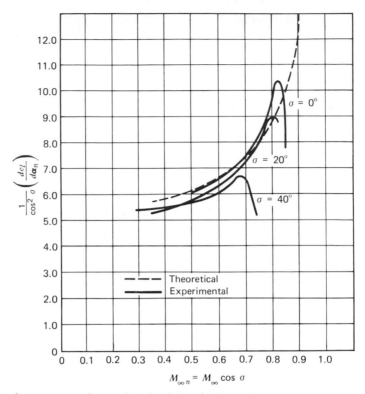

Fig. 6 Experiment versus theory for chordwise $dc_l/d\alpha_n$ of swept wing in compressible flow; $c_l = L'/\frac{1}{2}\rho_\infty V_{\infty n}^2$, α_n measured in plane normal to leading edge, theoretical curve follows Prandtl-Glauert theory. See text for reason for poor agreement at high $M_{\infty n}$ (Lippisch and Beushausen, 1946).

$M_{n\infty \text{ cr}} = M_{\infty \text{ cr}} \cos \sigma$ and is governed by the curve in Fig. 12.5 (Eq. 12.15) with the abscissa replaced by $M_{\infty n}$.

Another test of the Prandtl-Glauert theory based on the same measurements used in Fig. 5 is shown in Fig. 6, where $dc_l/d\alpha_n$ is plotted against $M_{\infty n}$. The definitions of the symbols are given in the figure caption. The curves are in good agreement, even for $40°$ sweepback, up to about $M_{\infty n} = 0.68$. Wind tunnel boundary interference, as affected by Mach number, is undoubtedly the main reason for the fact that the $M_{\infty n}$ at which the measured curves break downward occurs earlier as σ increases. As was pointed out in Section 10.12, at test Mach numbers near unity blockage of the stream and reflection of waves from the tunnel boundary will introduce large errors in test results unless a very small model is used in a large tunnel or the tunnel walls are porous. The test results of Fig. 6 were obtained in a relatively small tunnel with parallel, solid walls; therefore, since at $M_{\infty n} = 0.68$, $M_\infty = 0.89$ at $\sigma = 40°$, one would expect compressibility effects to increase with σ near "blockage conditions." Nevertheless the range of relatively good agreement indicates that chordwise aerodynamic characteristics are functions only of $M_{\infty n}$, and that the Prandtl-Glauert law is valid over a practical range of sweepback and angle of attack.

13.4 Design Rules for Wing–Fuselage Combinations

When a wing and other appendages are added to a fuselage, mutual interference effects, as introduced in Chapter 4 for incompressible flow, become important. These effects become especially important at high subsonic and transonic speeds and their treatment generally involves combinations of three interrelated concepts. These are (1) the Prandtl-Glauert-Goethert (PGG) rule described in Section 13.2, (2) the Area Rule (one aspect of which is the "waisting" of the fuselage) and its refinements by means of which flow interference effects among the fuselage and the wing and other appendages are minimized, and (3) corrections for viscous effects determined by theory and experiment as described in succeeding chapters. In this section, we restrict our treatment to item (2) because its main features can readily be described in terms of *physical* principles treated in earlier chapters. Mathematical formulas will be presented only insofar as they demonstrate the important physical features. Applications to transonic and supersonic designs will be treated separately.

TRANSONIC CONFIGURATIONS

One must first recognize that the flow equation applicable to slightly perturbed transonic flows (Eq. 11.11 for three-dimensional flow):

$$\left[(M_\infty^2 - 1) + (\gamma + 1) M_\infty^2 \frac{\phi_x}{V_\infty} \right] \phi_{xx} - \phi_{yy} - \phi_{zz} = 0 \qquad (11.11)$$

is nonlinear, since the second term in brackets involves the product $\phi_x \phi_{xx}$; thus, since superposition of flows is not permitted, special methods, such as the "Area Rule," are necessary to treat problems in transonic flow.

For transonic flows, the Area Rule may be stated as follows:

Within the limitations of small perturbation theory, at a given transonic Mach number, aircraft with the same longitudinal distribution of cross-sectional area, including fuselage, wings and all appendages, will, at zero lift, have the same wave drags.

This rule is a limiting case of a general theorem for supersonic flows derived by Hayes (1947) (e.g., see Lomax and Heaslet, 1956); it is applicable to near-sonic flows about body + wing, and so forth, provided the flows conform with small perturbation concepts.

The physical phenomena on which the rule is based are illustrated in Fig. 7, which shows a fuselage and swept wing configuration cut by plane AB, normal to the

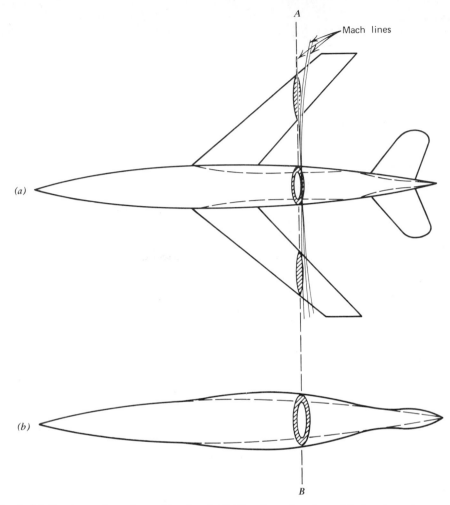

Fig. 7 (*a*) Fuselage-wing-tail combination (solid lines) cut by plane *AB* showing wing areas intercepted and fuselage waisted (broken line) by amount equal to sum of intercepted areas of wings. Light inclined curves near *AB* represent waves generated by the intercepted surfaces. (*b*) Original fuselage plus intercepted areas of wing and tail at each station (equivalent body).

fuselage axis. The flow is near-sonic so that, as is indicated in the figure, the surfaces of propagation of the waves generated by those surface elements in supersonic flow regions intercepted by the cutting plane are nearly coincident with that plane. Therefore each surface element of the cross section of the wings (and other appendages) intercepted by the cutting plane may affect the flow inclination in the region of the body very near the cutting plane. It follows that the contribution to the wave drag from each of the cross sections in a supersonic region intercepted by the cutting plane will be approximately the same regardless of its location in the

plane. Figure 7*b* illustrates the rule; the cross-sectional area of the wings intercepted by the cutting plane at each axial station is equal at that station to the area
of the annulus added to the fuselage to form an "equivalent body." The wing (and
all other appendages), as well as the equivalent body, must of course *each* satisfy
the restrictions imposed by the small perturbation analysis.

  The application of Hayes' rule to transonic aircraft was not appreciated until
1952 when Whitcomb (1956) carried out tests, some of which are reproduced in
Fig. 8; his rationale for the tests was based on physical reasoning essentially that

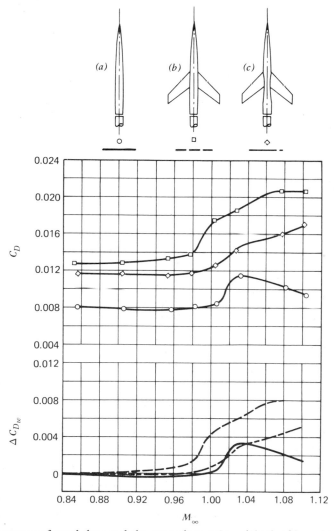

Fig. 8 Measurements of total drag and the wave drag ΔC_{D_w} (obtained by subtracting viscous
drag from total) at zero lift. In (*c*) the fuselage is waisted by the volume of the wings (Whitcomb, 1956, courtesy of NASA).

given above. Shown at the top of the figure are (a) a body of revolution, (b) the same body with swept wings, and (c) the body, waisted in accordance with the Area Rule, with the swept wings attached.

The three configurations were tested at an angle of attack of $0°$ (zero lift) from Mach .84 to 1.10; the plot of C_D gives the measured drag of the three configurations, C_{D_w} is found by calculating the skin friction drag for each combination and subtracting its coefficient from C_D. Thus the C_{D_w} curves in Fig. 8 represent the wave drags only. We see that when the wings were simply attached to the original body, the steep increase in wave drag began at about Mach 0.98, but by the application of the area rule the drag rise began at a significantly higher Mach number; at Mach 1.02, for instance, waisting the body halved the wave drag. The other compressibility effects would be similarly delayed.

An important inference can be drawn from the physical discussion above, that is, if at a given station the cross-sectional area distribution of the *equivalent* body changes rapidly with x, the waves generated will be strong enough to contribute disproportionately to the wave drag; this contribution may be decreased by smoothing the equivalent body even, if necessary, by increasing its cross section.

A striking example of the effect is illustrated in Fig. 9 which gives results of tests showing the effect of extending the "cab" of the Boeing 747 (Goodmanson and Gratzer, 1973). The results show that $M_{\infty cr}$ is delayed by smoothing the area distribution by fairing the fuselage-cab juncture. The effect of the fairing is negligible until $M_{\infty cr}$ is reached for the *unfaired* juncture; then the fairing delays M_∞ at which waves are generated. Thus the fairing causes an increase in $M_{\infty cr}$, as illustrated in Fig. 9, at $0.3 < C_L < 0.5$. Since the wind tunnel data are based on the measured *total* drag, and since viscous and vortex drags were not changed significantly by the modification, the drag decrease at the high subsonic Mach numbers comprises almost entirely the decrease in wave drag resulting from the fairing of the juncture.

Streamline contouring modification of the area rule to realize the maximum gain at a design lift coefficient requires that the waisting of the fuselage be apportioned between the regions above and below the wing. Figure 10*a* shows an infinite yawed wing generating lift. The pressure distributions will cause the streamlines at the edges of the boundary layers to be deflected upwind on the upper surface and downwind on the lower; for a finite wing these deflections will be increased by the spanwise flows due to the trailing vortices. The flow interference caused by the fuselage in the presence of these differential flow deflections can be compensated for by apportioning the waisting of the fuselage in the manner shown in Fig. 10*b*, that is, by waisting the sections above the wing root more than those below (Goodmanson and Gratzer, 1973, and Lock and Bridgewater, 1967).

The extreme sensitivity of the details of contour shapes to the generation of shock waves and areas of flow separation in near-sonic flows have motivated considerable recent research toward refinements of the above methods. Some of these

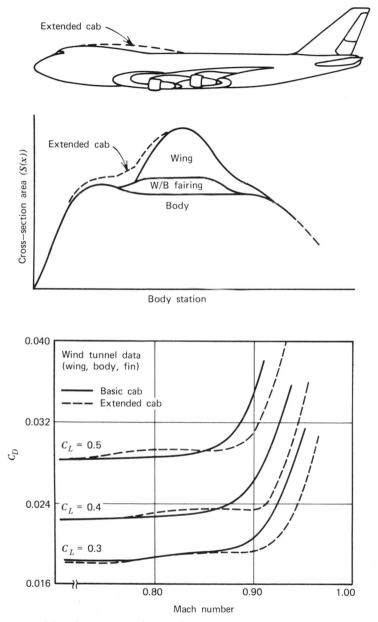

Fig. 9 Effect on $S(x)$ and on measured drag of Boeing 747 due to fuselage modification (Goodmanson and Gratzer, 1973, courtesy of the Boeing Company).

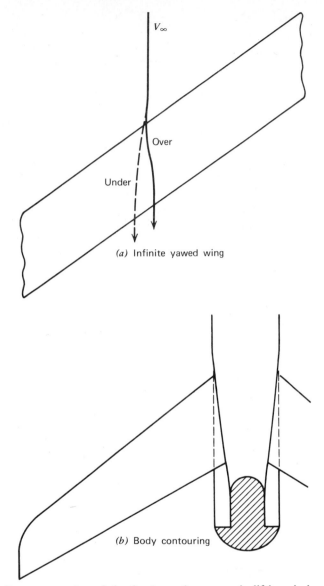

(a) Infinite yawed wing

(b) Body contouring

Fig. 10 Streamline countouring of the fuselage of a transonic lifting airplane, to correct for flow deflections on wing surfaces. Shaded area is cross section of fuselage at trailing edge of wing.

involve modifications of the wing as well as the fuselage sections in regions near the wing-root (see Lock and Bridgewater, 1967).

One can discern the effect of some of these refinements in Fig. 11, which is a photograph of the top surface of a model of a transonic airplane; the surface was coated to show the streamlines. The smooth streamlines near the wing-fuselage

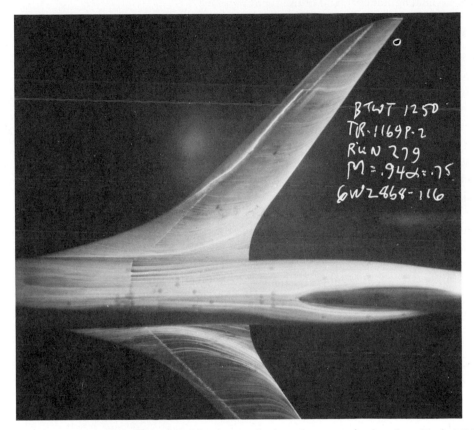

Fig. 11 Top view of a model of the fuselage and wing of a transonic aircraft at Mach 0.94, $\alpha = 0.75°$. The surface has been coated to render the streamlines visible. Through proper design of the fuselage-wing juncture, root stall was avoided. (Goodmanson and Gratzer, 1973, Courtesy of The Boeing Company.)

juncture show the effect of what appears to be near-optimum waisting and planform modification to avoid flow interferences that could cause flow separation and shock waves in this region. The shocks shown farther out on the wing are apparently not of sufficient intensity to cause separation.

It must be emphasized that these methods require the selection of a "design Mach number" at which the compressibility effects will be minimized, under the constraints of other factors such as viscous and structural effects.

The "Swing-Wing" planform proposed by Jones (1972) in which the entire wing may be rotated as a unit to provide the optimum cross-sectional area distribution for a given transonic Mach number is a promising method for minimizing the wave drag. Much of the current effort is directed toward overcoming mechanical difficulties but aerodynamically this configuration offers the flexibility to accommodate to the landing, takeoff, and cruise flight regimes.

SUPERSONIC CONFIGURATIONS

It was shown in Chapter 11 that application of small perturbation theory in completely supersonic flow about "slender" bodies, leads to the linear equation

$$(M_\infty^2 - 1)\,\phi_{xx} - \phi_{yy} - \phi_{zz} = 0 \tag{11.15}$$

Consistent with this equation, an approach based on an analysis by von Kármán (1935) leads to a reliable *quantitative* determination of the zero-lift wave drag of a slender body in a uniform supersonic flow. The method uses Hayes' Area Rule, which applies to supersonic flows, but which was utilized above for a qualitative discussion of the wave drag of transonic configurations.

In the description that follows, much of the mathematical analyses, which are quite involved, are omitted, since they do not appear to be necessary for an exposition of the physical concepts involved. It will be shown, however, that the results, some of them surprisingly simple and effective, play an indispensible role in the methods, treated subsequently, for the aerodynamic design of supersonic aircraft. Thus, readers, who are mathematically inclined, will hopefully be motivated to master the details of the analyses developed elsewhere (e.g., von Kármán, 1935; Liepmann and Roshko, 1957).

Von Kármán represented the flow about an axisymmetric body by the superposition of a uniform supersonic flow and a continuous supersonic source distribution along a line parallel to the flow, as shown in Fig. 12. Using the formula for the source given in Eq. 11.16, he wrote the velocity potential for the flow

$$\Phi = V_\infty x + \phi = V_\infty x + \int_0^{(x-\lambda r)} f(\xi)\,\frac{d\xi}{\sqrt{(x-\xi)^2 - \lambda^2 r^2}} \tag{21}$$

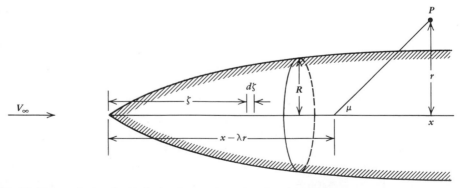

Fig. 12 Nomenclature for analysis of wave drag of body of revolution.

where $\lambda^2 = M_\infty^2 - 1$ and $f(\xi)\,d\xi = 2\pi$ times the source strength along $d\xi$ (the strength of a single source at the origin was designated by C in Eq. 11.16). The integral is the perturbation velocity potential, ϕ, at $P(x, r)$, for the source distribution on the x axis. $\sqrt{(x - \xi)^2 - \lambda^2 r^2}$ is the so-called hyperbolic radius, h, at point P at (x, r) and $r^2 = y^2 + z^2$. We note that in the linearized supersonic flow the source makes no contribution to the flow upstream of its downstream Mach cone with semiapex angle equal to that of the free stream Mach angle. The integral extends from the (sharp) nose of the body at $x = 0$ to $\xi = x - \lambda r$ the apex of the Mach cone passing through P.

The immediate problem to be solved is to determine the source density distribution $f(x)$ such that the surface of the body is a streamline of the superimposed flows. With perturbation velocity components

$$u_x = \frac{\partial \phi}{\partial x}; \qquad u_r = \frac{\partial \phi}{\partial r} \tag{22}$$

the boundary condition at the body states that the surface of the body must be a streamline of the flow, that is,

$$\left(\frac{u_r}{V_\infty + u_x}\right)_{r=R} = \frac{dR}{dx} \tag{23}$$

where R is the radius of the body. For this linearized flow, quadratic terms in perturbations are neglected, and, substituting from Eq. 22, Eq. 23 becomes

$$\left(\frac{\partial \phi}{\partial r}\right)_{r=R} = V_\infty \frac{dR}{dx} \tag{24}$$

The wave drag of the body is found by use of a method first pointed out by Max Munk for use in airship design. For the solution of the problem in supersonic flow, the method states that the wave drag of the line source is equal to the rate at which downstream momentum is transferred across the boundaries of a circular cylinder of radius r_1 enclosing the source. To illustrate, in Fig. 13, $\rho u_r \cdot 2\pi r_1\,dx$ is the rate at which mass is transferred outward through the cylindrical element $2\pi r_1\,dx$, and multiplying this rate of mass flow by u_x gives the rate at which the x momentum is transferred outward. This quantity when integrated over the length of the cylinder is the reaction to the wave drag of the source distribution representing the body.

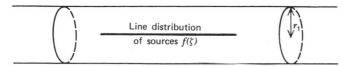

Line distribution
of sources $f(\xi)$

Fig. 13 Cylindrical control surface enclosing a line source.

Thus

$$D = -2\pi r_1 \rho \int_{-\infty}^{\infty} (u_r u_x)_{r=r_1} \, dx \tag{25}$$

The value of this integral is found to be independent of r_1 in linearized flow.

Returning to the source distribution of Fig. 12, von Kármán showed that the source density $f(x)$ of Eq. 21 is related to the area distribution of the body $S(x)$,

$$f(x) = S' \frac{V_\infty}{2\pi} \tag{26}$$

where $S' = dS/dx$. He also discerned that the equation for the wave drag can be expressed in a form identical with that of the integral representation of the induced drag of a wing in incompressible flow (Eq. 6.35).

Sears (1947) exploited this analogy to arrive at a simple formula to calculate the wave drag of the axisymmetric body. He showed that if, using Eq. 26, we expand $f(x)$ in a Fourier sine series

$$f(x) = \frac{V_\infty}{2\pi} \frac{dS}{dx} = \frac{V_\infty l}{2} \sum_{n=1}^{\infty} A_n \sin n\,\theta \tag{27}$$

where l is the length of the body, and θ, designated in Fig. 14 $(0 \leqslant \theta \leqslant \pi)$, is the angle defined by the circle drawn with radius $l/2$. The curve designated S' in Fig. 14 refers to the slope of the area distribution of the equivalent body of Fig. 7b . For given $S(x)$ and l, the Fourier coefficients, A_n, are determined by the method of Section 5.6 (Eqs. 5.25 and 5.26); x and θ are related by the equation $x = \frac{1}{2} l (1 - \cos \theta)$, which, except that l replaces c, is exactly the equation defined

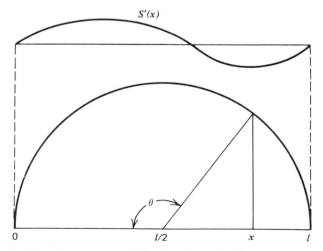

Fig. 14 $S'(x)$ and relation between x and θ for Fourier expansion.

in Fig. 5.6 (Section 5.6) for thin airfoil theory, and, with b replacing l, is that defined in Section 6.5 for finite wing theory. After considerable analysis Sears obtained the solution of Eq. 25 for the wave drag of the configuration of Fig. 7a in the forms

$$D_w = \frac{\pi^3}{4} \frac{\rho V_\infty^2 l^2}{2} \sum_{n=1}^{\infty} n A_n^2$$

$$C_{D_w} \equiv \frac{D_w}{q_\infty l^2} = \frac{1}{4} \pi^3 \sum_{n=1}^{\infty} n A_n^2$$

(28)

It was pointed out by Jones in 1953 (Jones, 1956) that the Area Rule, as formulated mathematically by Hayes, is applicable to the *quantitative* determination of the zero lift wave drag of supersonic configurations. As described above for slender fuselage-appendage configurations in transonic flow the wave drag is determined by the areas intercepted by the Mach planes, that is, by the Mach cones for *sonic* flow.

For supersonic flows however there will of course be an infinite number of Mach planes, inclined at the Mach angle, μ, each cutting the fuselage axis at a different *azimuthal* angle χ. Figure 15 shows a view of an airplane with a side view of a cutting plane inclined at angle $\mu = \sin^{-1}(1/M_\infty)$ to the axis at the point X. Its

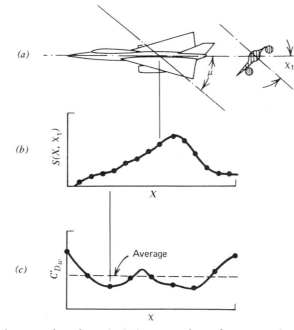

Fig. 15 Steps in the procedure for calculating wave drag of a supersonic aircraft at zero lift (Carlson and Harris, 1969, courtesy of NASA).

equation is

$$X = x - \lambda y \cos \chi - \lambda z \sin \chi$$

For a given M_∞ and constant χ the equation describes a set of parallel planes inclined at the Mach angle μ and intersecting the axis at values of $x = X$; with X constant, varying χ defines a set of planes enveloping the Mach cone with apex at $x = X$. The diagram at the right of (a) shows the intercepted area of the cutting plane at an azimuthal angle χ_1. Varying X varies the location of the apex of the Mach cone, and, since all of components of the aircraft are "slender" the waves generated at all surface points intersected by the cone propagate along surfaces nearly coincident with the Mach cone, just as those generated in a near-sonic flow nearly coincide with the normal plane in Fig. 7. Thus the flow interference between the waves generated at neighboring axial sections may be neglected.

A calculation procedure for determining the aerodynamic characteristics of a supersonic aircraft is given by Carlson and Harris (1969). The wave drag at zero lift is described by means of Fig. 15. The area intercepted by the Mach plane at $x = X$, $\chi = \chi_1$ is designated $s(X, \chi_1)$. The area

$$S = s \sin \mu \tag{29}$$

is then the projection of s on the plane normal to the axis; it represents the area intercepted by normal planes cutting the equivalent (slender) body of revolution for $\chi = \chi_1$. Figure 15b is a plot of $S(X, \chi_1)$ versus X. As was done for transonic flow, the Fourier coefficients are determined by writing (see Eq. 27 and Fig. 14)

$$S'(X, \chi_1) = \pi l_1 \sum A_{n1} \sin n\theta_1 \tag{30}$$

where l_1 and A_{n1} and θ_1 are respectively the lengths and Fourier coefficients corresponding to χ_1; in general, the length of the equivalent body l_1 will vary with the azimuth angle χ of the cutting plane. We designate by D'_w the drag of the equivalent body for a given χ_1. Then from Eq. 28,

$$D'_w(\chi_1) = \frac{\pi^3}{4} \frac{\rho_\infty V_\infty^2 l^2}{2} \sum n A_{n1}^2$$

$$C'_{D_w}(\chi_1) = \frac{D'_w(\chi_1)}{q_\infty l^2} = \sum \frac{1}{4} \pi^3 n A_{n1}^2 \tag{31}$$

For the calculation of Fig. 15, the above process is repeated for several values of χ_1 and the resulting variation of C'_{D_w} with χ is shown in (c). The total wave drag, C_{D_w}, which is the average C'_{D_w} for all azimuth angles is

$$C_{D_w} = \frac{1}{2\pi} \int_0^{2\pi} C'_{D_w} \, d\chi \tag{32}$$

The design calculations are simplified and extended to some lifting body configurations by Lomax and Heaslet (1956).

The drag due to lift, or, vortex drag, as determined by the Carlson-Harris method involves dividing the wing surface into a large number of elements, as shown in Fig. 16. At the "control point" of each panel a vortex, represented by the solution of the linearized flow equation designated in Eqs. 11.16 is placed. The method of simultaneous algebraic equations, described in Sections 4.13, 5.10, and 6.13 for

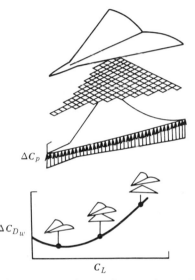

Fig. 16 Network of vortex sheets on a wing surface used to calculate wave drag increment ΔC_{D_w} due to lift. (Carlson and Harris, 1969, courtesy of NASA.)

incompressible flow, is used to satisfy the boundary condition requiring $\phi_n = 0$ at the surface of each element. Solution of the simultaneous equations yields the strength of the vorticity at each element and, by integration, the lift and moment coefficients. The drag is found for each element as the downstream component of the lift and the effect of twist and warp are taken into account by a method involving the interference between the warped and twisted wing and a planar wing of the same planform. The plot of Fig. 16 shows the vortex drag coefficient ΔC_{D_v} versus C_L.

Representative comparisons among the relative values of the three drag contributions are given in Fig. 17. In this figure "zero volume" assumes zero thickness lifting surfaces and "near field" refers to the fact that this drag due to lift is determined from downwash *at* the surface, "far field" refers, in effect to wave drag determination by the area rule, and, as indicated, skin friction drag is determined by the separate contribution calculated for each of the components of the aircraft.

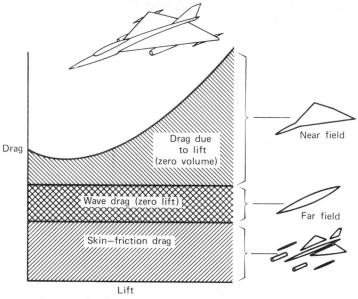

Fig. 17 Representative contributions to the drag of a supersonic aircraft (Carlson and Harris, 1969, courtesy of NASA).

13.5 Concluding Remarks

We have attempted to give an exposition of some important physical concepts that guide the design of high speed aircraft *insofar as they can be described on the basis of small disturbance theories.*

Three aspects must be remembered. (1) Since the small-disturbance equations are based on first-order effects, while the flow phenomena are influenced by higher-order effects to an extent depending mainly on the slenderness of the configuration and on angle of attack, the designer works back and forth between theory and experiment to seek optimum configurations. (2) All of the methods involve selection of a design Mach number, and the calculations do not indicate the characteristics under off-design conditions. (3) The methods described make no provision for viscous effects, such as effects of boundary layer thickness, transition from laminar to turbulent layers, flow separation, and wake and jet effects. Considering these effects the agreement of the predictions with experiment are truly remarkable, even though they are limited to small angles of attack of slender bodies over restricted ranges of Mach number.

CHAPTER 14

The Dynamics of Viscous Fluids

14.1 Introduction

The objective of Chapters 14 through 19 is to give the reader an understanding of the role played by viscosity in determining the flow of fluids. The exact equations of conservation of mass, momentum, and energy, which were derived for inviscid flow in Chapters 3 and 8, are derived, including viscosity effects, in Appendix B. The attendant complications are so great that few exact solutions of technical importance exist. However, remarkably good approximate solutions to a large number of important engineering problems have been obtained through simplification of the equations according to the circumstances of the particular application being treated.

Many of the solutions that have been obtained depend on a physical insight for their mathematical formulation. This insight is necessary to the solution of engineering problems, and we shall therefore stress the physical reasoning leading to the simplified equations.

The analyses of the preceding chapters apply to a good approximation to the flow outside the boundary layer and the wake behind the body. Then, as long as the angle of attack is small, we may expect the perfect-fluid analysis to give a good approximation to the pressure distribution and therefore to the lift and moment acting on an airfoil. This reasoning is not completely self-evident because it presumes that the pressure at the outer edge of the boundary layer is the same as that at the solid surface. The presumption is justified in this chapter on intuitive grounds.

As soon as the angle of attack becomes large enough to cause the airfoil to *stall*, the perfect-fluid analysis is no longer of any value and we must rely almost entirely on empirical results. The stall is synonymous with flow separation, which is described in Chapters 15, 18, and 19.

Two types of drag, *form* or *pressure drag* and *skin friction*, are encountered in the flow of a viscous incompressible fluid past a body. Form or pressure drag results from the separation of the flow from the body, for example, as occurs with a stalled airfoil. As a result of flow separation, the configuration of the streamlines and hence the pressure distribution over the body are altered considerably from what they are for perfect-fluid flow. The integration over the surface of the downstream component of the pressure forces on the elements of the surface gives the form drag. This integration neglects the skin friction because the skin friction acts tangentially to the surface. A *streamline body* is defined as a body for which the major contribution to the drag is skin friction; a body for which the major contribution is form drag is defined as a *bluff body*. An airfoil, for instance, is a streamline body at low angles of attack, since skin friction accounts for 80 to 90 percent of the drag. At angles of attack for which the airfoil is stalled, 80 to 90 percent of the drag is form drag, and so in this case the airfoil is a bluff body.

In taking up the factors affecting the type of flow and hence the shearing stress, we deal constantly with *similarity parameters*, one of which is the Reynolds number, $Re = Vl/\nu$ derived in Appendix A. The Reynolds number is actually a *class* of similarity parameters, depending upon the particular characteristic length l which is used in its definition. The chord of an airfoil, the distance from the leading edge to the point of transition from laminar to turbulent flow in the boundary layer, the thickness of the boundary layer, and the height of roughness elements are some of the characteristic lengths; each is important for a different aspect of the flow. Turbulence in the incident airstream and curvature of the surface introduce other parameters. The reader should constantly keep in mind that in all viscous-flow problems a number of similarity parameters govern the phenomena observed. Much of the work in aerodynamics has for its object the identification and the evaluation of the effects of those parameters which govern the particular aspects of flow being considered.

The two distinct types of flow—laminar and turbulent—are discussed in the following chapters. These flows are distinguished from each other according to the physical mechanism of the stresses. In laminar flow, solutions to problems can be obtained by more or less straightforward simplifications of the conservation equation. In turbulent flow, however, the number of variables outnumbers the equations, so great dependence is placed on dimensional reasoning and on hypotheses suggested by experimental results.

An extensive compilation of measurements of drag over a wide range of speeds is given by Hoerner (1965).

14.2 The No-Slip Condition

It was pointed out in Chapter 1 that the distinguishing feature of the flow of a viscous fluid around a body is the fact that at the fluid-solid interface no relative motion exists between the fluid and the body; that is, the *no-slip condition* prevails at a solid surface. The difference between the flow of viscid and inviscid fluids is therefore manifested in the boundary conditions. For inviscid fluids, the boundary condition at a solid surface is the vanishing of the velocity component normal to the surface. For a viscous fluid, however, the boundary condition must be that the *total* velocity vanishes at the surface.*

*No direct experimental check of the no-slip condition exists. Its acceptance rests rather on the excellent agreement between the theory employing this condition and experiment. A discussion of the condition is given in Goldstein (1938). In low-density flow such as exists at altitudes above about one hundred thousand meters, the mean free path of molecules is relatively large and the no-slip condition no longer obtains.

14.3 The Boundary Layer

The boundary layer is defined as the layer adjacent to a body within which the major effects of viscosity are concentrated. Intuitively, we would expect that the alteration to the flow caused by the no-slip condition will decrease as we move out from the surface and hence that the effect will not be detectable beyond a certain distance. In other words, outside of the boundary layer, the flow of a viscous fluid will resemble closely that of an inviscid fluid.

The justification for applying the results of perfect fluid analyses to viscous flows was provided by Ludwig Prandtl in 1904. He postulated that for fluids of small viscosity, the effects of viscosity on the flow around streamline bodies are concentrated in a *thin* boundary layer. The limitation of Prandtl's hypothesis to fluids of small viscosity is broad enough to include gases as well as "watery" fluids.

We must, of course, specify a characteristic dimension *with respect to which* the boundary layer is thin. On an airfoil, for instance, at speeds of significance in aeronautics, the boundary layer will vary from practically zero thickness near the leading edge to a few percent of the chord at the trailing edge. Then, the characteristic length with respect to which the boundary layer is thin is the distance from the forward stagnation point of the body to the point being considered.

The most important deduction from Prandtl's hypothesis of a thin boundary layer is that *the pressure change through the boundary layer is essentially zero*. This deduction is justified here on intuitive grounds. Figure 1a shows schematically the streamlines and velocity distributions in the boundary layer along a flat plate. The boundary layer thickness, designated by δ, is small everywhere, and hence $d\delta/dx$ will also be small. The streamlines will therefore be only very slightly curved and the radius of curvature will be large. From the equilibrium condition (Section 3.6)

$$\frac{\partial p}{\partial y} = \frac{\rho u^2}{R}$$

and it follows that $\partial p/\partial y$ will be negligible. If the surface is curved, as shown in Fig. 1b, the conclusion is still valid. Experiment and theory indicate that $\partial p/\partial y$ may be neglected even over surfaces of quite small radius of curvature.

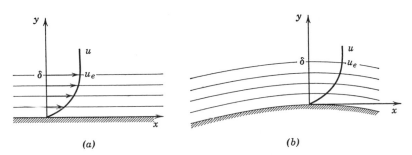

(a)   (b)

Fig. 1 Coordinate systems for boundary layer equation.

Since the pressure is nearly constant through the boundary layer, the *stagnation pressure*, p_0, will, by virtue of Bernoulli's equation $p + \frac{1}{2}\rho V^2 = p_0$ (Section 3.5), vary from p at $y = 0$ to $p + \frac{1}{2}\rho u_e^2$ at $y = \delta$. Then, by Section 3.6,

$$\frac{\partial p_0}{\partial n} = \frac{\partial p_0}{\partial y} = - (\text{curl}_z \ \mathbf{V}) \, \rho V$$

and since $\partial p_0/\partial y$ is different from zero, $\text{curl}_z \ \mathbf{V}$ has a finite value; that is, the boundary layer is a field of rotational flow. This conclusion also follows from the definition of $\text{curl}_z \ \mathbf{V}$ in two dimensions:

$$\text{curl}_z \ \mathbf{V} = \frac{\partial v}{\partial x} - \frac{\partial u}{\partial y}$$

Since δ is small everywhere, the component of velocity normal to the surface, v, must be small everywhere, and therefore $|\partial v/\partial x| \ll |\partial u/\partial y|$. Hence the flow is rotational throughout the boundary layer.

With δ small and $\partial p/\partial y$ negligible, it follows that the pressure distribution over a streamlined body is very nearly that calculated for an inviscid flow. Therefore, the lift and moment acting on wings and bodies may be calculated, as in previous chapters, by integrating the pressure distribution derived on the basis of inviscid flow.

For instance, in Section 5.5, we found that for a thin airfoil at a small angle of attack in an inviscid flow $\partial c_l/\partial \alpha = 2\pi$ and the aerodynamic center is at the $\frac{1}{4}$ chord point. Actually, for airfoils up to about 15 percent thickness to chord ratio, $\partial c_l/\partial \alpha = 2\pi\eta$, where η has values between 0.9 and 1.0, depending on the camber and thickness distribution; and the aerodynamic center of most airfoils is 1 to 2 percent ahead of the $\frac{1}{4}$ chord point. We see, therefore, that the effect of viscosity on these aerodynamic characteristics is not large.

The inviscid flow analyses cannot, of course, predict the frictional drag of bodies. The following chapters are devoted to the determination of the frictional drag and the factors that affect it.

Deductions based on the hypothesis of a thin boundary layer break down in regions of "flow separation" such as occur at a high angle of attack. These limitations are described in Section 15.4.

14.4   Viscous Stresses

The analysis of the boundary layer must rest on an understanding of the viscous stress. The approximate derivations of Chapter 1 yielded expressions for the coefficient of viscosity μ and the shearing stress τ in terms of the properties of the fluid and of the flow. These expressions are

$$\mu = \tfrac{1}{3}\rho c L$$

$$\tau = \mu \frac{\partial u}{\partial y} \qquad (1)$$

where c is the average velocity of the molecules,* L is a length associated with the mean free path of the molecules between collisions, u is the ordered velocity, and y is the coordinate normal to the flow. It was concluded there that μ is independent of the pressure and proportional to the square root of the absolute temperature.

The concept of a shearing stress, which must constantly be kept in mind, is that of a rate of transfer of downstream momentum in the direction lateral to the flow. This transfer is accomplished by the random motion of the molecules, effecting a continuous exchange of momentum between faster- and slower-moving layers of the flow. Equation 1 refers to a molecular process, but the process is qualitatively unchanged when the flow becomes turbulent, the only difference being that in a turbulent flow relatively large masses of fluid carry out the momentum transport. Regardless of the type of flow, we may describe the shearing stress between two layers of fluid as a transfer phenomenon; one is a molecular transfer, the other a turbulent transfer.

Figure 1 shows a schematic diagram of a boundary layer in which the velocity u varies from zero at the surface to the free-stream value. The shearing stress at the surface $\tau_w = \mu(\partial u/\partial y)_w$ is the skin friction (force per unit area) exerted by the fluid on the surface in the tangential direction. The shearing stress then varies continuously throughout the boundary layer from τ_w at $y = 0$ to zero at $y = \delta$.

Momentum transfer also takes place between adjacent fluid elements *along* a streamline through the motion of molecules across the interface. The resulting stress is called a normal viscous stress and is proportional to $\mu \partial u/\partial x$. Since the boundary layer is thin and its thickness changes only slowly with x, $|\partial u/\partial x| \ll |\partial u/\partial y|$ and we therefore neglect the viscous normal stress compared with the viscous shearing stress.

14.5 Boundary Layer Equation of Motion

In Section 3.4, Newton's second law of motion was applied to a fluid element acted upon only by pressure forces and gravity. Euler's equation for the equilibrium of an inviscid fluid resulted. This law, applied to a fluid element of mass $\rho \Delta x \Delta y \Delta z$, may be written

$$\rho \Delta x \Delta y \Delta z \, \frac{\mathcal{D}V}{\mathcal{D}t} = F \qquad (2)$$

*In the more accurate derivation of μ, where λ is the mean free path between collisions, $\mu = 0.49\,\rho c \lambda$.

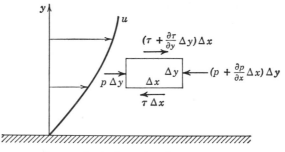

Fig. 2 Forces acting on an element in the boundary layer.

The left side of Eq. 2 represents the rate of change of the momentum of a fluid particle of mass $\rho \Delta x \Delta y \Delta z$, and the equation expresses the conservation of momentum of the fluid particle.

The equations expressing conservation of momentum in a viscous compressible fluid are derived in detail in Appendix B. They are also simplified to apply to boundary layers. In this section, we derive *directly* the approximate equation for flow in the two-dimensional boundary layer. The equation obtained is identical with that found by the more rigorous analysis of Appendix B.

We consider the forces acting on an element in a two-dimensional boundary layer as shown in Fig. 2. The element is of unit thickness in the z direction. The sum of the pressure and shear forces on the element gives the right side of Eq. 2, that is,

$$\rho \Delta x \Delta y \frac{\mathcal{D}u}{\mathcal{D}t} = \left(-\frac{\partial p}{\partial x} + \frac{\partial \tau}{\partial y} \right) \Delta x \Delta y$$

After dividing by $\Delta x \Delta y$, expanding the left side and substituting for τ from Eqs. 1, we get

$$\rho \left(\frac{\partial u}{\partial t} + u \frac{\partial u}{\partial x} + v \frac{\partial u}{\partial y} \right) = -\frac{\partial p}{\partial x} + \frac{\partial}{\partial y} \left(\mu \frac{\partial u}{\partial y} \right) \tag{3}$$

This equation is the *boundary layer equation of motion*. Then Eq. 3 and the continuity equation

$$\frac{\partial \rho}{\partial t} + \frac{\partial}{\partial x} (\rho u) + \frac{\partial}{\partial y} (\rho v) = 0 \tag{4}$$

which, for incompressible flow, reduces to

$$\frac{\partial u}{\partial x} + \frac{\partial v}{\partial y} = 0 \tag{5}$$

are the equations available for the solution of boundary layer problems.

For an incompressible flow, the variables are u, v, and p, but we have only two equations, Eqs. 3 and 5, to evaluate the three variables. The missing equation expresses conservation of the y component of momentum, the terms of which become negligible through the assumption that $\partial p/\partial y = 0$ within the boundary layer. Since we have only two equations to determine three variables, one of the variables must be given or must be determined independently. The pressure can be determined independently since, as was pointed out in Section 14.3, setting $\partial p/\partial y = 0$, and postulating a thin boundary layer enables us to use the methods of previous chapters for inviscid flows to find the pressure distribution over the body. After we have determined $p = p(x)$ for a particular body, Eqs. 3 and 5 and appropriate boundary conditions determine the velocity distributions in the boundary layer.

14.6 Similarity in Incompressible Flows

In this section, we demonstrate the importance of Reynolds number in comparing flows about geometrically similar bodies. We could carry out the demonstration with the boundary layer equation, but since the conclusion applies to the entire flow field, we shall use the exact equations of motion. These are the *Navier-Stokes equations* derived in Appendix B; they are the Euler equations of Section 3.4 with the addition of the terms describing the viscous forces on an element. The equations and the continuity equations for incompressible flow are

$$\rho \frac{\mathcal{D}u}{\mathcal{D}t} = -\frac{\partial p}{\partial x} + \mu \nabla^2 u$$

$$\rho \frac{\mathcal{D}v}{\mathcal{D}t} = -\frac{\partial p}{\partial y} + \mu \nabla^2 v$$

$$\rho \frac{\mathcal{D}w}{\mathcal{D}t} = -\frac{\partial p}{\partial z} + \mu \nabla^2 w \qquad (6)$$

$$\frac{\partial u}{\partial x} + \frac{\partial v}{\partial y} + \frac{\partial w}{\partial z} = 0$$

where

$$\nabla^2 \equiv \frac{\partial^2}{\partial x^2} + \frac{\partial^2}{\partial y^2} + \frac{\partial^2}{\partial z^2}$$

Let V be a representative velocity (say, that at a great distance from the body) and L be a characteristic length (say, the length of the body). Then Eqs. 6 can be made dimensionless by introducing the following dimensionless variables:

$$x' = \frac{x}{L}, \quad y' = \frac{y}{L}, \quad z' = \frac{z}{L}, \quad t' = \frac{Vt}{L}$$

$$u' = \frac{u}{V}, \quad v' = \frac{v}{V}, \quad w' = \frac{w}{V}, \quad p' = \frac{p}{\rho V^2}$$

Equations 6 then become

$$\frac{\mathcal{D}u'}{\mathcal{D}t'} = -\frac{\partial p'}{\partial x'} + \frac{1}{Re}\nabla'^2 u'$$

$$\frac{\mathcal{D}v'}{\mathcal{D}t'} = -\frac{\partial p'}{\partial y'} + \frac{1}{Re}\nabla'^2 v'$$

$$\frac{\mathcal{D}w'}{\mathcal{D}t'} = -\frac{\partial p'}{\partial z'} + \frac{1}{Re}\nabla'^2 w'$$

$$\frac{\partial u'}{\partial x'} + \frac{\partial v'}{\partial y'} + \frac{\partial w'}{\partial z'} = 0$$

(7)

where $Re = VL/\nu$. Now, given two geometrically similar bodies immersed in a moving fluid, Eqs. 7 show that the equations of motion for the two flows are identical, *provided* that the Reynolds numbers are the same. Also, the nondimensional boundary conditions for the two flows will be identical.

Therefore, *flows about geometrically similar bodies at the same Reynolds number are completely similar in the sense that u', v', w', and p' are, respectively, the same functions of x', y', z', and t' for the various flows.* Similarity of the bodies must involve not only the shapes but also the roughness; the flows must also be similar as regards turbulence (see Chapter 17).

Another important generalization results when we let the Reynolds number become infinite. Equations 7 then reduce to Euler's equation (Section 3.4) for a perfect fluid. Therefore, *the flow of a perfect fluid is identical with that of a viscous fluid at infinite Reynolds number.* It is important to realize that this is true only in the limit; a finite increase in the Reynolds number does not necessarily improve the agreement between a viscous flow and the flow of a perfect fluid.

In Chapter 16, it will be shown that for a compressible fluid with heat transfer several more similarity parameters must be taken into account.

14.7   Physical Interpretation of Reynolds Number

In most ranges of the dimensionless parameters in fluid flows, the Reynolds number is by far the most important dimensionless parameter. This number can be derived in physical terms by considering only the physical dimensions of those properties that determine the ratio of the inertia to the viscous forces on a volume of fluid; in so doing we completely disregard numerical factors.

Consider for example steady flow in a boundary layer of thickness δ; the governing equations are the Navier-Stokes equations, that is, the first three of Eqs. 6. The inertia stress terms, those on the left sides $\rho u \partial u/\partial x$, and so forth, have the dimensional form

$$\frac{\rho V^2}{\delta}$$

where V is the velocity outside the layer. The viscous stress terms, expressed by the second group of terms on the right side of the equation, $\mu \partial^2 u/\partial x^2$, and so forth, have the dimensional form

$$\frac{\mu V}{\delta^2}$$

and the ratio between the two stresses is

$$Re_\delta = \frac{\rho V \delta}{\mu} = \frac{V\delta}{\nu} \tag{8}$$

termed the boundary layer Reynolds number. If Re_δ is small (e.g. small V and δ, large ν) the viscous forces overshadow the inertia forces on a fluid element, and vice versa.

As is mentioned above, the length, δ, in Eq. 8 is simply a characteristic length and is chosen to describe a particular flow or an aspect of the flow. For flow past wings the chord is generally chosen, though use of the Reynolds number to compare flows implies that they are geometrically similar.

In general for low Reynolds number flows, such as the motion of fog droplets, the flight of small insects, or the flow of highly viscous fluids, the viscous forces are so much greater than the inertia terms that the latter may be neglected. Then Eqs. 6 and 7 become linear and superposition of simple flows to treat complicated ones is valid.

CHAPTER
15

Incompressible Laminar Flow in Tubes and Boundary Layers

15.1 Introduction

In the previous chapter we derived the equations necessary to the solution of the incompressible boundary layer problem. In this chapter, we first solve the problem of the steady, incompressible, viscid flow in a tube far from the entrance. Next we solve the boundary layer problem for flow along a flat plate oriented parallel to a flow.

The solutions of these two problems illustrate the importance of the Reynolds number; in fact, all the important quantities in which we are interested, when they are expressed in nondimensional form, are functions only of the Reynolds number. Later sections deal with the momentum relations within the boundary layer and with the flow associated with a pressure gradient. The Kármán-Pohlhausen method of analysis for boundary layer flow in a streamwise pressure gradient is described.

The results given in this chapter are remarkably good approximations even at relatively high speeds, but a *rigorous* treatment of the boundary layer at high speeds requires that variations of density and temperature be taken into account. The two additional equations needed are the equation of state and the equation expressing conservation of energy in the boundary layer. The latter will be derived in Chapter 16 preliminary to considerations of the compressible boundary layer.

15.2 Laminar Flow in a Tube

Consider incompressible flow in the conduit of Fig. 1. The conduit is straight, and, for simplicity of the analysis, a circular cross section is taken. The boundary layer will begin with zero thickness at the entrance and will grow with distance along the tube. At a large distance from the entrance, the outer edge of the boundary layer will have reached the center of the tube, and still further downstream, say for $x \geq x_1$, all velocity distributions will be identical. Then, by definition, for $x > x_1$, the flow is *fully developed*.*

If we sum the forces on the cylindrical element shown in Fig. 1, we have

$$F = \pi r^2 (p_1 - p_2) - 2\pi r l \tau \tag{1}$$

---

*Boussinesq calculated a formula $x_1 = 0.26 \, r Re$, where x_1 is the distance from the entrance to the point where the flow is fully developed in a tube of circular cross section, r is the radius of the tube, $Re = u_m r / \nu$, and u_m is the mean velocity in the tube. See Goldstein (1938, p. 299ff).

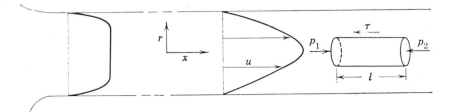

Fig. 1 Fully developed flow in a tube.

Since the element does not accelerate in a fully developed flow, $F = 0$; that is, the pressure force on the x faces of any element balance the frictional forces acting on the boundaries parallel to the flow. Further, with the origin at the center of the tube, $\tau = -\mu \, du/dr$, so that Eq. 1 becomes

$$\frac{du}{dr} = -\frac{r}{2\mu} \frac{p_1 - p_2}{l} \tag{2}$$

l is not specified and no other quantity in the equation except $p_1 - p_2$ depends on l. Therefore, $-(p_1 - p_2)/l = dp/dx$ = constant. Equation 2 is integrated to obtain

$$u = \frac{r^2}{4\mu} \frac{dp}{dx} + \text{constant}$$

The constant is evaluated by use of the boundary condition, $u = 0$ at $r = a$. The equation for the velocity distribution in fully developed flow then becomes

$$u = -\frac{1}{4\mu} \frac{dp}{dx} (a^2 - r^2) \tag{3}$$

The shape of the velocity profile is seen to be a paraboloid.* We next obtain the relations between the significant dimensionless parameters. These are the Reynolds number and pressure drop coefficient, defined as

$$Re = \frac{2au_m}{\nu}; \qquad \gamma = \frac{\tau_0}{\frac{1}{2} \rho u_m^2} = \frac{-\mu (du/dr)_{r=a}}{\frac{1}{2} \rho u_m^2} \tag{4}$$

where u_m is the mean velocity, that is, $u_m = Q/\pi a^2 \rho$ where Q is the mass flow per second through the tube. Since, for a paraboloid, the mean ordinate is half the maximum, at $r = 0$, $u = 2u_m$, and Eq. 3 gives

$$u_m = -\frac{a^2}{8\mu} \frac{dp}{dx}$$

*Equation 3 may also be obtained as an exact solution of the equation of motion derived in Appendix B.

Then

$$\gamma = -\frac{\frac{1}{2}\,a(dp/dx)}{\frac{1}{2}\,\rho u_m^2} = \frac{a(dp/dx)}{\rho(a^2/8\mu)\,(dp/dx)\,u_m} \tag{5}$$

$$\gamma = \frac{16}{Re} \tag{6}$$

The hyperbola γ versus Re describes the *scale effect* of Reynolds number on the pressure drop coefficient. The analysis applies only to *laminar flow;* at a critical Reynolds number dependent on factors to be taken up in Chapter 17, *transition* from laminar to *turbulent flow* occurs and time dependent terms associated with the unsteady character of the flow would have to be included if a rigorous solution were to be obtained.

15.3 Laminar Boundary Layer Along a Flat Plate

The solution of Eq. 14.3 for the steady flow of an incompressible viscous fluid along a flat plate, as shown in Fig. 2, was obtained by Blasius in 1908. For this

Fig. 2 The boundary layer on a flat plate.

case $\partial u/\partial t = \partial p/\partial x = 0$, and the equations of motion and continuity (Eqs. 14.3 and 14.5) become

$$u\,\frac{\partial u}{\partial x} + v\,\frac{\partial u}{\partial y} = \nu\,\frac{\partial^2 u}{\partial y^2}$$

$$\frac{\partial u}{\partial x} + \frac{\partial v}{\partial y} = 0 \tag{7}$$

To solve these equations, we need two boundary conditions for the first and one for the second. The conditions are

$$\text{at } y = 0: \qquad u = v = 0$$

$$\text{at } y = \infty: \qquad u = u_e \tag{8}$$

They express the physical conditions that there is no slip at the boundary ($u = 0$ at $y = 0$), that the boundary is a streamline ($v = 0$ at $y = 0$), and that the horizontal flow is unaffected at infinity ($u = u_e$ at $y = \infty$).

We see that in Eqs. 7 we have two equations to determine the two unknown variables. In order to get a single unknown variable and a single equation, we introduce the *stream function*, which was defined and discussed in Section 2.6. The stream function ψ is defined as that function of x and y such that

$$u = \frac{\partial \psi}{\partial y}, \qquad v = -\frac{\partial \psi}{\partial x} \qquad (9)$$

It is clear that the stream function defined in this way satisfies identically the continuity equation.

We now introduce ψ as a function of x and y such that the equation of motion of Eqs. 7 reduces to an ordinary differential equation, that is, the two independent variables are reduced to one. The reason for seeking an ordinary differential equation is that no general methods exist for solving partial differential equations of the type of the equation of motion. The most distressing feature of the equation is that it is nonlinear in the dependent variable ψ, as is evident if Eqs. 9 are substituted in Eqs. 7.

We therefore seek to express the equation of motion in terms of a single independent variable η, a function of x and y; that is, the equation of motion will be expressed in a form in which neither x nor y appears explicitly. Blasius found that if the new variable η were made proportional to $y/\sqrt{x}$, an ordinary differential equation resulted.*

It is, in general, most convenient to work with dimensionless quantities, and accordingly we define

$$\eta = \frac{y}{2}\left(\frac{u_e}{\nu x}\right)^{1/2}, \qquad \psi = (\nu u_e x)^{1/2} f(\eta) \qquad (10)$$

Here η is dimensionless and ψ has the dimensions: velocity $\times$ length. We next determine, by means of Eqs. 9 and 10, the terms in Eqs. 7. Differentiations with respect to η are denoted by primes. Then,

$$u = \frac{1}{2}u_e f', \qquad \frac{\partial u}{\partial x} = -\frac{1}{4}\frac{u_e}{x}\eta f''$$

$$\frac{\partial u}{\partial y} = \frac{u_e}{4}\left(\frac{u_e}{\nu x}\right)^{1/2} f'', \qquad \frac{\partial^2 u}{\partial y^2} = \frac{u_e}{8}\left(\frac{u_e}{\nu x}\right) f''' \qquad (11)$$

$$v = \frac{1}{2}\left(\frac{u_e \nu}{x}\right)^{1/2}(\eta f' - f)$$

When these values are substituted in the first of Eqs. 7 the result is the differential equation

$$f''' + ff'' = 0 \qquad (12)$$

---

*The *order-of-magnitude* analysis of Appendix B justifies the choice of $y/\sqrt{x}$ as the independent variable.

and the boundary conditions, Eqs. 8, become

$$\text{at } \eta = 0: \qquad f = f' = 0$$
$$\text{at } \eta = \infty: \qquad f' = 2 \tag{13}$$

and the solution $f(\eta)$ will, by Eqs. 11, enable the determination of u and v. The uniqueness of the solution has not been proved, but comparison with experiment has shown that the solution given is the one that describes the flow for the case considered. The differential equation, Eq. 12, appears simple; on the contrary, it is nonlinear and quite difficult. No closed solution has been found and, thus, solution by series is resorted to. We assume a solution of the form

$$f = A_0 + A_1 \eta + \frac{A_2}{2!} \eta^2 + \frac{A_3}{3!} \eta^3 + \cdots + \frac{A_n}{n!} \eta^n + \cdots$$

When the first two boundary conditions are applied to f, we find that $A_0 = A_1 = 0$. After substituting the series for f into Eq. 12, we get

$$A_3 + A_4 \eta + \frac{A_5}{2!} \eta^2 + \cdots + \left(\frac{A_2}{2!} \eta^2 + \frac{A_3}{3!} \eta^3 + \cdots \right) \left(A_2 + A_3 \eta + \frac{A_4}{2!} \eta^2 + \cdots \right) = 0$$

The multiplication is carried out and the coefficients of like powers of η are collected. Then,

$$A_3 + A_4 \eta + \left(\frac{A_2^2}{2!} + \frac{A_5}{2!} \right) \eta^2 + \cdots = 0$$

Since this equation must hold for all values of η, the coefficients of every power of η must vanish. Hence,

$$A_3 = A_4 = 0, \quad A_2^2 + A_5 = 0, \text{ etc.}$$

Then all terms can be expressed as functions of η and A_2:

$$f = \frac{A_2 \eta^2}{2!} - \frac{A_2^2 \eta^5}{5!} + \frac{11 A_2^3 \eta^8}{8!} - \frac{375 A_2^4 \eta^{11}}{11!} + \cdots \tag{14}$$

Equation 14 satisfies the first two boundary conditions of Eqs. 13, and the third will be used to determine A_2.

To accomplish this we write $f(\eta)$ in the equivalent form

$$f = A_2^{1/3} \left[\frac{(A_2^{1/3} \eta)^2}{2!} - \frac{(A_2^{1/3} \eta)^5}{5!} + \frac{11(A_2^{1/3} \eta)^8}{8!} - \frac{375(A_2^{1/3} \eta)^{11}}{11!} + \cdots \right]$$

$$\equiv A_2^{1/3} g(\Gamma)$$

where $\Gamma = A_2^{1/3} \eta$. The boundary condition (Eqs. 13) to be satisfied at $\eta = \infty$ is

$$\lim_{\eta \to \infty} f' = 2$$

which may be written

$$\lim_{\Gamma \to \infty} [A_2^{2/3} g'(\Gamma)] = 2$$

where the prime refers to differentiation with respect to Γ. But, for $A_2 > 0$ when $\eta \to \infty$, $\Gamma \to \infty$, and we may write instead of the above

$$\lim_{\eta \to \infty} [g'(\eta)] = \frac{2}{A_2^{2/3}}$$

or

$$A_2 = \left[\frac{2}{\lim_{\eta \to \infty} g'(\eta)} \right]^{3/2}$$

The right-hand side of this equation is plotted as a function of η, and A_2 can be determined to any desired approximation. Goldstein (1938) found that $A_2 = 1.32824$. The quantities f, f', and f'' are plotted in Fig. 3 for this value of A_2.

The solution shows that the value of u does not reach u_e until $\eta = \infty$, that is, at $y = \infty$. However, at $\eta = 2.6$, $u/u_e = 0.994$; therefore, if we choose the edge of the boundary layer $(y = \delta)$ as the point where u is within 1 percent of u_e, we get, from Eqs. 10,

$$\delta = 5.2 \sqrt{\frac{\nu x}{u_e}} = \frac{5.2 \, x}{\sqrt{Re_x}} \tag{15}$$

where $Re_x = u_e x/\nu$.

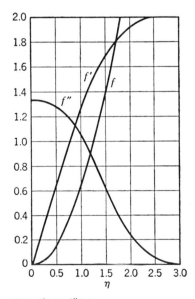

Fig. 3 The Blasius functions $f(\eta), f'(\eta), f''(\eta)$.

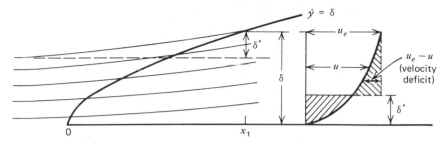

Fig. 4 Schematic representation of δ^*, the displacement thickness of the boundary layer at $x = x_1$ from the leading edge of a flat plate.

Since the definition of the boundary layer thickness, δ, is arbitrary, we define a *displacement thickness* δ^* as illustrated for flow along a flat plate in Fig. 4. We see that δ^* at $x = x_1$ is the amount by which the streamline entering the boundary layer at that point has been displaced outward by the retardation of the flow in the boundary layer. The velocity profile shown at the right illustrates that, since the two cross-hatched areas are equal, the displacement thickness is given by the integral

$$\delta^* = \int_0^\infty \left(1 - \frac{u}{u_e} \right) dy \tag{16}$$

in which δ^* is indicated at the surface instead of at the edge of the boundary layer.

We now calculate δ^*, which, according to Eqs. 11 and 16, is given by

$$\delta^* = \int_0^\infty \left(1 - \frac{u}{u_e} \right) dy = \left(\frac{\nu x}{u_e} \right)^{1/2} \int_0^\infty (2 - f')\, d\eta$$

$$= (\nu x / u_e)^{1/2} \, [2\eta - f]_0^\infty$$

$$= (\nu x / u_e)^{1/2} \, \lim_{\eta \to \infty} (2\eta - f)$$

Since, from Eqs. 13, $f'(\infty) = 2$, the solution for Eq. 12 which must hold for η large is $f = 2\eta + \beta$, where β is a constant; that is, $\lim_{\eta \to \infty} (2\eta - f) = -\beta$. β can be determined from a solution of Eq. 12 by successive approximation. (See Durand, 1943, Vol. 3, p. 87.) The result is $\beta = -1.7208$; that is,

$$\delta^* = \frac{1.7208\, x}{\sqrt{Re_x}} \tag{17}$$

The skin-friction coefficient $c_f = \tau_0 / \tfrac{1}{2} \rho u_e^2$ is calculated as follows:

$$\tau_0 = \mu \left(\frac{\partial u}{\partial y} \right)_{y=0} = \frac{\mu}{2}\, u_e f''(0) \frac{1}{2} \left(\frac{u_e}{\nu x} \right)^{1/2}$$

$$= \frac{1}{4}\, \mu A_2 u_e \left(\frac{u_e}{\nu x} \right)^{1/2}$$

Then

$$c_f = \frac{A_2}{2}\left(\frac{\nu}{u_e x}\right)^{1/2} = \frac{0.664}{\sqrt{Re_x}} \tag{18}$$

The average skin-friction coefficient C_f for one side of the flat plate of unit width of length l is given by

$$C_f = \int_0^l \frac{\tau_0\,dx}{\frac{1}{2}\rho u_e^2 l} = \frac{1.328}{\sqrt{Re_l}} \tag{19}$$

where $Re_l = u_e l/\nu$. Figures 5 and 6 show excellent agreement between theory and experiment for the velocity profile and for the local skin friction coefficient.

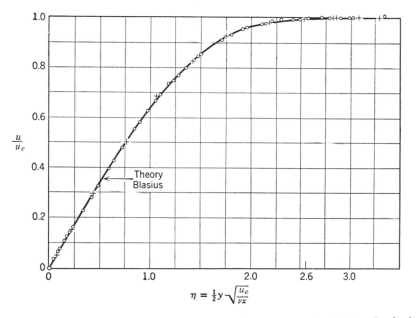

$$\eta = \tfrac{1}{2}y\sqrt{\frac{u_e}{\nu x}}$$

Fig. 5 Comparison between theoretical and experimental velocity distributions in the laminar boundary layer on a flat plate. Experiments by Nikuradse (1942) cover Reynolds number range 1.08×10^5 to 7.28×10^5.

As is well known, the above solution is valid only below a certain Reynolds number, the value of which is dependent on various influences (see Chapter 17). At higher Reynolds numbers, the flow in the boundary layer becomes *turbulent* and the equation of motion describing the flow must strictly include the transient term $\partial u/\partial t$. The turbulent boundary layer is discussed in Chapter 17. Figure 6 includes a comparison between skin-friction coefficients for laminar and for turbulent flow in the boundary layer.

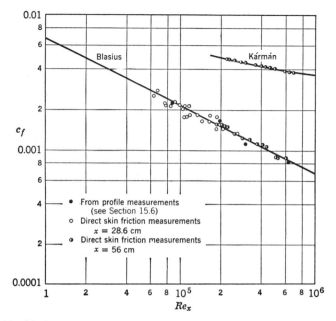

Fig. 6 Local skin friction in incompressible flow. The lower curve refers to laminar flow, the upper to turbulent. (Dhawan, 1953, Courtesy NASA.)

15.4 Flow with Pressure Gradient—Separation

The equation of motion for flow in the boundary layer will be analyzed to show some effects, on the flow, of the sign of the pressure gradient along the surface. Some of the facts about flow separation, of which a common example is the stalling of wings, can also be deduced.

Equation 14.3 gives some indication of the effect of pressure gradient on the velocity profile. If we make use of the boundary condition $u = v = 0$ at $y = 0$, the left side of the equation vanishes and we have (for μ = constant)

$$\mu \left(\frac{\partial^2 u}{\partial y^2} \right)_{y=0} = \frac{\partial p}{\partial x} \tag{20}$$

For the problem of a flat plate parallel to the flow, taken up in the previous section, $\partial p/\partial x = 0$, and hence, from Eq. 20, the profile has an inflection point at the surface. For $\partial p/\partial x < 0$, $(\partial^2 u/\partial y^2)_{y=0} < 0$; that is, the slope, $\partial u/\partial y$, decreases as y increases near the surface. Since $\partial u/\partial y = 0$ at $y = \delta$, we may expect that the decrease which begins at the surface will be *monotonic* to the edge of the boundary layer; in other words, the slope decreases with increasing y at every point within the layer. However, for $\partial p/\partial x > 0$, $(\partial^2 u/\partial y^2)_{y=0} > 0$; that is, $\partial u/\partial y$ *increases* as y increases near the surface. Since $\partial u/\partial y$ must be zero at $y = \delta$, $\partial^2 u/\partial y^2$ must go through zero somewhere within the boundary layer; in other words, an inflection point appears in the profile.

The physical reason for the appearance of an inflection point in the velocity profile in an *adverse pressure gradient* ($\partial p / \partial x > 0$) lies in the retarding effect, on the flow, of the upstream force associated with the adverse pressure gradient. The resulting loss of momentum of the fluid is especially noticeable near the surface where the velocity is low; hence $\partial u / \partial y$ near $y = 0$ becomes smaller and smaller the longer the distance over which the adverse gradient persists. This effect is shown diagrammatically in Fig. 7. Then, at some distance downstream of the pressure minimum, we reach a point where $(\partial u / \partial y)_{y=0} = 0$, and beyond this point the direction of flow reverses near the surface. The point where $(\partial u / \partial y)_{y=0} = 0$ is the *separation point*. The greater the adverse pressure gradient the shorter will be the distance from point A, the pressure minimum, to the separation point B.

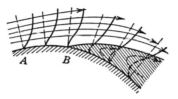

Fig. 7 Schematic velocity distributions in the vicinity of a separation point. A is the point of minimum pressure; B is the separation point.

Immediately downstream of the separation point, the schematic streamlines near the surface in Fig. 7 show a strong curvature, which must be associated with a strong pressure gradient normal to the surface. Accordingly, the reasoning of Section 14.3, which led to the conclusion that $\partial p / \partial y \cong 0$ through the boundary layer, can no longer be valid at, or downstream of, the separation point. The assumption which was the foundation of that analysis, that the boundary layer is thin compared with a characteristic dimension of the body, therefore breaks down, and it would be necessary to use the complete equations of motion to describe the flow. This analysis has not been carried out. An additional complication arises from the experimental observation that the flow downstream of the separation point is, except at very low Reynolds numbers, quite unsteady. The unsteadiness is confined to a region, extending downstream of the body, called the *turbulent wake*.

It will be seen from the above description that flow separation is a phenomenon associated with the boundary layer and therefore does not occur in an inviscid-fluid flow. The salient facts are that (1) separation occurs only in an adverse pressure gradient and then only if the adverse pressure gradient persists over a great enough length, the length being greater the more gentle the gradient; and (2) at and near the separation point, $\partial p / \partial y \neq 0$ near the surface.

Wherever the boundary layer is thin, Prandtl's hypothesis (Section 14.3) permits us to use the pefect-fluid analysis as a good approximation to the pressure distribution over a body in a viscous flow. But we can no longer employ this principle if an appreciable area of the body is in a region of separated flow; under these circumstances, theoretical and experimental pressure distributions will diverge widely.

The stalling of an airfoil at high angles of attack with its accompanying loss in lift and increase in drag is a separation phenomenon and is completely unpredictable by inviscid-fluid theory.

We are concerned here only with a description of flow separation; the point at which separation occurs is affected by many factors, among them the state of the boundary layer—whether it is laminar or turbulent. It will be pointed out in Section 18.11 that the higher velocities near the surface for a turbulent boundary layer enable it to drive further against an adverse pressure gradient than can the laminar layer; consequently, the effect of changing the boundary layer from laminar to turbulent is to move the separation point rearward.

15.5 Similarity in Boundary Layer Flows

The general condition for similarity of incompressible viscid flows about geometrically similar bodies was found to be equality of the Reynolds numbers (Section 14.6). Examples of the importance of Reynolds number are found in Sections 15.2 and 15.3, where it is shown that the resistance coefficient for flow in a tube and the skin-friction coefficient on a flat plate are determined by the Reynolds number.

"Similar" solutions of the boundary layer equations are those which yield scale factors that reduce all velocity profiles to a single curve. For instance, the Blasius solution of Section 15.3 predicts that $u/u_e = f'(\eta)$ where $\eta = \frac{1}{2}\, y/\sqrt{u_e/\nu x}$. Figure 5 shows excellent agreement with experiment.

A general class of similar profiles was found by Falkner and Skan (1930). They found that if u_e varies according to the law

$$u_e(x) = u_1 x^m \tag{21}$$

and

$$\psi(x, y) = \sqrt{\frac{2\nu u_1}{m+1}}\, x^{(m+1)/2}\, f(\xi)$$

$$\xi = y\sqrt{\frac{m+1}{2}\frac{u_e}{\nu x}} = y\sqrt{\frac{m+1}{2}\frac{u_1}{\nu}}\, x^{(m-1)/2} \tag{22}$$

the boundary layer equations, Eq. 14.3 with constant u and Eq. 14.5, reduce to

$$f''' + ff'' + \beta(1 - f'^2) = 0 \tag{23}$$

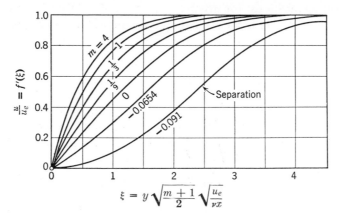

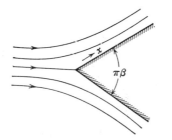

$$\xi = y\sqrt{\frac{m+1}{2}}\sqrt{\frac{u_e}{\nu x}}$$

Fig. 8 Velocity profiles for various values of m in Eq. 21.

Fig. 9 Streamlines for potential flow described by Eqs. 21 and 24.

where

$$\beta = \frac{2m}{m+1} \tag{24}$$

We see that for $m = 0$ ($\beta = 0$), Eq. 23 reduces to the Blasius equation Eq. 12, and ψ, ξ, and $u = u_e f'$ differ only by numerical factors from their counterparts in the Blasius analysis.

Equation 23 shows that similar boundary layer profiles exist everywhere on a body for which the velocity outside the boundary layer can be represented by Eq. 21 with constant m.* The profiles are shown in Fig. 8 for various values of m. The curve for $m = 0$ is the same as that shown in Fig. 5. We note that for $m = -0.091$ the flow is constantly on the verge of separation.

Equation 21 represents the velocity at the surface of a wedge which turns an inviscid flow through an angle $\beta\pi/2$ and therefore represents physically the flow past a wedge as shown in Fig. 9. The flows corresponding to $m > 0$, however, occur in

*Other "similar solutions" are given by Schlichting (1968, Chap. 8).

many practical applications; for instance, for $m = 1$ ($\beta = 1$) the boundary layer profile is that which occurs near the stagnation point of a two-dimensional body.

15.6 The von Kármán Integral Relation

Although solutions of the boundary layer equations for particular pressure distributions are useful, their practical application is limited to specific body shapes. A useful application of the momentum theorem, derived in Chapter 3, was devised by von Kármán (1921) to solve the boundary layer problem for any given pressure distribution. The method will be described here and its uses and limitations will be pointed out.

Consider a two-dimensional region (Fig. 10) bounded by a solid surface, the line $y = \delta$, and two parallel lines perpendicular to the solid surface. We analyze the

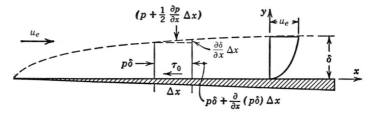

Fig. 10 Forces acting on a section of the boundary layer.

forces acting on the fluid according to the momentum theorem of Section 3.7, which may be stated: *the total rate of increase of momentum within a region is equal in both magnitude and direction to the force acting on the boundary of the region.* Consider the component of momentum parallel to the surface. A friction force $\tau_0 \Delta x$ is acting on the surface, and pressure forces are acting on the other three sides. Since body forces are neglected, the total downstream force acting on the boundaries of the element of length Δx is

$$- \tau_0 \, \Delta x + p\delta - \left[p\delta + \frac{\partial}{\partial x} (p\delta) \, \Delta x \right] + \left(p + \frac{1}{2} \frac{\partial p}{\partial x} \Delta x \right) \frac{\partial \delta}{\partial x} \, \Delta x$$

The last term of this expression is the downstream component of the mean pressure force acting on the sloping boundary $y = \delta$. Since Δx is small, we may neglect the term involving $(\Delta x)^2$, and the above expression for the downstream force simplifies to

$$\left(- \tau_0 - \delta \, \frac{\partial p}{\partial x} \right) \Delta x \tag{25}$$

To arrive at the momentum flux consider first the flux of mass through the region. The various contributions are:

$$\text{Mass entering at left per second} = \int_0^\delta \rho u \, dy$$

$$\text{Mass leaving at right per second} = \int_0^\delta \rho u \, dy + \frac{\partial}{\partial x}\left(\int_0^\delta \rho u \, dy\right)\Delta x$$

Then, from continuity, the mass entering the sloping face must equal the difference between these two values. Thus,

$$\text{Mass entering sloping face per second} = \frac{\partial}{\partial x}\left(\int_0^\delta \rho u \, dy\right)\Delta x$$

These expressions are used to find the flux of momentum, that is, the excess momentum leaving at the right over that entering at the left per second. The flux through the parallel faces is

$$\frac{\partial}{\partial x}\left(\int_0^\delta \rho u^2 \, dy\right)\Delta x \tag{26}$$

and the momentum entering the sloping face per second is the rate at which mass enters the sloping face, given above, multiplied by the free-stream velocity u_e. Thus the momentum enters the sloping face at the rate

$$u_e \frac{\partial}{\partial x}\left(\int_0^\delta \rho u \, dy\right)\Delta x \tag{27}$$

The momentum theorem is expressed by means of Eqs. 25, 26, and 27 and a term representing the time rate of increase of momentum within the element. Then

$$\int_0^\delta \frac{\partial}{\partial t}(\rho u)\, dy + \frac{\partial}{\partial x}\left(\int_0^\delta \rho u^2 \, dy\right) - u_e \frac{\partial}{\partial x}\left(\int_0^\delta \rho u \, dy\right) = -\tau_0 - \delta\frac{\partial p}{\partial x} \tag{28}$$

This is the one form of the *von Kármán integral relation*. It is applicable to an unsteady, compressible, viscid flow. We shall now particularize Eq. 28 to treat the incompressible boundary layer.

Equation 28 is put in a more convenient form by introducing the displacement thickness δ^* defined in Section 15.3 and the *momentum thickness* θ of the boundary layer. For an incompressible flow these quantities are defined by the relations

$$\delta^* = \int_0^\delta \left(1 - \frac{u}{u_e}\right)dy \tag{29}$$

$$\theta = \int_0^{\delta} \frac{u}{u_e} \left(1 - \frac{u}{u_e} \right) dy \tag{30}$$

In Section 15.3, δ^* was interpreted in terms of the velocity deficit in the boundary layer. Likewise, θ is a length associated with the momentum deficit that the air has suffered because of friction. To see this, consider the expression $\rho u (u_e - u) \, dy$, which is the momentum deficit of the mass $\rho u \, dy$ passing through the layer dy per second, relative to its momentum at velocity u_e. If this quantity is divided by ρu_e^2 and integrated through the boundary layer, we get Eq. 30, which then defines a length associated with the total momentum deficit in the boundary layer.

We may put the pressure term in a more usable form by means of the equation of motion for flow outside the boundary layer. Thus,

$$-\frac{\partial p}{\partial x} = \rho \left(\frac{\partial u_e}{\partial t} + u_e \frac{\partial u_e}{\partial x} \right)$$

and, after integrating from 0 to δ, this equation may be written

$$-\delta \frac{\partial p}{\partial x} = \int_0^{\delta} \rho \frac{\partial u_e}{\partial t} \, dy + \frac{\partial u_e}{\partial x} \int_0^{\delta} \rho u_e \, dy \tag{31}$$

Also, the last term on the left in Eq. 28 may be written

$$u_e \frac{\partial}{\partial x} \int_0^{\delta} \rho u \, dy = \frac{\partial}{\partial x} \left(u_e \int_0^{\delta} \rho u \, dy \right) - \frac{\partial u_e}{\partial x} \int_0^{\delta} \rho u \, dy \tag{32}$$

After substituting Eqs. 31 and 32 in Eq. 28,

$$\tau_0 = \frac{\partial}{\partial x} \left[\int_0^{\delta} \rho (u_e u - u^2) \, dy \right] - \frac{\partial u_e}{\partial x} \int_0^{\delta} \rho u \, dy + \frac{\partial u_e}{\partial x} \int_0^{\delta} \rho u_e \, dy$$

$$- \int_0^{\delta} \rho \frac{\partial}{\partial t} (u - u_e) \, dy \tag{33}$$

and Eqs. 29 and 30 enable us to put this formula in the form

$$\tau_0 = \rho \frac{\partial}{\partial x} (u_e^2 \theta) + \rho u_e \frac{\partial u_e}{\partial x} \delta^* + \rho \frac{\partial}{\partial t} (u_e \delta^*) \tag{34}$$

This equation has been the basis for many investigations of the incompressible boundary layer. It is derived here on physical grounds; it could as well be obtained as a first integral of the equation of motion as derived in Section 14.5.

Equation 34 provides a practical means for determining the distribution of shearing stress over a body. If velocity profiles in the boundary layer are measured at various stations, then the quantities occurring in the equation and their derivatives can be determined approximately and the shearing stress calculated. This method has been followed to determine the distribution of shearing stress over the surface of an airfoil.

15.7   The Pohlhausen Analysis of the Boundary Layer

The von Kármán integral relation was applied by Pohlhausen to an approximate investigation of the laminar boundary layer over bodies in steady incompressible flow when the pressure distribution is known. The method involves the evaluation of a polynomial with undetermined coefficients for the velocity distribution. The coefficients are determined as functions of the boundary layer thickness by means of boundary conditions, and Eq. 34 yields a differential equation for finding the boundary layer thickness as a function of x.

Since the velocity profile in an adverse pressure gradient has an inflection point (Section 15.4), we must employ at least a third-degree polynomial to represent it. In order to provide for a slightly better fit, Pohlhausen introduced the fourth-degree expression

$$\frac{u}{u_e} = A \left(\frac{y}{\delta}\right) + B \left(\frac{y}{\delta}\right)^2 + C \left(\frac{y}{\delta}\right)^3 + D \left(\frac{y}{\delta}\right)^4 \tag{35}$$

where A, B, C, D are undetermined coefficients. The method of solution consists first of expressing these coefficients as functions of δ by means of the boundary conditions, and second, of determining δ as a function of x by means of the von Kármán integral relation. The boundary conditions are

$$\text{at } y = 0: \ u = 0, \quad \mu \frac{\partial^2 u}{\partial y^2} = \frac{\partial p}{\partial x}$$
$$\tag{36}$$
$$\text{at } y = \delta: \ u = u_e, \quad \frac{\partial u}{\partial y} = 0, \quad \frac{\partial^2 u}{\partial y^2} = 0$$

where the second boundary condition at $y = 0$ is Eq. 20, and $\partial p/\partial x$ is considered to be given as a function of x. The three boundary conditions at $y = \delta$ may be considered simply as a defintion of δ. They are approximate; we have seen from the Blasius analysis of Section 15.3 that these conditions are exact only at $y = \infty$. Equation 35 already staisfies the condition $u = 0$ at $y = 0$, and so the remaining four serve to determine A, B, C, and D as functions of δ.

If the conditions of Eqs. 36 are applied to Eq. 35, we obtain the following equations:

$$A + B + C + D = 1$$
$$A + 2B + 3C + 4D = 0$$
$$2B + 6C + 12D = 0 \tag{37}$$
$$B = \frac{\delta^2}{2\mu u_e} \frac{\partial p}{\partial x}$$

By Bernoulli's equation for the flow at the edge of the boundary layer, the last of Eqs. 37 may be written

$$B = -\frac{\delta^2}{2\nu}\frac{du_e}{dx} = -\frac{\lambda}{2} \tag{38}$$

where

$$\lambda = \frac{\delta^2}{\nu}\frac{du_e}{dx} \tag{39}$$

is known as the *Pohlhausen parameter*. It is dimensionless and is a function of x. We solve Eqs. 37 for the constants A, C, and D in terms of λ and obtain

$$A = 2 + \frac{\lambda}{6}, \quad C = -2 + \frac{\lambda}{2}, \quad D = 1 - \frac{\lambda}{6} \tag{40}$$

and a particular boundary layer problem resolves itself into finding λ as a function of x; once λ is known, we can plot the velocity profiles by means of Eqs. 35 and 40. Velocity profiles for various values of λ are plotted in Fig. 11.

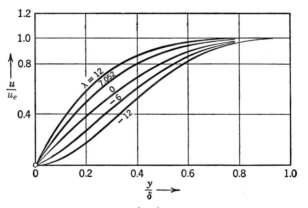

Fig. 11 Velocity profiles given by the von Kármán-Pohlhausen method.

The parameter λ has special significance with regard to the separation point, where the condition is, according to Section 15.4, $(\partial u/\partial y)_{y=0} = 0$. When this condition is applied to Eq. 35 we get $A = 0$ or

$$\lambda = -12 \tag{41}$$

as the condition for flow separation. The velocity profile for $\lambda = -12$ is shown in Fig. 11.

Further, $\lambda = 0$ at the minimum pressure point. If values of λ are substituted in

Eq. 35, we get values of $u/u_e > 1$ if $\lambda > +12$. Since this type of profile is unreasonable from a physical standpoint, we conclude that

$$-12 < \lambda < 12 \tag{42}$$

Pohlhausen determined λ as a function of x for the problem of the flow past a circular cylinder. He utilized the von Kármán integral relation of Eq. 34, which, when restricted to steady flow, may be written

$$\frac{\tau_0}{\rho} = u_e^2 \frac{d\theta}{dx} + (2\theta + \delta^*) u_e \frac{du_e}{dx} \tag{43}$$

This expression leads to an ordinary differential equation in δ^2/ν by the following process: Eq. 35 with the coefficients given in Eqs. 38 and 40 is substituted in Eqs. 29 and 30. The expressions for δ^* and θ so determined and

$$\tau_0 = \mu \left(\frac{\partial u}{\partial y}\right)_{y=0} = \frac{\mu A u_e}{\delta} = \frac{\mu u_e}{\delta} \left(2 + \frac{\lambda}{6}\right)$$

are substituted in Eq. 43. After some manipulation, an ordinary differential equation with δ^2/ν as the dependent variable results. The equation involves u_e, u_e', and u_e'', where the primes indicate differentations with respect to x, and so the solution of a particular problem requires that the pressure distribution over the surface be known.

The above method has been applied to determine the boundary layer development of bodies of various shapes. The results have been compared with more exact methods for specific cases, such as the flat plate problem of Section 15.3, the flow in the region of the two-dimensional stagnation point, and the flow past a circular cylinder.*

The various comparisons indicate that the Kármán-Pohlhausen method gives an excellent representation for the displacement and momentum thicknesses of the boundary layer and the skin-friction coefficient everywhere except near the separation point of the flow. A concomitant of this observation is the fact that the method predicts satisfactorily the location of the separation point only if the adverse pressure gradient is severe. On the other hand, on streamline shapes with gradual adverse gradients the actual separation point is in general well downstream of the location predicted by the von Kármán-Pohlhausen method.

One of the greatest uses of the method is to show trends of various effects, such as Mach number and local irregularities in pressure distributions.

*See Schlichting (1968, pp. 201–204). Holstein and Bohlen (Schlichting, 1968, pp. 192–201) have devised a form of the von Kármán-Pohlhausen method that is particularly well suited to practical calculations.

15.8 Three-Dimensional Boundary Layers

The two-dimensional boundary layers taken up in previous sections are characterized by a pressure gradient parallel to the flow direction. However, in many applications, such as the yawed cylindrical body in Fig. 12, the pressure gradient along
the surface makes an angle with the flow direction outside the boundary layer. The
component normal to the streamlines, designated by $\partial p / \partial n$, will cause the streamlines to curve in the planes tangent to the cylinder surface. The streamline curva-

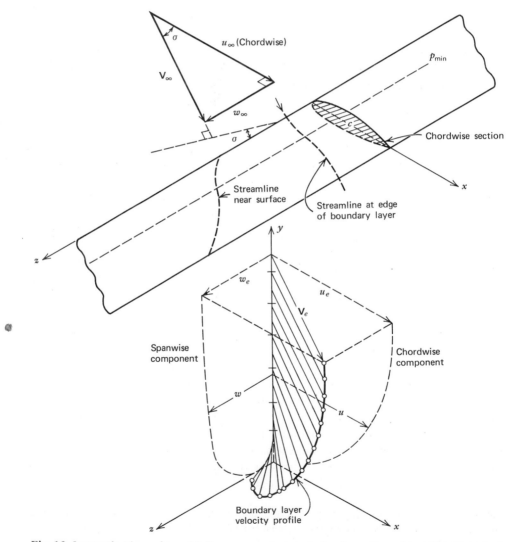

Fig. 12 Isometric view of an airfoil at yaw angle σ and the three-dimensional boundary layer
formed.

ture will be governed by the condition for equilibrium between the pressure and centrifugal forces on a fluid element (see Section 3.6),

$$\frac{\partial p}{\partial n} = \frac{\rho V^2}{R}$$

where V is the total velocity and R is the radius of curvature of the streamline. Since p is independent of y through the boundary layer, so will $\partial p/\partial n$ be constant. It follows that R will be proportional to V^2, that is, the radius of curvature will be small near the surface and will increase with distance from the surface. This three-dimensional boundary layer is shown schematically in the magnified region in Fig. 12. Typical streamlines very near the surface and near the outer edge of the boundary layer are also shown schematically.

The direction of the curvature of the streamlines can be rationalized as follows. As the streamline approaches the cylinder, it will bend toward the direction of the most rapid pressure decrease, that is, toward the normal to the line of minimum pressure. As the streamline *leaves* the pressure "trough," it will bend *away* from the direction of the most rapid *increase* in pressure. Therefore, there will be an inflection in the streamlines as they cross the line of minimum pressure. It follows also, as is shown in Fig. 12, that the pressure gradients will deflect the slow-moving air near the surface more than that near the edge of the boundary layer. We might also reason that the shapes of the streamlines will be qualitatively comparable to the paths of balls that traverse a shallow trough cut in a table at an angle to the initial direction of the roll.

Analytically the flow in the boundary layer of a yawed *infinite* cylinder may be treated with respect to the axes designated in Fig. 12. The exact equations of motion and continuity for incompressible flow are derived in Appendix B and were used in the similarity analysis of Section 14.6. They are

$$\rho \frac{\mathcal{D}u}{\mathcal{D}t} = -\frac{\partial p}{\partial x} + \mu \nabla^2 u$$

$$\rho \frac{\mathcal{D}v}{\mathcal{D}t} = -\frac{\partial p}{\partial y} + \mu \nabla^2 v \qquad (44)$$

$$\rho \frac{\mathcal{D}w}{\mathcal{D}t} = -\frac{\partial p}{\partial z} + \mu \nabla^2 w$$

$$\frac{\partial u}{\partial x} + \frac{\partial v}{\partial y} + \frac{\partial w}{\partial z} = 0$$

With the axes designated in Fig. 12, we see that none of the flow properties vary with z; that is, all derivatives with respect to z vanish. With this simplification, the first two equations and continuity contain only u, v, and p as functions of x, y, and t. The boundary conditions are also independent of z. Therefore, these equations may be solved for u, v, and p, and these solutions substituted in the third of Eqs. 44 enable us to solve for w.

It is shown in Appendix B that the boundary layer equation, Eq. 14.3 is a simplified form of the first equation of motion and that all of the terms in the second equation are negligible. Then, with p determined from inviscid flow theory, the u and v components in the boundary layer may be found from the boundary layer equation

$$\frac{\partial u}{\partial t} + u \frac{\partial u}{\partial x} + v \frac{\partial u}{\partial y} = -\frac{1}{\rho} \frac{\partial p}{\partial x} + v \frac{\partial^2 u}{\partial y^2} \tag{45}$$

and continuity

$$\frac{\partial u}{\partial x} + \frac{\partial v}{\partial y} = 0 \tag{46}$$

To solve for the spanwise component in the boundary layer, the third of Eqs. 44 simplifies to a form identical with Eq. 45 for u, except that $\partial p/\partial z = 0$. Then

$$\frac{\partial w}{\partial t} + u \frac{\partial w}{\partial x} + v \frac{\partial w}{\partial y} = v \frac{\partial^2 w}{\partial y^2} \tag{47}$$

As was pointed out above, Eqs. 45 and 46 may be solved for u and v and substituting these in Eq. 47 enables us to solve for w.

The above method of solution described is exact for *incompressible* flow over a *yawed* infinite cylinder. For compressible flow, the energy equation (Chapter 16) that must be used to obtain u and v cannot be made independent of w. For a finite yawed cylinder, for example, a finite wing, $\partial( )/\partial z$ will not be zero near the tips and w cannot be eliminated from the first two of Eqs. 44. For details, see Moore (1956).

CHAPTER 16

Laminar Boundary Layer in Compressible Flow

16.1 Introduction

This chapter discusses the effects of compressibility on boundary layer phenomena. The density and temperature, hitherto constant, now become variables; therefore, we need two new relations in addition to the equations of motion and continuity to solve the boundary layer problem. One of these relations is the equation of state, $p = \rho R T$, and the other is the equation expressing conservation of energy. This latter equation, derived for an adiabatic process in Chapter 8, was found to be $c_p T + \frac{1}{2} V^2 = c_p T_0$. When we consider the flow in the boundary layer, we need the *general* form of the energy equation, although the above simple form is found to be a useful approximation when there is no heat transfer to the wall.

The variation of temperature through a high-speed boundary layer brings with it a variation not only in density but also in viscosity and heat transfer coefficients. In Section 1.8, we found that the thermal conductivity and viscosity coefficient for a gas are theoretically connected by the relationship $k \sim c_p \mu$ and that $\mu \sim \sqrt{T}$. The first of these relations is found to hold quite closely, but in the second the viscosity coefficient actually varies more nearly according to the 0.76 power of the temperature.

A new parameter of considerable significance is the *Prandtl number*, $c_p \mu / k$, which, according to the previous paragraph, should be a constant for a given fluid. Actually its variation is small; values for air are given in Table 5 at the end of the book. We show later that the value of the Prandtl number is a measure of the degree to which effectively adiabatic conditions prevail in the boundary layer, and therefore of the limits of variation of the stagnation temperature within the layer.

The buoyancy forces resulting from density variations are neglected in the applications with which we deal. This simplification is justified because in high-speed flow the convection currents resulting from the buoyancy forces will invariably be small compared with the pressure gradient along the surface.

The velocity and temperature profiles through the boundary layer are described and the relation between skin-friction and heat transfer coefficients are given. The combined effects of Reynolds and Mach numbers are shown.

The occurrence of flow separation at supersonic speeds is complicated by the presence of shock waves. Whenever a shock wave intersects a surface, there will be a tendency for flow separation because the pressure is always greater on the downstream side of a shock (adverse gradient). Stalling of airfoils and flow separation in channels will be described.

We began the previous chapter on incompressible viscid flow by analyzing fully developed flow. When we realize that the effect of friction is to convert directed energy into heat, it becomes clear that fully developed flow of a gas can only be realized approximately, since as the gas becomes heated its density decreases and the flow never reaches an equilibrium distribution. In fact, as was shown in Section 9.8, the effect of friction is to accelerate a subsonic flow and to decelerate a super-sonic flow. Only a few of the simpler analyses that serve to illustrate the important concepts will be reproduced in this chapter.*

16.2 Conservation of Energy in the Boundary Layer

The conservation of energy principle applied to a group of particles of fixed identity has been given by Eq. 8.13. If this equation is specialized to the two-dimensional element $\rho \Delta x \Delta y$ shown in Fig. 1, we have

$$\rho \Delta x \Delta y \, \frac{de}{dt} = \frac{\delta}{\delta t} \iint_{\hat{S}_1} \rho (q - w) \, d\hat{S}_1 \qquad (1)$$

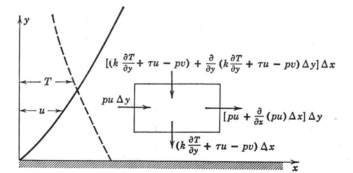

Fig. 1 Energy balance in the boundary layer.

where q and w represent the heat and work transfers across the surface $\hat{S}_1$ and the internal energy e, given by Eq. 8.19,[†] represents the average value for element $\rho \Delta x \Delta y$, moving with velocity u. Then

$$e = c_v T + \tfrac{1}{2} u^2 + \text{constant} \qquad (2)$$

*For further details, the reader should consult Schlichting (1968), Pai (1956), and Howarth (1953).

†Gravitational energy, $-gz$, has been neglected. Departing from the nomenclature of Chapter 8, the letter u in the following material refers to the x component of velocity and not to the intrinsic energy.

In Fig. 1, it is assumed that $\partial T/\partial y \gg \partial T/\partial x$. Therefore, we may neglect heat transfer through the ends of the element compared with that through the top and bottom faces. Further, we adopt the boundary layer approximations $\partial p/\partial y = 0$ and $\partial u/\partial y \gg \partial u/\partial x$.

Heat is transferred through the bottom face at the rate $-k(\partial T/\partial y)\Delta x$ and through the top face at the rate

$$\left[-k\,\frac{\partial T}{\partial y} + \frac{\partial}{\partial y}\left(-k\,\frac{\partial T}{\partial y} \right)\Delta y \right]\Delta x$$

and the heat transfer rate is

$$\frac{\partial}{\partial y}\left(k\,\frac{\partial T}{\partial y} \right)\Delta x \Delta y \tag{3}$$

Work performed by the pressure stress p and shearing stress τ is transferred across the horizontal and vertical faces as indicated in Fig. 1. The net rate of work transfer is

$$+\left[\frac{\partial}{\partial y}\,(\tau u) - \frac{\partial}{\partial x}\,(pu) - \frac{\partial}{\partial y}\,(pv) \right]\Delta x \Delta y \tag{4}$$

After substituting Eqs. 2, 3, and 4 into Eq. 1, we obtain

$$\rho\,\frac{d}{dt}\left(c_v T + \frac{u^2}{2} \right) = \frac{\partial}{\partial y}\left(k\,\frac{\partial T}{\partial y} \right) + \frac{\partial}{\partial y}\,(\tau u) - \frac{\partial}{\partial x}\,(pu) - \frac{\partial}{\partial y}\,(pv) \tag{5}$$

Equation 5 expresses the law: the rate of increase of total energy is equal to the sum of the rate at which heat is conducted into the element and the rate at which the stresses at the boundaries do work on the element. It is convenient to express the left side of Eq. 5 in terms of the enthalpy $c_p T + \frac{1}{2}\,u^2$,

$$\rho\,\frac{d}{dt}\left(c_v T + \frac{u^2}{2} \right) = \rho\,\frac{d}{dt}\left(c_p T - \frac{p}{\rho} + \frac{u^2}{2} \right)$$

$$= \rho\,\frac{d}{dt}\left(c_p T + \frac{u^2}{2} \right) - \frac{dp}{dt} + \frac{p}{\rho}\,\frac{d\rho}{dt} \tag{6}$$

After substituting from the continuity equation $\rho\,\mathrm{div}\,\mathbf{V} = -d\rho/dt$ in the last term and expanding the last two terms, Eq. 6 becomes

$$\rho\,\frac{d}{dt}\left(c_v T + \frac{u^2}{2} \right) = \rho\,\frac{d}{dt}\left(c_p T + \frac{u^2}{2} \right) - \frac{\partial}{\partial x}\,(pu) - \frac{\partial}{\partial y}\,(pv)$$

The last two terms of this equation cancel the last two terms of Eq. 5 which now becomes, after substituting $\tau = \mu \partial u/\partial y$,

$$\rho\,\frac{d}{dt}\left(c_p T + \frac{u^2}{2} \right) = \frac{\partial}{\partial y}\left(k\,\frac{\partial T}{\partial y} \right) + u\,\frac{\partial}{\partial y}\left(\mu\,\frac{\partial u}{\partial y} \right) + \mu\left(\frac{\partial u}{\partial y} \right)^2 \tag{7}$$

We can simplify this equation further by noting that if we multiply the boundary layer momentum equation (Eq. 14.3) for steady flow by u and subtract it from Eq. 7 we get, with c_p = constant,

$$\rho c_p \frac{dT}{dt} = u \frac{\partial p}{\partial x} + \frac{\partial}{\partial y}\left(k \frac{\partial T}{\partial y}\right) + \mu \left(\frac{\partial u}{\partial y}\right)^2 \tag{8}$$

This equation is the boundary layer energy equation in the form in which it is usually used.

The left side of Eq. 8 represents the rate of increase of enthalpy per unit volume as the element moves with the flow. The three terms on the right side represent, respectively, the rate at which work is done by the pressure forces, the rate at which heat is transferred through the sides of the element, and the rate at which viscous stresses dissipate energy of the ordered motion into heat (or enthalpy).

The equations governing the compressible laminar boundary layer are Eq. 8 above and the equations of motion and continuity, Eqs. 14.3 and 14.4, respectively. The latter are

$$\rho \left(u \frac{\partial u}{\partial x} + v \frac{\partial u}{\partial y}\right) = -\frac{\partial p}{\partial x} + \frac{\partial}{\partial y}\left(\mu \frac{\partial u}{\partial y}\right)$$

$$\frac{\partial}{\partial x}(\rho u) + \frac{\partial}{\partial y}(\rho v) = 0 \tag{9}$$

The boundary conditions are

$$\text{at } y = 0: \quad u = v = 0, \quad T = T_w(x)$$

$$\text{at } y = \infty: \quad u = u_e(x), \quad T = T_e(x) \tag{10}$$

In addition p, ρ, and T are connected by the equation of state $p = \rho R T$.

Various simplifications of these equations leading to results of practical value are taken up in subsequent sections.

16.3 Rotation and Entropy Gradient in the Boundary Layer

We shall show here that the rotation in the boundary layer is associated with an entropy gradient normal to the streamlines. From Eq. 8.33, the differential of the entropy is given by

$$dS = \frac{\delta q}{T} = d(\ln T^{c_v} - \ln \rho^R)$$

which becomes

$$dS = c_v \frac{dT}{T} - R \frac{d\rho}{\rho}$$

After introducing $p = \rho R T$

$$\frac{\partial S}{\partial y} = \frac{c_p}{T} \frac{\partial T}{\partial y} - \frac{R}{p} \frac{\partial p}{\partial y}$$

We set $\partial p / \partial y = 0$ for boundary layer flow and substitute for $\partial T / \partial y$ from

$$c_p T = c_p T_0 - \frac{u^2}{2}$$

Then

$$\frac{\partial S}{\partial y} = - \frac{u}{T} \frac{\partial u}{\partial y} + \frac{c_p}{T} \frac{\partial T_0}{\partial y}$$

In the boundary layer $\partial v / \partial x \simeq 0$ so that $\partial u / \partial y \simeq \mathrm{curl}_z \mathbf{V}$. We shall see later that $\partial T_0 / \partial y$ is small in the boundary layer over an insulated surface. Then, an approximate relation between entropy gradient and vorticity ω_z in the boundary layer is

$$\frac{\partial S}{\partial y} \simeq \frac{u}{T} \mathrm{curl}_z \mathbf{V} = \frac{u}{T} \omega_z$$

This relation, with y measured normal to the streamlines, also holds approximately in the free stream. It is known as Crocco's relation and shows that the entropy gradient along a detached shock (Section 10.7) is associated with a vorticity in the flow downstream of the shock.

16.4 Similarity Considerations for Compressible Boundary Layers

We showed in Section 14.6 that the condition for similarity of the incompressible flows around geometrically similar bodies is that the Reynolds numbers be identical. We show here what additional parameters are required for a compressible flow. We follow the same procedure as in Section 14.6, introducing the reference quantities L (length), U (velocity), ρ_1 (density), μ_1 (viscosity), and T_1 (temperature); and setting

$$u' = \frac{u}{U}, \quad v' = \frac{v}{U}, \quad x' = \frac{x}{L}, \quad y' = \frac{y}{L}, \quad t' = \frac{tU}{L}, \quad \mu' = \frac{\mu}{\mu_1}, \quad \rho' = \frac{\rho}{\rho_1}, \quad T' = \frac{T}{T_1},$$

$$p' = \frac{p}{\rho_1 U^2}, \quad Re = \frac{\rho_1 U L}{\mu_1} \quad (11)$$

Now, if the temperature variation through the boundary layer is not too great, we may for the purposes of the similarity analysis take μ and k as constants. Then the equations of motion reduce to those given in Section 14.6 for incompressible flow.

To find the additional parameters introduced by compressibility, we use Eqs. 11 to nondimensionalize the approximate form of the energy equation, Eq. 8. We obtain

$$\frac{\rho_1 U T_1 c_p}{L}\, \rho' \, \frac{dT'}{dt'} = \frac{\rho_1 U^3}{L}\, u'\, \frac{\partial p'}{\partial x'} + \frac{kT_1}{L^2}\, \frac{\partial^2 T'}{\partial y'^2} + \frac{\mu U^2}{L^2}\left(\frac{\partial u'}{\partial y'}\right)^2$$

In terms of dimensionless parameters this equation becomes

$$\rho'\, \frac{dT'}{dt'} = (\gamma - 1) M^2 u'\, \frac{\partial p'}{\partial x'} + \frac{1}{PrRe}\, \frac{\partial^2 T'}{\partial y'^2} + \frac{(\gamma - 1) M^2}{Re}\left(\frac{\partial u'}{\partial y'}\right)^2 \qquad (12)$$

where $Pr = c_p \mu / k$ is defined as the *Prandtl number*.

We see from Eq. 12 that the similarity of steady compressible boundary layer flows requires identical values of $(\gamma - 1) M^2$, Pr, and Re. Another parameter which enters, through the boundary conditions, is the *Nusselt number*, derived in Appendix A.

$$Nu = \frac{hL}{k}$$

where h is the rate of heat transfer per unit area per unit temperature difference and L is a characteristic length.

The solution of a particular compressible boundary layer problem may therefore be expressed formally as

$$f(c_f, Re, Nu, Pr, M, \gamma) = 0 \qquad (13)$$

where c_f is the skin friction coefficient $\tau_w / \frac{1}{2}\rho_1 U^2$. Fortunately γ and Pr are only weak functions of the temperature and they may be taken as constant for a wide range of applications. Hence, for most practical purposes a boundary layer problem reduced to the functional relationship:

$$f(c_f, M, Nu, Re) = 0 \qquad (14)$$

The solutions discussed in the remainder of this chapter are in the form of Eq. 14.

16.5 Solutions of the Energy Equation for Prandtl Number Unity

Many applications of boundary layer theory to compressible fluids yield good engineering approximations for air if the Prandtl number

$$Pr = \frac{c_p \mu}{k}$$

is set to unity. We shall in this section show that the problem of the compressible boundary layer is thus greatly simplified. Further, we shall demonstrate that the physical reason for the resulting simplification is that a Prandtl number of unity implies effectively adiabatic flow at every point in the boundary layer.

A physical interpretation of the Prandtl number will help in understanding its role in boundary layer theory. μ represents the rate of momentum transfer per unit area per unit velocity gradient and k/c_p represents the rate of heat transfer per unit area per unit enthalpy gradient. Then the Prandtl number represents the ratio between these two rates of transfer.

We can arrive at another physical interpretation by considering Eq. 7:

$$\rho \frac{d}{dt}\left(c_p T + \frac{u^2}{2}\right) = \frac{\partial}{\partial y}\left(k \frac{\partial T}{\partial y}\right) + u \frac{\partial}{\partial y}\left(\mu \frac{\partial u}{\partial y}\right) + \mu\left(\frac{\partial u}{\partial y}\right)^2 \tag{7}$$

The first term on the right-hand side of Eq. 7 represents heat transferred into the fluid element by conduction. The second and third terms correspond to energy generated through shear at the boundaries of the element and viscous dissipation within. If the heat transferred out of the element is just equal to the total heat generated through the action of viscosity, the right-hand side of Eq. 7 is zero. After making this assumption, the integral of Eq. 7 along a streamline can be written

$$c_p T + \frac{u^2}{2} = \text{constant} = c_p T_0 \tag{15}$$

Equation 15 indicates that the quantity $c_p T + \frac{1}{2} u^2$ is constant along a streamline. For the usual case of uniform flow upstream, the stagnation temperature has the same value on every streamline, and it may be concluded that $c_p T + \frac{1}{2} u^2$ is constant throughout the boundary layer.

The significance of this conclusion in terms of the Prandtl number may be seen by assuming the temperature gradient within the boundary layer to be small enough so that μ and k may be taken as constants. Then Eq. 7 may be written

$$\frac{d}{dt}\left(c_p T + \frac{u^2}{2}\right) = \frac{k}{\rho c_p} \frac{\partial^2}{\partial y^2}\left(c_p T + \frac{u^2}{2} Pr\right) \tag{16}$$

Equation 15 is a solution of Eq. 16, regardless of the velocity distribution, only if the Prandtl number is unity. Therefore, within the restrictions that μ and k are constant, the above analysis shows that Prandtl number unity implies effectively adiabatic conditions in the sense that the heat generated within an element by viscous work is transferred out of the element by conduction.

Equation 15 is but one of the solutions to Eq. 16 for Prandtl number unity. To determine the properties of the solution, we differentiate Eq. 15 and get

$$c_p \frac{\partial T}{\partial y} = -u \frac{\partial u}{\partial y} \tag{17}$$

We see from this equation that $\partial T/\partial y = 0$ where $u = 0 (y = 0)$ and where $\partial u/\partial y = 0 (y = \delta)$.

We conclude, therefore, that Prandtl number unity implies that for flow over an insulated surface, defined by $(\partial T/\partial y)_w = 0$, the energy equation reduces to Eq. 15

and that the velocity and temperature boundary layers have the same thickness. Note that this conclusion involves no restrictions on the pressure gradient in the flow direction.

We shall now find the corresponding relation between velocity and temperature with heat transfer at the wall. Here we make the approximation that $\partial p/\partial x = 0$. Then Eq. 8, with Prandtl number unity, is

$$\rho \left(u \frac{\partial T}{\partial x} + v \frac{\partial T}{\partial y} \right) = \frac{\partial}{\partial y} \left(\mu \frac{\partial T}{\partial y} \right) + \frac{\mu}{c_p} \left(\frac{\partial u}{\partial y} \right)^2 \tag{18}$$

We now assume that

$$T = A + Bu + Cu^2 \tag{19}$$

Equation 19 is substituted for T in Eq. 18 and the constants A, B, and C are evaluated by means of the boundary conditions:

$$\text{at } y = 0: \ u = 0, \quad T = T_w$$
$$\text{at } y = \infty: \ u = u_e, \quad T = T_e \tag{20}$$

Equation 18 with T from Eq. 19 is

$$\rho B \left(u \frac{\partial u}{\partial x} + v \frac{\partial u}{\partial y} \right) + 2\rho Cu \left(u \frac{\partial u}{\partial x} + v \frac{\partial u}{\partial y} \right)$$

$$= \frac{\partial}{\partial y} \left[\mu \left(B \frac{\partial u}{\partial y} + 2Cu \frac{\partial u}{\partial y} \right) \right] + \frac{\mu}{c_p} \left(\frac{\partial u}{\partial y} \right)^2 \tag{21}$$

The momentum equation (Eq. 9) with $\partial p/\partial x = 0$ is

$$\rho \left(u \frac{\partial u}{\partial x} + v \frac{\partial u}{\partial y} \right) = \frac{\partial}{\partial y} \left(\mu \frac{\partial u}{\partial y} \right) \tag{22}$$

We multiply Eq. 22 by B and subtract it from Eq. 21. Then, after expanding the term $\partial [2Cu(\partial u/\partial y)]/\partial y$, Eq. 21 becomes

$$2C\rho u \left(u \frac{\partial u}{\partial x} + v \frac{\partial u}{\partial y} \right) = 2Cu \frac{\partial}{\partial y} \left(\mu \frac{\partial u}{\partial y} \right) + 2C\mu \left(\frac{\partial u}{\partial y} \right)^2 + \frac{\mu}{c_p} \left(\frac{\partial u}{\partial y} \right)^2 \tag{23}$$

Now, if we multiply Eq. 22 by $2Cu$ and subtract from Eq. 23 we get

$$C = - \frac{1}{2c_p} \tag{24}$$

Then, after using the first boundary condition of Eqs. 20 to give $A = T_w$, Eq. 19 with Eq. 24 becomes

$$T = T_w + Bu - \frac{u^2}{2c_p}$$

and, with B evaluated by use of the second condition of Eqs. 20,

$$T = T_w + \left(\frac{T_e - T_w}{u_e} + \frac{u_e}{2c_p}\right) u - \frac{u^2}{2c_p} \tag{25}$$

or

$$\frac{T}{T_e} = \frac{T_w}{T_e} + \left(1 - \frac{T_w}{T_e}\right)\frac{u}{u_e} + \frac{u}{u_e}\left(1 - \frac{u}{u_e}\right)\frac{u_e^2}{2c_p T_e}$$

But

$$2c_p T_e = \frac{2\gamma R T_e}{\gamma - 1} = \frac{2 a_e^2}{\gamma - 1}$$

where a_e is the speed of sound outside the boundary layer. Then, with $M_e = u_e/a_e$, Eq. 25 finally becomes

$$\frac{T}{T_e} = \frac{T_w}{T_e} + \left(1 - \frac{T_w}{T_e}\right)\frac{u}{u_e} + \frac{\gamma - 1}{2}M_e^2\left(1 - \frac{u}{u_e}\right)\frac{u}{u_e} \tag{26}$$

This equation is generally referred to as Crocco's form of the energy equation. It applies strictly for $\partial p/\partial x = 0$, and $Pr = 1$, but k and μ need not be constants.

If we differentiate Eq. 26, we get a relation between the temperature and velocity gradients,

$$\frac{\partial T}{\partial y} = \left[\left(1 - \frac{T_w}{T_e}\right) + \frac{\gamma - 1}{2}M_e^2\left(1 - 2\frac{u}{u_e}\right)\right]\frac{T_e}{u_e}\frac{\partial u}{\partial y} \tag{27}$$

and at $y = 0\,(u = 0)$ and with $c_p\mu = k$, the relation between heat transfer and skin friction at the wall is

$$k\left(\frac{\partial T}{\partial y}\right)_w = \left[\left(1 - \frac{T_w}{T_e}\right) + \frac{\gamma - 1}{2}M_e^2\right]\frac{T_e c_p}{u_e}\mu\left(\frac{\partial u}{\partial y}\right)_w \tag{28}$$

We concluded from Eq. 15 that for an insulated plate Prandtl number unity implies that the temperature and velocity boundary layers are of equal thickness. Equation 28 permits us to broaden this conclusion to include arbitrary heat transfer at the wall at zero pressure gradient.

Since the Prandtl number for air is near unity, the above solutions (Eqs. 15 or 26) of the energy equation give satisfactory results for many boundary layer problems in aerodynamics. Before use can be made of these solutions, however, it is necessary to find the velocity distributions by solving the momentum equation.

16.6 Temperature Recovery Factor

The *adiabatic, recovery,* or *equilibrium temperature*, designated by T_{ad}, is the temperature of the wall in an airflow in which there is no heat transfer to the wall.

Mathematically $T_w = T_{ad}$ for $(\partial T/\partial y)_w = 0$. From the previous section (Eq. 15), we see that when the Prandtl number is unity, $T_{ad} = T_0$, the stagnation temperature in the outside flow. The correction to this result caused by deviations of the Prandtl number from unity will be investigated in this section. The problem of determining the recovery temperature is generally referred to as the "thermometer problem."

The flat plate thermometer problem was solved by Pohlhausen (1921) under the restrictions that the Mach number of the flow is low enough so that we may take ρ, μ, and k as constants. The governing equations are then Eqs. 8, 9, and 10. With $\partial p/\partial x = 0$, these are

$$u \frac{\partial u}{\partial x} + v \frac{\partial u}{\partial y} = \nu \frac{\partial^2 u}{\partial y^2}$$

$$\frac{\partial u}{\partial x} + \frac{\partial v}{\partial y} = 0 \tag{29}$$

$$\rho c_p \left(u \frac{\partial T}{\partial x} + v \frac{\partial T}{\partial y} \right) = k \frac{\partial^2 T}{\partial y^2} + \mu \left(\frac{\partial u}{\partial y} \right)^2$$

The boundary conditions are

$$\text{at } y = 0: \ u = v = 0, \quad \frac{\partial T}{\partial y} = 0 \tag{30}$$

$$\text{at } y = \infty: u = u_e, \quad T = T_e$$

The first two equations of Eqs. 29 and the boundary conditions on the velocities in Eqs. 30 constitute the Blasius problem solved in Section 15.3. The basic independent variable there was:

$$\eta = \frac{1}{2} \sqrt{\frac{u_e}{\nu x}} \, y$$

and

$$u = \frac{u_e}{2} f'(\eta); \quad v = \frac{1}{2} \sqrt{\frac{u_e \nu}{x}} [\eta f'(\eta) - f(\eta)] \tag{31}$$

We use these expressions for u and v and try to express the third equation of Eqs. 29 as an ordinary differential equation. We first introduce the dimensionless variable θ by writing

$$T = T_e + \frac{u_e^2}{2 c_p} \theta(\eta) \tag{32}$$

and the boundary conditions on the temperature in Eqs. 30 then become

$$\theta'(0) = 0 \quad \text{and} \quad \theta(\infty) = 0 \tag{33}$$

Then the adiabatic wall, or recovery, temperature is given by

$$T_r = T_{adw} = T_e + \frac{u_e^2}{2c_p} \theta(0)$$

When the expressions for u, v, and T in Eqs. 31 and 32 are substituted in the third equation of Eqs. 29, we obtain, after canceling common terms.

$$\theta'' + Pr f\theta' + 0.5 \, Pr f''^2 = 0 \tag{34}$$

After substituting values of f and f'' as given in Fig. 15.3, Pohlhausen found an approximate solution of Eq. 34. The recovery factor, which we shall call r, was found to be

$$r = \theta(0) = \frac{T_r - T_e}{u_e^2/2c_p} = \sqrt{Pr} = 0.845 \text{ (for air)} \tag{35}$$

Figure 2 is a plot of $\theta(\eta)$ through the boundary layer.

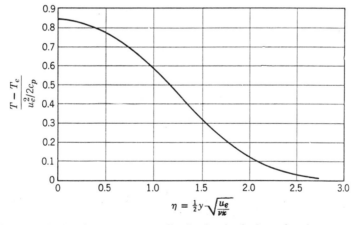

Fig. 2 Pohlhausen solution for temperature distribution in the boundary layer.

Equation 35 states that the air near the wall has lost stagnation enthalpy. Then conservation of energy demands, since we are dealing with an adiabatic process, that somewhere in the boundary layer the air must have gained stagnation enthalpy. This increase in stagnation enthalpy is shown clearly in the curves of Fig. 3 (Van Driest, 1952).

Accurate calculations by Van Driest (1952) show only small deviations of the laminar recovery factor from $\sqrt{Pr}$ up to Mach numbers of at least 8.

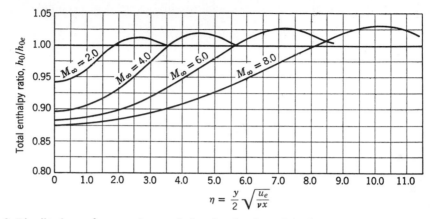

Fig. 3 Distribution of stagnation enthalpy in the boundary layer on an insulated plate (Van Driest, 1952, courtesy of NACA).

Some experimental results on a cone at various Mach numbers are shown in Fig. 4. The boundary layer at the low Reynolds numbers is laminar and the temperature recovery factor is 0.84 to 0.855 over the range of Mach numbers 1.79 to 4.5. Reference to Eq. 35 shows that the agreement between theory and experiment is remarkably good. Results from various laboratories give values of the laminar recovery factor of 0.85 ± 0.01 for Mach numbers between 1.2 and 6.

The curves in Fig. 4 begin to rise steeply at Reynolds numbers for which transition to the turbulent boundary layer occurs. These portions will be referred to in the following two chapters.

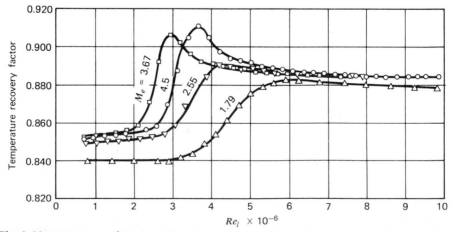

Fig. 4 Measurements of temperature recovery factor $[\theta(0)$ of Eq. 35] on 5° cone. (Measurements by Mack 1954; courtesy of NASA).

16.7 Heat Transfer Versus Skin Friction

Pohlhausen (1921) also calculated the heat transfer to a flat plate at constant temperature T_w with the same restrictions as applied in calculating the recovery factor in Section 16.6 ($\partial p/\partial x = 0$ and ρ, μ, and k constants); in addition, the velocities are assumed to be low enough so that the dissipation term $\mu(\partial u/\partial y)^2$ in the energy equation can be neglected.

The governing equations, Eqs. 8, 9, and 10, become

$$u \frac{\partial u}{\partial x} + v \frac{\partial u}{\partial y} = \nu \frac{\partial^2 u}{\partial y^2}$$

$$\frac{\partial u}{\partial x} + \frac{\partial v}{\partial y} = 0 \tag{36}$$

$$\rho c_p \left(u \frac{\partial T}{\partial x} + v \frac{\partial T}{\partial y} \right) = k \frac{\partial^2 T}{\partial y^2}$$

with the boundary conditions

$$\text{at } y = 0: \ u = v = 0, \quad T = T_w$$
$$\text{at } y = \infty; \ u = u_e, \qquad T = T_e \tag{37}$$

As in the previous section, we see that the first two equations of Eqs. 36 and the boundary conditions on the velocities in Eqs. 37 constitute the Blasius problem solved in Section 15.3. Then with η, u, and v as given in Eqs. 31 and a new variable

$$\beta(\eta) = \frac{T_w - T}{T_w - T_e} \tag{38}$$

we attempt to obtain an ordinary differential equation from the last equation of Eqs. 36. The derivatives occurring are

$$\frac{\partial T}{\partial x} = \frac{\eta}{2x} (T_w - T_e) \beta'$$

$$\frac{\partial T}{\partial y} = -\frac{1}{2} \sqrt{\frac{u_e}{\nu x}} (T_w - T_e) \beta' \tag{39}$$

$$\frac{\partial^2 T}{\partial y^2} = -\frac{1}{4} \frac{u_e}{\nu x} (T_w - T_e) \beta''$$

The boundary conditions on the temperature, Eqs. 37, become

$$\text{at } \eta = 0: \ \beta = 0$$
$$\text{at } \eta = \infty: \ \beta = 1 \tag{40}$$

We now substitute into the last equation of Eqs. 36 the expressions for u and v from Eqs. 31 and for the temperature derivatives from Eqs. 39. Then the equation becomes, after factoring common terms,

$$\beta'' + Pr f \beta' = 0 \tag{41}$$

where Pr is the Prandtl number $c_p \mu / k$ and f is the Blasius function of the previous section (see Fig. 15.3). Equation 41 is a linear ordinary differential equation in β and with Pr = constant; its solution is of the form

$$\beta' = \alpha \exp\left(-Pr \int_0^\eta f \, d\eta\right) \tag{42}$$

where α is the integration constant. After integrating Eq. 42 and applying the boundary condition $\beta = 0$ at $\eta = 0$ (Eq. 40), we get

$$\beta = \alpha \int_0^\eta \exp\left(-Pr \int_0^\eta f \, d\eta\right) d\eta \tag{43}$$

α is evaluated by the boundary condition $\beta = 1$ at $\eta = \infty$. Thus

$$\alpha = \left[\int_0^\infty \exp\left(-Pr \int_0^\eta f \, d\eta\right) d\eta\right]^{-1} \tag{44}$$

Pohlhausen substituted the Blasius function f shown in Fig. 15.3 and found, approximately,

$$\alpha = 0.664 \, Pr^{1/3} \tag{45}$$

The heat transfer coefficient is found by evaluating Q, the rate at which heat is transferred from a plate of width b and length l. We may write

$$Q = -kb \int_0^l \left(\frac{\partial T}{\partial y}\right)_w dx \tag{46}$$

After substituting for $\partial T / \partial y$ from Eq. 39, noting from Eq. 43 that $\beta' = \alpha$ at $y = \eta = 0$ and integrating,

$$Q = kb\alpha(T_w - T_e)\sqrt{\frac{u_e l}{\nu}} \tag{47}$$

The Nusselt number, Nu, derived in Appendix A and mentioned as one of the similarity parameters for boundary layer flow in Section 16.4, is given by

$$Nu = \frac{hL}{k} = \frac{L}{k} \frac{Q}{S(T_w - T_e)} \tag{48}$$

where L is a characteristic length, h is the rate of heat transfer per unit area per unit temperature difference, and S is the area of the plate $(S = lb)$. If we take the length of the plate, l, as the characteristic length and substitute for Q from Eq. 47 and for α from Eq. 45, Eq. 48 becomes

$$Nu = 0.664 \, Pr^{1/3} \, Re^{1/2} \tag{49}$$

where

$$Re = \frac{u_e l}{\nu}$$

Another dimensionless heat transfer coefficient, called the *Stanton number*, St, is defined as

$$St = \frac{h}{\rho c_p u_e} = \frac{Q}{\rho c_p S u_e (T_w - T_e)} \tag{50}$$

When we are dealing with the relation between heat transfer and skin friction, the Stanton number proves to be a convenient similarity parameter. Thus, after substituting for Q and for Nu,

$$St = \frac{Nu}{Pr \, Re} = \frac{0.664}{Pr^{2/3} \, Re^{1/2}} \tag{51}$$

If we compare Eq. 51 with the expression for the average skin-friction coefficient, C_f from Eq. 15.19, we may write

$$St = Pr^{-2/3} \, C_f / 2 \tag{52}$$

If we define the local Stanton number

$$st = \frac{q}{\rho c_p u_e (T_w - T_e)}$$

the analysis similar to that leading to Eq. 52 gives

$$st = Pr^{-2/3} \, c_f / 2 \tag{53}$$

Since the Prandtl number for air does not vary greatly, we have in Eqs. 52 and 53 remarkably simple relations between local and average heat transfer and skin-friction coefficients for a flat plate.

In spite of the approximations inherent in Eqs. 36 the agreement between Eqs. 52 and 53 and experiment is good up to reasonably high Mach numbers (see Chapman and Rubesin, 1949).

16.8   Velocity and Temperature Profiles and Skin Friction

In the previous sections, the approximations made amounted to a neglect of the effect of compressibility on the velocity profile. Although the results so obtained are applicable to some significant problems, their use is limited to moderate Mach numbers. Solutions applicable to high Mach numbers must take into account altera-tions to the velocity profile resulting from variations of μ and ρ with temperature (and therefore with y).

Many solutions of Eqs. 8, 9, and 10 have been obtained for specific variations of T_w, u_e, and T_e with x, for various Prandtl numbers, and for various relations be-tween μ and k and T. The analyses, compared with that for incompressible flow, are complicated considerably by the introduction of the new variables and equa-tions. The details of the solutions are beyond the scope of this book. Therefore, only the results of some of the studies that illustrate the important concepts are described here.

In practically all of the analyses, the Prandtl number is assumed constant, so that $k \sim c_p\mu$, but the variation of μ with T takes several forms. The most accurate relation is expressed by the Sutherland equation

$$\frac{\mu}{\mu_1} = \frac{T_1 + 120}{T + 120} \left(\frac{T}{T_1}\right)^{3/2} \tag{54}$$

where T_1 is a reference temperature. In analytical solutions the expression

$$\frac{\mu}{\mu_1} = C \left(\frac{T}{T_1}\right)^{\omega} \tag{54a}$$

where C is near unity and is a weak function of the temperature, and $0.5 < \omega < 1.0$, yields good approximations over a wide range of temperatures.

The general features of the velocity and temperature profiles in a compressible boundary layer are shown qualitatively in Fig. 5a. The effect of Mach number on

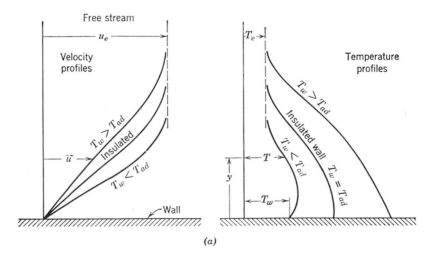

(a)

Fig. 5a Qualitative effect of wall temperature on temperature and velocity profiles in a boundary layer.

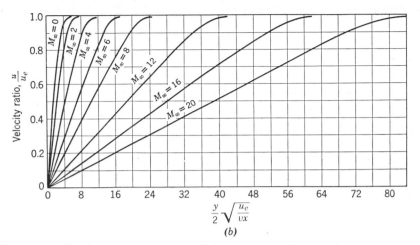

(b)

Fig. 5b Velocity distributions on an insulated plate at various Mach numbers.

an insulated flat plate is shown in Fig. 5b and Fig. 6 shows the dependence of the total skin friction on Mach and Reynolds numbers and temperature ratio (Van Driest, 1952).

It was pointed out by Tifford (1950) on the basis of earlier work by Busemann that remarkably good approximations to the skin friction on a flat plate can be obtained if the formulas are expressed in terms of the gas properties at the wall instead of those at the edge of the boundary layer. Thus, by use of Fig. 6, we may show that for $M_e \leqq 5$ and $0.25 < T_w/T_e < 5$,

$$C_{fw}\sqrt{Re_w}/1.328 \leqq 1.073 \tag{55}$$

where $C_{fw} = \int_0^l \tau_w\,dx/\frac{1}{2}\rho_w u_e^2 l$, $Re_w = u_e l/\nu_w$ (see Problem 16.8.1.).

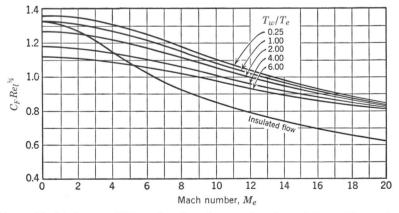

Fig. 6 Mean skin-friction coefficient for flat plate as function of Reynolds number, Mach number, and temperature ratio. (Van Driest, 1952, courtesy of NACA.)

16.9   Effects of Pressure Gradient

The effects of compressibility on the boundary layer in a favorable pressure gradient introduce complications in the solution of the equations, but no new concepts are involved.   On a surface which is warped so that the flow is turned away from its original direction, expansion waves form and the accompanying favorable pressure gradient accelerates the boundary layer flow.   If expansion waves intersect the body, a similar accelerating effect is experienced by the flow.

On the other hand, if the body is warped so that the flow turns *into* itself, compression waves will appear and these will tend to form an envelope as shown in the photograph of Fig. 7.   The envelope is a shock with intensity varying along its length so that there will be an entropy gradient (see Section 10.6) normal to the streamlines in the downstream flow.   Further, an adverse pressure gradient will exist on the concave wall with the resulting tendency for flow separation.   However, if the curvature continues as shown, there will be a tendency for the flow to *reattach* after separating.

The behavior of the boundary layer on a body in a supersonic flow in an adverse

Fig. 7 Shadowgraph of a projectile with concave nose travelling at Mach 2 in a firing range. Waves generated at knurled surface illustrate the formation of a shock along the envelope of the weak compressions formed along the concave surface; the apparent distortion of the nose is an optical effect. (Courtesy of U.S. Army Ballistics Research Laboratory.)

pressure gradient resolves itself into some phase of *shock wave–boundary layer interaction*. Considerable theoretical and experimental work has been carried out on two types of interaction: (1) the shock from a wedge intersecting the boundary layer on a flat plate, and (2) the flow over a sharp step.

When a shock wave intersects the boundary layer, its strength decreases steadily as it proceeds into the layer, and it becomes a Mach line at the streamline where the flow is sonic. The high pressure behind the wave provides a steep adverse pressure gradient which makes itself felt upstream through the subsonic portion of the layer. *Transition* to turbulent boundary layer (Chapter 17) or flow separation may result, depending on the intensity of the adverse gradient, that is, on the intensity of the shock.

Intuitively, it is logical that the thicker the subsonic portion of the boundary layer, the farther upstream the effects of the adverse gradient will be felt. Also, $\partial u/\partial y$ near $y = 0$ will be small for a thick subsonic portion, and hence a small adverse gradient (small shock intensity) will suffice to cause flow separation. In general, a laminar boundary layer will have a thicker subsonic portion than will the

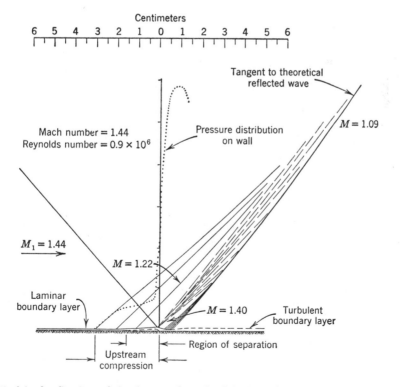

Fig. 8 Model of reflection of shock wave from laminar boundary layer. Expansion waves are shown by broken lines. (Liepmann, Roshko and Dhawan, 1952, courtesy of NASA.)

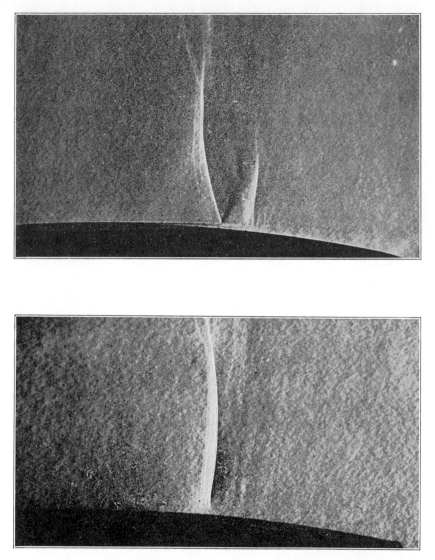

Fig. 9 Schlieren photographs of the shock wave configuration on the upper surface of an airfoil at a free-stream Mach number of 0.843. For the upper photograph the Reynolds number is 10^6 and the boundary layer is laminar upstream of the shock as indicated by the sharp line that occurs near the edge of the layer; the shock triggers transition to a turbulent layer the edge of which is ill-defined. For the lower the Reynolds number is 2×10^6 and the boundary layer is turbulent. Measurements at GALCIT. (Courtesy of H. W. Liepmann.)

turbulent layer. Since this is the only fact we need know to rationalize the differences between the interaction of a shock with a laminar or turbulent layer, both types will be discussed here. It is recommended that the student reread this section after studying the turbulent boundary layer.

In Fig. 8 (Liepmann, Roshko, and Dhawan, 1952), the incident shock, generated by a $4.5°$ wedge, intersects a laminar boundary layer on a flat plate. The shock wave configuration was identified by schlieren photographs, the boundary layer was studied with a total-head tube, and the pressure distribution was measured at the surface of the plate. The figure shows that the pressure rise has propagated upstream a distance of 3 cm or about 50 times the thickness of the boundary layer. The resulting adverse pressure gradient causes a thickening of the laminar layer followed by a "bubble" of separated flow and finally by transition to a turbulent boundary layer. The waves associated with the thickening and subsequent thinning of the layer are shown in the figure. The final reflected wave is parallel to the direction computed for reflection from a plane surface in an inviscid fluid. The initial Mach number given is measured; the others are computed from the wave configuration.

If the boundary layer is turbulent, the pressure rise propagates only about one-tenth as far upstream as for the laminar layer. In the experiments described, no flow separation was observed when the boundary layer was turbulent.

It was pointed out in Chapter 12 that for an airfoil in transonic flow the pressure recovery over the upper surface is generally accomplished by means of a shock wave. The interaction between this shock wave and the boundary layer has been the subject of many theoretical and experimental investigations. Figure 9 shows two schlieren photographs of the flow over an airfoil at a free-stream Mach number of 0.843; in the upper picture the boundary layer upstream of the shock wave is laminar; in the lower it is turbulent. In both pictures the flow direction is from left to right.

The intensity of the shock wave required to cause permanent flow separation depends on whether the boundary layer is laminar or turbulent, on the boundary layer thickness, on the pressure gradient in the flow immediately upstream, on whether the line of intersection is normal or inclined to the flow, and on the curvature of the surface. One obvious characteristic of importance is the inertia of the flow $\frac{1}{2}\rho u^2$ near the surface, since the shock must be intense enough to reverse the flow. However, if the surface is concave, as in Fig. 7, the shock must be quite intense to cause permenent separations. On the other hand, for a convex surface such as the upper surface of an airfoil, permanent separation will result from a relatively much weaker shock intersection. A thorough study of these factors was made by Gadd (1961).

CHAPTER
17

Flow Instabilities and Transition from Laminar to Turbulent Flow

17.1 Introduction

In this chapter we describe the circumstances under which a steady (i.e., laminar) shear flow becomes unstable and develops into the characteristically unsteady turbulent flow, with significant changes in flow properties, such as skin friction and heat transfer rate, location of flow separation, rate of growth of the boundary layer, etc. These changes occur because the turbulent flow is characterized by a "churning" motion, which generates a manyfold increase in the transfer (i.e., the coefficients of momentum, heat, and mass transfer) properties.

In most practical applications of flow phenomena, the flow is turbulent. Thus, a knowledge of the circumstances under which the transition from laminar to turbulent flow occurs is critical in the design of fluid machinery.

In the preceding two chapters we have treated the laminar boundary layer, represented as the "steady state solution" of the equations expressing conservation of mass, momentum, and energy in the flow near a solid surface. A laminar boundary layer is almost always realized near the forward stagnation point of a body, but farther back on the body a small disturbance, caused by an irregularity in the external flow, shock intersection, pressure pulses or surface roughness, can trigger the transition. As one would expect, on the basis of our previous studies of similarity parameters, the magnitude of a Reynolds number $U\delta/\nu$, where δ is the thickness of the laminar layer, will have a significant effect in determining whether a given flow disturbance at that location will trigger transition. The details vary, but if the boundary layer Reynolds number exceeds its critical value for a given flow, the layer becomes unstable, in the sense that a small disturbance generates imbalances in the forces acting on the fluid elements, causing the disturbance to grow as it proceeds downstream in the flow. A wave motion occurs, the amplitude of which grows as it propagates downstream and, unless some stabilizing influence intervenes, transition to a turbulent boundary layer follows. If all flow disturbances could be prevented it would be possible theoretically to maintain a laminar boundary layer over the entire surface at all Reynolds numbers; this perfectly smooth flow over a perfectly smooth body appears however to be unattainable.

This chapter treats the circumstances affecting the occurrence of instability and transition. A brief description of a turbulent flow field and the methods for measuring turbulence is followed by a description of transition and discussions of the factors affecting transition, methods of detecting transition, and of the flow around spheres and cylinders as affected by Reynolds number. The latter topic is included because it provides a spectacular example of the effect of boundary layer transition on the drag of bluff bodies.

A thorough understanding of the transition phenomenon must rest on a detailed knowledge of turbulent flow. A brief introduction to this subject is given in Section 17.2, and it is recommended that the student reread this chapter after studying Chapter 18.

17.2 Description of Turbulent Flow

The qualitative description of turbulence will be based on Fig. 1, which is sketched from photographs of the surface of a water tank on which powdered mica has been sprinkled. The grid shown at the right is being drawn through the tank at constant speed and the lines shown are the path lines of fluid particles. Immediately behind

Fig. 1 Path lines of the fluid particles behind a grid moving from left to right at uniform velocity through a fluid.

the grid, the wake of each rod, in the form of a regular succession of vortices, can be seen. This regular succession of vortices is called the *von Kármán vortex street* and will be described in Section 17.11. The wakes of the individual rods gradually merge into each other, until at the left, a turbulent field is shown in which the turbulence is *homogeneous*; that is, the turbulence in one part of the field has no average features distinguishing it from that in any other part.

The distance between the rods, called the *mesh length*, is a characteristic length of the turbulent field; 15 or 20 mesh lengths downstream of the grid, homogeneity is approximately established. If a speed-measuring instrument,* moving with the

---

*An excellent physical description of turbulence and methods of measurement are given by Bradshaw, 1971.

The hot-wire which was utilized for the measurements to be described here (and which would provide a trace similar to that of Fig. 2) utilizes platinum or tungsten wire 2 to 5×10^{-3} mm in diameter and 1 to 3 mm in length. The wire is heated to about $100°$ Celsius above room temperature. The velocity fluctuations in turbulent flow cause fluctuations in temperature and voltage drop which after amplification are displayed on an oscillograph record (such as is shown in Fig. 2), or the mean square value of the fluctuations is registered. With different hot-wire units and electronic equipment the three components of the fluctuation velocity are measured.

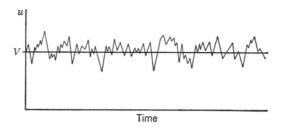

Fig. 2 Schematic representation of the velocity fluctuations in a turbulent flow.

grid, were placed at any cross-stream location in the stream at about 20 mesh lengths behind the grid, the record would look qualitatively like the trace in Fig. 2. The mean speed would be the speed of the grid, but at any particular instant the speed may deviate appreciably from the mean. These fluctuations have no definite frequency, and their amplitudes may have values anywhere between very small and fairly large values. The velocity distribution in space associated with these fluctuations is a *turbulent field*, such as exists in the free-stream of a wind tunnel, or in the atmosphere. The turbulent field is three dimensional in that the fluctuations from a mean value have components u_1, v_1, and w_1, respectively in the x, y, and z directions.

Some physical aspects of the turbulent field are as follows. A force is required to draw the grid of Fig. 1 through the fluid; a small part of the force is skin friction on the rods but the major portion is a "form drag" that stems from a change in the pressure distribution around the individual rods because of flow separation and the formation of the wakes. Thermodynamically, the system has suffered an increase in entropy because some of the directed energy has been transferred to turbulence; viscous decay in turn transforms the energy in the turbulent field into heat. If the grid were drawn through a homogeneous fluid the entropy increase would be evidenced as a drop in stagnation pressure through the grid; since Fig. 1 represents a water surface the entropy increase is evidenced by a drop in the water level as the grid passes a point.

A quantitative measure of turbulence is taken as the root-mean-square value of the fluctuations. Thus

$$\sigma = \frac{100}{V} \sqrt{\frac{1}{T} \int_0^T \frac{1}{3} (u_1^2 + v_1^2 + w_1^2)\, dt} \tag{1}$$

where T is large compared with the duration of any excursion, $\sqrt{u_1^2 + v_1^2 + w_1^2}$, from the mean speed V (Fig. 2), is a statistical measure of the magnitude of the turbulence in per cent of the mean speed.

The magnitude of the turbulence, σ, will determine its *diffusing* effect, that is, the rate at which a drop of coloring matter, for instance, will spread throughout the flow behind the grid in Fig. 1. It follows intuitively that the greater the disturbance behind the grid (large σ) the greater will be the rate at which the coloring matter will spread throughout the flow. If, on the other hand, the color were added to the water ahead of the grid, it would diffuse throughout the flow at an extremely slow rate. Ahead of the grid, the coloring is spread by molecular diffusion; behind, turbulence plays the major role in the diffusion. It follows that turbulent diffusion tends to destroy gradients in any property, whether it be color, momentum, heat, or density of particles in suspension.

Compare now a laminar boundary layer profile just before, with that just after, transition. The mean velocity profile of the turbulent layer is drawn through the band of instantaneous profiles superimposed on a mean motion. The intense mixing effect of the turbulence will tend to flatten the mean velocity profile, but it cannot carry this effect to the wall because of the no-slip condition there. Therefore, the effect of the turbulence will be qualitatively as shown in Fig. 3; the velocity gradient becomes smaller in the outer region and greater near the wall.

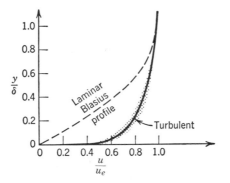

Fig. 3 Comparison between laminar and turbulent mean velocity profiles for the same boundary layer thickness. An *instantaneous* turbulent profile could be anywhere within the shaded region.

The change from the laminar profile to the mean velocity profile for turbulent flow takes place over a downstream interval which is called the *transition region*. The practical importance of transition and therefore of the factors that affect it are threefold: (1) since $(\partial u/\partial y)_w$ is greater for the turbulent than for the laminar layer, the shearing stress $\tau_w = \mu(\partial u/\partial y)_w$ will increase greatly through the transition region (see Fig. 15.6); (2) reference to Section 16.7 indicates that there will be a corresponding increase in heat transfer rate at the wall; (3) flow separation will be delayed because $(\partial u/\partial y)_w$ is greater in the turbulent layer.

17.3 Reynolds Experiment

The classical experiments of Osborne Reynolds (1883) demonstrated the fact that under certain circumstances the flow in a tube changes from laminar to turbulent over a given region of the tube. The experimental arrangement involved a water

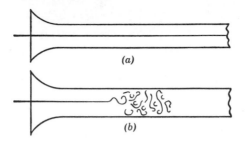

Fig. 4 Schematic representation of Reynolds' experiment.

tank and an outlet through a small tube, at the end of which was a stopcock for varying the speed of the water through the tube. The junction of the tube with the tank was nicely rounded, and a filament of colored fluid was introduced at the mouth. When the speed of the water was low, the filament remained distinct throughout the entire length of the tube, as shown in Fig. 4a; when the speed was increased, the filament broke up at a given point and diffused throughout the cross section, as shown in Fig. 4b. Reynolds identified a governing parameter as $U_m d/\nu$, where U_m is the mean velocity through the tube of diameter d, and this number has since been known as the Reynolds number.

Reynolds found that transition occurred at Reynolds numbers between 2000 and 13,000, depending upon the smoothness of the entry conditions. When extreme care is taken to obtain smooth flow, the transition can be delayed to Reynolds numbers as high as 40,000; on the other hand, a value of 2000 appears to be about the lowest value obtainable regardless of how rough the entrance conditions are made. The fact that the transition Reynolds number can be varied by disturbing the flow indicates that the transition Reynolds number is affected by the turbulence in the stream.

17.4 Tollmien-Schlichting Instability and Transition

The actual mechanism of transition, though far from completely understood, has been greatly illuminated by recent investigations, both theoretical and experimental. Tollmien, and later Schlichting, Lin, and others (see Lin, 1955) showed that for Reynolds numbers, $Re_{\delta *} = u_e \delta * / \nu$, above a definite minimum value, disturbances in a certain band of frequencies will tend to grow with time (the flow process is generally referred to as the T-S instability). Hot-wire records made by Schubauer and Skramstad (1947) at various distances behind the leading edge of a flat plate are shown in Fig. 5. These show the sequence of events following the generation of a disturbance in the laminar boundary layer. The oscillations, which represent fluctuations in the wind speed at 0.6 mm from the surface, are seen to grow as the distance from the leading edge increases. At $x = 1.83$ m, however, some irregularities

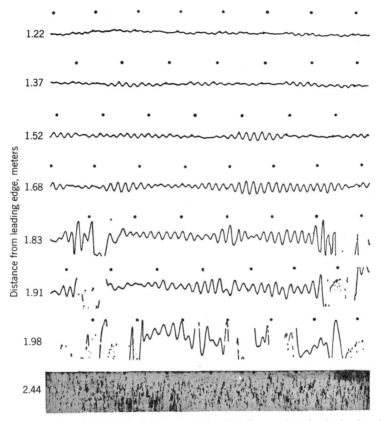

Fig. 5 Oscillograms by Schubauer and Skramstad showing fluctuations in the laminar boundary layer on a flat plate. Distance from surface = 0.6 mm, u_e = 24.4 m/sec, time interval between dots = 0.033 sec. (Courtesy of National Bureau of Standards.)

occur in the waves. These are "bursts" of high-frequency fluctuations ordinarily associated with turbulent flow. The bursts become more frequent and of longer duration with increasing x until, at 2.44 m, the entire record is turbulent.

The T-S disturbances shown in Fig. 5 are two-dimensional; the theoretical stream-lines are shown in Fig. 6. This representation is for obvious reasons termed the "cat's eye diagram." With suitable flow visualization techniques the cat's eyes can be photographed, their duration depending on their rate of amplification; they lose their identity as soon as the bursts of turbulence occur. The streamlines of Fig. 6 describe a wave motion that travels (propagates) downstream at about $u_e/3$ where u_e is the speed at the outer edge of the boundary layer.

The Schubauer-Skramstad experiments are in excellent agreement with the theory as shown in Fig. 7. In that diagram, $\beta \nu / u_e^2$ is plotted against $Re_{\delta^*} = u_e \delta^* / \nu$, where β is the frequency of the fluctuations and δ^* is the displacement thickness of the

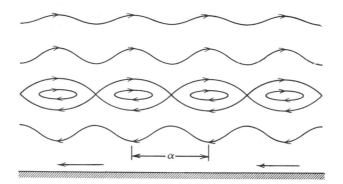

Fig. 6 Streamline pattern of T-S disturbances in laminar boundary layer on a flat plate (Lin, 1945). The streamlines shown are those seen by an observer moving downstream at the speed of propagation of the disturbance. This speed of propagation is about one-third of the flow speed at the edge of the boundary layer. The wavelength of the disturbance is designated α.

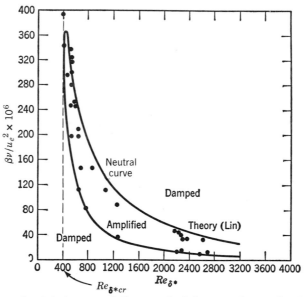

Fig. 7 Measurements by Schubauer and Skramstad of the neutral curve for the velocity fluctuations in the laminar boundary layer on a flat plate compared with the theoretical curve by Lin (see Lin, 1955).

laminar layer. For a given u_e and ν, disturbances of a given frequency are damped or amplified according to whether the calculated values of $\beta\nu/u_e^2$ and Re_{δ^*} define a point outside or within the loop designated "neutral curve." For instance, in Fig. 5 at $x = 1.83$ m, $\beta = 189$ hz $= 1190$ rad/sec. To place the point on Fig. 7, we calculate $\beta\nu/u_e^2 = 29 \times 10^{-6}$; $Re_x = 3.1 \times 10^6$ so that by Eq. 15.17 $\delta^* = 1.72x/\sqrt{Re_x} =$

1.8 mm and Re_{δ^*} = 3020. This point lies well within the loop in Fig. 7 indicating that the disturbance is being amplified. In fact, for the observed value of $\beta \nu / u_e^2$, amplification began at Re_{δ^*} = 1300, that is, for δ^* = 0.77 mm or x = 1.2 m, just upstream of the first trace in Fig. 5.

Another feature of Fig. 7 is the indicated existence of a vertical tangent to the neutral curve; in other words, there is a minimum critical Reynolds number, $Re_{\delta^* cr}$, below which all *infinitesimal* disturbances are damped. The term infinitesimal is emphasized because the theory is not applicable to finite disturbances. For the flat plate $Re_{\delta^* cr}$ has a value variously calculated to be between 420 and 575, but *finite* disturbances can be amplified at lower Re_{δ^*}.

The instability indicated by the growth of the disturbances is only a first step in the transition process. The theoretical predictions are based on a solution of the linearized equation of motion and therefore do not provide for any distortion of the sinusoidal disturbances shown in Fig. 5, much less for the appearance of the bursts of turbulence.

Emmons (1951) observed the transition on flow along a water table. He saw spots later identified as the high frequency bursts shown in Fig. 5 at $x \geqslant 1.83$ m; these grew in size and more appeared as x increased until they merged to the fully turbulent state representing the trace at x = 2.44 m in Fig. 5. The details of the flow in the spots was measured by Schubauer and Klebanoff (1955) and are shown in Fig. 8; the turbulent spot is generated by an electric spark and the details of its

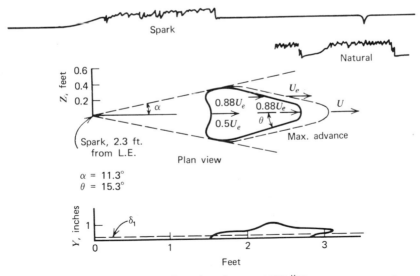

Fig. 8 Turbulent spot initiated by electric spark between needle electrode and surface. Oscillograms with 1/60-sec timing dots shown above, time progression from left to right, upper showing spark discharge on right and spot passage on left, lower showing natural transition (Schubauer and Klebanoff, 1955; courtesy of NACA).

growth and internal structure were measured as it was carried downstream by the flow. It appears that there is a sudden transition to a fully turbulent boundary layer at the boundary of the spot.

The turbulent spots, as for the appearance of the initial instability waves, are generated in practice by turbulence in the outside stream, pressure pulses, or surface roughness. It appears that when the instability waves have grown to critical amplitude, a slight further amplification has an explosive effect in generating the high-frequency fluctuations characteristic of turbulent flow. The range of validity of the small disturbance theory leading to the results in Figs. 6 and 7 is obviously exceeded when the bursts occur.

Transition may occur as a result of "transverse contamination" from the edges of the surface or behind a roughness element. The turbulent boundary layer grows laterally to subtend an angle of about $18°$ behind the roughness element and at about $9°$ from the leading edge at the wall.

17.5   Factors Affecting Transition by T-S Instability

The scope of the treatment of transition given here is necessarily sketchy; several comprehensive treatises are available, among them the cassettes and accompanying texts by Morkovin on transition and Mack on stability (1972), and the account by Stuart (1971) on nonlinear stability.

In general any influence that decreases the critical Reynolds number indicated in Fig. 7 also increases the size of the amplification loop and hastens transition. The various influences are discussed in the following paragraphs.

PRESSURE GRADIENT

The stabilizing effect of a favorable pressure gradient on transition, and the destabilizing effect of an adverse gradient, are illustrated by the Schubauer–Skramstad results shown in Fig. 9. A curved plate with the pressure distribution shown at the left was used. The figure shows a favorable pressure gradient ($\partial p/\partial x < 0$) over the forward portion and an unfavorable gradient over the rear portion. We see from the traces that disturbances which amplify in a region of zero pressure gradient are damped in a favorable gradient; when the gradient becomes adverse, they are again amplified.

Schlichting and Ulrich (see Schlichting, 1968) illustrated, by means of the neutral stability loops (Fig. 10), the effect of pressure gradient on the stability of the boundary layer. The Pohlhausen parameter λ (Section 15.7) is the dimensionless parameter; $\lambda > 0$ designates a favorable pressure gradient, and conversely. We see that the behavior of the traces in Fig. 9 could be predicted from the curves of Fig. 10; that is, when the pressure gradient is favorable, the $Re_{\delta*cr}$ is reduced and the area of the instability loop is much decreased, and conversely for the adverse gradient. One would accordingly expect transition to be delayed in a favorable pressure gradient and hastened in an adverse.

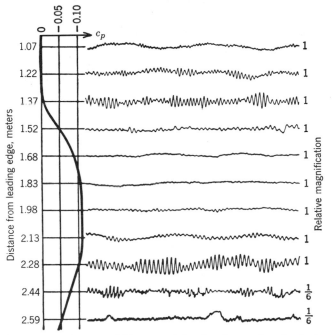

Fig. 9 Oscillograms by Schubauer and Skramstad showing fluctuations in the laminar boundary layer on a surface with the pressure distribution indicated at the left. Distance from surface = 0.53 mm, u_e = 29 m/sec. (Courtesy of National Bureau of Standards.)

The above behavior illustrates the theoretical result (for incompressible flow only) that velocity profiles with an inflection point have a low $Re_{\delta*cr}$; that is, they tend to be unstable. Profiles with inflection points are shown in Fig. 15.11 for $\lambda < 0$, and Fig. 10 shows that if $\lambda < 0$, $Re_{\delta*cr}$ is low and the neutral loops enclose large unstable regions.

The low-drag feature of the laminar boundary layer can be exploited by maintaining a favorable pressure gradient over as much of the surface as possible. To this end, NACA developed the "laminar flow" or low-drag families of airfoils; the drag characteristics of one of the airfoils is shown in Fig. 11 (see Abbott and von Doenhoff, 1949). Figure 11a shows the 65_3-018 airfoil; the meanings of the numbers are as follows:

> 6: series or family designation,
> 5: maximum thickness is at $0.5c$,
> 3: the drag is a minimum over a range $\Delta c_l = 0.3$,
> 0: $c_l = 0$ is the "design lift coefficient"[†],
> 18: the maximum thickness is $0.18c$.

[†]The "design lift coefficient" has the following meaning. The camber of the mean chord line around which the thickness distribution is "draped," is chosen so that its slope at leading edge is parallel to V_∞ for the design lift coefficient; this coefficient is generally chosen in accordance with the cruise condition of the aircraft. Abbott and von Doenhoff (1949) give characteristics of these airfoils for several mean lines.

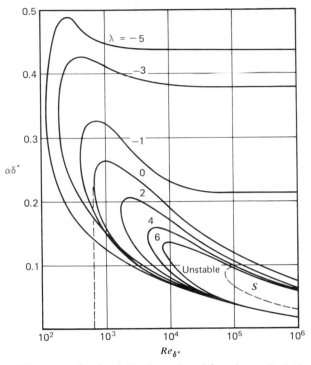

Fig. 10 Neutral stability loops for favorable ($\lambda > 0$) and for adverse ($\lambda < 0$) pressure gradients on a flat plate [$\lambda = (\delta^2/\nu)\,du_e/dx$]; α is the wavelength of the disturbance and δ^* is the displacement thickness. Verticle broken line designates Re_{δ^*cr} for $\lambda = 0$; curve labeled S refers to "asymptotic suction" surface for $\lambda = 0$. See Schlichting (1968, p. 471; courtesy of McGraw-Hill.)

The plot of $(V/V_\infty)^2$ is related to that of the pressure coefficient as follows:

$$\left(\frac{V}{V_\infty}\right)^2 = \left(\frac{V}{V_\infty}\right)^2 - 1 + 1 = \frac{p - p_\infty}{q_\infty} + 1$$

$$= C_p + 1$$

Thus the figure shows that the pressure gradient is favorable over the forward $0.45c$ to $0.47c$ for lift coefficients up to at least 0.32. The stability loops indicate therefore that *if precautions are taken relative to the other effects listed below*, the boundary layer will tend to be laminar over nearly half the chord. That this effect does in fact occur is indicated by the drag coefficient curve shown in Fig. 11*b* where the "low-drag bucket" in the c_d versus c_l curve for the smooth airfoil indicates an extensive area of laminar boundary layer over the range $-0.3 < c_l < 0.3$. The addition of roughness near the leading edge triggers early transition so that the boundary layer is turbulent over most of the airfoil surface; as a consequence the drag is increased and the "bucket" is destroyed.

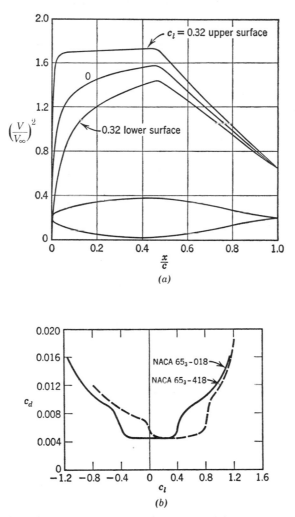

Fig. 11 Characteristics of NACA 65_3-018 and 65_3-418 airfoils. (a) Pressure distribution $(u_e^2/V_\infty^2 = c_p + 1)$ at $c_l = 0$ and 0.32. (b) Measured c_l versus c_d at $Re = 6 \times 10^6$ for smooth airfoil. (Abbott and von Doenhoff, 1949, courtesy of NACA.)

Experimental studies described previously (see Figs. 16.8 and 16.9a) indicate that the sudden pressure increase associated with the intersection of even a weak compression shock with a laminar boundary layer is often sufficient to trigger transition.

The opposite effect, that is, "reverse transition" or "laminarization" of a turbulent boundary layer, can occur in a highly favorable pressure gradient (Sternberg, 1954; Back, Cuffel, and Messier, 1969). The accompanying reduction in skin friction and heat transfer can have important effects in rocket nozzles, for instance.

The experiments of Back et al. indicate that when the parameter

$$\left(\frac{\nu_e}{u_e^2}\right)\frac{du_e}{dx} > 2 \times 10^{-6}$$

significant laminarization occurs.

SUCTION

The application of suction at a surface (1) decreases the boundary layer thickness and (2) causes the velocity profile to become more full, that is, the greater the suction the more the profile deviates from one with an inflection point. Both effects are stabilizing, that is, for the "asymptotic suction profile" described in Problem 15.3.4 has a minimum critical Reynolds number of around 70,000; its stability loop is designated by S in Fig. 10. The profile has a value of the shape parameter $H = \delta^*/\theta$ of 2, which has the maximum Re_{δ^*cr} in Fig. 12, in which the trends of various influences are indicated as functions of H.

Thus, theory indicates that through the use of suction a laminar boundary layer could be realized over a large portion of the surface of an aircraft. The prospect of the resulting large decrease in drag has prompted many theoretical and experimental investigations (see, e.g., Pfenninger, 1965; Pfenninger and Reed, 1966). While the potential of the application has not to date been realized, largely because of surface roughness, noise, and three-dimensional flow effects (see below), some gains have been made and work continues in many laboratories.

HEAT

It was pointed out that Fig. 10 shows that the effect of an inflection point in the laminar-boundary layer velocity profile is to increase appreciably the rate of amplification of disturbances. We can show quite simply that if a surface is heated the velocity profile develops an inflection point. This analysis is similar to that of Section 15.4, which shows the effect of an adverse pressure gradient on the velocity profile.

It was shown in Chapter 1 that the viscosity coefficient μ of a gas is theoretically proportional to the square root of the absolute temperature. If a temperature gradient exists in the y direction (normal to the surface) μ will vary with y, and the boundary layer equation of motion becomes

$$\rho\left(u\frac{\partial u}{\partial x} + v\frac{\partial u}{\partial y}\right) = -\frac{\partial p}{\partial x} + \frac{\partial}{\partial y}\left(\mu\frac{\partial u}{\partial y}\right) \tag{2}$$

At the surface $u = v = 0$ and Eq. 2 becomes

$$\frac{\partial p}{\partial x} = \left[\frac{\partial}{\partial y}\left(\mu\frac{\partial u}{\partial y}\right)\right]_{y=0}$$

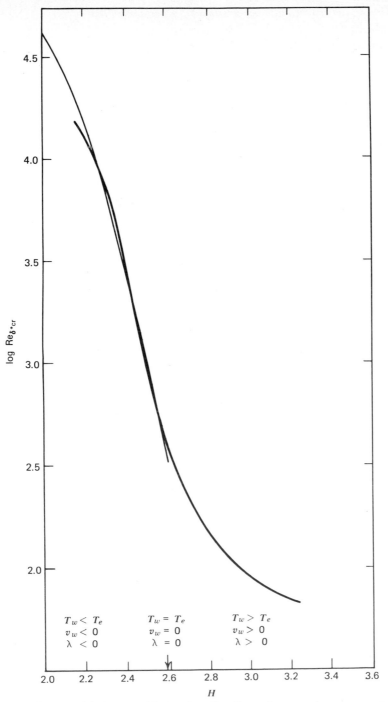

Fig. 12 Minimum critical Reynolds number (log Re_{δ^*cr}) versus shape parameter $H = \delta^*/\theta$; T_w = surface temperature, T_e = temperature at $y = \delta$, v_w = velocity normal to surface ($v_w < 0$ for suction), $\lambda = (\delta^2/\nu)\, du_e/dx$. $H = 2.58$ for isothermal laminar flow over impervious surface at $dp/dx = 0$. Effects of plate temperature, suction, and λ are indicated qualitatively. Theoretical curve, drawn through data from many sources, adapted from Stuart (1963). Courtesy of Oxford University Press.

which may be written

$$\left(\frac{\partial^2 u}{\partial y^2}\right)_{y=0} = \frac{1}{\mu}\left[\frac{\partial p}{\partial x} - \left(\frac{\partial \mu}{\partial y} \cdot \frac{\partial u}{\partial y}\right)_{y=0}\right] \tag{3}$$

This result will be discussed in the light of the curve shown in Fig. 12. Consider first a heated flat plate $(\partial p/\partial x = 0)$. Then the temperature and therefore μ will decrease with y, i.e., $(\partial\mu/\partial y)_{y=0} < 0$. Since $(\partial u/\partial y)_{y=0} > 0$, Eq. 3 shows that for this case $(\partial^2 u/\partial y^2)_{y=0} > 0$. But near the outer edge of the boundary layer $\partial^2 u/\partial y^2 < 0$. Therefore, at some point within the layer $\partial^2 u/\partial y^2 = 0$; that is, the velocity profile has an inflection point. This behavior is shown in Fig. 16.5a.

As indicated earlier, a velocity profile with an inflection point is unstable. This is borne out by the low values of Re_{δ^*cr} in Fig. 12, at $H > 2.58$, corresponding to an inflection point profile in an isothermal flow over a surface with an adverse pressure gradient. The above results indicate that the effect of heating a surface will tend to increase H and thus decrease Re_{δ^*cr}.

The converse, of course, holds as well, that is, cooling the surface is stabilizing.

The effect is shown in the measurements at Mach 3 plotted in Fig. 13 for transition on the surface of a cone. While there are uncertainties in the "conversion" to an equivalent flat plate (see Mack, 1975) the trend of the curve is considered reliable, except perhaps under conditions of extreme cooling, as is pointed out below in the paragraph on "Roughness".

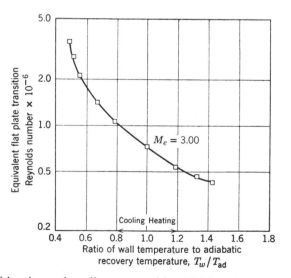

Fig. 13 Effect of heating and cooling on transition. Measurements made on cone and converted to "equivalent flat plate" (Jack and Diaconis, 1955; courtesy of NACA).

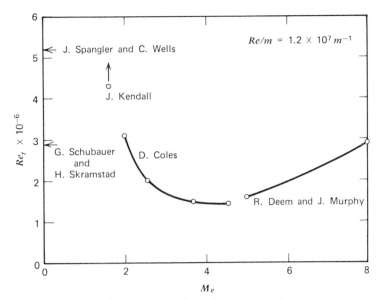

Fig. 14 Effect of Mach number on start-of transition Reynolds number as measured on flat plates in wind tunnels. The Schubauer and Skramstad curve applies to variation of turbulence alone. Mack (1975) courtesy of AIAA.

COMPRESSIBILITY

Theoretical studies and numerical techniques have recently yielded fairly complete solutions of the stability equations for compressible flow (Mack, 1975). However the experimental investigations carried out have been beset by extreme difficulties, the two most important of which are the high noise levels in high-speed and, especially, supersonic tunnels, particularly those with turbulent wall boundary layers, and the still obscure and controversial effects of "unit Reynolds number," V_∞/ν, that cause wide disparity, particularly among measurements in different wind tunnels. According to Mack, 1975, the most reliable measurements for the onset of transition on flat plates as a function of "unit" Mach number at $Re/m = 1.2 \times 10^7$ m^{-1} are those reproduced in Fig. 14. The trends indicated in these measurements are also found in most other measurements as well as stability theories, that is, the initial effect of compressibility is to decrease the critical and transition Reynolds numbers to a minimum between Mach 3 and 5, followed by an increase at higher Mach numbers.

TURBULENCE AND NOISE

The separate effects of pressure disturbances and turbulence were studied by Schubauer and Skramstad (1947). They also demonstrated that tones with frequency near that of the Tollmien instability waves of a boundary layer will excite instability and hasten transition; however, only recently have wind tunnels been available that enabled measurement of the separate effects of pressure and velocity fluctuations.

The experiments of Spangler and Wells (1968), carried out in a "quiet" wind tunnel in which sound levels of various magnitudes could be introduced (compared in Fig. 15 with those of Schubauer and Skramstad, 1947), indicate that the relative magnitudes and frequencies of the pressure and velocity fluctuations affect significantly the transition Reynolds numbers in flow along a flat plate. Without going into details of the various experiments it is clear that the acoustic frequencies imposed on the flow (27 to 82 cps) have a tremendous effect on transition at a given turbulence level as measured by a hot wire.

Tests indicate that the turbulence in the atmosphere is essentially zero insofar as any effect on boundary layer transition is concerned; however, Fig. 15 indicates that the noise and vibration generated by the engines can trigger transition of the boundary layers on an aircraft.

These and many other experiments have been analyzed by Loehrke, Morkovin, and

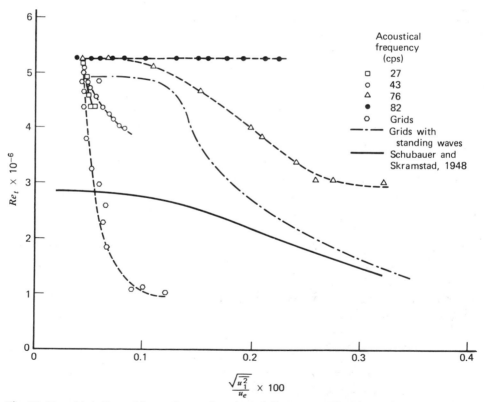

Fig. 15 Transition Reynolds number as function of free-stream disturbance intensity. ($u_e/v = 7.9 \times 10^5$/m; $u_e = 11.7$ m/sec.) (From Spangler and Wells, 1968; courtesy of AIAA.)

Fejer (1975); they are able to bracket the effects of the many variables. Since all the effects are nonlinear, their effects are not superposable and any one of them may, if large enough, exert the *governing* effect.

ROUGHNESS

The effect of roughness will, of course, be qualitatively in the same direction as that of increasing disturbance in the main flow, whether caused by turbulence or sound. Any of the factors introduces disturbances in the laminar flow and the transition Reynolds number will tend to decrease. Although there are several types of roughness—single-, two-, or three-dimensional roughness elements, distributed sand roughness, etc.—their general effect on transition is represented in Fig. 16.

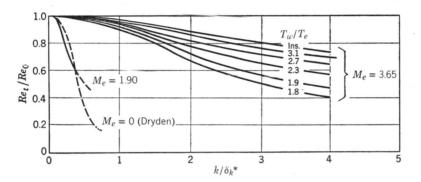

Fig. 16 Effects of roughness, surface temperature, and Mach number on transition on a cone at supersonic speeds (Van Driest and Boison, 1957) with the effect of roughness on a flat plate at low speeds (Dryden, 1959). Re_0 is the transition Reynolds number $u_e \delta^*/\nu$ for the smooth surface; Re_t, that for the surface with sand roughness of height k; T_w is the surface temperature; T_e and M_e refer to the main flow; δ_k^* is the displacement height of the boundary layer thickness at transition.

The experimental results of Fig. 16 show however that *cooling* a *rough* surface in the supersonic flows moves transition *forward* whereas it was stated earlier that cooling a *smooth* plate has a *stabilizing* effect. The reason for this seeming paradox probably lies in the circumstance that, as the surface is cooled, the kinematic viscosity in the immediate vicinity of the surface decreases and the Reynolds number of the flow past a roughness element therefore increases. Associated with this increase in Reynolds number, there is an increase in the magnitude of the disturbance in momentum (ρu_1) in the wake of the roughness element. The result would be a decrease in the transition Reynolds number. We may conclude then (as was mentioned above) that the validity of the results shown in Fig. 13 is limited to smooth surfaces, and, if a surface is rough enough, the indicated effect of surface cooling may be reversed.

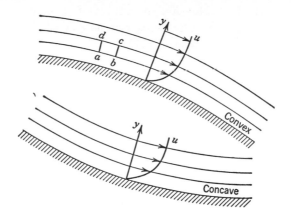

Fig. 17 Schematic representation of the flow along concave and convex surfaces.

17.6 Centrifugal Instability and Transition

Transition due to streamwise curvature takes place in an entirely different way from that due to the factors previously discussed. The cause can be illustrated simply by means of Fig. 17, which represents boundary layer profiles for flow over curved surfaces. Consider a small element of fluid $abcd$; it must be in equilibrium under the actions of pressure and centrifugal forces. Equating these forces,

$$\left(-\frac{\partial p}{\partial n}\right) \cdot dr \cdot r\,d\theta = \rho r\,dr\,d\theta \cdot \frac{u^2}{r}$$

or

$$-\frac{\partial p}{\partial n} = \frac{\rho u^2}{r} \tag{4}$$

where r is the radius of curvature (positive for concave upward) and u is the velocity. For equilibrium, $\partial p/\partial n$ will be positive for a convex and negative for a concave surface; that is, the pressure force will be toward the surface for convex curvature and away from the surface for concave curvature. For the convex surface, if a small element of fluid is disturbed (by turbulence or roughness) so that it moves outward, its velocity will be less than that of the surrounding air, and hence the centrifugal force tending to carry it farther out will be less than the pressure force tending to return it to its layer of origin. Thus, it will tend to return to its layer of origin; in other words, the effect of convex curvature is to stabilize the flow. For the concave flow, if a particle is displaced outward, the centrifugal force is less than that on the surrounding air because its velocity is less; hence the pressure force, which is just sufficient to balance the centrifugal force on the surrounding air, will carry the particle farther from its layer of origin. Therefore, concave curvature has a destabilizing effect, whereas convex curvature has a stabilizing effect. Although the above analysis is somewhat rough, the conclusion agrees qualitatively with the theoretical investigation by Goertler.

Goertler (see Lin, 1955) found that laminar flow over a concave surface is unstable if the parameter, $Re_\theta \sqrt{\theta/r}$, where θ is the momentum thickness of the boundary layer, exceeds 0.57. Experimentally, Liepmann (see Lin, 1955) found that the parameter had values between 6 and 9 at transition, depending on the turbulence level in the free stream. There is a considerable discrepancy between theoretical and experimental values of the parameter, but it must be remembered that the theoretical value refers to instability and therefore must be somewhat lower than the experimental value, which refers to transition.

An essential difference exists between the Tollmien waves shown in Fig. 6 and the disturbances found by Goertler, shown in Fig. 18. The latter type consists essentially of vortices oriented in the stream direction; they are similar to the disturbances found by G. I. Taylor (see Lin, 1955) in his investigation of the instability of the flow between rotating cylinders.

Some of the most important flows subject to centrifugal instability are those involving three-dimensional boundary layers (Fig. 15.12) in which the plane of curvature of the streamlines is essentially parallel to the surface and the magnitude of the curvature varies with distance from the surface. Some of those of interest in aircraft design concern the boundary layers on swept wings and on the blades of propellers, helicopters and fan rotors. These suffer a combination of T-S and Goertler or centrifugal instability; if we consider outward or inward displacement of a fluid element of the three-dimensional boundary layer of Fig. 15.12 the curvature will be different and there will be a net centrifugal force generated tending to carry the element farther from the streamline from which it was displaced. Thus, instability of a boundary layer will be enhanced by sweepback as well as by the centrifugal forces acting on a displaced particle on a rotating blade.

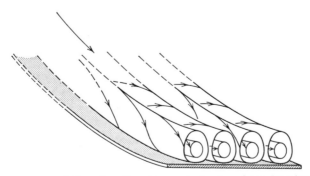

Fig. 18 Streamline pattern of Goertler disturbances in laminar boundary layer on a concave surface.

17.7   Stratified Flows

There are many instabilities of parallel flows that have important applications in aeronautics. Many of these are treated by Yih (1969) and by Stuart (1963).

One of these, the effect of temperature gradients across a boundary layer, has been mentioned above in connection with boundary-layer stability; its most familiar manifestation in the atmosphere is governed by the dimensionless parameter termed the Richardson number:

$$\frac{-(g/\rho)(d\rho/dy)}{(du/dy)^2} \tag{5}$$

where $-gd\rho/dy$ is the restoring force on unit volume of a gas displaced in the y direction in a density gradient $d\rho/dy$; unit volume of fluid has inertia per unit volume $\rho[u + (du/dy)\,dy]^2$, so that, *relative to the fluid* at its layer of origin where the velocity is u, the inertia is $\rho(du/dy)^2$. Expression 5 is then the ratio between the restoring force of gravity and inertia of the element per unit volume; thus the higher this ratio the more stable will be the flow. Figure 19 shows fair agreement between measurements by Reichardt and the theoretical neutral curve by Schlichting (1968, p. 493), in that the turbulent flows occur in the unstable region and most of the laminar flows are in the stable region.

"Helmholtz instability" is another common occurrence, forming "gravity waves"

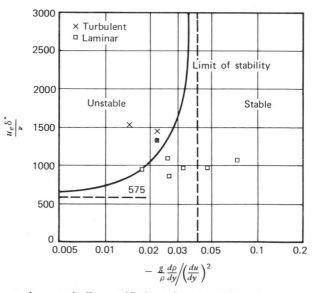

Fig. 19 Measurements in a vertically stratified gas flow by Reichardt compared with stability limit curve calculated by Schlichting. Points indicate whether the flow was laminar or turbulent (Schlichting, 1968; courtesy of McGraw-Hill).

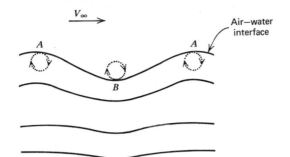

Fig. 20 Waves at an air-water interface with relative velocity V_∞. The circles represent particle paths as the waves propagate to the right.

at an air-water interface in relative motion (Prandtl, 1952, pp. 88–92). Figure 20 shows air flowing over a water surface on which small waves have formed; the relative velocity will be high at the crests and low in the troughs of the waves. By Bernoulli's theorem, therefore, the pressure will be high in the troughs and low at the crests thus tending to cause the crests to rise and the troughs to deepen. The effect therefore is to tend to increase the amplitude of the waves.

The process however reaches a limit. As the waves propagate to the right the water particles near the surface move approximately in the circles shown by the dotted lines, thus steepening the wave form in a way resembling superficially that in which finite compression waves in a gas steepen into shock waves (the two processes are physically quite different). Finally the water waves "break" and form the familiar "white caps."

These gravity waves can also form in the atmosphere in steep density gradients. It has been hypothesized by Bekofske and Liu, 1972, that the "breaking" of these waves generates the "clear-air-turbulence" (CAT) often encountered by aircraft.

The Taylor–Goertler instability is also often visible in the atmosphere in the form of parallel lines of clouds. Qualitatively these denote the line vortices similar to those shown in Fig. 18; the water vapor in warm moist air, carried aloft by convection, condenses and forms lines of clouds while the cold currents carried downward are warmed.

17.8 Stability of Vortex Sheets

In Section 5.2, the vortex sheet was introduced to represent the mean line of an airfoil; however, there the shape of the mean line was fixed and so therefore was the shape of the vortex sheet. Now we consider instead the vortex sheet that develops at the boundary between two flows of different velocities (e.g., at the edge of a jet), as indicated schematically in Fig. 21. When a small disturbance occurs the velocities normal to the sheet induced by the vorticity elements making up the

Fig. 21 Instability of the vortex sheet between streams of different velocities.

sheet cause the sheet to roll up and resemble the build up of water waves. Eventually, depending on the Reynolds number, transition to a turbulent shear flow occurs and any regularity becomes more and more difficult to find. In the next chapter, the turbulent "mixing region" formed after transition will be described briefly for jets and for wakes.

17.9   The Transition Phenomenon

The difficulty of predicting transition, even assuming we knew thoroughly the effects of the many separate influences, has been pointed out but reemphasis seems in order. It constitutes, along with turbulence itself, one of the most intriguing problems in physics, and therefore, in engineering. We must know the effects of the several factors but also we must live with them long enough to develop a "feeling" for the steps to take to avoid transition, or to trigger transition (e.g., to avoid flow separation) to achieve optimum performance in a specific application.

In aircraft we know that at landing and takeoff, the angles of attack are high enough so that the high adverse pressure gradients and (at takeoff) the noise are likely to be the governing factors; as a result the boundary layer will be turbulent over most of the aircraft. At cruising speeds extraordinary precautions in the way of providing a smooth surface and large areas of favorable pressure gradient, with perhaps suction at critical points may be necessary to attain appreciable areas of laminar flow. In transonic or supersonic flow it is almost certain that if shocks intersect the surface they will trigger transition (see Figs. 16.8 and 16.9*a*). Thus interference between flows about the various components of the aircraft must be analyzed (perhaps by the source and doublet sheet method of Chapters 4 and 5) to avoid as far as possible unfavorable effects; on the contrary *favorable* interference effects are also possible, and devoutly to be sought.

A transition rule, found by Smith (1956) after examination of many measurements, is that transition occurs when the amplification of a disturbance, whether it be of the T-S or Goertler type, reaches e^9. The rule has been shown to be fairly reliable on smooth surfaces with pressure gradients.

Once the transition point is estimated the laminar and turbulent skin friction and form and, at transonic and supersonic speeds, wave drag can be estimated. Turbulent skin friction is treated in the next chapter.

17.10 Methods for Experimentally Detecting Transition

Some understanding of the difference between laminar and turbulent boundary layers can be gained from a consideration of the methods that have been devised for locating transition. See for instance Bradfield (1964).

1. *Hot-Wire Anemometer*

The difference between hot-wire records in the laminar and turbulent boundary layers is demonstrated in Fig. 5 and is described in Section 17.4. Laser methods would give similar indications.

2. *Total Head Near Surface*

If a total head tube of small dimensions is moved upstream near the surface, when it passes from the turbulent to the laminar layer, there will be an appreciable drop in the total head. The reason for this drop is that the velocity very near the surface, and hence the total head, is much less in a laminar layer than in a turbulent. This method was demonstrated in flight by Jones (1938); Fig. 22 shows some typical results.

3. *Stethoscope*

If a standard medical stethoscope is applied to short tubes leading from a total-head tube in the boundary layer, irregular pulses shown in the transition region in the hot-wire records of Fig. 5 can be detected. A steady noise is heard when the total-head tube is in the turbulent boundary layer.

4. *Sublimation*

This method utilizes a coating of a volatile substance on the surface; the rate of sublimation depends upon whether the boundary layer is laminar or turbulent. The

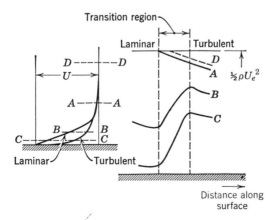

Fig. 22 Schematic representation of the variation of stagnation pressure through the transition region at various heights above the surface.

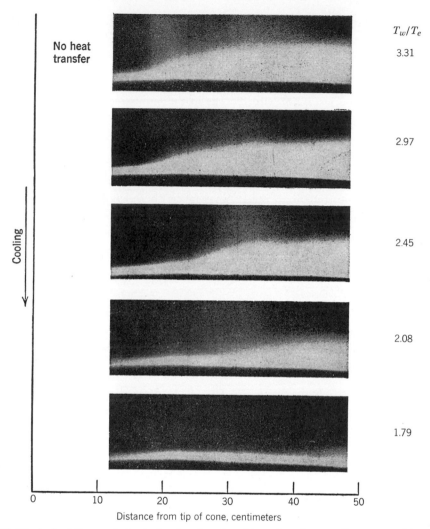

Fig. 23 Typical schlieren photographs showing the effect of surface cooling on transition on a smooth $10°$ cone for $M_e = 3.65$. (Van Driest and Boison, 1957.)

greater velocity gradient at the surface in the turbulent layer causes a higher rate of sublimation than if the layer is laminar. Consequently, behind the transition region the surface will clean more rapidly than ahead, and the line of demarcation marks the transition point. One application utilizes a mixture of ether and camphor which is sprayed on the surface; as soon as it has dried, the windstream is started and is continued until the camphor has sublimated from the turbulent area.

5. *Surface Temperature*

Figure 16.4 shows a marked increase of the temperature recovery factor in the transition region; thus measurements of the equilibrium temperature of an insulated surface provide a practical means for determining the transition region. For flows of short duration, the surface will not reach the equilibrium temperature, but since the skin friction is markedly greater in the turbulent region than in the laminar (Fig. 15.6 and Fig. 18.5), the rate of heat transfer to the surface will be much greater in the transition region (see Eq. 16.53).

6. *Schlieren Photographs*

Magnified schlieren photographs, sensitive to the density gradients normal to the surface, provide an accurate means of locating transition. Figure 23 reproduces a series of photographs with various degrees of surface cooling.

17.11   Flow Around Spheres and Circular Cylinders

The flow around circular cylinders and spheres are described here to illustrate the marked effects of roughness, turbulence; and pressure gradient on flow separation; the reason being that the flow regimes initiated are interesting and of real practical significance. Figure 24 is the curve of drag coefficient of a circular cylinder $C_D = \mathrm{drag}/q_\infty d$ versus Reynolds number $V_\infty d/\nu$, where d is the cylinder diameter. This is an experimental curve with data from various laboratories.

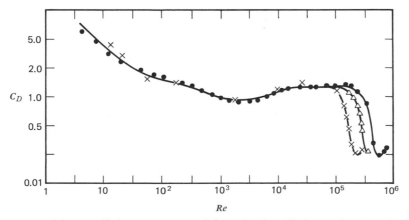

Fig. 24 Measured drag coefficient, $C_D = D/q_\infty d$, for a circular cylinder as a function of Reynolds number, $Re = V_\infty d/\nu$.

Flow visualization techniques enable identification of five flow configurations, shown schematically in Fig. 25, each characteristic of a specific range of Reynolds numbers.

1. The "Stokes range" extends up to a Reynolds number of about 5. The flow (Fig. 25*a*) resembles closely that of an inviscid fluid, in that no separation is evident. The drag coefficient varies approximately inversely with the Reynolds number.

2. The "symmetrical wake range," extending from the Stokes range to about $Re = 10$, is characterized by a "captive" vortex pair in the wake as shown in Fig. 25*b*. The boundary layer is laminar forward of the flow separation point, which is near $90°$.

3. The breakup of the symmetrical wake of (2) begins at around $Re = 10$, when a dissymmetry in the flow or a surface irregularity causes one of the vortices to be dislodged and carried downstream with the external flow. The resulting flow is unsymmetrical and a short time later the second vortex is shed while a new vortex is being formed to replace the one shed first.

4. Thus the "vortex street range" (first analyzed by von Kármán), depicted in Fig. 25*d*, is generated; it extends over a Reynolds number range from around 50 to

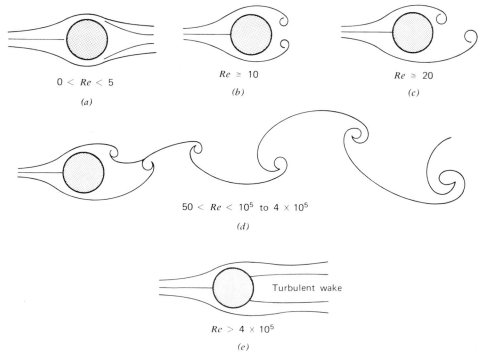

$0 < Re < 5$

(a)

$Re \gtrsim 10$

(b)

$Re \gtrsim 20$

(c)

$50 < Re < 10^5$ to 4×10^5

(d)

Turbulent wake

$Re > 4 \times 10^5$

(e)

Fig. 25 Schematic streamlines of the flow around a circular cylinder in various Reynolds number ranges.

4×10^5. The wake is characterized by a *Strouhal* number, nd/V_∞, where n is the shedding frequency, of about 0.21; the details of the flow have been studied extensively by many researchers, among them Kovasznay (1940) and Roshko (1954). The boundary layer is laminar until the flow separates at about $82°$ from the forward stagnation point.

5. The vortex street range ends at the Reynolds number at which stream turbulence, or surface roughness or heating, are intense enough to trigger transition to a turbulent boundary layer upstream of the $82°$ station, that is, forward of the laminar separation point; this condition occurs at 2 to 4×10^5. The turbulent boundary layer drives the separation point rearward to over $100°$; the wake decreases in width and the resulting decrease in form drag causes the drag coefficient to drop to around a third of its former value. Thus, the tests of Fig. 24 show that the higher the turbulence the lower the Reynolds number at which the sharp drop in C_D occurs. After the drop, C_D rises slowly with increasing Reynolds number, as shown in Fig. 24.

Spheres exhibit flow phenomena very similar to those about cylinder except that, instead of a vortex street the flow exhibits, a succession of *spiral vortices* of opposite circulation over a Reynolds number range comparable to that of the von Kármán vortex street behind a circular cylinder. The C_D versus Re curve is shown in Fig. 26

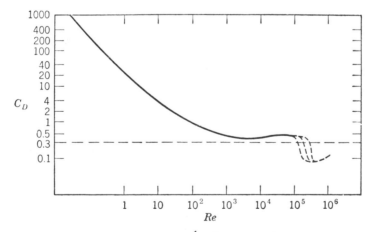

Fig. 26 Drag coefficient of spheres, $C_D = D/\frac{1}{4}\pi d^2 q_\infty$ as a function of Reynolds number $Re = V_\infty d/\nu$.

where the location of the rapid drag decrease is a function of the same variables as for the circular cylinder flow. The Reynolds number shortly after the break in the drag coefficient curve begins, that is, at $C_D = 0.3$ is defined by Dryden and Kuethe (see Goldstein, 1938, p. 500) as the "critical Reynolds number" of the sphere. The pressure distributions for Reynolds numbers below 1.6×10^5 and above 4.3×10^5 are shown in Fig. 27. In both instances, the flow separates shortly after the ad-

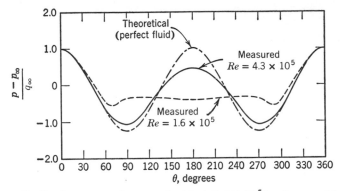

Fig. 27 Pressure distribution over a sphere at subcritical (1.6×10^5) and supercritical (4.3×10^5) Reynolds number compared with inviscid theory.

verse pressure begins, about $76°$ for the "subcritical" ($Re = 1.6 \times 10^5$) and about $110°$ for the "supercritical" condition.

The von Kármán vortex street is itself unstable as is shown in Fig. 28 behind a flat plate with a laminar boundary layer. The smooth laminar flow first forms the

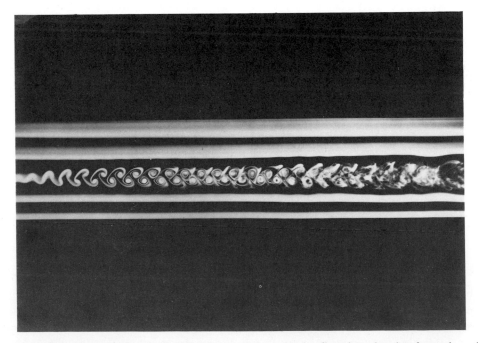

Fig. 28 Photograph of smoke filaments in the wake behind a flat plate showing formation of the vortex street and its dissolution. (Courtesy Prof. F. N. M. Brown, the University of Notre Dame)

vortex street which itself gradually loses its regularity until it finally almost literally explodes into a turbulent wake. This explosion phenomenon is analogous to that for transition the boundary layers treated earlier.

17.12   Concluding Remarks

The instabilities described in this section and the resulting changes in flow characteristics are only a few of those that occur in technology and in nature. They are of course inherent in the differential equations (the Navier-Stokes equations) that govern the flow of the fluid. The equations are derived in detail in Appendix B for compressible viscous fluids, but their peculiar nature can be discussed in terms of the *dimensionless* form of the equations in one of its simplest forms, that of the incompressible laminar boundary layer, Eq. 15.7. That form is

$$u' \frac{\partial u'}{\partial x'} + v' \frac{\partial u'}{\partial y'} = -\frac{\partial p'}{\partial x'} + \frac{1}{Re} \frac{\partial^2 u'}{\partial y'^2} \qquad (6)$$

where the lengths are nondimensionalized with a characteristic length L, the velocities with V_∞ and $p' = p/\rho \, V_\infty^2$.

As was shown in Section 14.6, the Reynolds number, $Re = V_\infty L/\nu$, is the single parameter in the equation. The results in the present chapter indicate that most, probably all, flow phenomena change their character completely above a critical Reynolds number; the instabilities are triggered when the coefficient of the highest order term, $\partial^2 u'/\partial y'^2$, in Eq. 6 falls below a critical value. As long as $1/Re$ is high the flow is orderly, in other words laminar and predictable, but when the velocity or characteristic length increases or the kinematic viscosity decreases sufficiently the transition to a turbulent, seemingly chaotic, state follows.

Equations of this type are termed *boundary layer type equations* and in general exhibit this spectacular behavior; furthermore, the higher the critical Reynolds number (lower $1/Re$ in Eq. 6) the more explosive the transition is likely to be when the critical value is exceeded.

The introduction of the many other fluid properties (compressibility, low densities, high temperatures, dissociation and ionization with resulting electrical conductivity, etc.) all modify the critical Reynolds number and add numerous new flow configurations and new dimensionless parameters, many of them with critical values of their own.

However, the "safe" ranges in which their variations have negligible effect (when they can be identified) make it possible to simplify and solve the equations for many applications.

CHAPTER 18

Turbulent Shear Flows

18.1 Introduction

Turbulent shear flows are of four types: flow through tubes and channels, boundary layers, jets, and wakes. These are categorized according to the characters of their boundaries, as indicated in Fig. 1. The tube or channel flow is wholly confined and mean velocities as well as turbulent fluctuations must vanish at the walls; as they must at the wall for the turbulent boundary layer. The irregular "free" boundaries of the outer edge of the boundary layer are sharply defined, but their shape changes with time, as do those for the wakes and jets in (c) and (d). Typical wake boundaries at a given instant are shown in Fig. 16.7 which is a shadowgraph of the wake behind a projectile moving through still air at Mach 2. The boundaries between the turbulent and the surrounding irrotational field are only slightly less sharp in an *incompressible* turbulent flow; in either case, if the body is held stationary in a moving stream, a measuring instrument such as a hot-wire is placed near the edge of the wake it will register high frequency fluctuations when a turbulent "bulge" passes and relatively steady flow during the remainder of the time.

In the chapters on laminar flow, we were able to solve some problems and to indicate in a straightforward manner a variety of solutions. When the flow becomes turbulent, however, as was pointed out in Chapter 17, the properties are time dependent and for a given set of boundary conditions, such as for boundary layer flow ($u = v = 0$ at $y = 0$, $u \to u_e$ as $y \to \infty$), there are an infinite number of solutions of the equations of motion. On the other hand, if the velocity components are averaged in the governing equations to treat the steady mean properties of a turbulent flow, new variables are introduced but no new equations are forthcoming. The hypotheses used for the solution of practical problems are therefore based on empirically determined factors.

Cebeci and Smith (1974) estimated that with computers currently available it would require thousands of years to solve the unsteady Navier-Stokes equations for the two-dimensional turbulent boundary layer. Such a solution would require a network of panel elements, each of them smaller than the smallest scale turbulence element, that is, the smallest value of $u\Delta t$, where Δt is the shortest duration of a deviation of the instantaneous velocity from the mean value.

The systematization of our knowledge of turbulent flow is therefore probably the most impressive example, at least outside of the "life sciences," of establishing some order and even solving many practical problems in a field in which the problems are not even "formulated," that is, the known equations, even for the *mean*

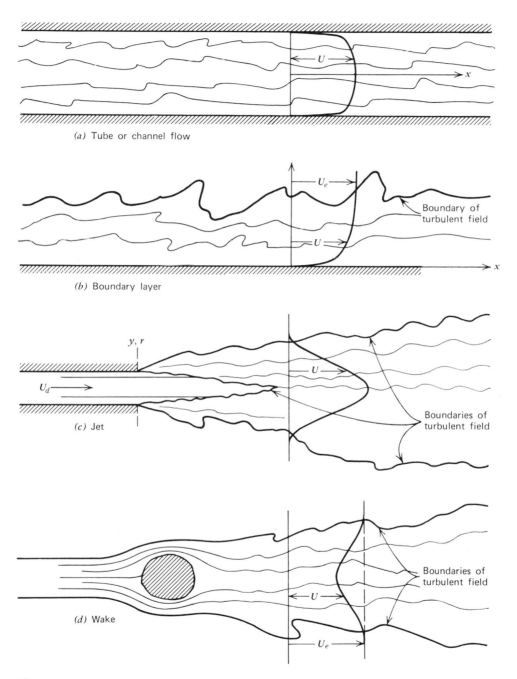

(a) Tube or channel flow

(b) Boundary layer

(c) Jet

(d) Wake

Fig. 1 Schematic instantaneous streamlines, mean velocity distributions and field boundaries of turbulent shear flows: (a) fully developed flow through tube or channel, (b) boundary layer, (c) jet, and (d) wake. *Note:* turbulent field boundaries are *not* streamlines.

flow, are fewer than the number of known variables. The state of our knowledge of turbulent flows is thus a synthesis of brilliant deductions from many carefully conducted experiments. Bradshaw (1975) gives a more comprehensive introduction to the physical aspects of the subject than can be included here.

In the previous chapter, we began the description of turbulence, in connection with the treatment of instability of laminar flows and transition. In the current chapter the description is continued and applied briefly to the turbulent shear flows shown in Fig. 1. The treatments depend to a large extent on "similarity analyses" for the form of the solutions and on experiment for numerical factors. Effects of roughness, pressure gradient, and compressibility are described. The analogy between skin friction and heat transfer is delineated.

18.2 Quantitative Description of Turbulence

The instantaneous velocity vector in a turbulent field will differ from the mean velocity vector in both magnitude and direction. If the x axis is taken in the direction of the mean velocity U, the velocity components at any particular instant are designated by $U + u_1$, v_1, and w_1, where u_1, v_1, and w_1 are, respectively, x, y, and z components of the instantaneous deviation of the velocity from its mean value. The values of these fluctuation components will vary in a random fashion with time about a zero value in the same way as the speed fluctuates about its mean value in Fig. 17.2. Hence u_1, v_1, and w_1 at a given point are functions of the time and are characterized by zero mean values; that is, the integral of the area between the trace in Fig. 17.2 and the line representing the mean value is zero. Mathematically,

$$\frac{1}{T} \int_0^T u_1 \, dt = \frac{1}{T} \int_0^T v_1 \, dt = \frac{1}{T} \int_0^T w_1 \, dt = 0 \tag{1}$$

where T is a time that is very long in comparison with the duration of any of the individual excursions from the mean value. In order to simplify the notation, a bar placed over a quantity signifies the mean value of the quantity with respect to the time, and Eq. 1 may be written $\overline{u_1} = \overline{v_1} = \overline{w_1} = 0$.

Theoretical analyses show that a significant quantity in a turbulent field is the root mean square value of the fluctuations defined in Eq. 17.1. In terms of the above notation,

$$\sigma = \frac{100}{U} \sqrt{\frac{\overline{u_1^2} + \overline{v_1^2} + \overline{w_1^2}}{3}} \tag{2}$$

In the main stream of a wind tunnel (outside of the boundary layer) the quantities $\overline{u_1^2}$, $\overline{v_1^2}$, and $\overline{w_1^2}$ are nearly equal, and so the definition generally employed for the turbulence number is

$$\sigma = \frac{100\sqrt{\overline{u_1^2}}}{U} = \frac{100}{U}\sqrt{\frac{1}{T}\int_0^T u_1^2\, dt} \tag{3}$$

In Section 17.5, the experimental results shown in Fig. 17.15 show that turbulence in the free stream, depending on its magnitude, on the Reynolds number, and on the frequency and intensity of the sound level, triggers transition from a laminar to a turbulent boundary layer. Fortunately, once the turbulent layer is established its properties are practically unaffected by the sound level; sound generation by turbulence is treated briefly in Section 19.6.

The above description of turbulence characterizes it as a *statistical phenomenon*, which means that the significant properties of a turbulent field are expressed in terms of the mean square value of the fluctuations rather than as instantaneous values.

When turbulent fluctuations are generated in a flow, as for instance by the grid in Fig. 17.1, some of the energy of the flow is transformed into turbulent energy. This energy loss by the main flow would be evidenced by a drop in stagnation pressure through the grid; in the water flow of Fig. 17.1 a drop in the water level would occur through the grid. Then the energy lost by the main stream must equal the mean kinetic energy of the turbulent fluctuations. This amount can be calculated by considering the mean kinetic energy of the fluctuations per unit volume, given by

$$\overline{\text{K.E.}} = \tfrac{1}{2}\rho\overline{[\{(U+u_1)^2 + (V+v_1)^2 + (W+w_1)^2\} - \{U^2 + V^2 + W^2\}]}$$

From the integral definition of mean values above it is clear that

$$\overline{a+b} = \overline{a} + \overline{b}$$

$$\overline{Uu_1} = U\overline{u_1} = 0, \text{ etc.}$$

Using these relations we get

$$\overline{\text{K.E.}} = \tfrac{1}{2}\rho(\overline{u_1^2} + \overline{v_1^2} + \overline{w_1^2})$$

Comparison with Eq. 2 shows that the square of the turbulence number, σ, is proportional to the ratio between the kinetic energy of the turbulence and that of the main stream.

As the disturbances caused by the grid are carried downstream the turbulent fluctuations gradually dissipate and their energy is transformed into heat, that is, into molecular agitation. This process, by which energy is drawn from the main stream and converted into heat, must be associated with an increase in entropy of the flow system (see Chapter 8).

We have so far encountered three aerodynamic phenomena in which the gas suffers an increase in entropy. They are, shock waves, the laminar boundary layer, and turbulence. In the laminar boundary layer and in the shock wave the energy of the main flow is transformed directly into random energy, or heat. In turbulence, however, energy is first drawn from the main flow into disturbances or eddies. The viscous stresses occurring within the eddies cause them to decay, and their energy is eventually transformed completely into heat. This process is responsible for the major portion of the entropy increase in the turbulent boundary layer.

18.3   The Turbulent Shearing Stress

Just as we identified the shearing stress in a laminar boundary layer (Sections 1.7 and 14.4) in terms of the momentum transferred by the random molecular motions, we shall now describe turbulent shearing stress in terms of the momentum transferred by the random "churning" motions in a turbulent flow. *The shearing stress is identified as the rate of cross-stream transfer of downstream momentum per unit area.* Its magnitude in a laminar flow in the x direction is $\tau_l = \mu \, du/dy$, where μ is the "molecular" viscosity coefficient; in a turbulent flow the main contribution to the shearing stress may be written

$$\tau_t = \rho \epsilon \, dU/dy \tag{4}$$

where ϵ, termed the *eddy viscosity* coefficient, is the turbulent analog for the molecular kinematic viscosity $\nu = \mu/\rho$, and dU/dy is the gradient of the mean velocity at a point. However, whereas ν is a property of the fluid, that is, proportional to the product of the mean free path and the average molecular speed, ϵ is proportional to the product of a speed, and a length, both associated, instead, with the *turbulence field*. The question of identifying the significant velocity and length is considered in the next section.

In the meantime, we derive an expression for the turbulent shearing stress in much the same way as was done for a laminar flow in Chapter 1. The transfer of downstream momentum takes place across the plane AB in Fig. 2 as a result of the exchange of momentum by the turbulent motions. The fluid elements crossing the plane AB from above carry, on the average, high momentum fluid to a low momentum layer, while those crossing from below carry low momentum fluid to a high momentum layer.

The shearing stress, the rate of transfer of x momentum across unit area of AB, is evaluated as follows: in Fig. 2, the vectors indicate that, at the instant considered, an element of fluid with x velocity equal to the mean value at $y + \Delta y$ crosses plane AB with vertical velocity $-v_1$; the rate of mass transfer across unit area of AB from above at that instant is therefore $-\rho v_1$. The x velocity component of the fluid ele-

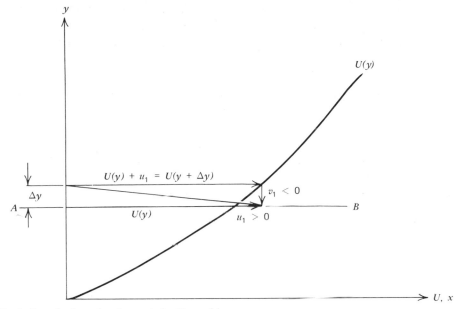

Fig. 2 Terminology for determining Reynolds stress.

ment at that instant is $U + u_1$ where U is the mean velocity at AB and u_1 is the x component of the instantaneous departure from U. The magnitude of the instantaneous rate of transfer of x momentum across AB will be the product of the rate of mass transfer per unit area $(-\rho v_1)$ and its x velocity increment u_1, that is, $-\rho u_1 v_1$. For an element crossing AB from *below* $\rho v_1 > 0$ and $u_1 < 0$ and the rate of momentum transfer is again $-\rho u_1 v_1$. The **turbulent shearing stress, the mean rate of momentum transfer per unit area,** is

$$\tau_t = -\rho \overline{u_1 v_1} \tag{5}$$

where the bar, as in the previous section designates the mean value over a relatively long time.

As indicated previously, the turbulent fluctuations are random in magnitude and direction, and it is therefore to be expected that a fluid element moving upward, for instance, will not *invariably* carry with it a *deficit* in momentum compared with that above plane AB. In other words, at any given instant $-\rho u_1 v_1$ may be positive or negative; its value will, however, be positive more often than negative, so that, if the mean value is taken over a long time, $-\rho \overline{u_1 v_1} > 0$.

The total shearing stress acting at any given point will be the sum of the mean laminar stress and the turbulent stress. Thus,

$$\tau = \tau_l + \tau_t = \mu \frac{\partial U}{\partial y} - \rho \overline{u_1 v_1} \tag{6}$$

Since the mean and turbulent velocity components must vanish at the wall, Eq. 6 reduces to $\tau_w = \mu(\partial U/\partial y)_w$ at $y = 0$. Except for very small values of y, however, the *turbulent* mixing is responsible for nearly all of the momentum transfer. Therefore, $\tau = -\rho\overline{u_1 v_1}$ is a good approximation to Eq. 6 *except* in the immediate vicinity of the wall.

Equation 6 is a rough approximation of the shearing stress in a mean two-dimensional shear flow. Even this approximate form however avails us nothing unless we can evaluate $-\rho\overline{u_1 v_1}$, generally termed the "Reynolds shear stress," or, simply, the **Reynolds stress**, in terms of the distributions of the *mean* properties of the flow. In the next section, we average the boundary layer equation for incompressible turbulent flow to verify Eq. 6 formally and we discuss hypotheses for obtaining expressions to derive approximate solutions to practical problems.

18.4 The Equations for Turbulent Shear Flows

We can derive the approximate form of the equations governing *two-dimensional* turbulent shear flows, that is, flow through tubes, in boundary layers, wakes, and jets by substituting fluctuating quantities for u, v, and p in the boundary layer equation derived for laminar flow in Chapter 14. As in Section 18.2, we write

$$u = U(x, y) + u_1(x, y, t); \qquad v = V(x, y) + v_1(x, y, t)$$
$$p = P(x) + p_1(x, y, t) \tag{7}$$

Thus U and V are mean values over time and the fluctuations are small, that is, $u_1 \ll U$, $v_1 \ll V$, $p_1 \ll P$. In reality, these fluctuations are functions of z as well, and w_1 exists as well as u_1 and v_1; also, near the wall, u_1 and v_1 reach values as high as $0.1U$. Nevertheless, the analysis that follows still gives some valid physical insights into the turbulent flow phenomena of importance in engineering; furthermore they act as a guide for experiments and a framework for dimensional analysis of turbulent flows.

After substituting Eqs. 7 in Eqs. 14.3 and 14.5 these equations become

$$\rho\left\{\frac{\partial}{\partial t}(U + u_1) + (U + u_1)\frac{\partial}{\partial x}(U + u_1) + (V + v_1)\frac{\partial}{\partial y}(U + u_1)\right\}$$

$$= -\frac{\partial}{\partial x}(P + p_1) + \frac{\partial}{\partial y}\left\{\mu\frac{\partial}{\partial y}(U + u_1)\right\}$$

$$\frac{\partial}{\partial x}(U + u_1) + \frac{\partial}{\partial y}(V + v_1) = 0$$

We are interested in the *mean* flow over time so we take the mean value of each term and the above equations become

$$\rho\left\{\frac{\partial U}{\partial t} + \overline{\frac{\partial u_1}{\partial t}} + U\frac{\partial U}{\partial x} + \overline{\frac{\partial}{\partial x}\left(\frac{u_1^2}{2}\right)} + \overline{\frac{\partial}{\partial x}(Uu_1)} + V\frac{\partial U}{\partial y} + \overline{V\frac{\partial u_1}{\partial y}} + \overline{v_1\frac{\partial U}{\partial y}}\right.$$

$$\left.+ \overline{v_1\frac{\partial u_1}{\partial y}}\right\} = -\frac{dP}{dx} - \overline{\frac{\partial p_1}{\partial x}} + \frac{\partial}{\partial y}\left(\mu\frac{\partial U}{\partial y}\right) + \overline{\frac{\partial}{\partial y}\left(\mu\frac{\partial u_1}{\partial y}\right)}$$

$$\frac{\partial U}{\partial x} + \frac{\partial V}{\partial y} + \overline{\frac{\partial u_1}{\partial x}} + \overline{\frac{\partial v_1}{\partial y}} = 0$$

We apply the rule expressed in Eq. 1 that the mean value of any term linear in a fluctuation component vanishes, and assume that the mean values of their derivatives vanish as well. Then,

$$\rho\left(U\frac{\partial U}{\partial x} + V\frac{\partial U}{\partial y}\right) = -\frac{dP}{dx} + \frac{\partial}{\partial y}\left(\mu\frac{\partial U}{\partial y} - \rho\overline{u_1 v_1}\right) - \frac{\partial}{\partial x}\left(\tfrac{1}{2}\rho\overline{u_1^2}\right)$$

$$\frac{\partial U}{\partial x} + \frac{\partial V}{\partial y} = 0$$

The last term of the first equation represents a turbulent *normal* stress on an element and if this stress is neglected in comparison with the preceding term, which is the *shearing* stress, the equation become

$$\rho\left(U\frac{\partial U}{\partial x} + V\frac{\partial U}{\partial y}\right) = -\frac{dP}{dx} + \frac{\partial}{\partial y}\left(\mu\frac{\partial U}{\partial y} - \rho\overline{u_1 v_1}\right) \tag{8}$$

$$\frac{\partial U}{\partial x} + \frac{\partial V}{\partial y} = 0 \tag{9}$$

We see that the term in parentheses on the right side of Eq. 8 represents the total shearing stress given in Eq. 6 and the meanings of the various terms are the same as those for the laminar equations, Eqs. 14.3 and 14.5.

The boundary conditions for instance for a boundary layer flow are:

$$\text{at } y = 0, \quad U = V = \overline{u_1 v_1} = 0$$

$$\text{at } y = y_e, \quad U = U_e \tag{10}$$

If we compare these with the corresponding expressions for laminar flow (Eqs. 15.8), we see that one boundary condition $\overline{u_1 v_1} = 0$ at $y = 0$ has been added. Therefore, mathematically and physically, the turbulent flow problem is not "formulated," since we have one more variable, $\overline{u_1 v_1}$, than we have equations. Only when the "closure" condition is formulated, that is, when a relation expressing $\overline{u_1 v_1}$ in terms of the mean velocities U and V, are straightforward solutions possible. Therefore, current solutions of the many engineering problems in turbulent flow are based on assumptions for $\overline{u_1 v_1}$ based on many experimental studies. Some of these are given in the next section.

As was mentioned above, only $\overline{u_1 v_1}$, the most important of the terms involving velocity fluctuations, is included in Eqs. 8 and 10. Neglected terms include $\overline{u_1^2}$, $\overline{v_1^2}$, $\overline{w_1^2}$, $\overline{v_1 w_1}$, and $\overline{u_1 w_1}$ and their derivatives. Thus, for a complete solution, we would need new equations relating these with each other and with the mean flow. Fortunately, for many of the problems that concern us here, $-\rho\overline{u_1 v_1}$ is by far the largest of the terms expressing the turbulent stresses.

18.5 Hypotheses for Reynolds Stress, $-\rho\overline{u_1 v_1}$

We are concerned here with hypotheses connecting Eq. 5 ($\tau_t = -\rho\overline{u_1 v_1}$) with the mean properties of a turbulent shear flow; the mean flow is in the x direction, as in Fig. 2, and its gradient is in the y direction.

In Fig. 2, the momentum transferred across the plane AB is proportional to $u_1 = (\Delta U/\Delta y)\Delta y$ or, in the limit, to $(dU/dy)\,dy$. However, as we have shown, unless the mean value $-\rho\overline{u_1 v_1} > 0$, no *net* momentum will be transferred from the high to the low momentum layers, and vice versa; therefore, on the average, $-v_1$ and u_1 must be *positively correlated*; that is, there must be a relation between them such that their average product is positive. This requirement is satisfied most simply by assuming that $-v_1$ is proportional to u_1, that is, to $\partial U/\partial y$. Accordingly, Prandtl wrote

$$-\rho\overline{u_1 v_1} = \rho l^2 \left(\frac{\partial U}{\partial y}\right)^2 \tag{11}$$

where l^2 is a proportionality factor; l has the dimensions of a length and was designated by Prandtl as the **mixing length**. It follows that Eq. 5 and the parentheses on the right in Eq. 8 may be written

$$\tau = \tau_l + \tau_t = \left(\mu + \rho l^2 \left|\frac{\partial U}{\partial y}\right|\right)\frac{\partial U}{\partial y} \tag{12}$$

where the absolute value of $\partial U/\partial y$ within the parentheses is specified because both the laminar and turbulent contributions to the shearing stress must have the same sign as the velocity gradient.

The second hypothesis for the Reynolds stress provides a more obvious analogy with that for laminar flow, where, for shear flows for which the velocity varies much more rapidly in the y (cross-stream) than in the x (downstream) direction,

$$\tau_l = \mu\,\frac{\partial u}{\partial y} = \rho\nu\,\frac{\partial U}{\partial y} \tag{13}$$

where $\nu = \mu/\rho$ is the kinematic viscosity. By analogy, we write for the Reynolds stress

$$\tau_t = -\rho\overline{u_1 v_1} = \rho\epsilon\,\frac{\partial U}{\partial y} \tag{14}$$

where ϵ is termed the eddy viscosity. It, like ν, has the dimensions of a velocity times a length and, from Eq. 12, in terms of the mixing length, $\epsilon = l^2|\partial U/\partial y|$. *Physically $\rho\epsilon$ is the mean rate of transfer of x momentum across unit area of the plane AB of Fig. 2 by turbulent fluctuations.* The expression for the *total* shearing stress corresponding to Eq. 12 is then

$$\tau = \tau_l + \tau_t = (\mu + \rho\epsilon)\frac{\partial U}{\partial y} \tag{15}$$

The discussion of these two hypotheses (Eqs. 11 and 14) will be continued in the next section along with the treatment of the laminar sublayer.

The dimensionless variables are brought to light by writing Eq. 8 in terms of the two hypotheses and reference constants V_∞ and L. With

$$U = \tilde{U}V_\infty; \quad V = \tilde{V}V_\infty; \quad P = \tilde{P}\rho V_\infty^2; \quad x = \tilde{x}L; \quad y = \tilde{y}L \tag{16}$$

Equations 12 and 15 are substituted in Eq. 8 to give the dimensionless equations:

$$\tilde{U}\frac{\partial\tilde{U}}{\partial\tilde{x}} + \tilde{V}\frac{\partial\tilde{U}}{\partial\tilde{y}} = -\frac{d\tilde{P}}{d\tilde{x}} + \frac{1}{Re}\frac{\partial}{\partial\tilde{y}}\left(1 + \frac{l^2|\partial U/\partial y|}{\nu}\right)\frac{\partial\tilde{U}}{\partial\tilde{y}} \tag{17}$$

and

$$\tilde{U}\frac{\partial\tilde{U}}{\partial\tilde{x}} + \tilde{V}\frac{\partial\tilde{U}}{\partial\tilde{y}} = -\frac{d\tilde{P}}{d\tilde{x}} + \frac{1}{Re}\frac{\partial}{\partial\tilde{y}}\left(1 + \frac{\epsilon}{\nu}\right)\frac{\partial\tilde{U}}{\partial\tilde{y}} \tag{18}$$

where $Re = V_\infty L/\nu$, V_∞ is a characteristic velocity, such as the free stream value, and L is a characteristic length, such as the length of a body in the flow or the chord of an airfoil. In analyses of specific problems it is generally advantageous, particularly for computer calculations, to work with dimensionless equations such as these.

Equations 17 and 18 emphasize the essential difference between laminar and mean turbulent flows. Whereas the properties of the incompressible laminar flows of Chapter 14 were functions of the Reynolds number only, their turbulent counterparts, Eqs. 17 and 18, depend as well on the ratio between the laminar and "effective" turbulent viscosities. This ratio is in general a function of the space coordinates *and* of the Reynolds number.

One of the questions that arises relative to the Reynolds stress is: just how efficient is the turbulence as a "transfer mechanism?" In Eq. 11, for instance, Prandtl assumed implicitly that $-v_1$ is proportional to u_1 at every instant, that is, that they are perfectly correlated. We can then define a "correlation factor"

$$Q'_{xy} = \frac{\overline{u_1 v_1}}{\sqrt{\overline{u_1^2}}\sqrt{\overline{v_1^2}}}$$

If we assume that u_1 and v_1 are, for instance, sine functions in phase with each other, we find $Q'_{xy} = 1$. Actually, measurements by Schubauer and Klebanoff (1951) show $Q'_{xy} \cong 0.44$ over the major portion of the boundary layer.

A physically more significant parameter is written

$$Q_{xy} = \frac{-\rho \overline{u_1 v_1}}{\overline{K.E.}}$$

where $\overline{K.E.}$ is the total kinetic energy in the turbulent fluctuations. The measurements (see Hinze, 1959, p. 492) show that $Q_{xy} \cong 0.32$ throughout most of the boundary layer. This number indicates that only about 32 percent of the energy generated in a turbulent shear flow is effective as a momentum transfer mechanism.

18.6   The Laminar or Viscous Sublayer

Since the boundary conditions (Eqs. 10) require that all velocities vanish at the wall, the Reynolds stress $-\rho \overline{u_1 v_1} \to 0$ as $y \to 0$. Thus the shearing stress at the wall is laminar and in terms of mixing length l of Eq. 12 and eddy viscosity ϵ of Eq. 14, the wall conditions are expressed by

$$l \text{ and } \epsilon \to 0 \text{ at a solid boundary}$$

Therefore, for turbulent flow in tubes or boundary layers, there is a thin layer, called the "laminar sublayer" (sometimes called the viscous sublayer), designated by δ_l, within which the shearing stress is predominantly laminar, that is, at

$$0 < y < \delta_l, \quad l^2 \left| \frac{\partial U}{\partial y} \right| \text{ and } \epsilon \ll \nu \tag{19}$$

The experimental results described later show that, while the velocity gradient within the laminar sublayer determines the skin friction, its thickness, δ_l, is only of the order of 1 percent of the total boundary layer thickness, δ. A quantitative value in terms of ν and τ will be chosen empirically in Eq. 20.

Outside of the laminar sublayer, the turbulence that generates the Reynolds stress increases very rapidly with distance through a "buffer" layer, so that, for the greater part of the shear layer, the eddy viscosity ϵ is very much larger than ν; just how much larger can be determined from measurements. Clauser (1956) found that in a turbulent boundary layer outside of the laminar sublayer, a rough approximate value is $\epsilon = 0.017 U_e \delta^* \cong$ constant, where δ^* is the displacement thickness of the turbulent layer, *independent of pressure gradient* and *roughness*. The reader may verify later by means of Eqs. 34 and 35, for the turbulent boundary layer one meter behind the leading edges of a flat plate, with $U_e = 50$ m/sec, that, for air, $\delta = 1.8$ cm, $\delta^* = 2.2$ mm and, from the above formula, $\epsilon = 1.87 \times 10^{-3}$ m²/sec; this value is $1.87 \times 10^{-3}/1.44 \times 10^{-5}$ = about 130 times larger than the molecular or laminar kinematic viscosity! So the turbulent boundary layer may be likened superficially to a viscous molasses-like fluid flowing over a thin film of a watery fluid, the latter corresponding to the laminar sublayer.

We arrive at the thickness of the laminar sublayer through experiment and dimensional reasoning on the basis of the physical factors that must affect the flow very near the surface; these factors are τ_w, ρ and μ. We decrease these to two by writing

$$U_\tau = \sqrt{\frac{\tau_w}{\rho}} \quad \text{and} \quad \nu = \frac{\mu}{\rho}$$

The first of these is termed the "friction velocity," simply because it has the dimensions of a velocity. Experiments indicate that the laminar sublayer is represented approximately by

$$\delta_l = \frac{4\nu}{U_\tau} \tag{20}$$

Results of analysis of experiments, to be described in following sections, indicate that the thickness δ_l constitutes only about 1 percent of the tube radius or boundary layer thickness; the importance of this thin layer however derives from the fact that it is the device for transferring skin friction, heat and other properties of the main flow to the surface of a body.

We shall treat the turbulent shear flows in the following sections in the order in which they are represented in Figure 1. The mixing lengths and eddy viscosities are expressed as functions of y as determined by application of dimensional considerations to measurements in the various flows.

18.7   Fully Developed Flow in Tubes and Channels

Fully developed flow, as described for laminar flow in Section 15.2 is achieved at points far from the entrance of a tube. It is characterized by the fact that all mean velocity profiles are identical and hence that $V = \partial U/\partial x = 0$. Thus, the equation of motion, Eq. 8, for flow in a two-dimensional channel reduces to

$$\frac{dP}{dx} = \frac{d\tau}{dy} \tag{21}$$

Equation 21 becomes, after integration and application of the boundary condition $\tau = \tau_w$ at $y = 0$,

$$\tau = \tau_w + y\,\frac{dP}{dx} \tag{22}$$

and since from symmetry $\tau = 0$ at $y = b/2$, where b is the breadth of the channel,

$$\tau_w = -\tfrac{1}{2}b\,\frac{dP}{dx} \tag{23}$$

We shall now consider, instead of a two-dimensional channel, a tube of circular cross section, as was analyzed in Section 15.2. The equilibrium of the pressure and shearing forces on an element (Eq. 15.1) gives

$$2\pi r \tau = -\pi r^2 \frac{dP}{dx}$$

where $dP/dx = (p_2' - p_1)/l$. Then,

$$\tau = -\frac{1}{2} r \frac{dP}{dx} \tag{24}$$

with the origin at the center of the tube. For this case the skin-friction coefficient γ at the wall, as in Section 15.2, is given by

$$\gamma = \frac{\tau_w}{\frac{1}{2}\rho U_m^2} = \frac{a}{\rho U_m^2}\left(-\frac{dP}{dx}\right) \tag{25}$$

where U_m is the mean speed over the cross section and a is the radius of the tube. By means of this formula it is particularly easy to determine, for a fluid of a given density, the skin-friction coefficient experimentally; it is necessary only to measure the discharge through the tube to get U_m and to measure dP/dx by means of pressure taps in the wall.

It will be noted that Eqs. 21 through 25 are valid for either laminar or turbulent flow. Figure 3 shows schematically a comparison between fully developed laminar and turbulent flow in a tube as given by experiment for the same mean velocity over the cross section. In both flows the shearing-stress distribution is linear, but in laminar flow its magnitude is much less. The shearing stress is, by Eq. 12, the sum of the laminar (τ_l) and the Reynolds (τ_t) stresses. As was pointed out in Section 18.3, τ_l represents the entire stress at the wall, since the no-slip condition requires that $\overline{u_1 v_1}$ be zero at that point. However, $-\rho \overline{u_1 v_1}$ far overshadows the laminar stress everywhere *except* in the immediate vicinity of the wall.

The method given below for finding the velocity distribution is justified, not by its rigor, but rather by the fact that it provides a systematic framework for the

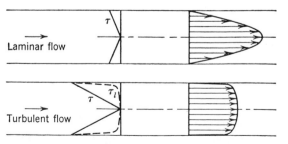

Fig. 3 Schematic representation of fully developed laminar and turbulent flow in a tube, for the same mean velocity.

analysis of experimental data. We first neglect the laminar shearing stress and, as a consequence, the velocity distribution cannot possibly hold in the immediate vicinity of the wall. Second, we set $\tau = \tau_w$ = constant. The consequence of the two assumptions is that any agreement between theory and experiment could be expected to be limited to a layer near, but not extending to, the wall, throughout which the shearing stress does not vary much from its value at the wall. The fact that the derived velocity distribution actually agrees with experiment over the entire tube, except in the immediate vicinity of the wall, is simply a fortunate circumstance.

The significant variables that must determine the mean velocity distribution are τ_w, ρ, μ, l, and y. However, experimental results indicate that, rather than consider τ_w as a separate variable, we should employ $\sqrt{\tau_w/\rho}$, which has the dimensions of a velocity and is called the *friction velocity* U_τ. Then the variables are U_τ, μ, ρ, l, and y. By the Π theorem (Appendix A), we can form only two dimensionless parameters from these five quantities. Accordingly, we may write

$$\frac{U}{U_\tau} = f\left(\frac{U_\tau y}{\nu}, \frac{l}{y}\right) \tag{26}$$

With $\tau = \tau_w$ we assume the simple relation l/y = constant = K. Then Eq. 11 becomes

$$U_\tau^2 = K^2 y^2 \left(\frac{dU}{dy}\right)^2 = \text{constant} \tag{27}$$

which integrates to give

$$U^+ = K^{-1} \ln y + \text{constant} \tag{28}$$

where $U^+ = U/U_\tau$. The integration constant and the functional dependence of Eq. 26 indicates that Eq. 28 may be expressed in the form

$$U^+ = A + \kappa^{-1} \ln y^+ \tag{29}$$

where $y^+ = yU_\tau/\nu$. The symbol κ was introduced by von Kármán (see Schlichting, 1968), who showed that κ has a nearly constant value of about 0.4 for all of the flows considered. The experimental results (see Schlichting, 1968), shown in Fig. 4,* are employed to evaluate A and κ. The straight line has the equation

$$U^+ = 5.5 + 5.75 \log_{10} y^+ \tag{30}$$

The agreement with experiment is good from y^+ = 30 to very near the center of the channel.

An example will illustrate how small y is for y^+ = 30 for fully developed pipe flow at a representative Reynolds number. Figure 5 shows faired curves of the measured pressure drop coefficient, $\gamma = \tau_w/\frac{1}{2}\rho U_m^2$, where U_m is the average value of the

---

*Nikuradse's measurements were made in a circular tube. More recent measurements by Laufer (1950) verify the general form of Eq. 29 for flow through a channel with height: width ratio of 12 : 1; the constants are, however, slightly different.

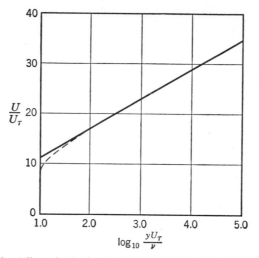

Fig. 4 Measurements by Nikuradse in fully developed turbulent flow in a tube are represented by the dashed curve for small values of $\log_{10}(yU_\tau/\nu)$ and by the straight line over the remainder of the range shown. The measurements cover the range of Reynolds numbers from 4×10^3 to 3.24×10^6. Courtesy of McGraw-Hill.

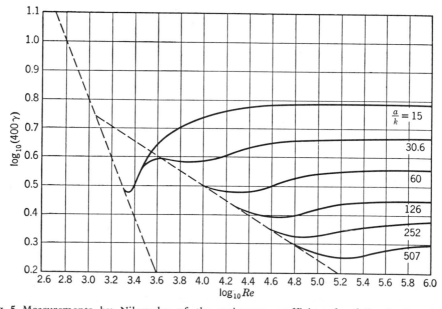

Fig. 5 Measurements by Nikuradse of the resistance coefficient for fully developed flow through a tube of radius a with various sizes of the roughness elements a/k are represented by the solid curves. The dashed lines represent theory for laminar and for turbulent flow in a smooth tube. Courtesy of McGraw-Hill.

mean velocity over the channel cross section, as a function of $\log_{10} Re$ (the effect of surface roughness represented by various values of a/k will be described below). We take $\log_{10} (400\gamma) = 0.5$ ($\gamma = 0.008$) at $Re = U_m a/\nu = 10^4$, so that if we take the mean velocity $U_m = 30$ m/sec the pipe diameter ($2a$) is 1 cm. Then $U_\tau = \sqrt{\frac{1}{2}\gamma U_m^2} = 1.9$ m/sec; for air these values give $y = 0.23$ mm for $y^+ = 30$. We see, therefore, that at $Re = 10^4$, Eq. 30 is a good approximation to the measurements of Fig. 4 over about 95 percent of the channel radius.

To determine the resistance coefficient, we observe from Eq. 29

$$\frac{U_c}{U_\tau} = A + \left(\frac{1}{\kappa}\right) \ln\left(\frac{U_\tau a}{\nu}\right)$$

where U_c is the velocity at the center $y = a$. Experiments over a wide range of Reynolds numbers indicate that

$$\frac{(U_c - U_m)}{U_\tau} = \text{constant}$$

where U_m is the mean velocity over the cross section of the tube. Hence, we may write

$$\frac{U_m}{U_\tau} = A' + B' \ln\left(\frac{aU_\tau}{\nu}\right)$$

or, if we write

$$\gamma = \frac{\tau_w}{\frac{1}{2}\rho U_m^2} = 2\frac{U_\tau^2}{U_m^2}; \qquad Re = \frac{2aU_m}{\nu}$$

we have

$$\sqrt{2}\gamma^{-1/2} = A_1 + B' \ln Re\gamma^{1/2}$$

The constants A_1 and B' are evaluated from experiments, giving

$$\frac{U_m}{U_\tau} = 0.29 + 5.66 \log_{10}\frac{2aU_\tau}{\nu}$$

$$\gamma^{-1/2} = -0.40 + 4.00 \log_{10} Re\gamma^{1/2}$$

(31)

for the resistance of smooth pipes.

The theory leading up to Eqs. 31 is, to be sure, rough and is qualitative to the extent that two empirically determined constants are required. However,

$$B' = \kappa^{-1} \ln 10$$

which gives a value for κ of 0.408, as compared with a value of 0.417 derived from the expression for the velocity distribution Eq. 30.

Figure 5 compares calculated pressure drop coefficients for fully developed turbulent and laminar flows (Eqs. 31 and Eq. 15.6) through smooth tubes of circular cross section. Included are curves based on measurements with various degrees of surface roughness; a/k is the ratio of the tube radius to k, the height of uniform sand grains glued to the surface. The flow is laminar for Reynolds numbers below about 2200; then transition occurs and for each roughness the curves approach a constant value of the coefficient. For all but the two highest roughness values the separate curves follow the turbulent curve for some distance before they branch off to approach their final values. These curves are interpreted on the hypothesis that the surface roughness does not contribute to the pressure drop on flow resistance unless the roughness elements project beyond the laminar sublayer. As is pointed out in the previous section, this sublayer thickness, $\delta_l = 4\nu/U_\tau$, is, in practical cases, less than 1 percent of the tube radius; this figure is then a measure of the "maximum allowable roughness" for minimum pressure drop in fully developed turbulent flow through a tube. The constant value approached for each roughness signifies that the Reynolds number has been reached at which the pressure drop along the tube is proportional to the square of the mean velocity. Commercially available pipes generally have roughness value considerably greater than the above minimum ($0.01a$) but the curves indicate that the penalty in increased pressure drop is small except at relatively high Reynolds numbers.

A physically more meaningful measure of roughness is expressed by Schlichting (1968). The roughness of a tube flow is identified by the following ranges of $k^+ = U_\tau k/\nu$, where k is the roughness height,

$$0 < k^+ < 5 \quad \text{hydraulically smooth}$$
$$5 < k^+ < 70 \text{ transition regime}$$
$$k^+ > 70 \text{ completely rough}$$

In the hydraulically smooth regime the roughness has no effect on γ and in the completely rough regime γ is a function of k/a only. In this latter regime the pressure drop is proportional to the product ρU_m^2. The results are closely related to roughness effects in boundary layers, as is pointed out in Section 18.9.

18.8   Incompressible Flow in Turbulent Boundary Layers

The analytical representation of the flow in a turbulent boundary layer differs from that in a tube in two related important aspects.

1. The most significant difference stems from the circumstance that for the boundary layer a free surface exists at the outer edge of the layer. The irregularity of the boundary as well as the sharp boundary between the rotational turbulent flow within the layer and the irrotational (potential) external flow are indicated in Fig. 6. The corresponding phenomenon in wakes is shown in the photographs of

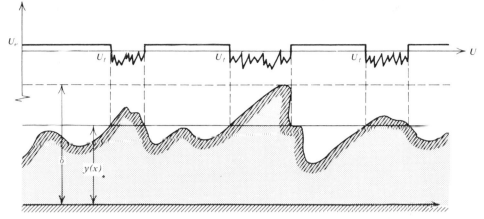

Fig. 6 Schematic instantaneous boundary of turbulent region (cross-hatched) in boundary layer on a surface. Instantaneous velocity distribution at station indicated is shown; mean velocity follows the log law (Eq. 36) within turbulent tongues, and is irrotational and approximately equal to U_e between the tongues.

Figs. 10.2 and 16.7. A vorticity or velocity measuring instrument located within the turbulent region not too far from the edge of the layer shows intermittent intervals of turbulent flow characterized by high frequency fluctuations (see Fig. 17.5) interspersed with smooth or slowly varying potential flow. To take account of this effect, not present in fully developed flow in tubes, an "intermittency function"

$$\gamma_K(y/\delta) = \frac{U_e - U}{U_e - U_t} \tag{32}$$

is defined, where U_t is the mean velocity during the turbulent intervals; it was measured first by Klebanoff (1955) for flow along a flat plate. He found that γ_K is unity for $0 < y/\delta \gtrsim 0.4$, that is, tongues of the external flow penetrate at irregular intervals as far as $y/\delta = 0.4$. Thus γ_K decreases continuously from 1 to zero in the range $0.4 < y/\delta < 1.0$. The tongues of potential flow gradually become entrained by molecular diffusion with the turbulence.

2. Schubauer (1954) noted (a) that within the tongues of potential flow the velocity is near that of the external flow and (b) that within the turbulent intervals the mean velocity was near that given by the logarithmic law of Eq. 30.

These circumstances and the search for a method of representing measured velocity profiles over a wide range of conditions led Sarnecki to formulate a new intermittency function, γ_S, shown in Fig. 7. While γ_S differs from Klebanoff's measured intermittency the two functions agree at the edges of the layer and both have roughly the same S shape, Sarnecki found that γ_S is very nearly a universal function, independent of Reynolds number and pressure gradient as discussed in the next section.

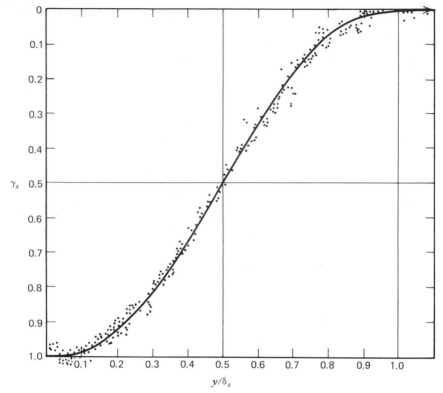

Fig. 7 "Intermittency correlation" found by Sarnecki (1967). Experimental points cover a wide range of conditions. Reproduced by courtesy of Great Britain Aeronautical Research Council.

Thompson (1967) used Sarnecki's function γ_S in a semiempirical representation of turbulent boundary layer profiles over a wide range of conditions.* He represented the velocity profiles by the equation

$$U(y) = \gamma_S U_t + (1 - \gamma_S) U_e \tag{33}$$

This equation expresses the velocity profile as a two parameter family of curves; in the first term U_t represents the effect of the wall by means of functional relationships $U_t^+ = f(y^+)$, the function being different in each of three "wall regions," as detailed below; the second parametric relationship is $\gamma_S(y/\delta)$ of Fig. 7, which measures the effect of the intermittent incursion of the external flow with velocity U_e. The three parts of the "wall region" are designated in Fig. 8; the functional relationships between U_t^+ and y^+ for each is given below.

*Coles (1968, see Cebeci and Smith, 1974) devised a method conceptually equivalent to that of Thompson; Coles identified the difference between the half-profile for channel flow with that for the boundary layer by evaluating a so-called "wake function" $h(y/\delta)$ which takes into account the intermittency of the outer flow.

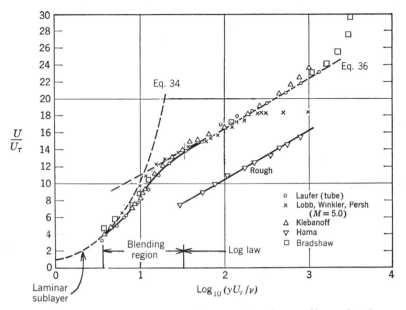

Fig. 8 Distribution of mean velocity near wall in turbulent flow. Effects of surface roughness and compressibility are indicated. The Lobb et al. data use values of ρ and μ at the wall. Measurements by Klebanoff and by Bradshaw were made in adverse pressure gradients.

1. The laminar sublayer where $\gamma_S = 0$ and $y^+ < 4$ (Eq. 20), the velocity distribution is assumed to be linear. Then, in this layer

$$U_t^+ = y^+ \tag{34}$$

2. The blending layer for $4 < y^+ < 30$ the empirical equation suggested by Galbraith and Head (1975) is approximately

$$U_t^+ = 4.2 - 5.7 \ln y^+ + 5.1(\ln y^+)^2 - 0.7(\ln y^+)^3 \tag{35}$$

where the coefficients have been rounded to one decimal place.
   3. The log-law region for $y^+ > 30$ in which

$$U_t^+ = 5.50 \log_{10} y^+ + 5.45 \tag{36}$$

Many analyses have been made leading to equations of the form of Eqs. 35 and 36. The coefficients vary slightly with the different analyses though all agree on the form of the relationships.
   The "mixing zone" described by the second term of Eq. 33, is the region in which the flow is comprised of irregular tongues of the turbulent and the external flow. As was mentioned above, Schubauer (1954) observed that the mean velocity is approximately that of the log law of Eq. 36, interspersed with regions where the

velocity is approximately U_e. Then, as y increases, this zone is characterized by increasing deviation from the log law.

In general, for incompressible flow past a smooth surface, the velocity plots of Fig. 8 are identical for $y < 0.15\ \delta$. This result is independent of the pressure gradient, since the Klebanoff and Bradshaw measurements were made in strong adverse pressure gradients. Thus the effect of the penetration of the tongues of external flow on the distributions is appreciable only in the outer 85 percent of the layer, as is indicated in the solution of Problem 18.11.1. The various curves in Fig. 8 will be discussed further in following sections.

It is clear from the above that the turbulent velocity profiles cannot be correlated throughout in terms of the single coordinate y^+; this feature is contrary to that of the laminar profile in which u/u_e can be described throughout in terms of the single parameter $y\sqrt{u_e/\nu x}$.

The application of Eqs. 33–36 and the function γ_S of Fig. 7, along with $U_\tau = U_e\sqrt{c_f/2}$ determine the velocity profile. These will be treated in the following sections.

18.9   Turbulent Boundary Layer on a Flat Plate

Some velocity distributions for flow along a flat plate are shown in Fig. 8 for smooth and *very* rough plates. The laminar sublayer, the blending layer, and the logarithimic region described, respectively, by Eqs. 34, 35, and 36 are designated. Near the end of the logarithmic region, where the intermittency term becomes appreciable the experimental points deviate from the logarithmic line to form the free-stream mixing region. In the tube flow, where no free-stream flow exists, that is, $\gamma_S = 1$ throughout, the points are seen to follow the logarithmic curve throughout.

The experimental points for the entire flat plate boundary layer profile, including those in Fig. 8 for incompressible flow, are plotted in Fig. 9, to define von Kármán's "velocity defect law." The law is expressed mathematically by

$$\frac{(U_e - U)}{U_\tau} = f\left(\frac{y}{\delta}\right) \tag{37}$$

where $f(y/\delta)$ is the same function for smooth and rough plates. This important feature of Fig. 9 is shown by the circumstance that experimental points of Fig. 8 are concentrated near $y/\delta = 0$ so that the intermittency phenomenon and the free stream govern the nondimensional velocity defect over the major portion of the layer, independent of surface roughness.

The various regimes for $k^+ = U_\tau k/\nu$ given in Section 18.7 are still a reasonable approximation for flow along a flat plate. As may be inferred from the velocity profiles of Fig. 12 and Section 18.11, since the velocities near the surface are lower in an adverse gradient, roughness makes a relatively smaller contribution to the drag than for a flat plate.

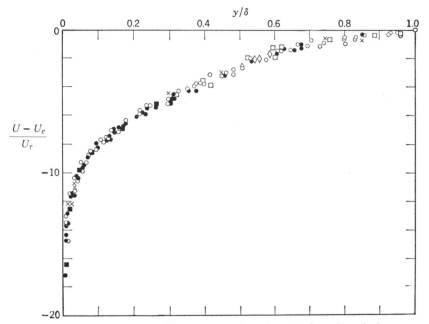

Fig. 9 Turbulent boundary layer profiles on smooth and rough flat plates in incompressible flow.

CALCULATION METHODS AND FORMULAS

By dimensional reasoning beginning with the velocity distribution in tubes von Kármán devised laws with empirical coefficients, respectively, for the local and average skin-friction coefficients on flat plates. These are

$$c_f^{-1/2} = 1.7 + 4.15 \log_{10} (Re_x c_f) \tag{38}$$

$$C_f^{-1/2} = 4.13 \log_{10} (Re_l C_f) \tag{39}$$

where l is the length of the plate, $c_f = \tau_w/q_e$, $C_f = (\int_0^l \tau_w \, dx)/lq_e$. Prandtl (see Durand, Div. 3, p. 153) derived another, easier to apply formula

$$C_f = 0.455(\log_{10} Re_l)^{-2.58} \tag{40}$$

The formulas for C_f give practically undistinguishable results. They are plotted in Fig. 10 along with that for a laminar boundary layer. The transition curve sketched in the figure represents the total skin-friction coefficient for the boundary layer on the flat plate of length l completely laminar up to $Re_l = 5 \times 10^5$. As the Reynolds number increases further, the transition point moves forward toward the leading edge and the skin-friction coefficient approaches asymptotically that for a completely turbulent boundary layer.

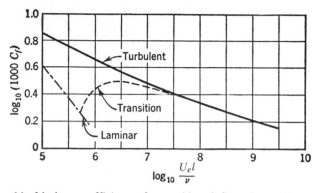

Fig. 10 Average skin-friction coefficient of one side of flat plate with completely laminar boundary layer, with completely turbulent boundary layer, and with transition occurring at a distance from the leading edge dependent on Reynolds number.

For purposes of calculation, Prandtl assumes that transition takes place at a point x and that in the turbulent layer behind that point the skin friction is the same as it would be if the layer were turbulent from the leading edge. Then the skin friction over the plate of length l is given by the total turbulent skin friction over the length l minus the correction term made up of the difference between the turbulent and laminar friction over the length x. The correction term is, accordingly, $\frac{1}{2}\rho U_e^2 x \cdot (C_{f_t} - C_{f_L})_{Re_x}$, where the coefficients of mean laminar and turbulent skin friction C_{f_L} and C_{f_t} are evaluated at Re_x. Then the coefficient of mean skin friction may be written

$$C_f = C_{f_t} - \frac{Re_x}{Re_l}(C_{f_t} - C_{f_L})_{Re_x}$$

or

$$C_f = C_{f_t} - \frac{A}{Re_l}$$

where $A = Re_x(C_{f_t} - C_{f_L})_{Re_x}$ is a function of Re_x. A transition curve for $A = 1800$ is shown in Fig. 11.

Convenient approximate formulas derived by Blasius from experimental results are

$$\frac{U}{U_e} = \left(\frac{y}{\delta}\right)^{1/7} \tag{41}$$

$$\delta = \frac{0.37x}{Re_x^{0.2}} \tag{42}$$

$$c_f = \frac{0.0592}{Re_x^{1/5}} = \frac{0.0256}{Re_\theta^{1/4}}$$

$$C_f = \frac{0.072}{Re_l^{1/5}}$$

(43)

for the range $5 \times 10^5 < Re_l < 10^7$. Use of Eqs. 42 and 43 assumes that the boundary layer is turbulent effectively from the leading edge; if the layer is laminar for an appreciable distance, x is measured from some point between the leading edge and the transition point.

18.10 Effects of Compressibility on Skin Friction

The first estimate of the effect of compressibility on the turbulent skin friction on a flat plate was made by von Kármán (1935). He assumed that Eq. 38 is valid for a compressible boundary layer provided that properties at the wall, ρ_w and μ_w are used in the calculation of C_f and Re.* He further assumed a Prandtl number $(c_p \mu/k)$ of unity and then calculated the ratio C_f/C_{fi}, where the subscript i refers to incompressible flow, as a function of the free-stream Mach number. The curve plotted in Fig. 12 was the result. Many other more involved calculations have been made, using some form of the mixing-length hypothesis of Section 18.5 and introducing compressibility by means of the turbulent form of the energy equation of Section 16.2. Those theoretical curves which bracket the available experimental results are shown in Fig. 11. The theoretical and experimental results in Fig. 11 refer to the insulated plate. Three different techniques were used to obtain the experimental results shown: (1) by analyzing measured velocity profiles by the von Kármán momentum integral taken up in Section 15.6; (2) by subtracting the measured total drags of bodies of different length; and (3) by measuring the drag on a small section of a flat plate suspended on a sensitive balance. There is some effect of Reynolds number in the measurements and in the theories but the major variation is with the Mach number.

Figure 8 provides some justification for the trend of von Kármán's assumption, since in plotting the measurements of Lobb et al. for a free-stream Mach number of 5.0, ρ_w and μ_w were used. Excellent agreement with the low-speed measurements is shown. A remarkable feature of these results is the relatively large part of the boundary layer that conforms approximately with the law of the wall. The edge of the boundary layer is at $\log_{10}(yU_\tau/\nu) = 2.4$, whereas in the low-speed boundary layer this point represents only 10 to 15 percent of the boundary layer thickness. Outside of the inner or laminar sublayer region, the velocity profile

*In Section 16.8 the significance of wall properties in *laminar* boundary layer calculations was pointed out.

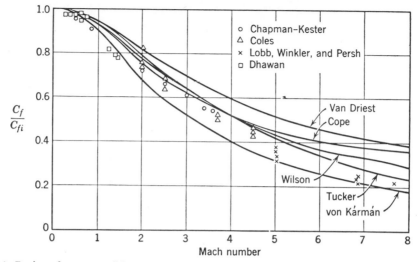

Fig. 11 Ratio of compressible to incompressible skin-friction coefficients for flat plate as function of Mach number of external flow.

closely satisfied a power law $U/U_e = (y/\delta)^{1/n}$ (see Eq. 41) where n decreased from 7 to 5.5 as the Mach number increased from 5 to 7.7.

A comprehensive account of methods for calculating compressible turbulent boundary layers with heat transfer is given by Cebeci and Smith (1974).

18.11   Effect of Adverse Pressure Gradient–Separation

The description of the turbulent boundary layer as given in Section 18.8 applies to all pressure gradients, but special difficulties are encountered when the pressure gradient is adverse and particularly, when flow separation is approached.

Some velocity distributions measured by Schubauer and Klebanoff (1951) on a two-dimensional curved body 8.5 m long are shown, along with the velocity distribution $U_e/U_{e\,\text{max}}$ in Fig. 12. We note that the boundary layer changes shape and thickens rapidly in the region of decreasing velocity, that is, in the adverse pressure gradient. The boundary layer was turbulent from the leading edge.

The important criterion governing *laminar* flow separation was found to be $\lambda = (\delta^2/\nu)\,\partial u_e/\partial x = -12$ (Section 15.7). However, the turbulent boundary layer is much more difficult to describe analytically and, while the factors in the above expression show the right trends, that is, increasing δ^2 and $-\partial u_e/\partial x$ and decreasing ν hasten separation, they can no longer form a single criterion. Qualitatively, how-

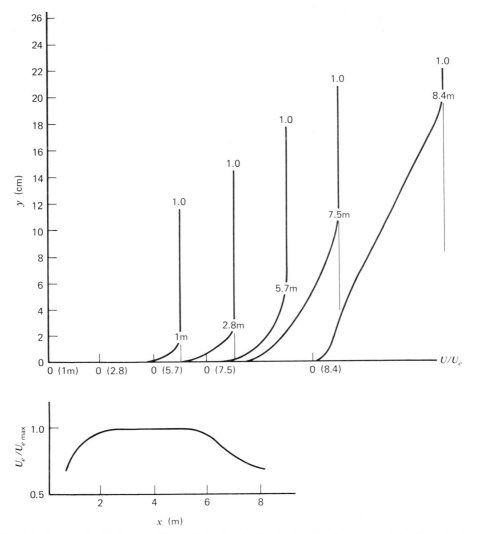

Fig. 12 Measured velocity profiles and longitudinal velocity distributions; origins for velocity profiles are indicated (Schubauer and Klebanoff, 1951, courtesy of NACA).

ever, we know that the high value of $(\partial U/\partial y)_w$ in the turbulent layer enables it to drive farther than could the laminar layer against a given adverse pressure gradient.

Several methods for locating the turbulent separation point have been developed. These are discussed in detail by Cebeci and Smith (1974). The methods follow one of two general procedures: (a) the boundary layer equation Eq. 17 or 18 with appropriate assumptions respectively for l or ϵ, and initial conditions, at the transition point, for instance; step-by-step solutions then follow the boundary layer to the separation point; (b) the von Kármán momentum integral relation derived in Sections 15.6 and 15.7. Following is a brief description of the second procedure: the momentum equation (15.43), with the notation for the turbulent boundary layer is

$$\frac{\tau_w}{\rho} = U_e^2 \frac{d\theta}{dx} + (2\theta + \delta^*) U_e \frac{dU_e}{dx}$$

where δ^* and θ, respectively, the displacement and momentum thicknessses of the boundary layer, are defined in Eqs. 15.29 and 15.30. A significant relation describing the turbulent boundary layer in a pressure gradient is

$$H = \frac{\delta^*}{\theta}$$

The starting point for the analysis of the effect of pressure gradient is the nondimensional form of this equation, that is

$$\frac{d\theta}{dx} = \frac{c_f}{2} - \frac{\theta}{U_e} (H + 2) \frac{dU_e}{dx} \tag{44}$$

For given initial values of θ, c_f, U_e, and dU_e/dx, H can be evaluated as a function of x by a step-by-step solution of Eq. 44. For the initial calculation, the known velocity profile yields θ and δ^*, $H = \delta^*/\theta$, and $Re_\theta = U_e\theta/\nu$; c_f is given by Fig. 13 in the form of curves of constant H on a c_f versus Re_θ plot. This plot, determined by Galbraith and Head (1975), on the basis of experimental data from many sources covers a wider range of parameters but is not greatly different from a compilation by Ludwieg and Tillman used in most other analyses.

The first step in the solution is to use initial values of c_f, θ, U_e, H, and dU_e/dx, at, for instance, the transition point, and to solve Eq. 44 to find θ at an $x + \Delta x$ where Δx is small. Thus, a new Re_θ is found, but the new values of c_f and H require another relation.

This relation was found by Head (1960) by means of a relation suggested by the increase in rate of volume flow in the boundary layer through its entrainment of the external flow. Head defines the new relation based on the increase in δ^* through entrainment of the external flow by the boundary layer. Since by definition

$$\delta^* = \int_0^\delta \left(1 - \frac{U}{U_e}\right) dy$$

$$U_e(\delta - \delta^*) = \int_0^\delta U \, dy$$

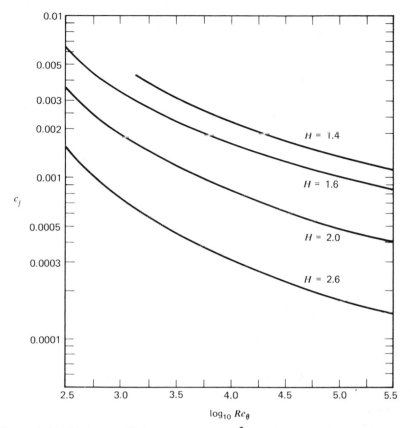

Fig. 13 Local skin-friction coefficient $c_f = 2(U_\tau/U_e)^2$ as a function of H and Re_θ. (Galbraith and Head, 1975, courtesy of Royal Aeronautical Society.)

Head assumes that the rate of entrainment is given by

$$\frac{d}{dx}[U_e(\delta - \delta^*)] = U_e F(H_{\delta-\delta^*}) \tag{45}$$

where $H_{\delta-\delta^*} \equiv (\delta - \delta^*)/\theta$ is another shape parameter, which Head finds can be expressed as a function of H, that is,

$$H_{\delta^*-\delta} = G(H)$$

The functions F and G are evaluated from experiments (see Cebeci and Smith, 1974, p. 167). He then expresses Eq. 45 in the finite difference form,

$$\Delta[U_e(\delta - \delta^*)]_0 = U_e G_1(H)\,\Delta x \tag{46}$$

H at $x + \Delta x$ is found from this relation and thus the step-by-step solution of Eq. 44 can be continued. It is found by the several analyses of the problem that separation

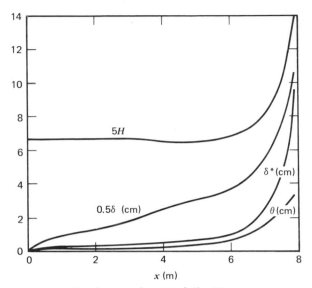

Fig. 14 Boundary layer properties for experiments of Fig. 12.

occurs at a value of H between 1.8 and 2.8. Figure 14 shows plots of the various quantities for the experiments of Schubauer and Klebanoff (1951), shown in Fig. 12. The curve $H(x)$ shows that the above range of predictions is not as vague as it seems, since H is shown to increase rapidly with x near the separation point.

Figure 15 compares the separation point on a given airfoil surface as predicted by four methods, designated by their authors (Cebeci, Mosinskis, and Smith, 1972). Separation actually occurs where the function $(1 - C_p)^{1/2}$ flattens; upstream of that point $(1 - C_p)^{1/2} = U_e/U_\infty$. The computation method of Cebeci and Smith (C-S) is perhaps a more versatile (and, understandably, more involved) method, in that it can incorporate effects of compressibility, suction, blowing, and heating, and is applicable to combinations of surfaces as well as bodies of revolution.

The effect of compressibility is of course complicated by the presence of shock waves generated in adverse pressure gradients in supersonic regions (see Fig. 16.7), and by shock wave-boundary layer interactions. These methods use, in addition to the equations above, the turbulent form of the energy equations such as Eq.16.7 with the expressions for turbulent viscosity and heat conduction coefficients (see Cebeci-Smith, 1974).

18.12 Reynolds Analogy—Heat Transfer and Temperature Recovery Factor

The relation between heat transfer and skin friction in the laminar boundary layer (Section 16.5) rests on the proportionality between the heat transfer and viscosity

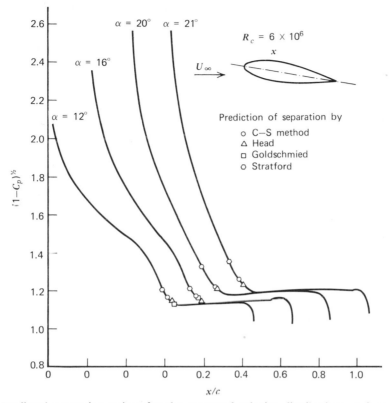

Fig. 15 Predicted separation points for the measured velocity distributions at four angles of attack on the NACA 65,2-421 airfoil.   The origin is displaced as designated for each curve. Flow separation is observed near the point where the distribution flattens. (Cebeci, Mosinskis and Smith, 1972, courtesy of AIAA.)

coefficients, that is, $k \sim c_p \mu$.   For the turbulent boundary layer, the Reynolds analogy provides the corresponding relation between skin friction and heat transfer.

The Reynolds analogy rests on the assumption that the mechanisms of turbulent transfer of heat and momentum are similar.   In Section 18.3, we showed that the mean rate of momentum transfer by turbulence is given by

$$\tau_t = -\rho \, \overline{u_1 v_1} \tag{47}$$

By the same reasoning as that used there, ρv_1 is the mass of air crossing the unit area of a plane normal to the temperature gradient, and it carries with it a temperature deficit or surplus of $(\partial T/\partial y) \, dy = \theta_1$.   Then the turbulent rate of heat transfer per unit area is

$$q_t = \rho c_p \overline{v_1 \theta_1} \tag{48}$$

Now, if we assume that the *rate of heat transfer per unit enthalpy gradient is equal to the rate of momentum transfer per unit momentum gradient*, we write,

$$\frac{\rho c_p \overline{v_1 \theta_1}}{c_p \partial T/\partial y} = \frac{-\rho \overline{u_1 v_1}}{\partial U/\partial y} \tag{49}$$

With $q_t = k_t \partial T/\partial y$ and $\tau_t = (\epsilon \rho) \partial U/\partial y$, Eq. 49 becomes

$$k_t = c_p \epsilon \rho = c_p \mu_t \tag{50}$$

that is, Eq. 50 implies a turbulent Prandtl number of unity.

At a given point in the boundary layer the temperature gradient, $\partial T/\partial y$, will be proportional to $T_w - T_e$, the change in temperature across the layer; we assume that, with the same proportionality factor, $\partial U/\partial y$ will be proportional to U_e. Then Eq. 49 becomes

$$\frac{\rho c_p \overline{v_1 \theta_1}}{c_p (T_w - T_e)} = \frac{-\rho \overline{u_1 v_1}}{U_e} \tag{51}$$

Then, from Eq. 51 and Eqs. 47 and 48, we form the dimensionless coefficients,

$$\frac{q_t}{\rho c_p U_e (T_w - T_e)} = \frac{\tau_t}{\rho U_e^2}$$

The left side is the local "Stanton number," st (Section 16.7), and the right side is $c_f/2$. Then the Reynolds analogy is expressed by

$$st = c_f/2 \tag{52}$$

More detailed studies by Rubesin (1953) predict, instead of Eq. 52,

$$st = 0.6 \, c_f \tag{53}$$

practically independent of Mach number up to about 5.

The effective Prandtl number of the turbulent boundary layer, comprising turbulent and laminar contributions, respectively, from the outer region and the laminar sublayer, determines the temperature recovery factor. Many measurements indicate that the recovery factor r for the turbulent boundary layer may be expressed in terms of the molecular Prandtl number as

$$r = Pr^{1/3} \cong 0.89 \tag{54}$$

for air over a wide range of Mach numbers.

Figure 16.4 shows experimental recovery factors for laminar and turbulent boundary layers. The agreement between the experimental data and Eq. 16.35 for laminar flow and Eq. 54 above for turbulent flow are seen to be good.

18.13 Free Turbulence

Turbulent jets and wakes differ from boundary layers in that *both* boundaries are unconfined and irregular, as indicated in Fig. 1; also, since there are no solid boundaries the flows are completely turbulent (at relatively low Reynolds numbers) in that there are no laminar sublayers and no buffer layers. In this treatment, we only indicate the general features of these flows, since the details are similar in physical concepts to those for turbulent boundary layers. The reader may find the details in Schlichting (1968), Townsend (1956), and Hinze (1959).

The flows, as for the turbulent boundary layer, are characterized by eddy viscosities for the turbulent flows considerably higher than the kinematic viscosities that govern their laminar counterparts. The size of the eddies responsible for the shearing stress are of the order of the width of the turbulent mixing region and these are responsible for the irregular boundaries indicated in Fig. 1.

Prandtl hypothesized that for free turbulent flows the eddy viscosity may be represented by the simple formula

$$\epsilon = \kappa_1 b (U_{max} - U_{min})$$

where κ_1 is constant and $2b$ is the width of the mixing region, or, since the boundaries are ill-defined, a more practical formula would be

$$\epsilon = \kappa b_{1/2} (U_{max} - U_{min}) \tag{55}$$

where $b_{1/2}$ is half the width of the region (within the mixing region) at the boundaries of which $U = \frac{1}{2}(U_{max} - U_{min})$. It has been found that when this formula is substituted into the boundary layer equation for the mean motion (Eq. 18) that the experiments show that κ is constant at any distance greater than around 15 or 20 diameters of a jet or of a lateral dimension of a body producing a wake. The values of ϵ are in practical cases several hundred times greater than their counterpart ν for laminar flow so that the rate of turbulent mixing is many times greater than the laminar. For a 1 m/sec circular jet for instance the eddy viscosity is equal to about 0.002 m^2/sec (see Schlichting, 1968) which is about 150ν and this is also about the ratio between the rate of spreading of the turbulent to that of the laminar circular jet. One has only to observe the plume of smoke from a cigarette in quiet surroundings to see first the waviness associated with instability, then the sudden breakup into a large diameter, highly irregular dispersion of the smoke particles.

As one might expect, from the violent flow shown in the photographs of the wakes of Figs. 10.2 and 16.7, that the critical Reynolds numbers for wakes and jets are quite low. Calculations (see Stuart, 1963) show, for instance, that for a plane jet the critical Reynolds number, based on the jet width, is about 5. Therefore, in most cases of practical importance free surface flows are turbulent.

CHAPTER
19

Boundary Layer Control

19.1 Introduction

As the uses aircraft and fluid machinery in general broaden to serve wider areas the need for effective boundary layer control (BLC) becomes more evident; increased use of vertical-short takeoff and landing aircraft (V-STOL), helicopters and ground effect machines (GEM) are examples. For conventional aircraft the aerodynamic objectives of BLC are to minimize drag at cruising speed, to maximize lift at landing and takeoff, and to minimize the excitation of vibration, noise, and unsteady flow effects in general.

This section gives only brief accounts of some of the geometric parameters of "naked" airfoils as they affect maximum lift and unsteady effects, and effects of some current devices to control, first, boundary layer transition, and second, maximum lift. Comprehensive treatments are given by Smith, 1975, and by Hoerner and Borst, 1975.

19.2 Airfoils in Incompressible Flow Without BLC

Figure 1 shows pressure distributions and schematic streamline diagrams at different angles of attack for a conventional airfoil shape (NACA 4412) with maximum thickness near the leading edge in two-dimensional flow. The streamline diagrams show the smooth flow, characteristic of that at low angles of attack, and that for stall near the leading edge, showing the instantaneous streamlines that characterize the unsteady wake. The pressure distribution at $\alpha = 24°$ indicates a decreased lift and a strong diving moment.

These airfoils (the NACA four digit series), whose major disadvantage is that their maximum thicknesses are relatively near the leading edge (around $0.3c$), still show some of the aerodynamic features common to all airfoils. As was pointed out in Chapter 5, these common features include a lift curve slope near 2π per radian and the aerodynamic center near the quarter chord point.

The pressure distributions show a minimum pressure point (maximum $(U_e/V_\infty)^2$) near the leading edge; as a result the adverse pressure gradient occurs early; thus (Section 17.5) boundary layer transition to the turbulent (high skin friction) condition will occur early.

Figure 2 shows the coefficient of total viscous drag, that is, the sum of skin friction and form drags, for a systematic series (the lower three vary in maximum thick-

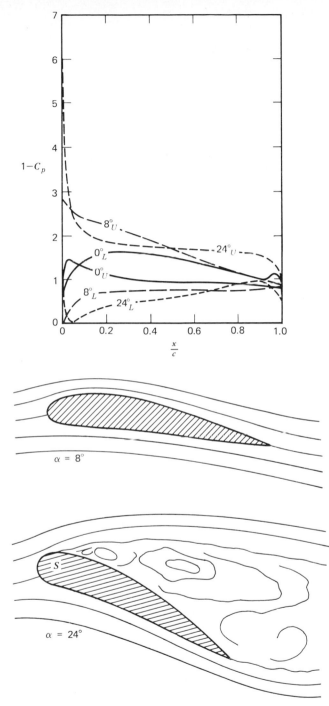

Fig. 1 Pressure distributions on an NACA 4412 airfoil at angles of attack of 0°, 8°, and 24°; the subscripts U and L designate the distributions on the upper and lower surfaces. Schematic streamlines at 8° and 24° are shown at right; at 24° the airfoil is stalled, the location of flow separation, designated by S, moves irregularly and the flow in the wake is highly unsteady. (Measurements by Pinkerton, 1936, reproduced by courtesy of NACA.)

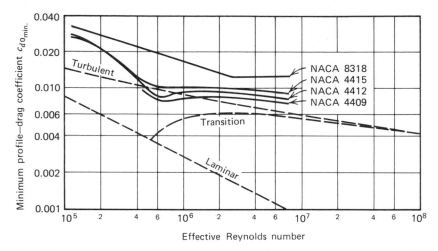

Fig. 2 NACA measured drag coefficients for various wing sections compared with theoretical skin friction for flat plate.

ness from $0.09c$ to $0.15c$ but all have $0.04c$ maximum camber located at $0.4c$). The broken curves apply to a flat plate with laminar and with turbulent boundary layers and with transition at the trailing edge at $Re = 5 \times 10^5$, and moving forward with increasing Reynolds number. The angles of attack are near their zero lift value. The main feature to be noted in this comparison is that the form drag increases with maximum thickness. The top curve, for $0.08c$ maximum camber and $0.18c$ thick indicates a large increase with camber as well, though that trend will not in general extend to airfoils of smaller, more practical degrees of camber.

A feature of this family of airfoils is that, for a given camber and maximum thickness, the "leading edge radius" is relatively large. We see, from Fig. 3, that over a considerable range of Reynolds numbers the larger leading edge radii, corresponding to the thicker sections, lead to higher maximum lift coefficients. This radius is 0.89 percent c for the 0009 shape, 1.58 percent c for the 0012 and 2.48 percent for the 0018.

It has been pointed out in previous sections that, beyond a certain angle of attack, the airfoil is no longer a streamline body but rather has the nature of a bluff body, with its large wake. At high angles of attack, the minimum pressure point on the upper surface of an airfoil is close to the leading edge, as shown in Fig. 1. Beyond the minimum pressure point, the adverse pressure gradient is steep and tends to cause transition and, at high angles of attack, flow separation. The nature of the stall, whether it is abrupt or gentle, depends on the nature of the boundary layer and its behavior in the presence of an adverse pressure gradient.

Ahead of the minimum pressure point, the large favorable pressure gradient tends to maintain a laminar boundary layer. Beyond the minimum pressure point, the steep adverse gradient can cause one of three effects, as indicated schematically in

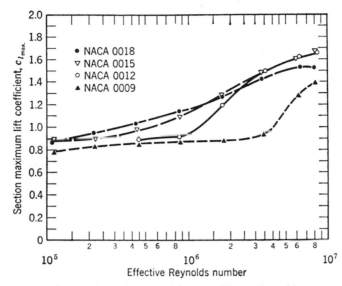

Fig. 3 Variation of maximum lift coefficient with Reynolds number. (Courtesy of NACA.)

Fig. 4: in A, the laminar boundary layer separates at point S if the Reynolds number is below a critical value (below about 4×10^6) for the 0009 section, and the lift drops rapidly to a relatively low value; in B the Reynolds number is above about 6×10^6 causing the separated flow for the 0009 section to rejoin the surface at

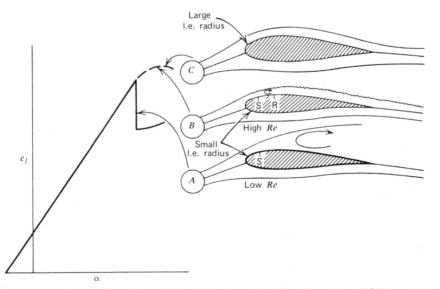

Fig. 4 Lift coefficient curves for small and for large leading edge radius airfoils at angles of attack near maximum lift.

point R to form a turbulent boundary layer (a "separation bubble" is formed between S and R) which can drive farther against the adverse pressure gradient;* in C for the thicker section the initial adverse gradient is not severe enough to cause separation at the low Reynolds numbers, so, instead, transition to a turbulent layer occurs at a low enough Reynolds number so the severe drop in lift does not occur. As indicated in Fig. 4, the c_l versus α curve is rounded near maximum lift for conditions B and C. For conditions B and C, the "permanent" flow separation begins near the trailing edge and moves steadily forward with increasing angle of attack; this is the desirable flow condition if the aircraft is to perform smoothly and predictably near maximum lift. One possibility, that of oscillation between conditions A and B, is devoutly to be avoided.

The effect of camber on maximum lift is related to the effects of slats and slots on the pressure distributions. The general effect is to increase the maximum lift coefficient, as can be seen in Fig. 6; Section 19.4 describes theory and experiment for a multicomponent configuration.

On highly swept leading edges such as delta planforms, "leading edge separation," a condition similar to B in Fig. 4 occurs. A vortex with axis along the leading edge forms, as shown in Fig. 5, and reinforces the trailing vortex of the wing. The vortex generates a region of low pressure near the leading edge which results in a significant contribution to the lift, as described by Legendre (1966).

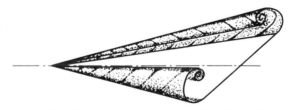

Fig. 5 Schematic drawing of leading edge vortex generated by delta wing.

19.3 BLC for Low Drag

As was discussed in Section 17.5, a large area of favorable pressure gradient, as achieved on the NACA 65 series airfoils shown in Fig. 17.11, is practically a necessity if a laminar boundary layer and its associated low skin friction is to be achieved over significant areas. These airfoils are distinguished by a favorable pressure gradient up to about $0.5c$ and, if the surface and the flow are smooth, by a "drag bucket," signifiying that the boundary layer is laminar up to $0.5c$ or beyond.

For the illustration used in Fig. 17.11, these "buckets" extend over a c_l range of about 0.3 (the subscript in the airfoil designation). The low maximum lift coeffi-

*Blick *et al* (1975) have shown that the permanent laminar separation of A can be avoided by introducing a serrated leading edge similar to that occurring on the feather along the leading edge of some owls' wings. Similar results are reported by Soderman (1973) (see Section 19.6).

cient associated with the small leading edge radius can, as will be shown, be obviated by proper boundary layer control surfaces at the leading edge. As will be described below the design of the airfoil shape aft of the maximum thickness location is an important factor in determining the maximum lift coefficient. It is noteworthy that airfoils such as those shown in Fig. 17.11 will, at low angles of attack, have a high critical Mach number (see Section 12.5). The reason becomes clear when we compare Fig. 17.11 with Fig. 1 and observe that for a given design lift coefficient the pressure peak disappears for the 65 series airfoil and therefore the critical Mach numbers will be much higher for the flat distribution of Fig. 17.11. This feature is utilized in transonic aircraft especially near the wing root (see Lock and Bridgewater, 1967).

The most promising combination for low drag aircraft appears to be a combination of a smooth surface, the largest possible area of favorable pressure gradient such as is achieved by the NACA 65 series or similar shapes described above and in Section 17.11, and localized or area suction to maintain a maximum area of laminar flow. The use of localized or area suction and/or auxiliary surfaces and/or slots tend to obviate flow separation. Pfenninger (1965), Lachmann (1961) and Thwaites (1960) present particularly thorough and novel studies of various methods as well as references to other significant researches.

Thwaites gives the framework for calculating the coefficient for the power used for suction, for the suction volume, and for the drag coefficients associated with the various losses.

19.4 BLC for High Lift

Both passive and active devices are used to achieve high lift, the latter being particularly applicable to V-STOL and GEM.

Among the passive devices a few are shown in Fig. 6, along with corresponding values of $c_{l_{max}}$. These achieve values up to 3.7 (that giving the highest value is not *quite* passive, since a suction slot is incorporated). As was pointed out in Section 5.8 (see Fig. 5.15), the greater the flap or slot deflection the greater will be the effective camber of the combination, the lower will be the angle of zero lift, and the greater will be the upward displacement of the lift curve; however the slope of the lift curve below the stall remains practically constant.

The increased effective camber with its resulting increase in $c_{l_{max}}$ may be achieved by deflecting slats and flaps or by a free hinged (Thwaites) flap or a downward directed jet near a *rounded* trailing portion of the airfoil.

Analysis of various slot-slat-flap combinations, as well as suction and blowing effects are reviewed and their potentialities analyzed by Smith, 1975, (see also Cebeci and Smith, 1974). They use panel methods similar to those described in Section 5.10. Goradia and Colwell (1975) also treat multiple surfaces including the

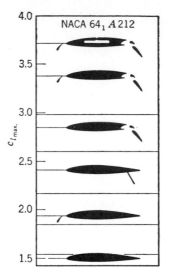

Fig. 6 Maximum lift coefficient for an airfoil with various high lift devices. (Courtesy of NACA.)

effects of ordinary and "confluent" boundary layers as designated in Fig. 7; as is indicated, the confluent boundary layer includes the effects of the turbulent layers shed by upstream surfaces. In their analysis of the pressure distribution the panel method used by Goradia and Colwell differs in two respects from the uniform source plus vorticity panel method of Section 5.10, in that they approximate the shapes of the solid surfaces plus boundary layer displacement by closed polygons of vortex panels only, each with linearly varying vorticity; the control points, where the boundary condition $\nabla\phi = V_n = 0$ is satisfied at the midpoints of the panels, as

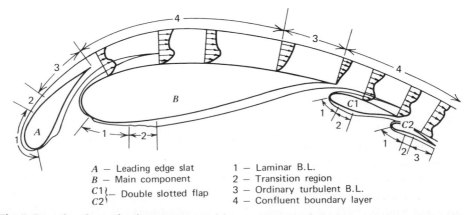

A — Leading edge slat   1 — Laminar B.L.
B — Main component   2 — Transition region
C1⎫— Double slotted flap   3 — Ordinary turbulent B.L.
C2⎭   4 — Confluent boundary layer

Fig. 7 Boundary-layer development on multicomponent aerofoil. (Goradia and Colwell, 1975. Courtesy of The Royal Aeronautical Society.)

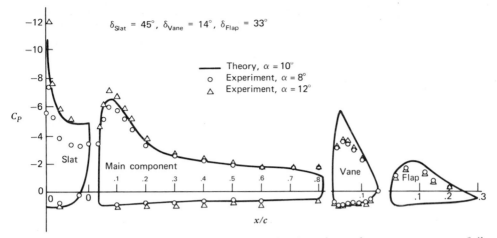

Fig. 8 Experimental and predicted pressure distributions for a four-component aerofoil. (Goradia and Colwell, 1975. Courtesy of The Royal Aeronautical Society.)

in Section 5.10. Figure 8 shows good agreement between the calculations and experiment for a four-component configuration.

The most extreme high lift devices are those used for V-STOL aircraft; some of these devices are described by Goodmanson and Gratzer (1973) and are shown in cross-section along with some test results in Fig. 9. These configurations utilize the jet flow directly, as in "vectored thrust," or, in combination with blowing near the leading edge just behind a slot to prevent separation before the flow reaches the flaps. The tests refer to four-engine configurations with the nondimensional parameters:

$$C_\mu = \frac{\text{(momentum flow rate through leading edge slot)}}{q_\infty S} \tag{1}$$

$$C_J = \frac{\text{(momentum flow rate of primary jet)}}{q_\infty S_J} \tag{2}$$

where S is the wing area, S_J is the cross-sectional area of the jet, and δ_F are flap angles. The test results shown in Fig. 9 refer to $C_J = 2$; calculated curves for 100% efficiency of the jet and for $C_J = 0$ are shown for comparison.

Other devices include moving belt surfaces and leading edge regions comprising rotating cylinders. In the latter case, if the cylinder surface moves on the order of six times V_∞, lift coefficients over 10 can be achieved.

19.5   Circulation Control

Circulation control uses tangential blowing to delay flow separation over a rounded trailing edge; recent developments (see Englar, 1975) indicate its potential

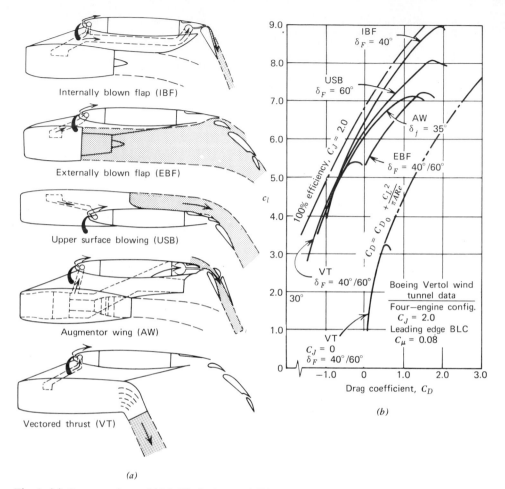

Fig. 9 (*a*) Cross sections of high lift devices and (*b*) results of tests at different values of C_μ and C_J (Eqs. 1 and 2). (Goodmanson and Graetzer, 1973. Courtesy of The Boeing Company.)

significance for helicopters and V-STOL aircraft. The device utilizes the "Coanda Effect," illustrated in Fig. 10. The jet with velocity V_j, Mach number M_j issuing from the slot of width h follows the curved surface contour approximately as would an inviscid fluid, until the balance between the pressure gradient normal to the surface and the centrifugal force exerted by the flow is destroyed by the thickening boundary layer and the dynamic pressure of the free stream.

Maximum lift coefficients achieved by a cylindrical Coanda surface-split flap configuration is shown in Fig. 11. In extreme cases the jet can be turned nearly $180°$ so that it opposes the external flow; the resulting combination of high lift and high drag is attractive for the landing of V-STOL aircraft.

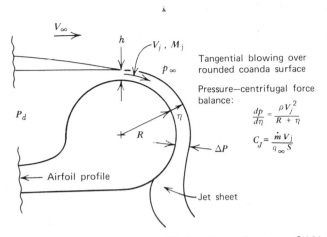

Fig. 10 Basic circulation control aerodynamics (Englar, 1975. Courtesy of AIAA.)

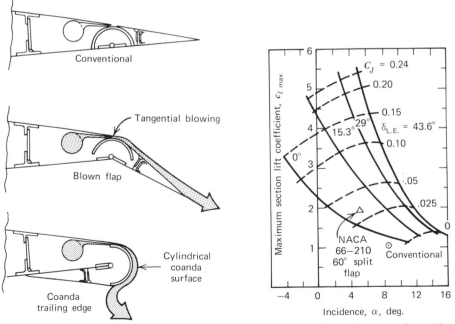

Fig. 11 (*a*) Circulation control modes of operation. Maximum lift coefficients of two-dimensional NACA 66-210 wing with circulation control with various C_J and leading edge droop (δ_{Le}). (Englar, 1975, courtesy of AIAA.)

19.6   Aerodynamic Noise

One of the most important environmental and engineering problems concern the sonic boom at very high speeds, and, at lower speeds as well, the noise associated with unsteady flow over lifting surfaces of helicopters, turbines, and the high-lift devices of V-STOL aircraft.

The framework for modern researches in sound generation and attenuation originated with the works of Lighthill (see Lighthill, 1961) and by works carried out and reviewed by Ribner (1968).

High-speed jets are one of the main sound sources, though their importance has been decreased somewhat by decreasing lateral velocity gradients near the jet exhaust through the introduction of high bypass turbine engines. Another source of sound is that arising through unsteady lift forces associated with unsteady wakes. When a vortex of a given circulation is shed at the trailing edge, as in the von Kármán vortex street, the circulation around the surface changes by an equal and opposite amount and with successive shedding the lift fluctuates; as a result, the surface radiates sound at the vortex shedding frequency. It appears that this source of sound can be attenuated by suppressing the vortex street.

Paterson, Vogt, Fink, and Munch (1973) found that, for an airfoil with a sharp trailing edge, sound measurements indicate that the von Kármán vortex street does not form unless the boundary layers on one or both surfaces are laminar. For a rough surface with blunt trailing edge, observations by Kuethe (1972) indicate that the vortex street forms regardless of the state of the boundary layer. However, the introduction of inclined surface waviness or "vortex generators" that disturbs sufficiently the two-dimensional character of the flow at the trailing edge was effective in suppressing the vortex street; it follows that the sound generated by the street should be attenuated significantly. Soderman (1973) found that the addition of serrations at the leading edge of a lifting surface also suppressed the vortex street and the attendant sound.

19.7   Three-Dimensional and Unsteady Flows

The analyses in this book are confined to two-dimensional steady flows in polar (axisymmetric flows) or rectangular coordinates. It is pointed out however that no new concepts are involved in extending the *panel* method to steady *three* dimensional flows; also for flow past yawed cylindrical bodies, for example, sweptback airfoils, suitable orientation of the coordinate system enables the chordwise and spanwise velocity components to be analyzed separately.

In spite of the fact that most flows of practical importance are three-dimensional with unsteady wakes, comparisons of two-dimensional theory with experiment are in many instances surprisingly accurate. Exceptions occur in regions where lateral or spanwise pressure gradients are too great to be ignored; examples include the flow near the boundaries between stalled and unstalled regions on a wing, such as occur near the tips of deflected slats or flaps and near wing tips where the lateral gradients generate large spanwise velocity gradients.

Currently, major effects are being made to avoid the generation of unsteadiness, such as through boundary layer control; the resulting increases in aerodynamic efficiency, and alleviation of structural vibration and noise are areas of great potential pay-off in the development of fluid machinery.

Appendix
A

Dimensional Analysis

If an algebraic equation expresses a relation among physical quantities, it can have meaning only if the terms involved are alike dimensionally. For example, two numbers may be equal, but if they represent unlike physical quantities they may not be equated. This requirement of dimensional homogeneity in physical equations is useful in determining the combinations in which the variables occur. Specifically, let it be required that all the terms in an equation be pure numbers. Then the variables involved may occur only in combinations that have zero dimensions. Any physical equation can be expressed in terms of dimensionless combinations of the variables. The formal statement of this fact is embodied in the Π theorem, which may be stated as follows: (a proof is given, for instance, by Durand, 1943).

Any function of N variables

$$f\{P_1, P_2, P_3, P_4, \cdots P_N\} = 0 \tag{1}$$

may be expressed in terms of $(N - K)$ Π products

$$f\{\Pi_1, \Pi_2, \Pi_3, \cdots \Pi_{N-K}\} = 0 \tag{2}$$

where each Π product is a dimensionless combination of an arbitrarily selected set of K dimensionally independent variables and one other; that is,

$$\Pi_1 = f\{P_1, P_2, \cdots P_K, P_{K+1}\}$$

$$\Pi_2 = f\{P_1, P_2, \cdots P_K, P_{K+2}\}$$

$$\cdots$$

$$\Pi_{N-K} = f\{P_1, P_2, \cdots P_K, P_N\}$$

K is equal to the number of fundamental dimensions required to describe the variables P. If the problem is one in mechanics, all quantities P may be expressed in terms of mass, length, and time, and $K = 3$. In thermodynamics, all quantities may be expressed in terms of mass, length, time, and temperature, and $K = 4$. The arbitrarily selected set of K variables may contain any of the quantities P_i with the restriction that the K set itself may not form a dimensionless combination.

To illustrate the application of the Π theorem to a problem in mechanics, we consider the force experienced by a body that is in motion through a fluid. We assume that the force will depend upon the following parameters:

$$F = f\{\rho, V, l, \mu, a\} \tag{3}$$

where the symbols and their dimensions are as tabulated.

Symbol	Name	Dimensions
F	Force	MLT^{-2}
ρ	Density	ML^{-3}
V	Velocity	LT^{-1}
l	A length characterizing the size of the body	L
μ	Coefficient of viscosity	$ML^{-1}T^{-1}$
a	Speed of sound	LT^{-1}

Let us write Eq. 3 in the form of Eq. 1:

$$g\{F, \rho, V, l, \mu, a\} = 0 \qquad (4)$$

There are six variables and three fundamental dimensions. Therefore, there are three Π products. If we choose ρ, V, and l as the K set, the Π products are

$$\Pi_1 = f_1\{F, \rho, V, l\}$$
$$\Pi_2 = f_2\{\mu, \rho, V, l\}$$
$$\Pi_3 = f_3\{a, \rho, V, l\}$$

The Π theorem guarantees that the Π products above can be made dimensionless. As an example, we find a dimensionless combination of the variables in Π_1, in the form $F\rho^a V^b l^c$. We write the quantity in terms of its dimensions

$$(MLT^{-2})(ML^{-3})^a(LT^{-1})^b(L)^c$$

The exponents of M, L, and T must be zero. This process leads to the three equations

$$1 + a = 0$$
$$1 - 3a + b + c = 0$$
$$-2 - b = 0$$

from which

$$a = -1$$
$$b = -2$$
$$c = -2$$

and Π_1 becomes

$$\Pi_1 = \frac{F}{\rho V^2 l^2}$$

Proceeding in the same manner with Π_2 and Π_3, we get

$$\Pi_2 = \frac{\rho V l}{\mu}$$

$$\Pi_3 = \frac{V}{a}$$

Then Eq. 4 may be written

$$f\left\{\frac{F}{\rho V^2 l^2}, \frac{\rho V l}{\mu}, \frac{V}{a}\right\} = 0 \tag{5}$$

To illustrate the application of the Π theorem to a problem that includes thermal effects, we consider the heat transferred between a solid and the surrounding fluid when the solid is in motion through the fluid. The heat per second transferred through the boundary layer is given by

$$\frac{\partial Q}{\partial t} = hA(\theta_2 - \theta_1) \tag{6}$$

where A is the area through which the heat is transferred and $\theta_2 - \theta_1$ is the temperature difference between the solid and the fluid. The symbol θ is used in this appendix to represent temperature because the symbol T is reserved for the dimension *time*. The constant h is the heat-transfer coefficient, and we assume that it will be a function of the following parameters:

$$h = f\{\rho, \mu, k, l, a, c, V\} \tag{7}$$

c is the specific heat of the fluid and is defined as the heat required to raise unit mass one degree.* k is the thermal conductivity of the fluid and is defined by the equation

$$\frac{\partial Q}{\partial t} = kA \frac{\partial \theta}{\partial s} \tag{8}$$

where the derivative of the temperature is in a direction normal to the area through which the heat is being transferred. Notice that the thermal conductivity is the proportionality constant in an equation that expresses the rate of heat transfer through a continuous medium, whereas the heat transfer coefficient is the proportionality constant in an equation that expresses the rate of heat transfer between

---

*The specific heats c_p and c_v have been defined in terms of the fluid characteristics in Section 8.2. The definition given here is equivalent.

two different media. The dimensions of h and k may be found from the defining Eqs. 6 and 8. The thermal parameters c, h, and k have the following dimensions:

Symbol	Name	Dimensions
c	Specific heat	$L^2 T^{-2} \theta^{-1}$
h	Heat transfer coefficient	$MT^{-3}\theta^{-1}$
k	Thermal conductivity	$MLT^{-3}\theta^{-1}$

Equation 7 contains eight parameters, and, since four dimensions are needed to describe the eight parameters, the number of Π products will be four. If we choose ρ, μ, k, and l as the K set, the Π products are

$$\Pi_1 = f\{h, \rho, \mu, k, l\} \qquad \Pi_3 = f\{c, \rho, \mu, k, l\}$$
$$\Pi_2 = f\{V, \rho, \mu, k, l\} \qquad \Pi_4 = f\{a, \rho, \mu, k, l\}$$

A dimensionless combination of the variables in Π_1, in the form $h\rho^a\mu^b k^c l^d$, is found by the method employed in the preceding example. We write the quantity in terms of its dimensions.

$$(MT^{-3}\theta^{-1})(ML^{-3})^a(ML^{-1}T^{-1})^b(MLT^{-3}\theta^{-1})^c L^d$$

The exponents of M, L, T, and θ must be zero. Thus we obtain the four equations

$$1 + a + b + c = 0$$
$$-3a - b + c + d = 0$$
$$-3 - b - 3c = 0$$
$$-1 - c = 0$$

from which

$$a = 0$$
$$b = 0$$
$$c = -1$$
$$d = 1$$

and Π_1 becomes

$$\Pi_1 = \frac{hl}{k}$$

A similar procedure for Π_2, Π_3, and Π_4 leads to the result

$$\Pi_2 = \frac{\rho Vl}{\mu}$$

$$\Pi_3 = \frac{c\mu}{k}$$

$$\Pi_4 = \frac{a\rho l}{\mu} = \Pi_2 \frac{a}{V}$$

Then Eq. 7 may be written

$$f\left\{\frac{bl}{k}, \frac{\rho Vl}{\mu}, \frac{c\mu}{k}, \frac{a}{V}\right\} = 0 \tag{9}$$

Two specific heats of a fluid are commonly used: the specific heat at constant pressure c_p and the specific heat at constant volume c_v. If each of these is included in Eq. 7, then both Π products that involve them will be of the form of Π_3. Therefore, one Π product could be taken as

$$\Pi_3 = \frac{c_p \mu}{k}$$

and the other as

$$\Pi_5 = \frac{c_v}{c_p} \Pi_3$$

In place of $c\mu/k$ in Eq. 9, we would write the two quantities γ and $c_p\mu/k$, where γ is the ratio of the specific heats.

Dimensional analysis has application in many other problems that arise in aeronautical engineering. It is commonly employed to isolate the important dimensionless parameters in propeller theory and stability problems of the airplane as a whole. Much of the intuitive reasoning that leads to the laws of skin friction variation in a turbulent boundary layer can be aided by dimensional considerations.

Appendix
B

Derivation of the Navier-Stokes and the Energy Equations

1 Introduction

It is the object here to develop the equations of motion* and energy for unsteady compressible viscous flow. Special forms of the general equations have been used in various sections of the book. To derive the general equations we follow a proce- dure analogous to that employed in the theory of elasticity and assume that, with respect to the *principal axes*, the stress is proportional to the rate of extension and to the rate of increase of specific volume of the fluid element.

Consider the stress τ exerted across a fluid surface A. If both shear and normal stresses are assumed to be acting, their resultant will be oblique to the surface. Let the surface be the oblique face of the small tetrahedron shown in Fig. 1. Then, if **n**

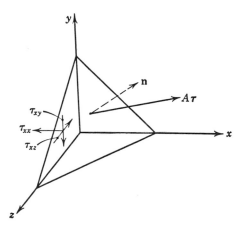

Fig. 1

is the unit normal to A, the areas of the tetrahedron faces that are normal to the x, y, and z coordinate axes will be $(\mathbf{n} \cdot \mathbf{i})A$, $(\mathbf{n} \cdot \mathbf{j})A$, and $(\mathbf{n} \cdot \mathbf{k})A$, respectively. The stress on the x face of the tetrahedron may be resolved into three components, of which one is normal to the face and is given the symbol τ_{xx} and the other two, τ_{xy} and τ_{xz}, are shearing stresses and lie in the x face. The notation for the stress com- ponents is as follows. The first subscript represents the face on which the stress acts, and the second subscript represents the direction in which the stress acts.

*The equations of motion were derived by Navier in France in 1827 and by Stokes in England in 1845.

The stresses on the x, y, and z faces of the tetrahedron are thus resolved into nine components.

For equilibrium of the tetrahedron

$$\left.\begin{array}{l} \boldsymbol{\tau} \cdot \mathbf{i} = (\mathbf{i}\tau_{xx} + \mathbf{j}\tau_{yx} + \mathbf{k}\tau_{zx}) \cdot \mathbf{n} = \mathbf{f}_1 \cdot \mathbf{n} \\[2mm] \boldsymbol{\tau} \cdot \mathbf{j} = (\mathbf{i}\tau_{xy} + \mathbf{j}\tau_{yy} + \mathbf{k}\tau_{zy}) \cdot \mathbf{n} = \mathbf{f}_2 \cdot \mathbf{n} \\[2mm] \boldsymbol{\tau} \cdot \mathbf{k} = (\mathbf{i}\tau_{xz} + \mathbf{j}\tau_{yz} + \mathbf{k}\tau_{zz}) \cdot \mathbf{n} = \mathbf{f}_3 \cdot \mathbf{n} \end{array}\right\} \tag{1}$$

Only three of the six shear stresses are independent, because conservation of angular momentum requires that the torque arising from the shear stresses acting on a fluid element be equal to the time rate of change of angular momentum of the element. If the law is applied to an element $\Delta x\, \Delta y\, \Delta z$, it is easily shown that in the limit, as $\Delta x\, \Delta y\, \Delta z$ approaches zero, we have*

$$\tau_{xy} = \tau_{yx}$$

$$\tau_{xz} = \tau_{zx}$$

$$\tau_{yz} = \tau_{zy}$$

It is shown in the next section that a set of axes can be found such that the shearing stresses vanish when referred to them. In deriving the equations of motion it is convenient to express the stresses in this *principal* system of coordinates. Then the basic assumption relating normal stresses to the extension derivatives is applied. A transformation back to the arbitrary coordinate system yields the six components of stress in terms of the rates of extension and strain of the fluid element. The rates of extension and strain of a fluid element have been discussed in Chapter 2.

2 Principal Axes

It will be shown that the stresses at a point can be represented by pure normal stresses when referred to the *principal axes*. This property will, for simplicity, be demonstrated in two dimensions; the conclusions hold for three dimensions as well. Consider the element ABC in Fig. 2. Both shear and normal stresses act on the x and y faces; the stresses on the inclined face are dictated by the equilibrium condition. If the area of the side BC is A and if X and Y are the stress components indicated,

$$AX = A \cos \alpha\, \tau_{xx} + A \sin \alpha\, \tau_{yx}$$

$$AY = A \sin \alpha\, \tau_{yy} + A \cos \alpha\, \tau_{xy}$$

---

*If these relations are used in Eqs. 1, then the vectors $\mathbf{f}_1$, $\mathbf{f}_2$, and $\mathbf{f}_3$ may be interpreted as the net forces per unit area on the x, y, and z faces, respectively.

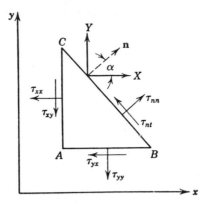

Fig. 2

We may solve for τ_{nn} and τ_{nt} on the BC face. These values are

$$\tau_{nn} = X \cos \alpha + Y \sin \alpha$$

$$= \tau_{xx} \cos^2 \alpha + \tau_{yy} \sin^2 \alpha + 2\tau_{xy} \sin \alpha \cos \alpha$$

$$\tau_{nt} = -X \sin \alpha + Y \cos \alpha$$

$$= \tau_{xy}(\cos^2 \alpha - \sin^2 \alpha) + (\tau_{yy} - \tau_{xx}) \sin \alpha \cos \alpha$$

Now, if we choose the angle α so that $\tau_{nt} = 0$, we get

$$\frac{\tau_{xy}}{\tau_{xx} - \tau_{yy}} = \frac{\sin \alpha \cos \alpha}{\cos^2 \alpha - \sin^2 \alpha} = \frac{1}{2} \tan 2\alpha \qquad (2)$$

and hence there will be two perpendicular directions such that $\tau_{nt} = 0$. These are the *principal directions*. The corresponding normal stresses are the *principal stresses*.

The extension of the foregoing analysis to three dimensions leads to the following conclusion. Through any point in a flow, three principal planes can be found over which the shearing stresses vanish. Since only normal stresses act on principal planes, there are no stresses acting which can cause a rate of strain of the fluid in these planes (see Section 2.7). We therefore set the rate of strain of the principal planes equal to zero. This property of principal planes is used to advantage in the next section in expressing the rates of strain in terms of the extension derivatives.

3 Transformation Equations

Consider two sets of orthogonal axes x, y, z and x', y', z' that are rotated with respect to each other. The primed set refers to the principal axes. The direction cosines connecting the two sets are given in the following table:

$$
\begin{array}{c|ccc}
 & x & y & z \\
\hline
x' & l_1 & m_1 & n_1 \\
y' & l_2 & m_2 & n_2 \\
z' & l_3 & m_3 & n_3
\end{array}
\tag{3}
$$

It is desired to transform the following quantities to the primed system of coordinates

$$
\frac{\partial u}{\partial x}, \qquad \gamma_x = \frac{\partial w}{\partial y} + \frac{\partial v}{\partial z}
$$

$$
\frac{\partial v}{\partial y}, \qquad \gamma_y = \frac{\partial u}{\partial z} + \frac{\partial w}{\partial x}
$$

$$
\frac{\partial w}{\partial z}, \qquad \gamma_z = \frac{\partial v}{\partial x} + \frac{\partial u}{\partial y}
$$

It was shown in Sections 2.4 and 2.7 that the quantities above represent the rates of extension and strain of a fluid element. The same combinations with the primes represent the rates of extension and strain referred to the principal axes. As explained in the last section,

$$
\gamma_x' = \gamma_y' = \gamma_z' = 0
\tag{4}
$$

To carry out the transformation, we may, for example, write

$$
\frac{\partial u}{\partial x} = \left(l_1 \frac{\partial}{\partial x'} + l_2 \frac{\partial}{\partial y'} + l_3 \frac{\partial}{\partial z'} \right) (l_1 u' + l_2 v' + l_3 w')
$$

and

$$
\gamma_x = \left(m_1 \frac{\partial}{\partial x'} + m_2 \frac{\partial}{\partial y'} + m_3 \frac{\partial}{\partial z'} \right) (n_1 u' + n_2 v' + n_3 w')
$$

$$
+ \left(n_1 \frac{\partial}{\partial x'} + n_2 \frac{\partial}{\partial y'} + n_3 \frac{\partial}{\partial z'} \right) (m_1 u' + m_2 v' + m_3 w')
$$

Similar expressions hold for the other rates of extension and strain. After we have performed the indicated operations and made use of Eqs. 4, the rates of extension and strain become

$$\left.\begin{aligned}
\frac{\partial u}{\partial x} &= l_1^2 \frac{\partial u'}{\partial x'} + l_2^2 \frac{\partial v'}{\partial y'} + l_3^2 \frac{\partial w'}{\partial z'} \\[2mm]
\frac{\partial v}{\partial y} &= m_1^2 \frac{\partial u'}{\partial x'} + m_2^2 \frac{\partial v'}{\partial y'} + m_3^2 \frac{\partial w'}{\partial z'} \\[2mm]
\frac{\partial w}{\partial z} &= n_1^2 \frac{\partial u'}{\partial x'} + n_2^2 \frac{\partial v'}{\partial y'} + n_3^2 \frac{\partial w'}{\partial z'}
\end{aligned}\right\} \tag{5}$$

$$\left.\begin{aligned}
\gamma_x &= 2\left(m_1 n_1 \frac{\partial u'}{\partial x'} + m_2 n_2 \frac{\partial v'}{\partial y'} + m_3 n_3 \frac{\partial w'}{\partial z'} \right) \\[2mm]
\gamma_y &= 2\left(n_1 l_1 \frac{\partial u'}{\partial x'} + n_2 l_2 \frac{\partial v'}{\partial y'} + n_3 l_3 \frac{\partial w'}{\partial z'} \right) \\[2mm]
\gamma_z &= 2\left(l_1 m_1 \frac{\partial u'}{\partial x'} + l_2 m_2 \frac{\partial v'}{\partial y'} + l_3 m_3 \frac{\partial w'}{\partial z'} \right)
\end{aligned}\right\} \tag{6}$$

Thus, the rates of strain referred to arbitrary axes are expressed in terms of the extension derivatives taken with respect to principal axes.

If the three relations in Eq. 5 are added, and the equation

$$l^2 + m^2 + n^2 = 1$$

is used, it appears that the sum of the three extension derivatives in the arbitrary coordinate system is equal to the sum of the three derivatives in the principal system. This is to be expected, because the sum of these derivatives is simply div **V**, which is invariant to the choice of coordinate systems.

4   The Stresses at a Point

The principal stresses at a point are denoted by τ_1, τ_2, and τ_3 normal, respectively, to the x', y', and z' planes. The plane *BCD* in Fig. 3 is taken normal to the x axis; according to the table (Eq. 3) the direction cosines of its normal are l_1, l_2, and l_3, and

$$A\tau_{xx} = \tau_1 l_1 A l_1 + \tau_2 l_2 A l_2 + \tau_3 l_3 A l_3$$

After analyzing in a similar way the normal stresses on planes normal to the y and z axes, we may write

$$\begin{aligned}
\tau_{xx} &= \tau_1 l_1^2 + \tau_2 l_2^2 + \tau_3 l_3^2 \\[2mm]
\tau_{yy} &= \tau_1 m_1^2 + \tau_2 m_2^2 + \tau_3 m_3^2 \\[2mm]
\tau_{zz} &= \tau_1 n_1^2 + \tau_2 n_2^2 + \tau_3 n_3^2
\end{aligned} \tag{7}$$

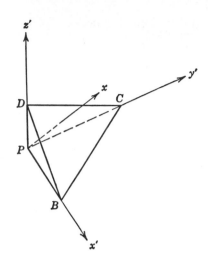

Fig. 3

To find τ_{xy} we observe that the y-direction has, respectively, the direction cosines m_1, m_2, and m_3 with respect to the x', y', and z' axes. Then,

$$A\tau_{xy} = \tau_1 l_1 Am_1 + \tau_2 l_2 Am_2 + \tau_3 l_3 Am_3$$

The other shearing stresses follow by analogy, so that

$$\tau_{yz} = \tau_1 m_1 n_1 + \tau_2 m_2 n_2 + \tau_3 m_3 n_3$$

$$\tau_{zx} = \tau_1 n_1 l_1 + \tau_2 n_2 l_2 + \tau_3 n_3 l_3 \qquad (8)$$

$$\tau_{xy} = \tau_1 l_1 m_1 + \tau_2 l_2 m_2 + \tau_3 l_3 m_3$$

For a fluid at rest or in uniform motion the three normal stresses in the principal directions τ_1, τ_2, and τ_3 all will have the same value. They will be equal to the static pressure in the fluid $-p$. If, however, the fluid has an arbitrary motion, the stresses τ_1, τ_2, and τ_3 will differ from the static pressure $-p$ by an amount dependent upon the extension derivatives $\partial u'/\partial x'$, $\partial v'/\partial y'$, and $\partial w'/\partial z'$. At this point it is necessary to make an assumption relating the normal stresses and the rates of extension. It is assumed that the stress and rates of extension are related in the following manner:*

*In the theory of elasticity the assumption for the principal stresses $\tau_i(i = 1, 2, 3)$ are

$$\tau_i = \lambda^*(e_1' + e_2' + e_3') + 2\mu^* e_i'$$

where λ^* is the bulk modulus, μ^* is the modulus of rigidity of the material, and e_1', e_2', and e_3' are the extensions along the principal axes. The familiar quantities, Poisson's ratio and Young's modulus, are algebraic functions of λ^* and μ^*. See, for instance, S. Timoshenko and J. N. Goodier (1951).

$$\tau_1 = -p + \lambda \left(\frac{\partial u'}{\partial x'} + \frac{\partial v'}{\partial y'} + \frac{\partial w'}{\partial z'} \right) + 2\mu \frac{\partial u'}{\partial x'}$$

$$\tau_2 = -p + \lambda \left(\frac{\partial u'}{\partial x'} + \frac{\partial v'}{\partial y'} + \frac{\partial w'}{\partial z'} \right) + 2\mu \frac{\partial v'}{\partial y'} \qquad (9)$$

$$\tau_3 = -p + \lambda \left(\frac{\partial u'}{\partial x'} + \frac{\partial v'}{\partial y'} + \frac{\partial w'}{\partial z'} \right) + 2\mu \frac{\partial w'}{\partial z'}$$

λ is the coefficient of viscosity that relates normal stress to div $\mathbf{V}$. From the equation of continuity

$$\text{div } \mathbf{V} = -\frac{1}{\rho} \frac{\mathcal{D}\rho}{\mathcal{D}t} \qquad (10)$$

and therefore div $\mathbf{V}$ may be though of as the time rate of change of specific volume of the fluid per unit specific volume. λ is of the nature of a *bulk modulus*. μ is the coefficient of viscosity that relates normal stress to the rates of extension. The factor 2 is inserted for convenience in the operations that follow.

To find the normal stresses in the original coordinate system, Eqs. 9 are substituted in Eqs. 7. The first of Eqs. 7 becomes

$$\tau_{xx} = (l_1^2 + l_2^2 + l_3^2)(-p + \lambda \text{ div } \mathbf{V}) + 2\mu \left(l_1^2 \frac{\partial u'}{\partial x'} + l_2^2 \frac{\partial v'}{\partial y'} + l_3^2 \frac{\partial w'}{\partial z'} \right)$$

But

$$l_1^2 + l_2^2 + l_3^2 = 1$$

and from the first of Eqs. 5

$$\frac{\partial u}{\partial x} = l_1^2 \frac{\partial u'}{\partial x'} + l_2^2 \frac{\partial v'}{\partial y'} + l_3^2 \frac{\partial w'}{\partial z'}$$

Similar operations with the second and third of Eqs. 7 lead to the following result for the normal stresses:

$$\tau_{xx} = -p + \lambda \text{ div } \mathbf{V} + 2\mu \frac{\partial u}{\partial x}$$

$$\tau_{yy} = -p + \lambda \text{ div } \mathbf{V} + 2\mu \frac{\partial v}{\partial y} \qquad (11)$$

$$\tau_{zz} = -p + \lambda \text{ div } \mathbf{V} + 2\mu \frac{\partial w}{\partial z}$$

The shearing stresses in the original coordinate system are found by substituting Eqs. 9 into Eqs. 8. The first of Eqs. 8 becomes

$$\tau_{yz} = (m_1 n_1 + m_2 n_2 + m_3 n_3)(-p + \lambda \operatorname{div} \mathbf{V})$$

$$+ 2\mu \left(m_1 n_1 \frac{\partial u'}{\partial x'} + m_2 n_2 \frac{\partial v'}{\partial y'} + m_3 n_3 \frac{\partial w'}{\partial z'} \right)$$

But

$$m_1 n_1 + m_2 n_2 + m_3 n_3 = 0$$

and from the first of Eqs. 6

$$\mu \gamma_x = 2\mu \left(m_1 n_1 \frac{\partial u'}{\partial x'} + m_2 n_2 \frac{\partial v'}{\partial y'} + m_3 n_3 \frac{\partial w'}{\partial z'} \right)$$

Therefore,

$$\tau_{yz} = \mu \gamma_x$$

Similar operations with the second and third relations in Eqs. 8 lead to the following result for the shear stresses:

$$\tau_{yz} = \mu \gamma_x = \mu \left(\frac{\partial w}{\partial y} + \frac{\partial v}{\partial z} \right)$$

$$\tau_{zx} = \mu \gamma_y = \mu \left(\frac{\partial u}{\partial z} + \frac{\partial w}{\partial x} \right) \qquad (12)$$

$$\tau_{xy} = \mu \gamma_z = \mu \left(\frac{\partial v}{\partial x} + \frac{\partial u}{\partial y} \right)$$

The viscosity coefficients λ and μ may be realted in the following manner. We add Eqs. 11 and obtain

$$\tau_{xx} + \tau_{yy} + \tau_{zz} = -3p + (3\lambda + 2\mu) \operatorname{div} \mathbf{V}$$

from the continuity equation

$$\operatorname{div} \mathbf{V} = -\frac{1}{\rho} \frac{\mathcal{D}\rho}{\mathcal{D}t}$$

so that

$$\tau_{xx} + \tau_{yy} + \tau_{zz} = -3p - (3\lambda + 2\mu) \frac{1}{\rho} \frac{\mathcal{D}\rho}{\mathcal{D}t} \qquad (13)$$

This equation states that the average of the normal stresses differs from the static pressure by a quantity proportional to the substantial derivative of the density.

Therefore, if we make the usual assumption that the pressure is a function only of the density and not of the rate of change of the density as the element moves through the flow, the coefficient of the last term in Eq. 13 must vanish; that is,

$$\lambda = -\tfrac{2}{3}\,\mu \tag{14}$$

The validity of this assumption is discussed elsewhere (e.g., Vincenti and Kruger, 1965). For air, measurements indicate that $-\lambda$ is of the same order as μ. Therefore, the error involved in using Eq. 14 could be appreciable only if $\mathcal{D}\rho/\mathcal{D}t$ is very large. This could happen within a shock wave.

5 Conservation of Momentum—Navier-Stokes Equations

Let a local region $\hat{R}$ in a fluid be bounded by a surface $\hat{S}$. According to the conservation of momentum principle (Section 3.7), the time rate of increase of momentum within $\hat{R}$ is equal to the rate at which momentum is flowing into $\hat{R}$ plus the forces acting on the fluid within $\hat{R}$. These forces are the force of gravity $\rho\mathbf{g}$ acting on each unit volume and the force applied to the fluid at the boundary by the surface stress $\boldsymbol{\tau}$. The mathematical expression for the conservation of momentum in the x direction is

$$\frac{\partial}{\partial t}\iiint_{\hat{R}} \rho u \, d\hat{R} = -\iint_{\hat{S}} (\rho\mathbf{V}\cdot\mathbf{n})\,u\,d\hat{S} + \iiint_{\hat{R}} \rho\mathbf{g}\cdot\mathbf{i}\,d\hat{R} + \iint_{\hat{S}} \boldsymbol{\tau}\cdot\mathbf{i}\,d\hat{S} \tag{15}$$

where $\mathbf{n}$ is the unit vector normal to $\hat{S}$. $\boldsymbol{\tau}\cdot\mathbf{i}$ is given by the first of Eqs. 1. The surface integrals in Eq. 15 are transformed to volume integrals by means of the divergence theorem (Section 2.4), and after a slight rearrangement Eq. 15 becomes

$$\iiint_{\hat{R}} \left[\frac{\partial\rho u}{\partial t} + \operatorname{div}\rho u\mathbf{V} - \rho\mathbf{g}\cdot\mathbf{i} - \operatorname{div}\,(\mathbf{i}\tau_{xx} + \mathbf{j}\tau_{yx} + \mathbf{k}\tau_{zx}) \right] d\hat{R} = 0 \tag{16}$$

Equation 16 is true for all regions no matter how small, and therefore the integrand must vanish. After expanding and regrouping the terms in the integrand of Eq. 16, we have

$$u\left(\rho\operatorname{div}\mathbf{V} + \frac{\mathcal{D}\rho}{\mathcal{D}t}\right) + \rho\frac{\mathcal{D}u}{\mathcal{D}t} = \rho X + \frac{\partial\tau_{xx}}{\partial x} + \frac{\partial\tau_{yx}}{\partial y} + \frac{\partial\tau_{zx}}{\partial z} \tag{17}$$

where X is the force of gravity per unit mass. According to the equation of continuity, the first term in Eq. 17 must vanish. What remains is the statement of conservation of momentum in the x direction. Conservation of momentum in the y and z directions are found in a similar fashion. The three equations may be written

$$\rho \frac{\mathcal{D}u}{\mathcal{D}t} = \rho X + \frac{\partial \tau_{xx}}{\partial x} + \frac{\partial \tau_{yx}}{\partial y} + \frac{\partial \tau_{zx}}{\partial z}$$

$$\rho \frac{\mathcal{D}v}{\mathcal{D}t} = \rho Y + \frac{\partial \tau_{xy}}{\partial x} + \frac{\partial \tau_{yy}}{\partial y} + \frac{\partial \tau_{zy}}{\partial z} \quad (18)$$

$$\rho \frac{\mathcal{D}w}{\mathcal{D}t} = \rho Z + \frac{\partial \tau_{xz}}{\partial x} + \frac{\partial \tau_{yz}}{\partial y} + \frac{\partial \tau_{zz}}{\partial z}$$

The above equations may be expressed in terms of the velocity derivatives by using Eqs. 11 and 12. We have, finally,

$$\rho \frac{\mathcal{D}u}{\mathcal{D}t} = \rho X - \frac{\partial p}{\partial x} + \frac{\partial}{\partial x} (\lambda \ \mathrm{div} \ \mathbf{V}) + \mathrm{div} \left(\mu \frac{\partial \mathbf{V}}{\partial x} \right) + \mathrm{div} \ (\mu \ \mathrm{grad} \ u)$$

$$\rho \frac{\mathcal{D}v}{\mathcal{D}t} = \rho Y - \frac{\partial p}{\partial y} + \frac{\partial}{\partial y} (\lambda \ \mathrm{div} \ \mathbf{V}) + \mathrm{div} \left(\mu \frac{\partial \mathbf{V}}{\partial y} \right) + \mathrm{div} \ (\mu \ \mathrm{grad} \ v) \quad (19)$$

$$\rho \frac{\mathcal{D}w}{\mathcal{D}t} = \rho Z - \frac{\partial p}{\partial z} + \frac{\partial}{\partial z} (\lambda \ \mathrm{div} \ \mathbf{V}) + \mathrm{div} \left(\mu \frac{\partial \mathbf{V}}{\partial z} \right) + \mathrm{div} \ (\mu \ \mathrm{grad} \ w)$$

For an incompressible fluid div $\mathbf{V} = 0$ everywhere and λ and μ are constant. Then Eqs. 19 reduce to

$$\rho \frac{\mathcal{D}u}{\mathcal{D}t} = \rho X - \frac{\partial p}{\partial x} + \mu \nabla^2 u$$

$$\rho \frac{\mathcal{D}v}{\mathcal{D}t} = \rho Y - \frac{\partial p}{\partial y} + \mu \nabla^2 v \quad (20)$$

$$\rho \frac{\mathcal{D}w}{\mathcal{D}t} = \rho Z - \frac{\partial p}{\partial z} + \mu \nabla^2 w$$

6   Conservation of Energy

The mathematical formulation of the conservation-of-energy principle follows the procedure of Section 8.3. Consider a local region $\hat{R}$ bounded by a surface $\hat{S}$. The time rate of increase of the energy per unit mass e within the region is equal to the time rate at which energy crosses the boundary $\hat{S}$ of the region, plus the rate at which heat is conducted into the region through $\hat{S}$, plus the time rate at which work is done on the fluid within $\hat{R}$ by the surface stresses $\boldsymbol{\tau}$. This may be written

$$\frac{\partial}{\partial t} \iiint_{\hat{R}} \rho e \ d\hat{R} = - \iint_{\hat{S}} (\rho \mathbf{V} \cdot \mathbf{n}) e \ d\hat{S} + \iint_{\hat{S}} k\mathbf{n} \cdot \nabla T \ d\hat{S} + \iint_{\hat{S}} \boldsymbol{\tau} \cdot \mathbf{V} \ d\hat{S} \quad (21)$$

$\mathbf{n} \cdot \nabla T$ in the integrand of the second integral on the right is the derivative of the temperature in a direction normal to $d\hat{S}$. k is the thermal conductivity. The second integral is an application of the heat conduction law given by Eq. 8 of Appendix A. The integrand of the third integral is the scalar product of a force per unit area and the velocity. This is the rate at which the surface stresses do work on the fluid within $\hat{R}$, and by means of Eqs. 1 it may be written

$$\boldsymbol{\tau} \cdot \mathbf{V} = (u\mathbf{f}_1 + v\mathbf{f}_2 + w\mathbf{f}_3) \cdot \mathbf{n} \tag{22}$$

With the help of Eq. 22, the three surface integrals of Eq. 21 may be converted to volume integrals by using the divergence theorem. After this transformation and rearrangement of terms, Eq. 21 becomes

$$\iiint_{\hat{R}} \left[\frac{\partial \rho e}{\partial t} + \operatorname{div}(\rho e \mathbf{V}) - \operatorname{div}(k \nabla T) - \operatorname{div}(u\mathbf{f}_1 + v\mathbf{f}_2 + w\mathbf{f}_3) \right] d\hat{R} = 0 \tag{23}$$

Because Eq. 23 is true for all regions no matter how small, the integrand must vanish. After expansion and rearrangement of terms, the integrand becomes

$$e\left(\frac{\mathfrak{D}\rho}{\mathfrak{D}t} + \rho \operatorname{div} \mathbf{V} \right) + \rho \frac{\mathfrak{D}e}{\mathfrak{D}t} - \operatorname{div}(k \nabla T) - (u\nabla \cdot \mathbf{f}_1 + v\nabla \cdot \mathbf{f}_2 + w\nabla \cdot \mathbf{f}_3)$$

$$- (\mathbf{f}_1 \cdot \nabla u + \mathbf{f}_2 \cdot \nabla v + \mathbf{f}_3 \cdot \nabla w) = 0 \tag{24}$$

The first expression vanishes because of continuity of the flow. The derivative of the second term is a substantial time rate of the total internal energy e which, according to the discussion of Section 8.4, is made up of three parts.[*]

$$e = \tilde{u} + \frac{V^2}{2} - \mathbf{g} \cdot \mathbf{r} \tag{25}$$

$\mathbf{r}$ is the displacement vector and $\mathbf{g} \cdot \mathbf{r}$ is the gravitational potential energy. $\tilde{u}$ and $V^2/2$ are the intrinsic energy and kinetic energy[†] respectively. The term $u\nabla \cdot \mathbf{f}_1$ in the first bracket of Eq. 24 may be expanded in the form

$$u\left(\frac{\partial \tau_{xx}}{\partial x} + \frac{\partial \tau_{yx}}{\partial y} + \frac{\partial \tau_{zx}}{\partial z} \right)$$

From the first of Eqs. 18 this is equivalent to

$$\rho \frac{\mathfrak{D}}{\mathfrak{D}t}\left(\frac{u^2}{2} \right) - \rho u X$$

---

[*]The x component of the velocity and the intrinsic energy both have been given the symbol u in the preceding chapters. To distinguish one from the other in this appendix, the symbol $\tilde{u}$ is adopted for intrinsic energy.

[†]$\mathbf{V}$ is the instantaneous velocity, and so the mean value of e would include the mean energy of turbulent fluctuations.

The terms $v \nabla \cdot \mathbf{f}_2$ and $w \nabla \cdot \mathbf{f}_3$ may be expanded in a similar manner. The sum becomes

$$u \nabla \cdot \mathbf{f}_1 + v \nabla \cdot \mathbf{f}_2 + w \nabla \cdot \mathbf{f}_3 = \rho \frac{\mathfrak{D}}{\mathfrak{D} t} \left(\frac{V^2}{2} \right) - \rho \mathbf{V} \cdot \mathbf{g} \qquad (26)$$

If Eqs. 25 and 26 are substituted into Eq. 24, there results

$$\rho \frac{\mathfrak{D} \tilde{u}}{\mathfrak{D} t} - \operatorname{div} (k \nabla T) - (\mathbf{f}_1 \cdot \nabla u + \mathbf{f}_2 \cdot \nabla v + \mathbf{f}_3 \cdot \nabla w) = 0 \qquad (27)$$

The last bracket is expanded using Eqs. 1 and then the stresses are written in terms of strains with the help of Eqs. 11 and 12. If $\lambda = -\frac{2}{3} \mu$, Eq. 27 finally becomes

$$\rho \frac{\mathfrak{D} \tilde{u}}{\mathfrak{D} t} = \operatorname{div} (k \nabla T) - p \operatorname{div} \mathbf{V} + \Phi \qquad (28)$$

where

$$\Phi = -\frac{2}{3} \mu (\operatorname{div} \mathbf{V})^2 + 2\mu \left[\left(\frac{\partial u}{\partial x} \right)^2 + \left(\frac{\partial v}{\partial y} \right)^2 + \left(\frac{\partial w}{\partial z} \right)^2 \right] + \mu(\gamma_x^2 + \gamma_y^2 + \gamma_z^2) \qquad (29)$$

Φ is the dissipation function and represents the time rate at which energy is being dissipated per unit volume through the action of viscosity.

If the specific heat c_v may be considered constant, then, following the discussion of Chapter 8, we may write

$$\frac{\mathfrak{D} \tilde{u}}{\mathfrak{D} t} = c_v \frac{\mathfrak{D} T}{\mathfrak{D} t}$$

Assuming the thermal conductivity k to be constant also, Eq. 28 specializes to

$$\rho c_v \frac{\mathfrak{D} T}{\mathfrak{D} t} = k \nabla^2 T - \rho \operatorname{div} \mathbf{V} + \Phi \qquad (30)$$

For incompressible viscous flows, Eq. 30 may be further specialized to

$$\rho c_v \frac{\mathfrak{D} T}{\mathfrak{D} t} = k \nabla^2 T + \Phi \qquad (31)$$

where the dissipation function Φ is given by Eq. 29 with the omission of the term containing $\operatorname{div} \mathbf{V}$.

A more useful form of the equation than Eq. 28 may be derived as follows. From Section 8.3

$$\tilde{u} = c_p T - \frac{p}{\rho} = h - \frac{p}{\rho}$$

where h is the enthalpy. Differentiating,

$$\frac{\mathcal{D}\tilde{u}}{\mathcal{D}t} = \frac{\mathcal{D}h}{\mathcal{D}t} - \frac{1}{\rho}\frac{\mathcal{D}p}{\mathcal{D}t} + \frac{p}{\rho^2}\frac{\mathcal{D}\rho}{\mathcal{D}t}$$

and, using continuity,

$$\rho\frac{\mathcal{D}\tilde{u}}{\mathcal{D}t} = \rho\frac{\mathcal{D}h}{\mathcal{D}t} - \frac{\mathcal{D}p}{\mathcal{D}t} - p \text{ div } \mathbf{V} \tag{32}$$

After equating Eqs. 28 and 32, we get

$$\rho\frac{\mathcal{D}h}{\mathcal{D}t} - \frac{\mathcal{D}p}{\mathcal{D}t} = \text{div } (k \text{ grad } T) + \Phi \tag{33}$$

7 Boundary Layer Equations

The application of a perfect-fluid analysis to problems in aerodynamics was justified when Prandtl in 1904 simplified the Navier-Stokes equations to the boundary layer equations by postulating that, for a fluid of small viscosity, such as air (or water), the viscosity will alter the flow around a streamline body only in the immediate vicinity of the surface. Outside of this layer, viscosity can be neglected and the flow is predicted to a high degree of accuracy by perfect-fluid analysis.

We consider a two-dimensional flow over a cylindrical surface with the coordinate system shown in Fig. 4. The curvature of the surface must not be too great, otherwise the centripetal terms must be included. The flow will be described by the first two momentum equations of Eqs. 19, the continuity equation, Eq. 10, and the energy equation, Eq. 33.

Simplification of these equations to obtain the boundary layer equations depends

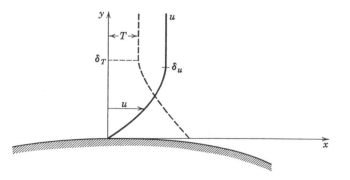

Fig. 4

on an *order-of-magnitude* analysis based on Prandtl's hypothesis that the effects of viscosity are confined to a *thin* boundary layer.

There will, in general, be velocity and temperature boundary layers, whose respective thicknesses are designated δ_u and δ_T. To examine the relative magnitudes of δ_u and δ_T we recall in Section 16.5 we showed that $\delta_u = \delta_T$ for a fluid with a Prandtl number of unity. Since for air the Prandtl number varies from about 0.68 to 0.75, the actual δ_u and δ_T are not far from equal and for purposes of the order-of-magnitude analysis, we designate both thicknesses by δ.

We consider the dimensionless boundary layer thickness $\delta' = \delta/L$ where L is a characteristic length. The sequence of orders of magnitude from small to large will be δ'^2, δ', 1, $1/\delta'$, $1/\delta'^2$. If L is taken as the distance from the leading edge, δ' will in many practical cases be less than 10^{-2}.

In determining the orders of magnitude of the various terms in the equations we shall neglect those effects which will not change the orders of magnitude of the various terms. For instance, if we neglect variations in density, we are in effect postulating that $(1/\rho)\,\mathcal{D}\rho/\mathcal{D}t$ in Eq. 10 will be of order *not greater* than that of either of the other two terms in the equation, $\partial u/\partial x$ and $\partial v/\partial y$. We would then be justified in the hypothesis that $\partial u/\partial x$ and $\partial v/\partial y$ will be of the same order of magnitude in the boundary layer.* Also, since $\mu \sim T$, approximately, very large temperature gradients can be accommodated without changing the order of magnitude of the terms involving viscosity. With these simplifications, the first two of Eqs. 19, Eq. 10, and Eq. 33 become

$$\rho\,\frac{\mathcal{D}u}{\mathcal{D}t} = -\frac{\partial p}{\partial x} + \mu \nabla^2 u$$

$$\rho\,\frac{\mathcal{D}v}{\mathcal{D}t} = -\frac{\partial p}{\partial y} + \mu \nabla^2 v \tag{34}$$

$$\text{div } \mathbf{V} = 0$$

$$\rho c_p\,\frac{\mathcal{D}T}{\mathcal{D}t} - \frac{\mathcal{D}p}{\mathcal{D}t} = k\nabla^2 T + \Phi$$

where Φ is given by Eq. 29.

We now nondimensionalize these equations by introducing

$$u' = \frac{u}{V}, \quad v' = \frac{v}{V}, \quad x' = \frac{x}{L}, \quad y' = \frac{y}{L}, \quad t' = \frac{tV}{L},$$

$$p' = \frac{p}{\rho_1 V^2}, \quad Re = \frac{VL}{\nu_1}, \quad T' = \frac{T}{T_1}, \quad M_1 = \frac{V}{a_1}, \quad Pr = \frac{\mu_1 c_p}{k}$$

*For flow through a normal shock, this hypothesis would be in error because there $\partial u/\partial x$ has very high values and $\partial v/\partial y = 0$.

where V and the subscript 1 refer to free-stream values. Equations 34 then become

$$\frac{\mathcal{D}u'}{\mathcal{D}t'} = -\frac{\partial p'}{\partial x'} + \frac{1}{Re}\nabla'^2 u'$$

$$\frac{\mathcal{D}v'}{\mathcal{D}t'} = -\frac{\partial p'}{\partial y'} + \frac{1}{Re}\nabla'^2 v' \qquad (36)$$

$$\text{div}'\,\mathbf{V} = 0$$

$$\frac{\mathcal{D}T'}{\mathcal{D}t'} - (\gamma-1)M_1^2\frac{\mathcal{D}p'}{\mathcal{D}t'} = \frac{1}{PrRe}\nabla'^2 T' + \frac{(\gamma-1)M_1^2}{Re}\frac{\Phi'}{\mu_1}$$

We postulate that u', $\partial^n u'/\partial x'^n$, $\partial u'/\partial t'$, $\partial T'/\partial t'$ and $\partial^n T'/\partial x'^n$ are at most of order unity, that is,

$$u' = O(1), \quad \frac{\partial^n u'}{\partial x'^n} = O(1), \quad \frac{\partial u'}{\partial t'} = O(1)$$

$$\frac{\partial T'}{\partial t'} = O(1), \quad \frac{\partial^n T'}{\partial x'^n} = O(1) \qquad (37)$$

These relations should be interpreted in the sense that the equations we shall derive from them become inapplicable to problems in which the u' or T' derivatives approach a larger order of magnitude than that postulated. We shall see later that the restrictions are not serious.

Now, from the continuity equation and Eqs. 37

$$\frac{\partial u'}{\partial x'} = -\frac{\partial v'}{\partial y'} = O(1)$$

and then

$$v' = \int_0^{\delta'}\frac{\partial v'}{\partial y'}\,dy' \simeq \int_0^{\delta'} O(1)\,dy = O(\delta') \qquad (38)$$

In carrying out the integration we have in effect substituted a mean value for $\partial v'/\partial y'$. This is permissible since we are interested only in finding which terms in the equations of motion are small enough to neglect. Equations 37 imply that differentiation with respect to x' or t' does not change the order of magnitude of a quantity, whereas Eq. 38 implies that differentiation with respect to y' increases the order of magnitude by one; that is, $v' = O(\delta')$ and $\partial v'/\partial y' = O(1)$. Then,

$$\frac{\partial u'}{\partial y'} = O\left(\frac{1}{\delta'}\right), \quad \frac{\partial^2 u'}{\partial y'^2} = O\left(\frac{1}{\delta'^2}\right)$$

$$\frac{\partial v'}{\partial x'} = O(\delta'), \quad \frac{\partial^2 v'}{\partial x'^2} = O(\delta')$$

$$\frac{\partial v'}{\partial t'} = O(\delta'), \quad \frac{\partial^2 v'}{\partial y'^2} = O\left(\frac{1}{\delta'}\right) \tag{39}$$

$$\frac{\partial T'}{\partial y'} = O\left(\frac{1}{\delta'}\right), \quad \frac{\partial^2 T'}{\partial y'^2} = O\left(\frac{1}{\delta'^2}\right)$$

The above differentiation rule was derived by means of the equation of continuity and therefore does not apply to the pressure. However, since $v' \ll u'$, Euler's equation applied at the edge of the boundary layer and Eqs. 37 gives

$$\frac{\partial p'_i}{\partial x'} \simeq -u'_e \frac{du'_e}{dx'} = O(1)$$

Then, by the same reasoning that led to Eqs. 37 we may write

$$\frac{\partial p'}{\partial x'} = O(1) \tag{40}$$

Equation 40 should be interpreted as were Eqs. 37 above; it simply places an upper limit on the order of magnitude of $\partial p'/\partial x'$. Equations 36 are rewritten in expanded form and the orders of magnitude of the terms (Eqs. 37–40) inserted:

$$
\overset{(1)}{\frac{\partial u'}{\partial t'}} + \overset{(1)\cdot(1)}{u'\frac{\partial u'}{\partial x'}} + \overset{(\delta')\cdot(1/\delta')}{v'\frac{\partial u'}{\partial y'}} = \overset{(1)}{-\frac{\partial p'}{\partial x'}} + \frac{1}{Re}\left(\overset{(?)}{\frac{\partial^2 u'}{\partial x'^2}} + \overset{(1/\delta'^2)}{\frac{\partial^2 u'}{\partial y'^2}}\right)
$$

$$
\overset{(\delta')}{\frac{\partial v'}{\partial t'}} + \overset{(1)\cdot(\delta')}{u'\frac{\partial v'}{\partial x'}} + \overset{(\delta')\cdot(1)}{v'\frac{\partial v'}{\partial y'}} = \overset{(?)}{-\frac{\partial p'}{\partial y'}} + \frac{1}{Re}\left(\overset{(\delta')}{\frac{\partial^2 v'}{\partial x'^2}} + \overset{(1/\delta')}{\frac{\partial^2 v'}{\partial y'^2}}\right)
$$

$$\tag{41}$$

$$
\overset{(1)}{\frac{\partial T'}{\partial t'}} + \overset{(1)\cdot(1)}{u'\frac{\partial T'}{\partial x'}} + \overset{(\delta')\cdot(1/\delta')}{v'\frac{\partial T'}{\partial y'}} - (\gamma-1)M_1^2\left(\overset{(1)}{\frac{\partial p'}{\partial t'}} + \overset{(1)\cdot(1)}{u'\frac{\partial p'}{\partial x'}} + \overset{(\delta')\cdot(?)}{v'\frac{\partial p'}{\partial y'}}\right)
$$

$$
= \frac{1}{PrRe}\left(\overset{(?)}{\frac{\partial^2 T'}{\partial x'^2}} + \overset{(1/\delta'^2)}{\frac{\partial^2 T'}{\partial y'^2}}\right) + \overset{(?)}{\frac{(\gamma-1)M_1^2}{Re}}\overset{(1/\delta'^2)}{\frac{\Phi'}{\mu_1}}
$$

Evaluating the first of these equations, we see (1) that $\partial^2 u'/\partial x'^2$ is negligible compared with $\partial^2 u'/\partial y'^2$; and (2) that, since all other terms are of order unity, the one involving the Reynolds number cannot be of larger order. Then, to make this term of order unity,

$$Re = O\left(\frac{1}{\delta'^2}\right) \tag{42}$$

This equation is substituted into the second of Eqs. 41, from which we see that all terms except $\partial p'/\partial y'$ are of order δ'. Therefore, this term cannot be of an order larger than δ', and we write

$$\frac{\partial p'}{\partial y'} = O(\delta') \tag{43}$$

In other words, *the pressure is approximately constant through the boundary layer.*

The following simplifications may be made in the third of Eqs. 41: by Eq. 43 we may neglect $v' \, \partial p'/\partial y'$; by reference to Eq. 29, all terms in Φ'/μ_1 are of order unity except $(\partial u'/\partial y')^2$, which is of order $1/\delta'^2$.

It was pointed out earlier that whereas the above simplifications apply strictly to incompressible flow only, the orders of magnitudes of the terms are not changed by compressibility effects except perhaps at very high Mach numbers and temperature ratios.

Then, on the basis of the simplifications in Eqs. 41, Eqs. 19 reduce to the following for the compressible two-dimensional boundary layer.

$$\rho \left(\frac{\partial u}{\partial t} + u \frac{\partial u}{\partial x} + v \frac{\partial u}{\partial y} \right) = - \frac{\partial p}{\partial x} + \frac{\partial}{\partial y} \left(\mu \frac{\partial u}{\partial y} \right)$$

$$\rho c_p \left(\frac{\partial T}{\partial t} + u \frac{\partial T}{\partial x} + v \frac{\partial T}{\partial y} \right) - \frac{\partial p}{\partial t} - u \frac{\partial p}{\partial x} = \frac{\partial}{\partial y} \left(k \frac{\partial T}{\partial y} \right) + \mu \left(\frac{\partial u}{\partial y} \right)^2 \tag{44}$$

The boundary layer equations derived in an approximate way in Sections 14.5 and 16.2, respectively are identical to Eqs. 44. They were solved for various conditions in Chapters 15 and 16.

The above order-of-magnitude analysis is a more reliable way of obtaining the equations than that used in the text, because here we can assess the approximate magnitude of the terms neglected as well as the circumstances under which they might be large enough to affect the solution.

We can gain from the solution of Section 15.3 some appreciation of the amount by which a term must change in order to change its order of magnitude. For instance, Eq. 15.15 gives

$$\frac{\delta}{x} = 5.2 \sqrt{\frac{\nu}{u_e x}}$$

With $x = 1$ m, $u_e = 100$ m/sec and $\nu = 1.44 \times 10^{-5}$ m²/sec for air, $\delta = 1.97 \times 10^{-3}$ m. Then, for the conditions assumed, the ratio between successive orders of magnitude is $1/1.97 \times 10^{-3} = 508$. With such a large ratio between successive orders of magnitude, we see that the approximations involved in using Eqs. 44 are not serious for a wide range of practical problems.

Problems

Section 1.3

1. Assume that of the $N/6$ molecules moving toward, and normal to, a surface, $n_1/6$ molecules have the speed c_1, $n_2/6$ have the speed c_2, and so on, where $n_1 + n_2 + \cdots = N$. Show that the pressure exerted on the surface is

$$p = \tfrac{1}{3}\rho\overline{v^2}$$

where

$$\overline{v^2} = \frac{(n_1 c_1^2 + n_2 c_2^2 + \cdots)}{N}$$

is called the "mean square molecular speed."

Section 1.4

1. Show that for an isothermal atmosphere the pressure distribution is described by

$$p = p_0 \exp\left(-gz/RT\right)$$

2. In an isentropic atmosphere the relationship between pressure and density is governed by

$$\frac{p}{p_0} = \left(\frac{\rho}{\rho_0}\right)^{\gamma}$$

where γ is the ratio of the specific heat of air at constant pressure to that at constant volume. Show that the pressure distribution in such an atmosphere is described by

$$p = p_0\left(1 - \frac{\gamma - 1}{\gamma}\frac{\rho_0}{p_0}gz\right)^{\gamma/(\gamma-1)}$$

Section 1.5

1. A hollow sphere (with vacuum inside) of outer radius r_0 and inner radius r_i is made of a material of specific gravity α. It is required that the sphere can stay anywhere in a stagnant tank of water when completely submerged. Show that the inner radius must have the value

$$r_i = r_0(1 - \alpha^{-1})^{1/3}$$

Section 1.6

1. A rectangular tunnel of height h is built on the bottom of a river of depth H. The pressure inside the tunnel is the same as that at the upper surface of the river. Show that for unit length of the tunnel the hydrostatic force acting on each side wall is $\rho g h(H - h/2)$ and the moment about the base of the wall is $\frac{1}{2}\rho g h^2(H - \frac{2}{3}h)$.

Section 1.9

1. For $V = 100$ mph and $l = 3$ ft, show that the Reynolds number and force coefficient (Eq. 18) have the same values in FPS and SI units. Calculate for both air and water under standard conditions. Tables 1 and 2 give conversion factors and numerical values.

Section 2.2

1. A two-dimensional pressure field is defined by the expression

$$p = x^2 y + y^2$$

Show that at the point (3, 2) the derivative of p in the direction of a line that makes an angle of $45°$ with the positive x axis has the value of 17.68.

2. For the pressure field described in Problem 1, show that the derivative at the point (3, 2) has the value of 15.5 in the direction of the curve

$$3y^2 - 4x = 0$$

At the same point, show that the direction for which the derivative is the maximum makes an angle of $47.25°$ with the positive x axis; and that the value of the derivative of p in this direction is 17.69.

3. A temperature field is described by the equation

$$T = x^3 y$$

Show that at the point (1, 2) the magnitude of grad T is 6.08 and the slope of the gradient line is 1/6. Show also that the equations of the gradient line and isotherm passing through the point (1, 2) are, respectively,

$$x^2 - 3y^2 + 11 = 0; \qquad x^3 y = 2$$

Section 2.3

1. A two-dimensional velocity field is described in terms of its Cartesian components

$$u = 2xy^2; \qquad v = 2x^2 y$$

Show that the equation of the streamline passing through the point $(1, 7)$ is

$$y^2 - x^2 = 48$$

2. In a flow field the absolute value of the velocity is constant along circles that are concentric about the origin. The streamlines are straight lines passing through the origin. Show that the velocity components in Cartesian coordinates are

$$u = \frac{xf(x^2 + y^2)}{\sqrt{x^2 + y^2}} \; ; \qquad v = \frac{yf(x^2 + y^2)}{\sqrt{x^2 + y^2}}$$

and, in polar coordinates,

$$u_r = f(r), \quad u_\theta = 0$$

3. Expand the function $y \cos x$ in a Taylor series about the point $(0, 0)$. Show that for small values of x and y

$$y \cos x \simeq y - \tfrac{1}{2}x^2 y$$

4. The absolute value of the velocity and the equation of the streamlines in a velocity field are given respectively by

$$|V| = \sqrt{x^2 + 2xy + 2y^2} \; ; \qquad y^2 + 2xy = \text{constant}$$

Show that $u = x + y$ and $v = -y$.

5. In the flow field of Problem 4, show that the rate of change of u is the maximum in the direction of a line that makes an angle of $45°$ with the positive x axis. Show that $|\text{grad } u| = \sqrt{2}$.

Section 2.4

1. In cylindrical polar coordinates (r, θ, z) shown in Fig. 3, show that the divergence of the velocity vector $\mathbf{V}$ is

$$\text{div } \mathbf{V} = \frac{\partial u_r}{\partial r} + \frac{1}{r}\left(u_r + \frac{\partial u_\theta}{\partial \theta}\right) + \frac{\partial u_z}{\partial z}$$

2. The streamlines of a two-dimensional velocity field are straight lines through the origin described by the equation

$$y = mx$$

The absolute value of the velocity varies according to the law

$$|V| = \frac{\Lambda}{2\pi r}$$

Using both Cartesian and polar coordinates, show that the value of div $\mathbf{V}$ at the point (1, 2) is zero. In general, what can be said about div $\mathbf{V}$ at all points in the field? What can be said about div $\mathbf{V}$ at the point (0, 0)?

3. In the *three-dimensional* flow from a point source the streamlines radiate in all directions from a point. If we designate the strength of the source by Λ' show that

$$|\mathbf{V}| = \Lambda'/4\pi r^2$$

Note the dimensions of Λ' compared with Λ (for a line source).

Section 2.5

1. Which of the following flows satisfy *conservation of mass* for the flow of an incompressible fluid?

(a) $u = -x^3 \sin y$
$\quad\ v = -3x^2 \cos y$

(b) $u = x^3 \sin y$
$\quad\ v = -3x^2 \cos y$

(c) $u_r = 2r \sin \theta \cos \theta$
$\quad\ u_\theta = -2r \sin^2 \theta$

(d) $|\mathbf{V}| = \dfrac{k}{r^2}$
$\quad\ x^2 + y^2 = c$ (streamlines)

2. For a certain flow field, the absolute value of the velocity and the equation of the streamlines are given respectively by

$$|\mathbf{V}| = f(r); \quad y = mx$$

Show that $f(r)$ must have the form for source flow (and no other) if the pattern is to satisfy conservation of mass for the flow of an incompressible fluid.

Hint: Apply the condition div $\mathbf{V} = 0$ and solve the resulting differential equation for $f(r)$.

3. A flow field is described by

$$|\mathbf{V}| = f(r); \quad x^2 + y^2 = c$$

What form must $f(r)$ have if *continuity* is to be satisfied? Explain your results.

4. A flow field is described by

$$|\mathbf{V}| = f(r, \theta); \quad x^2 + y^2 = c$$

Is there any function $f(r, \theta)$ for which this field satisfies continuity? Explain your answer.

Section 2.6

1. The stream function of a two-dimensional incompressible flow is given by the equation

$$\psi = x^2 + 2y$$

(a) Show that the velocity vector at the point $(2, 3)$ makes an angle of $-63.4°$ with the positive x axis and its magnitude is 4.47.

(b) At the point $(2, 3)$, show that the velocity component in the direction that makes an angle of $30°$ with the positive x axis has the value of -0.266.

2. The existence of a stream function depends on the flow satisfying *continuity*. Therefore, any velocity field derived from a stream function automatically satisfies *continuity*. Prove the latter statement.

3. An incompressible two-dimensional flow is described by the stream function

$$\psi = x^2 + y^3$$

Write the equation of the streamline that passes through the point $(2, 1)$. Show that the magnitude of $\mathbf{V}$ at $(2, 1)$ is equal to the absolute value of grad ψ at $(2, 1)$. Show that the direction of the velocity is perpendicular to the direction of grad ψ.

4. A two-dimensional incompressible flow is described by the velocity components

$$u = 2x; \qquad v = -6x - 2y$$

Does the flow satisfy continuity? If so find the stream function.

Section 2.7

1. The streamlines of a certain flow are concentric circles about the origin, and the absolute value of the velocity varies according to the law

$$|\mathbf{V}| = kr^n$$

Show that the angular speed of any fluid element in the flow is described by

$$\epsilon_z = \tfrac{1}{2}k(n + 1)r^{n-1}$$

and that the corresponding rate of strain of the element is $\gamma_z = -2k/r^2$ if $\epsilon_z = 0$.

2. Consider a "boundary layer" of thickness δ at a distance x $(x \gg \delta)$ from the leading edge of a flat plate. Assume the velocity components are

$$u = 2u_e \left[\left(\frac{y}{\delta} \right) - \frac{1}{2} \left(\frac{y}{\delta} \right)^2 \right], \quad v = 0$$

Find ϵ_z and γ_z for $0 < y < \delta$.

Section 2.8

1. In two-dimensional polar coordinates, show that the curl of the velocity vector **V** is

$$\text{curl}_z \, \mathbf{V} = \frac{\partial u_\theta}{\partial r} + \frac{1}{r}\left(u_\theta - \frac{\partial u_r}{\partial \theta}\right)$$

Hint: Apply Eq. 33 to a region formed by two circular arcs and two radius vectors.

2. When $n = 1$ for the flow of Problem 2.7.1, show that the circulation along a circular path of radius r with center at the origin has the value $2\pi k r^2$.

3. The stream function of a two-dimensional incompressible flow is given by

$$\psi = \frac{\Gamma}{2\pi}\ln r$$

Show that the circulation about a closed path enclosing the origin is Γ and is independent of r. Find the circulation for a specific radius.

Section 2.9

1. Does the irrotational velocity field of a two-dimensional vortex flow satisfy *continuity*?

Section 2.10

1. The components of **V** are given by the expressions

$$u = x^2 + y^2; \qquad v = 2xy^2$$

Show that the integral along the path **s** between the points $(0, 0)$ and $(1, 2)$ of the component of **V** in the direction of **s** has the value of
(a) 17/3 if **s** is a straight line.
(b) 83/15 if **s** is a parabola with vertex at the origin and opening to the right.
(c) 17/3 if **s** is a portion of the x axis and a straight line perpendicular to it.

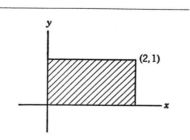

Prob. 2.10.2, page 450

2. The absolute value of the velocity and the equation of the streamlines in a two-dimensional velocity field are given by the expressions

$$|\mathbf{V}| = \sqrt{x^2 + 4xy + 5y^2}; \qquad y^2 + xy = c$$

By two different methods, show that the value of the integral of curl$_z$ **V** over the surface shown is -4.

Section 2.11

1. A velocity field is described by the following components:

$$u = 2xy; \qquad v = x^2 + 1$$

Beginning with these expressions, show that the velocity potential is given by

$$\phi = x^2 y + y$$

2. Does a potential exist for the field described by

$$u = 3xy; \qquad v = x^2 + 1$$

Show that every field that is derived from a potential is necessarily irrotational.

3. A two-dimensional velocity field is described in the following manner:

$$u = x^2 - y^2 + x; \qquad v = -(2xy + y)$$

Show that this field satisfies *continuity* and is also irrotational. Beginning with these components show that the velocity potential is given by

$$\phi = \tfrac{1}{3}x^3 + \tfrac{1}{2}(x^2 - y^2) - xy^2$$

Show that the line integral of **V** · d**s** along a straight line connecting $(0, 0)$ and $(2, 3)$ has the value of $-107/6$. Check this answer by computing the value of the line integral directly from the potential.

4. A two-dimensional potential field is described in the following manner:

$$\phi = \frac{x^3}{3} - x^2 - xy^2 + y^2$$

Show that the velocity component in the direction of the path $x^2 y = -4$ at the point $(2, -1)$ has the value of $1/\sqrt{2}$.

5. If a velocity field has a potential, are we guaranteed that the conservation of mass principle is satisfied?

6. Show that the two fields described below are identical.

$$\psi = 2xy + y; \qquad \phi = x^2 + x - y^2$$

Section 2.12

1. Show that the velocity induced at the center of a circular vortex filament of strength Γ and radius R is $\Gamma/2R$.

2. A rectangular vortex filament of strength $4\pi \times 10^4$ m²/sec is shown in the figure below. Show that the magnitude of the velocity at point A induced by 0.01 m of filament at point B is 12.5 m/sec. What is the direction of that induced velocity? What is the velocity induced at point A by 0.01 m of filament at C?

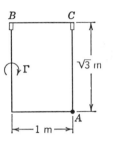

Prob. 2.12.2

Section 3.2

1. Flow in a right angle corner is described by the stream function

$$\psi = Axy$$

where A is a constant. (See example Section 2.2.)
Show that as the fluid elements move along the streamlines their speed increases at the rate

$$\frac{\mathcal{D}V}{\mathcal{D}t} = \frac{A^2(x^2 - y^2)}{r}$$

where r is the radial distance from the origin. (What is the curve along which $\mathcal{D}V/\mathcal{D}t$ varies inversely as r?)

2. Show that in a steady flow the acceleration of a fluid element at a distance r from a *line* source is $-\Lambda^2/4\pi^2r^3$, and that for a *point* source (see Problem 2.4.3) the value is $-\Lambda'^2/8\pi^2r^5$. Distinguish between acceleration and $\mathcal{D}V/\mathcal{D}t$.

3. For flow along elliptical streamlines described by $\psi = ax^2 + by^2$, show that the stream speed of a fluid element increases at the rate

$$\frac{\mathcal{D}V}{\mathcal{D}t} = \frac{4xyab(a - b)}{(a^2x^2 + b^2y^2)^{1/2}}$$

For what values of a and b or the relation between them does $\mathcal{D}V/\mathcal{D}t = 0$? What are the shapes of the streamlines for these flows?

4. Show that the acceleration of a fluid particle in the steady two-dimensional field described by $\psi = xy + y^2$ is $\mathbf{i}x + \mathbf{j}y$.

5. Show that the acceleration of a fluid particle whose motion is described by

$$|\mathbf{V}| = \Omega r; \qquad x^2 + y^2 = k \text{ (streamlines)}$$

is in the radial direction and has the magnitude $-\Omega^2 r$. Compare your answer with the centripetal acceleration of a rigid particle moving along a curved path.

6. Consider a velocity field described in the following manner:

$$u = xy + 20t; \qquad v = x - \tfrac{1}{2}y^2 + t^2$$

Show that the acceleration in the x-direction of a fluid particle is $x^2 + x(\tfrac{1}{2}y^2 + t^2) + 20yt + 20$.

7. The velocity potential of a steady flow field is given by the expression

$$\phi = 2xy + y$$

The temperature is the following function of the field coordinates:

$$T = x^2 + 3xy + 2$$

show that the time rate of change of temperature of a fluid element as it passes through the point (2, 3) is 108.

Section 3.3

1. When a **pitot tube** as shown is placed in a creek with its lower open end facing upstream, water rises in the vertical portion to a height of 5 cm above the water surface. Show that the flow velocity V is 0.99 m/sec.

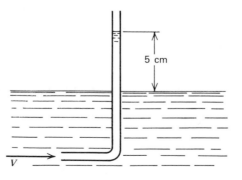

Prob. 3.3.1

2. An airfoil is traveling through sea level air at a speed of 180 km/hr. Show that the pressure at a stagnation point on the airfoil is 1533 N/m² above the atmospheric pressure. Consider a point on the airfoil where the velocity of the air relative to the airfoil is 60 m/sec. Show that the pressure at that point is 674 N/m² below the atmospheric value.

3. Sea level air is being drawn into a vacuum tank through a duct as shown in the accompanying diagram. The static pressure at station AA in the duct measures 9.33×10^4 N/m². Show that the velocity at station AA is 115 m/sec. Assume an incompressible nonviscous flow.

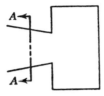

Prob. 3.3.3

4. In Fig. *a* below, sea level air is being drawn into a vacuum tank through a duct. In Fig. *b*, the airfoil is moving through sea level air at a speed of 30 m/sec. In both cases, the relative velocity between airfoil and air at point A is 60 m/sec and the air is incompressible. Show that the static pressure at point A in the first case is 2207 N/m² below, while that in the second case is 1655 N/m² below the atmospheric pressure.

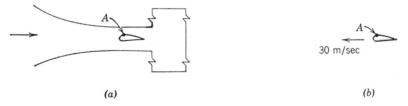

(a)   (b)

Prob. 3.3.4

Section 3.4

1. Show that with gravity neglected, the pressure gradient for vortex flow is $\rho \Gamma^2/4\pi^2 r^3$ in the radial direction.

2. A two-dimensional incompressible flow is described by the stream function $\psi = x^2 - y^2$. Show that $\nabla p/\rho$ at the point (1, 2) is $-4\mathbf{i} - 8\mathbf{j}$.

Section 3.6

1. A fluid is rotating as a solid body according to the law

$$|\mathbf{V}| = \Omega r$$

Show that the difference in total head is $11\rho\Omega^2$ between streamlines whose radii are 5 m and 6 m, respectively. The density of the fluid is ρ.

Section 3.7

1. Derive Euler's equation (Eqs. 13) from the momentum integrals (Eqs. 30) by letting $\hat{R} \to 0$. *Note:* Use the divergence theorem (Section 2.4) to express the surface integrals of Eqs. 30 as volume integrals, then let $\hat{R} \to 0$. (In Appendix B, this transformation is carried out for a viscous fluid.)

2. Air at a pressure of 1.08×10^5 N/m^2 and a velocity of 30 m/sec enters the tank shown in the figure below at A and B. The entrance areas are each 5 cm^2. The air discharges at atmospheric pressure at C through an area of 10 cm^2. Steady conditions are assumed at the entrances and exit, and the density of the air may be considered to have the sea level value everywhere. Show that the reactions are $R_1 = 3.86$ N and $R_2 = 6.64$ N, respectively, in the directions shown in the figure.

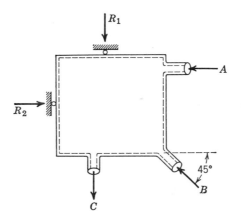

Prob. 3.7.2

3. The velocity distribution is measured at a distance of 0.4 chords behind an airfoil. It is found that the wake half-width $b/2 = 0.1$ chords and the velocity dis-

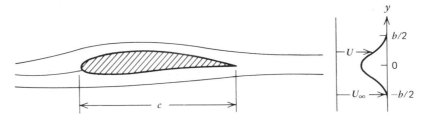

Prob. 3.7.3

tribution in the wake is given approximately by the formula

$$U = U_\infty \left[1 - 0.83 \cos^2 \left(\frac{\pi y}{b}\right)\right]$$

Show that the drag coefficient of the airfoil is $c_d = D'/\frac{1}{2}\rho U_\infty^2 c = 0.063$.

4. A fire engine with a high velocity nozzle is shown schematically below. The nozzle exit diameter is 7.5 cm and the water flow rate is 8000 l/min. The pressure at the nozzle exit is atmospheric. The diameter of the inlet pipe is 30 cm and the inlet pressure is 2×10^5 N/m^2.

Determine the components of the force **F**, in addition to the weight of the engine, required to hold the fire engine in place. Express your results in Newtons. Water and air are at standard conditions.

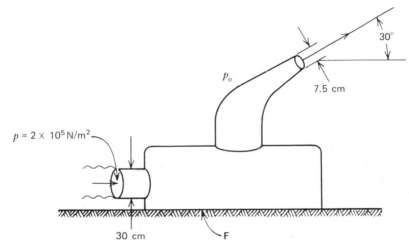

Prob. 3.7.4

Section 4.2

1. Let $G(x, y)$ be a solution of the two-dimensional Laplace equation. Show that $G(x, y)$ may represent the velocity potential or stream function of a two-dimensional, nonviscous, incompressible flow.

2. State clearly what boundary conditions are required to describe a *particular* solution of Laplace's equation.

Section 4.4

1. Show that at $\theta = 67°47'$ the wind speed on the cliff is the same as that far upstream.

2. Superimpose a vortex flow and a source flow with centers at the origin. Show that the result is a spiral flow in which the velocity is everywhere inclined at the same angle to the radius vector from the origin, the angle being $\tan^{-1}(-\Gamma/\Lambda)$.

Section 4.5

1. A source and sink of equal strength Λ, separated by a distance $2l$, are located symmetrically about the origin on a line that makes an angle of $45°$ with the positive x axis. Show that the stream function of the doublet, formed from this source-sink pair by letting $l \to 0$ while keeping $2l\Lambda = \kappa$, is

$$\psi = -\frac{\kappa}{2\pi} \cdot \frac{1}{\sqrt{2}} \frac{\sin\theta - \cos\theta}{r} = -\frac{\kappa}{2\pi} \cdot \frac{1}{\sqrt{2}} \frac{y - x}{x^2 + y^2}$$

2. Show that a source-sink pair (source and sink of equal strength) when viewed from infinity looks like a doublet.

3. A vortex of strength Γ is located at $(0, y_0)$ and a vortex of strength $-\Gamma$ is located at $(0, -y_0)$. We let y_0 approach zero while keeping the product Γy_0 a constant. Show that the stream function for the resulting flow has the same form as that for a source-sink doublet on the x axis.

Section 4.6

1. A source of strength Λ located on the x axis at $x = -a$ and a sink of the same strength located at $x = +a$ are in the presence of a uniform stream of speed V_∞ in the direction of the positive x axis. Plot the streamlines and show that the equations of the streamlines that contain the stagnation points are

$$y = 0; \qquad x^2 + y^2 - a^2 = \frac{2ay}{\tan(2\pi V_\infty y/\Lambda)}$$

Plot the body shape for $a = V_\infty = \Lambda/2\pi = 1$.

2. Plot the polar pressure coefficient distribution about the circular cylinder described in this section.

Section 4.7

1. Show that the pressure coefficient on the surface of a circular cylinder of radius a in a uniform stream, with a circulation Γ around the cylinder, has the form

$$C_p = 1 - 4\sin^2\theta \left(1 + \frac{\Gamma}{4\pi a V_\infty \sin\theta}\right)^2$$

2. A circular cylinder of 1 m radius is moving in the direction of the negative x axis with a velocity of 10 m/sec. The circulation around the cylinder is 20π m^2/sec. Plot the polar pressure coefficient distribution about the cylinder.

Section 4.8

1. A velocity field is described by the stream function

$$\psi = 100y\left(1 - \frac{25}{r^2}\right) + \frac{628}{2\pi}\ln\frac{r}{5}$$

Find (a) the shape of the zero streamline, (b) the locations of the stagnation points, (c) the circulation around the body, (d) the velocity at ∞, and (e) the force acting on the body.

2. Show that the pressure coefficient in the flow field of Problem 4.8.1 at the point $(6, -1)$ is 0.885.

3. A uniform stream is flowing past a vortex at the origin. Using the momentum theorem, show that the force per unit length on the vortex is

$$\mathbf{F} = \rho\mathbf{V}_\infty \times \mathbf{\Gamma}$$

where $\mathbf{\Gamma}$ is a vector formed from the circulation by using right-hand rule.

Hint: In this problem the inner boundary may be taken as a point at the origin.

Section 4.12

1. Show that if the source in Fig. 17 is not constrained, it would move away from the wall at an initial speed of $\Lambda/4\pi a$.

2. Show that the stream function of the flow shown in Fig. 20, obtained by placing a doublet of strength κ in a uniform flow of speed V_∞ at a height a above a plane wall, has the expression

$$\psi = V_\infty y - \frac{\kappa}{2\pi}\left[\frac{y - a}{x^2 + (y - a)^2} + \frac{y + a}{x^2 + (y + a)^2}\right]$$

Show that on the closed body formed by $\psi = 0$, there is a force of magnitude $\rho\kappa^2/8\pi a^3$ per unit spanwise length acting in the direction toward the wall.

Hint: Compute the force acting on the plane wall.

Section 4.13

1. Consider a source panel of length l and of uniform strength λ per unit length lying on the y axis. Show that the velocity potential at a point (x, y) caused by the source panel is

$$\phi(x, y) = \frac{\lambda}{4\pi}\int_{y_1}^{y_1+l}\ln\left[x^2 + (y - y_0)^2\right]\,dy_0$$

where y_1 is the distance of the lower end of the panel from the x axis. Then verify that the velocity components at that point are

$$u(x, y) = \frac{\lambda}{2\pi} \left[\tan^{-1} \left(\frac{y - y_1}{x} \right) - \tan^{-1} \left(\frac{y - y_1 - l}{x} \right) \right]$$

$$v(x, y) = \frac{\lambda}{4\pi} \{ \ln [x^2 + (y - y_1)^2] - \ln [x^2 + (y - y_1 - l)^2] \}$$

From the above relations show that on the surface of this panel except at the end points, the normal velocity induced by the source distribution is $\lambda/2$; and that the induced tangential velocity at the midpoint is zero.

2. For the source panel arrangement around a circular cylinder shown in Fig. 23a, verify that for $i = 4$, the $j = 6$ term in the summation of Eq. 33 is $0.4018 \lambda_6/2\pi$.

Section 5.3

1. Prove that the velocity induced in the region surrounding a doubly infinite vortex sheet satisfies the equation of continuity everywhere.

Section 5.5

1. Show that the distribution

$$\gamma = \frac{2\alpha V_\infty (1 + \cos \theta)}{\sin \theta}$$

satisfies both the Kutta condition and the condition of parallel flow at the boundary.

2. Plot the γ/V_∞ distribution versus chord for lift coefficients of 0.1, 0.5, and 1.0. Explain your results at the leading edge. How is the parameter γ/V_∞ related to the pressure coefficient

$$\Delta C_p = \frac{(p_U - p_L)}{\frac{1}{2}\rho V_\infty^2}$$

3. Show that in order to develop a sectional-lift coefficient of 0.5, a symmetrical airfoil must fly at an angle of attack of 4.55 degrees. Through what point on the airfoil does the line of action of the lift act?

4. Using the method of Section 5.5, show that the expression for the moment about a point $\frac{3}{4}$ chord behind the leading edge of a symmetrical airfoil is $\frac{1}{2}\pi\alpha\rho V_\infty^2 c^2$. Verify this result by using the known fact that the center of pressure is at the $\frac{1}{4}$ chord point for all angles of attack.

Section 5.6

1. An airfoil has a mean camber line that has the shape of a circular arc (constant radius of curvature). The maximum mean camber is kc, where k is a constant and c is the chord. The free-stream velocity is V_∞, and the angle of attack is α. Under the assumption that $k \ll 1$, show that the γ distribution is approximately

$$\gamma = 2V_\infty \left(\alpha \, \frac{1 + \cos \theta}{\sin \theta} + 4k \sin \theta \right)$$

Section 5.7

1. For the circular-arc airfoil described in Problem 5.6.1, show that the angle of zero lift is $-2k$ radians and that the moment coefficient about the aerodynamic center is $-\pi k$.

2. A mean camber line with a reflexed trailing edge must have a point of inflection, and, therefore, the simplest equation that can describe it is a cubic. Four boundary conditions are required to determine a cubic. Two of them are the condition of zero camber at the leading and trailing edges. Therefore, if a reflexed mean camber line is represented by a cubic, the equation will contain two arbitrary constants and may be written in the dimensionless form

$$\bar{z} = a[(b - 1)\bar{x}^3 - b\bar{x}^2 + \bar{x}]$$

where $\bar{z} = z/c$ and $\bar{x} = x/c$, c being the chord.

Show that $b = 7/3$ and $c_{mac} = a\pi/24$ if the angle of zero lift of the airfoil is zero. Show also that c_{mac} vanishes when $b = 15/7$. Plot the mean camber line in each case.

Section 5.8

1. Consider the double flapped airfoil shown in the figure; $x_{h_1} = 0.2c$ and $x_{h_2} = 0.8c$ deflected respectively at angles η_1 and η_2. We will show that for proper values

Prob. 5.8.1

of η_1 and η_2 this mean line will resemble roughly a "reflex" profile; for a given $\eta_1 > 0$ (simulating positive camber) the sign and magnitude of η_2 will govern c_{mac}. Show that $\eta_2 = -.25\eta_1$ is the condition that $c_{mac} = 0$ for the combination. Show that with $\eta_2 = 0$, $\eta_1 > 0$, $\Delta c_{mac} < 0$, that is, a diving moment acts, but a very slightly negative η_2 will make the c_{mac} of the combination vanish. The opposite effect is shown for a downward deflection of a flap in Problem 5.8.2.

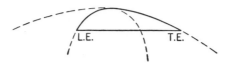

Prob. 5.8.2

2. The mean camber line of an airfoil consists of two parabolas joined at their vertices as shown. The maximum mean camber is 4 percent of the chord, and the position of the maximum mean camber behind the leading edge is 20 percent of the chord.

Show that this airfoil has the characteristics that $c_{mac} = -0.074$ and $\alpha_{L0} = -0.063$ radians.

When the geometric angle of attack is $3°$, show that the lift coefficient is 0.722, and the center of pressure is $0.352c$ behind the leading edge.

At the same angle of attack, a flap that is 15 percent of the entire chord of the airfoil is deflected $2°$ downward. Show that the lift coefficient becomes 0.828 and the center of pressure moves to a point which is $0.365c$ behind the leading edge.

Section 6.2

1. The bound vortex AB shown in the accompanying figure is in the presence of a sea level stream of speed equal to 100 m/sec. The total force on the vortex is 10,000 N. At the end of 1/5 sec, the vortex formation in the fluid is shown in the figure. Using the law of Biot and Savart, show that the induced downward velocity at the point E is 0.22 m/sec.

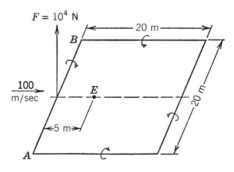

Prob. 6.2.1

Section 6.3

1. Show that on a lifting line of span b and of constant circulation Γ, the drag induced by the starting vortex filament alone is expressed by

$$D = \frac{\rho \Gamma^2}{2\pi} \left[\sqrt{1 + \left(\frac{b}{l}\right)^2} - 1 \right]$$

where l is the distance between the starting vortex and the lifting line. In finite-wing calculations, this drag is neglected. Why?

2. The downwash on the tail resulting from the wing wake is almost twice as great as the downwash on the wing resulting from the wing wake. Why?

3. Show that the downwash velocity, induced at the wing tip (point A) by unit area of the vortex sheet that forms the wake, has a magnitude

$$\frac{4}{(17)^{3/2}} \frac{L'_s}{\pi \rho V_\infty b^3}$$

The unit area is downstream of the lifting line by a distance b and off the centerline of the wing by a distance $b/4$. Assume that the vorticity is constant over the unit area and equal to the value at the center of the area. The lift distribution varies linearly from root to tip according to the equation

$$L' = L'_s\left[1 - \frac{1}{2}\left(\frac{|y|}{b/2}\right)\right]$$

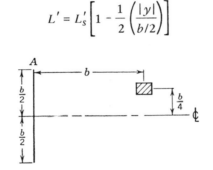

Prob. 6.3.3

Section 6.4

1. For the lift distribution described in Problem 6.3.3, show that the induced angle of attack at the spanwise station y_0, which is neither at the root nor at the tips of the wing, has the expression

$$\alpha_i(y_0) = \frac{L'_s}{4\pi \rho V_\infty^2 b} \ln \frac{y_0^2}{(b/2)^2 - y_0^2}$$

Section 6.5

1. A wing with an elliptical planform is flying through sea level air at a speed of 45 m/sec. The wing loading $W/S = 1000$ N/m^2. The wing is untwisted and has the same section from root to tip. The lift curve slope of the section m_0 is 5.7. The span of the wing is 10 m, and the aspect ratio is 5. Show that the sectional-lift and induced-drag coefficients are 0.806 and 0.041, respectively; and that the effective, induced, and absolute angles of attack are constant along the span and

have respectively the values $2.94°$, $8.1°$, and $11.04°$. Show that the power required to overcome the induced drag of this wing is 46,200 Nm/sec.

2. Two identical wings having the same weight and flying at the same speed are arranged in tandem. Which wing requires the greater horsepower? *Hint:* Consider the velocities induced by one wing on the other.

Section 6.6

1. Show that for a symmetrically loaded wing, A_n in Eq. 26 vanish for even values of n.

2. Show that in Eq. 38

$$\tau = \frac{\pi c}{4\bar{c}} \frac{\sum\limits_{n=1}^{\infty} nA_n \sin n\theta / \sin \theta}{\sum\limits_{n=1}^{\infty} A_n \sin n\theta} - 1$$

where $\bar{c}$ is the mean chord.

3. Show that an untwisted rectangular wing of aspect ratio 10 has the approximate characteristics that $C_L = 5.0412\alpha$ and $C_{D_i} = 0.8742\alpha^2$. What is your conclusion after comparing the result with that computed in the example of Section 6.6 for the rectangular wing of aspect ratio 6?

Section 6.7

1. Verify the curves in Figs. 8b and 9b by correcting those in Figs. 8a and 9a to $R = 5$.

Section 6.8

1. The equation (45) for a twisted wing is solved twice. At a station midway between the root and tip, the local-lift coefficient is found to be 0.8 when the wing-lift coefficient is 0.85. For the second solution, the local-lift coefficient is found to be 0.5 when the wing-lift coefficient is 0.52. The wing loading W/S is 1000 N/m². Show that the local-lift coefficient at this station is 1.68 when the wing is flying at a speed of 30 m/sec through sea level air.

2. Show that the induced-drag coefficient is the sum of a constant term, a term proportional to C_L, and a term proportional to C_L^2.

3. A wing having the properties described below is flying through sea level air at a speed of 300 km/hr. The wing loading $W/S = 770$ N/m². Plot c_{l_b} and c'_{l_a}

versus span. Plot c_l and c_{d_i} versus span. Solve for at least four values of A_n. Notice that for a symmetrically loaded wing A_n vanish for even values of n.

Wing Properties
Taper ratio = tip chord/root chord = 0.6.
Aspect ratio = 7.
Twist = $0°$ at root and increases linearly to $3°$ at tip (geometric washout).
Root section NACA 23012: $\alpha_{L0} = -1.2°$, $m_0 = 0.1$/deg.
Tip section NACA 2412: $\alpha_{L0} = -2°$, $m_0 = 0.098$/deg.
Assume a linear variation in properties from root to tip.

Section 6.10

1. For the wing described in Problem 6.8.3, assume that the trailing edge is perpendicular to the plane of symmetry. Show that the aerodynamic center of the wing is 1.09 m behind the leading edge of the root section, the moment coefficient about the aerodynamic center is -0.0373, and the angle of zero lift of the wing is $-0.213°$ measured from the chord line of the root section.

If the wing weighs 28,600 N, show that the induced drag is 113.5 N when it is flying through sea level air at 300 km/hr.

2. For a linearly tapered symmetrical wing of taper ratio λ that has a constant value of the sectional c_{mac}, show that the second integral on the right-hand side of Eq. 57 has the expression

$$\frac{4c_{mac}(1 + \lambda + \lambda^2)}{3(1 + \lambda)^2}$$

Section 7.2

1. Show that for a compressible fluid,

$$\text{div } \mathbf{V} = \frac{1}{v}\frac{\mathcal{D}v}{\mathcal{D}t}$$

where v is the specific volume, and the term on the right-hand side of the above equation is called the *dilatation*. Interpret the physical meaning of div $\mathbf{V}$ from the point of view of this equation.

Section 7.3

1. Verify that the velocity field represented by Eqs. 9 is irrotational.

Section 8.4

1. Air enters a tank at a speed of 100 m/sec and leaves it at 200 m/sec. If no heat is added to and no work is done by the air, show that the temperature of the air at the exit is $15°C$ below that at the entrance.

2. Air enters a machine at $373°$K with a speed of 200 m/sec, and leaves it at the temperature of the standard sea-level atmosphere. Show that in order to have the machine deliver 100,000 Nm/kg of air without any heat input, the exit air speed is 103.3 m/sec; and the exit speed becomes 459 m/sec when the machine is idling.

3. Two jets of air of equal mass flow rate are mixing thoroughly before entering a large reservoir. One jet is at $400°$K and 100 m/sec and the other is at $200°$K and 300 m/sec. In the absence of heat addition or work done, show that the temperature of the air in the reservoir is $324.9°$K.

4. Sea level air flowing at 500 m/sec is slowed down to 300 m/sec by a shock wave. Show that the temperature of the air behind the shock is $367.7°$K.

Section 8.7

1. Show that the Bernoulli's equation for an isothermal flow of a compressible fluid, derived by integrating Eq. 44, has the form

$$\tfrac{1}{2}V^2 + RT \ln \rho = \text{constant}$$

Section 8.8

1. When air is released adiabatically from a tire, the temperature of the air at the nozzle exit is $37°$C below that inside the tire. Show that the exit speed of air is 272.6 m/sec.

2. A stream of air drawn from a reservoir is flowing through an irreversible adiabatic process into a second reservoir in which the pressure is half of that in the first. Show that the entropy difference between the two reservoirs is 198.92 Nm/kg $°$K.

Section 9.2

1. Newton assumed that the process for very small discontinuities was *isothermal* and found an expression for speed of sound in the form

$$a = \sqrt{RT}$$

Verify this result and show that at sea level the indicated speed of sound is 287 m/sec, which is low compared with the correct (isentropic) value given by Eq. 6.

2. Sea level air is drawn from the atmosphere through a duct and into a vacuum tank. If the air remains a perfect gas at all temperatures, and if it can be expanded reversibly, show that the maximum speed that the air can attain is 760 m/sec. Show that if a maximum air speed of 1000 m/sec is desired, the air must be heated to $498°$K before entering the duct.

Section 9.3

1. A surface temperature of $500°K$ is recorded for a missile that is flying at an altitude of 15 km. Assume that the conditions at the surface are the same as those at a stagnation point. Show that the speed of the missile is 754 m/sec.

Note: The no-slip condition at the surface of a body requires that the flow velocity be zero at the surface. The full stagnation temperature is not reached, however, because the fluid is not brought to rest adiabatically. See Section 16.6 for the discussions.

2. Equation 5 shows that the flow of an incompressible fluid corresponds to that of a fluid with $\gamma = \infty$. Show that, by letting $\gamma \to \infty$, Eqs. 14, 16, and 21 reduce to the familiar expressions for incompressible flow.

3. An intermittent wind tunnel is designed for a Mach number of 4 at the test section. The tunnel operates by sucking air from the atmosphere through a duct and into a vacuum tank. The tunnel is located in Boulder, Colorado (altitude 1650 m, $\rho = 1.044$ kg/m^3), and the flow is assumed isentropic. Show that the density at the test section is 0.029 kg/m^3.

Section 9.4

1. An airplane is flying through sea level air at $M = 0.6$ (204 m/sec; 734 km/hr). Show that at the head of a pitot tube that is directed into the stream a pressure of 129,234 N/m^2 will be recorded. Show also that if the recorded total pressure were used to calculate the flight speed from Bernoulli's equation for *incompressible* fluids, that speed would be 768 km/hr. What is the conclusion from the result?

Section 9.5

1. Show that in a one-dimensional channel at a section where $dA/ds = 0$ but $M \neq 1$, the quantities dV/ds, $d\rho/ds$, and dp/ds must be all zero.

Section 9.6

1. Near a Mach number of unity, small variations in the free cross-sectional area of a wind tunnel (tunnel cross-sectional area minus model cross-sectional area) cause large variations in the flow parameters. Show that corresponding to a 1 per cent change in free area at test Mach numbers of 1.1, 1.2, 1.5, and 2.0, the percentages of change in Mach number are, respectively, 5.91, 2.93, 1.16, and 0.60.

2. Sea level air is being drawn isentropically through a duct into a vacuum tank. The cross-sectional areas of the duct at the mouth, at the throat, and at the entrance to the vacuum tank are 2 m^2, 1 m^2, and 4 m^2, respectively. Show that the maximum amount of air that can be drawn into the vacuum tank is 241 kg/sec.

3. Show that if the maximum flow rate is to be attained for the duct-tank configuration of Problem 2, the pressure in the vacuum tank must be 3020 N/m^2.

4. The geometry of a Laval nozzle is shown in the accompanying figure. The cross-sectional diameters vary linearly from the reservoir to the throat and from the throat to the exit. The throat area A_t = 1 m^2. The ratio of the reservoir area

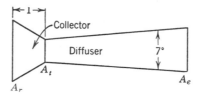

Prob. 9.6.4

to the throat area (A_r/A_t) = 20. The ratio of the exit area to the throat area (A_e/A_t) = 4. The length of the collector L_c = 1 m. The diffusion angle is 7°.
   (a) For a Mach number of unity at the throat, plot the distribution of M and p/p_0 along the tube.
   (b) For a Mach number of 0.5 at the throat, plot the distribution of M and p/p_0 along the tube.

5. Consider a manned orbital laboratory (M.O.L.) circling the earth at an altitude such that the external pressure is very nearly zero. A small meteorite punches a hole in the wall of the laboratory and the air within the laboratory starts to leak out. The astronauts within the laboratory initially are not wearing pressure suits. Problem: determine the time the astronauts have to put their pressure suits on. Assume:

1. Size of M.O.L.—Diameter = 3.66 m
               Length = 12.2 m;
2. Size of hole in M.O.L. wall—5.9 cm diameter;
3. Gas in M.O.L. ~ air, γ = 1.4;
4. Initial pressure in M.O.L. ~ pressure equivalent to that at 2000 m altitude;
5. Pressure at which astronauts fail to function in a rational manner without oxygen at 12,000 m altitude;

Prob. 9.6.5

6. Sonic flow through the hole in the M.O.L. wall;
7. The air within the tank follows a reversible isothermal process;
8. The air in flowing through the hole in the wall follows an isentropic process;
9. The temperature within the M.O.L. is 15°C.

Section 9.8

1. Sea level air moving at a speed of 170 m/sec enters a constant area duct in which heat is added at the rate of 1×10^5 Nm/kg. Show that for the air after heat addition, temperature is 365°K, pressure is 7980 N/m^2, density is 0.76 kg/m^3, and velocity is 274 m/sec.

2. In Problem 9.8.1, show that the heat rate that will produce thermal choking is 1.35×10^5 Nm/kg.

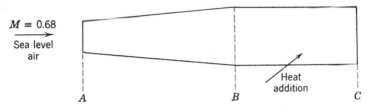

Prob. 9.8.3

3. In the tube shown, sea level air enters at $M = 0.68$ and reaches a value of $M = 0.25$ at the exit of the diffuser (station B). The entrance area is 1 m^2.
   (a) Assuming no dissipative losses in the diffuser, show that the area of station B is 2.16 m^2. Will the area be larger or smaller if losses are present?
   (b) Show that assuming no losses, the static pressure at station B is 132,500 N/m^2 and the density there is 1.48 kg/m^3. If losses are present, will the stagnation pressure rise or fall from station A to station B? Will the stagnation density rise or fall? Why?
   (c) Heat is added at $M = 0.25$ between station B and station C until thermal choking occurs. Show that the heat added is 8.36×10^5 Nm/kg and the stagnation temperature at station C is 1121°K.

Section 10.2

1. Sea level air moving at a Mach number of 1.5 is turned isentropically in an expansion direction through an angle of 5°. By treating M as a constant in Eqs. 7 and 10, show that the corresponding velocity increase is 7.81 percent and pressure drop is 24.6 percent. How can this result be improved?

2. Using the approximate velocity obtained in Problem 10.2.1 and the energy equation (9.9), show that the Mach number of the flow after turning through 5° is approximately 1.68.

3. In Problem 10.2.1, the air is turned in an expansion direction through an angle of 15°. By dividing the turn into three 5° increments, show that the approximate Mach number after turning is 2.03.

Section 10.3

1. Sea level air moving at a Mach number of 1.5 is turned isentropically in a *compressive* direction through an angle of 5°. Following the approximation procedure used in Problems 10.2.1 and 10.2.2, show that the approximate Mach number of the flow after turning is 1.34. Why is your answer approximate?

2. A small-amplitude wave on a *shallow* water surface is propagating in the direction of the negative x axis at a constant speed u. The accompanying figure shows the picture as seen by an observer moving with the wave front. With the effect of surface tension ignored, the pressure is a constant at all points on the water surface.
   (a) Using Eq. 3.19, the Bernoulli's equation in differential form, show that $u\, du = -g\, dh$.
   (b) Show that for small increments, conservation of mass requires that $h\, du = -u\, dh$. From the above two equations, show that the propagation speed of a shallow-water surface wave of small amplitude has the expression $u = \sqrt{gh}$. This wave phenomenon is the basis for the "surface analogy" used in practice to identify the two-dimensional wave configurations around complicated shapes.

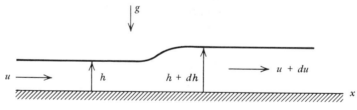

Prob. 10.3.2

Section 10.4

1. Sea level air moving at a Mach number of 1.5 is turned isentropically in a *compressive* direction through an angle of 5°. Find the exact Mach number of the flow after turning, and compare your answer with that obtained in Problem 10.3.1.

2. For the flow of Problem 10.4.1, show that after 5° of turning, the static values of the pressure, density, and temperature are, respectively, 129,000 N/m², 1.45 kg/m³, and 309°K.

3. Sea level air moving at a Mach number of 1.5 is turned isentropically in an *expansive* direction through an angle of 15°. Find the exact Mach number at the

end of $5°$ turning and at the end of $15°$ turning. Compare your answers with those obtained in Problems 10.2.2 and 10.2.3.

4. For the flow of Problem 10.4.3, show that after $15°$ of turning, the static values of the pressure, density, and temperature are, respectively, 46,100 N/m², 0.698 kg/m³, and $230°$K.

Section 10.5

1. Using the procedure of Section 10.5, derive the following relation among u_1, u_2, and v_2 (see Fig. 9).

$$v_2^2 = (u_1 - u_2)^2 \frac{u_1 u_2 - a^{*2}}{2u_1^2/(\gamma + 1) - u_1 u_2 + a^{*2}}$$

A plot of the above equation using the dimensionless variables $u_1^* = u_1/a^*$, $u_2^* = u_2/a^*$, and $v_2^* = v_2/a^*$ is the shock polar diagram illustrated below. The polar has been drawn for a particular value of u_1^*.

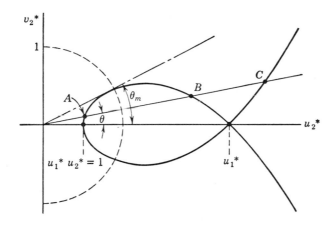

Prob. 10.5.1

Interpret the points A, B, and C corresponding to the turning angle θ. Interpret the two points on the polar corresponding to $\theta = 0$ in terms of the waves produced. How does the maximum turning angle θ_m vary with u_1^*?

Section 10.6

1. Using a value of γ of 1.4, show that the ratios of the flow parameters across a shock wave in terms of the normal Mach number may be written

$$\frac{p_2}{p_1} = \frac{7M_{1n}^2 - 1}{6}$$

$$\frac{\rho_2}{\rho_1} = \frac{6M_{1n}^2}{M_{1n}^2 + 5}$$

$$\frac{T_2}{T_1} = \frac{(7M_{1n}^2 - 1)(M_{1n}^2 + 5)}{36M_{1n}^2}$$

$$\frac{p_2^0}{p_1^0} = \left(\frac{6}{7M_{1n}^2 - 1}\right)^{2.5} \left(\frac{6M_{1n}^2}{M_{1n}^2 + 5}\right)^{3.5}$$

2. A body with a conical nose is traveling through sea level air at a Mach number of 2. The shock wave at the nose is observed to make an angle of $50°$ with the flow direction. Show that the pressure, density, temperature, and total head immediately downstream of the shock are, respectively, 260,000 N/m^2, 2.345 kg/m^3, $386°$K, and 730,000 N/m^2.

3. Other conditions remaining unchanged, what is the influence of altitude on the quantities computed in Problem 10.6.2?

4. Verify that, for $(p_2/p_1) < 1$, Eqs. 37, 38, and 39 indicate a decrease in entropy across an expansion shock. Thus, the existence of an expansion shock is precluded by the second law of thermodynamics.

Section 10.7

1. Show that if $(V_1/a^*) > 1$, then V_1/a_1 is also greater than unity; that is, show that $(V_1/a^*) > 1$ means that the flow ahead of a shock wave is supersonic.

2. Sea level air at a Mach number of 3 passes through a normal shock wave. Show that in passing through the shock, the static pressure of the air rises 945,000 N/m^2 and the flow speed decreases from 1020 m/sec to 265 m/sec. What do you think of the normal shock wave as a pressure-recovery device? Reserve your judgment until you have worked Problem 10.7.3.

3. Let the stream of Problem 10.7.2 be compressed between the initial and final speeds isentropically. What is the pressure rise? This is the maximum pressure recovery that can be obtained.

Section 10.8

1. Sea level air moving at a Mach number of 1.5 is turned in a compressive direction through an angle of $5°$. Show that the Mach number of the flow after turning is 1.33. What is your conclusion after comparing the present result with that of Problem 10.3.1?

2. A wedge having a total vertex angle of $60°$ is traveling at a Mach number of 3 at an altitude of 15 km. Show that downstream of the shock the static values of the pressure, density, and temperature are, respectively, 76,100 N/m², 0.613 kg/m³, and $433°K$; and that the stagnation values of those quantities are, respectively, 76,100 N/m², 0.613 kg/m³, and $433°K$. Show also that 44.28 percent of the stagnation pressure is lost across the shock wave.

3. If the total vertex angle of the wedge is $30°$ instead of $60°$ in Problem 10.8.2, show that the loss in stagnation pressure across the shock is 10.11 percent.

4. Show that in order to maintain an oblique shock attached to the nose, the minimum speed of the wedge of Problem 10.8.2 is approximately 738 m/sec.

5. Just above the critical speed at which the shock detaches from the nose, show that the static pressure immediately behind the shock of Problem 10.8.4 is 69,600 N/m² and the wave angle is $64.5°$. What is the pressure acting on the surface of the wedge? Show that the stagnation pressure loss in crossing the shock is 82,200 N/m².

Section 10.9

1. Show that the cone of a cone angle of $45°$ will have the same critical speed as the wedge of Problem 10.8.4; and that at this critical speed the pressure at the surface of the cone is 80,600 N/m².

2. Show that the pressure immediately downstream of the shock of Problem 10.9.1 is 75,500 N/m²; and that the direction of the flow there makes an angle of $29°$ with the cone axis.

Section 10.10

1. Sea level air is being drawn into a reservoir through a duct as shown in the accompanying figure. The cross-sectional areas at the mouth, at the throat, and at the entrance to the vacuum tank are 2 m², 1 m², and 4 m², respectively. By schlieren photography a normal shock is detected at a position in the duct where the cross-sectional area is 3 m². Show that the pressure in the reservoir is 9750 N/m².

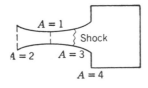

$A = 1$

Shock

$A = 2$ $A = 3$

$A = 4$

Prob. 10.10.1

Section 10.11

1. A shock wave which makes an angle of $30°$ with the upstream flow is produced by holding a wedge in a stream of sea level air traveling at a Mach number of 3. The wave strikes a plane surface as shown in Fig. 21. Show that the reflected shock wave makes an angle of $23°$ with the wall.

2. Show that the incident shock wave of Problem 10.11.1 will not be reflected if the wall is deflected through $13°$ in a clockwise direction at the point of incidence.

Section 11.2

1. Combine the basic equations of steady compressible flow theory into the form

$$\mathbf{V} \cdot \frac{d\mathbf{V}}{dt} = a^2 \operatorname{div} \mathbf{V}$$

Using the relation curl $\mathbf{V} = 0$, show that the Cartesian form of the above is Eq. 6 of Section 11.2.

Section 11.3

1. Prove that the linearized potential equation for three-dimensional steady compressible flow

$$(1 - M_\infty^2) \frac{\partial^2 \phi}{\partial x^2} + \frac{\partial^2 \phi}{\partial y^2} + \frac{\partial^2 \phi}{\partial z^2} = 0$$

is a valid approximation, provided:

$$M_\infty^2 \frac{u'}{V_\infty} \ll 1 \qquad M_\infty^2 \frac{v'}{V_\infty} \ll 1 \qquad M_\infty^2 \frac{w'}{V_\infty} \ll 1 \qquad \frac{M_\infty^2}{M_\infty^2 - 1} \frac{u'}{V_\infty} \ll 1$$

Section 11.4

1. Verify that Eqs. 16 satisfy Eq. 15.

Section 11.5

1. From the development of Section 11.5, the perturbation density appears only in combination with perturbation velocities, and therefore in the linear problem it can make no contribution to the pressure coefficient. With this in mind, show that the pressure coefficient could have been derived from Bernoulli's equation for incompressible flow:

$$p_\infty + \tfrac{1}{2} \rho_\infty V_\infty^2 = p + \tfrac{1}{2} \rho_\infty V^2$$

Section 12.2

1. The geometric boundary condition for the lifting and nonlifting problems determines the sign relationships of $\partial \phi / \partial y$ at $y = 0+$ and $y = 0-$. Show that the pres-

sure coefficient for the lifting and nonlifting problems plays a similar role for the derivative $\partial\phi/\partial x$ at $y = 0+$ and $y = 0-$. Can anything be said about the signs of $\partial\phi/\partial z$ at $y = 0+$ and $y = 0-$ for the lifting and nonlifting problems?

Section 12.3

1. The pressure coefficient at a point on a two-dimensional airfoil is -0.5 at a very low Mach number. By using the linearized theory, show that the pressure coefficient at that point has the value of -0.578 at $M_\infty = 0.5$, and becomes -0.834 at $M_\infty = 0.8$.

Section 12.4

1. A two-dimensional airfoil is so oriented that its point of minimum pressure occurs on the lower surface. At a free stream Mach number of 0.3, the pressure coefficient at this point is -0.782. Using the Prandtl-Glauert rule, show that the critical Mach number of the airfoil is approximately 0.65.

Section 12.6

1. A wing of infinite span with a symmetrical diamond-shaped section is traveling to the left through sea level air at a Mach number of 2. The maximum thickness to chord ratio is 0.15 and the angle of attack is $2°$. Using shock-expansion theory, show that the pressure at point B on the airfoil indicated in the figure is 52,400 N/m^2.

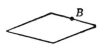

Prob. 12.6.1

2. Assume the wing of Problem 12.6, 1 is at zero angle of attack. Show that the pressure at point B obtained by using shock-expansion theory of Chapter 10 is 61,000 N/m^2 and that obtained by using linearized theory is 52,630 N/m^2.

3. A two-dimensional flat plate is flying at an altitude of 6 km and a Mach number of 2. The angle of attack is $10°$. Show that the pressure difference between the lower and upper surfaces is 53,500 N/m^2 based on shock-expansion method, and is 53,000 N/m^2 based on linearized theory.

4. An infinite wing whose symmetrical cross section is composed of two circular arcs is flying at a Mach number of 3. The angle of attack is zero degrees and the maximum thickness to chord ratio is 0.2. Neglecting viscous drag, show that the drag coefficient is 0.075 based on linearized theory.

5. The two-dimensional airfoil shown in the accompanying diagram is traveling at a Mach number of 3 and at an angle of attack of $2°$. The thickness to chord

Prob. 12.6.5

ratio is 0.1, and the maximum thickness occurs 30 percent of the chord down-stream from the leading edge. Using the linearized theory, show that the moment coefficient about the aerodynamic center is −0.0352, the center of pressure is at 1.214 c, and the drag coefficient is 0.0354. Show that the angle of zero lift is zero degrees.

6. Using linearized theory, compute the lift per unit span acting on and the circulation around a flat plate flying at a small angle of attack with a supersonic speed. Show that the Kutta-Joukowski theorem, $L' = \rho_\infty V_\infty \Gamma$, derived under the assumption of incompressible flow, still holds for the supersonic lifting body just described.

7. Repeat the same computations as those in Problem 12.6.6 for a supersonic flow at a Mach number of 2 past a flat plate at an angle of attack of $20°$, using shock-expansion method. Show that the Kutta-Joukowski theorem becomes invalid in this case. What does the result suggest?

8. An uncambered diamond-shaped airfoil section of chord c has a maximum thickness t which occurs at distance ac from the leading edge. Show that when flying at a supersonic speed, the wave drag coefficient of the airfoil due to the thickness envelope alone is

$$(c_d)_t = \frac{m_0}{4a(1-a)} \left(\frac{t}{c}\right)^2$$

From this result show that the drag is minimum for $a = \frac{1}{2}$.

Section 13.2

1. A rectangular wing of aspect ratio 10 is flying at a Mach number of 0.6. Show that the approximate value of $dC_L/d\alpha$ is 6.04. Compare the result with that of Prob. 6.6.3, which applied to the same wing in incompressible flow.

Section 13.3

1. Show that for

$$M_\infty > \sqrt{1 + \left(\frac{2c}{3b}\right)^2}$$

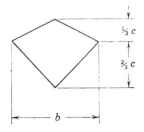

Prob. 13.3.1

the wing shown in the figure has supersonic leading edges, that is, $M_{\infty n} > 1$ at the leading edges; and that for

$$M_\infty < \sqrt{1 + \left(\frac{4c}{3b}\right)^2}$$

it has subsonic trailing edges, that is, $M_{\infty n} < 1$ at the trailing edges.

2. An airfoil at a given altitude has a critical Mach number of 0.7. Show that a sweep of $39°$ would be required to obtain a critical Mach number of 0.9 at the same altitude.

Section 13.4

1. The radius of a body of revolution is described by the equation

$$\frac{r}{r_0} = \left[1 - \left(1 - 2\frac{x}{l}\right)^2\right]^{1/2}$$

where r_0 is the maximum radius and l the total length of the body. Show that the Fourier coefficients A_n in Eq. 27 vanish for odd values of n; and that for even values of n, they have the values shown in the following order:

$$\frac{8}{\pi}\left(\frac{r_0}{l}\right)^2 \left[\frac{4}{3}, \frac{8}{15}, \frac{12}{35}, \frac{16}{63}, \frac{20}{99}, \cdots\right]$$

Show that the wave-drag coefficient of this body flying at a transonic speed can be expressed in the form of a summation

$$32\pi\left(\frac{r_0}{l}\right)^4 \sum_{m=1}^{\infty} m\left(\frac{4m}{4m^2 - 1}\right)^2$$

2. Show that by integrating Eq. 27 the area distribution of a body can be expressed in terms of Fourier coefficients in the following form:

$$S(\theta) = \frac{1}{4}\pi l^2 \left\{ A_1 \left(\theta - \frac{1}{2}\sin 2\theta \right) + \sum_{n=2}^{\infty} A_n \left[\frac{\sin (n-1)\theta}{n-1} - \frac{\sin (n+1)\theta}{n+1} \right] \right\}$$

A further integration shows that the total volume of the body is

$$V = \frac{1}{8}\pi^2 l^3 (A_1 + \frac{1}{2}A_2)$$

Section 14.4

1. From kinetic considerations similar to those in Section 1.7 for the derivation of the coefficient of viscosity, show that the thermal conductivity k is related to the specific heat of a gas c by the equation $k = \mu c$.

Section 14.5

1. Consider a boundary layer of thickness δ on a two-dimensional surface with a radius of curvature R. The fluid is air, and the velocity distribution in the boundary layer is given by

$$\frac{u}{u_e} = 2\left(\frac{y}{\delta}\right) - \left(\frac{y}{\delta}\right)^2$$

Assume that the streamlines in the boundary layer have the same curvature as the surface. Set up the equilibrium condition for the pressure and centrifugal forces and integrate to show that the change in pressure through the boundary layer is

$$\Delta p = \frac{8}{15}\frac{\delta}{R} \rho u_e^2$$

For $\delta = 0.01$ m, $R = 0.3$ m, $u_e = 100$ m/sec, and with standard sea level conditions at the edge of the boundary layer, show that the change of pressure through the boundary layer is 218 N/m^2 (which is small compared with the pressure at the edge).

2. Show that for a steady flow through a normal shock the momentum equation is

$$\rho u \frac{du}{dx} = -\frac{dp}{dx} + \frac{d}{dx}\left(\sigma \frac{du}{dx}\right)$$

where σ is a viscosity coefficient.

Section 14.6

1. Consider a two-dimensional flow in the xy plane. By differentiating and subtracting the two equations of motion (Eqs. 6) to eliminate the pressure, one obtains

$$\frac{\mathcal{D}\omega}{\mathcal{D}t} = \nu \nabla^2 \omega$$

where ω is the z component of the vorticity derived in Section 2.8. Assuming a perfect fluid, interpret the above equation in terms of the Helmholtz' third vortex theorem given in Section 2.13.

The equation for the diffusion of heat in a two-dimensional, inviscid, incompressible flow field is (see Eq. 31, Appendix B)

$$\rho c_v \frac{\mathcal{D}T}{\mathcal{D}t} = k \nabla^2 T$$

where T is the temperature, c_v is the specific heat at constant volume, and k is the thermal conductivity. In this equation the density and thermal conductivity are assumed constant. The analogy between the diffusion of heat and of vorticity in a two-dimensional flow field is evident from a comparison of the two equations.

Section 14.7

1. A wing of 2-m chord is flying through sea level air at a speed of 50 m/sec. A microorganism of a characteristic length of 10 microns (10×10^{-6} m) is swimming at a speed of 20 microns/sec through water whose kinematic viscosity is 1.15×10^{-6} m^2/sec. Show that the characteristic Reynolds numbers of these two flows are 6.88×10^6 and 1.74×10^{-4}, respectively.

Section 15.2

1. Show that the fundamental equation for Poiseuille flow in a tube may also be derived by deleting the proper terms in the complete Navier-Stokes equations for incompressible flow given in Section 14.6.

2. Show that the velocity distribution for fully developed flow between two stationary concentric cylinders of radii a and b, caused by an axial pressure gradient, is given by

$$u_z = -\frac{1}{4\mu} \frac{dp}{dz} \left[(a^2 - b^2) \frac{\ln (r/b)}{\ln (a/b)} - (r^2 - b^2) \right]$$

Derive the result both by means of the equations of this section and by means of the method employed in Problem 15.2.1.

3. If the entire flow between the two cylinders of Problem 15.2.2 is caused by moving the inner cylinder parallel to the axis with a velocity U relative to the outer, show that the velocity distribution is given by

$$u_z = U \frac{\ln (a/r)}{\ln (a/b)}$$

4. Are the solutions of Problems 15.2.2 and 15.2.3 additive? Why? If so, how can the flow represented by the sum of the two solutions be produced?

5. From the Navier-Stokes equations given in Section 14.6, show that fully developed flow in a pipe of any cross section is

$$\frac{\partial^2 u}{\partial y^2} + \frac{\partial^2 u}{\partial z^2} = \frac{1}{\mu} \frac{\partial p}{\partial x}$$

Show that this equation is identical with that for the deflection u of a diaphragm stretched over the pipe cross section with a pressure difference across it.

Section 15.3

1. We wish to calculate the drag of one side of triangular flat plate, shown in the figure, for a laminar boundary layer in an airflow at sea level conditions parallel to the surface. Write the expression for the drag of the element $dxdz$ shown and integrate in two equivalent ways: (1) integrate first with respect to x to find the drag of the strip of width dz, then with respect to z to the boundary; (2) integrate first with respect to z to find the drag of the strip of width dx, then with respect to x. *Answer:* 10.7 N. Calculate C_f = force/qS, where S = 4.5 m^2, the area of the plate. Compare with C_f calculated by Eq. 15.19 for a plate 1 m wide and 4.5 m long. Give reasons for the discrepancy, if there is any.

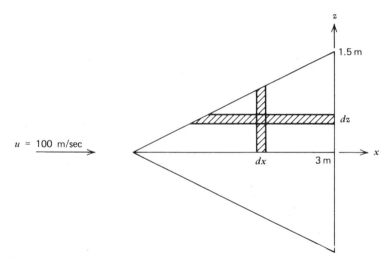

Prob. 15.3.1

2. Substitute $\xi = y/x^n$, $\psi = (\nu u_e x)^{1/2} f(\xi)$ into Eqs. 7 and show that $n = \frac{1}{2}$ is the condition that neither x nor y appear explicitly in the resulting differential equation.

3. Show that outside the laminar boundary layer along a flat plate, there is a velocity component normal to the plate of the magnitude

$$(v)_{y \to \infty} = 0.8604 \frac{u_e}{\sqrt{Re_x}}$$

which is not zero but is varying with the distance from the leading edge. It shows that in specifying the boundary conditions (Eqs. 8), nothing can be said about the vertical velocity at the edge of the boundary layer.

4. A semi-infinite flat plate is aligned with a flow of sea level air traveling at 30 m/sec. Show that at a station 1 m downstream from the leading edge, the boundary layer thickness is 3.6 mm, the skin friction is 0.254 N/m^2, and the vertical velocity at the edge of the boundary layer is 0.018 m/sec.

5. Consider the boundary layer on a flat plate with zero pressure gradient through which suction is applied uniformly, that is, the suction velocity v_0 is constant along the plate. The boundary conditions are then

$$\text{at } y = 0: \quad u = 0, v = -v_0$$

$$\text{at } y = \infty: \quad u = u_e$$

Show that a particular solution of Eqs. 7 satisfying these boundary conditions is

$$u = u_e \left[1 - \exp\left(-v_0 y/\nu\right)\right]; \quad v = -v_0$$

The solution is the "asymptotic suction profile," because this solution will only be approached at large distance from the leading edge of the plate. We return to this solution in Section 17.5 where its stability is discussed in connection with transition to a turbulent boundary layer.

Show that for such a velocity profile the displacement thickness is ν/v_0, and that the skin friction is $\rho u_e v_0$ (which is independent of the coefficient of viscosity).

Section 15.4

1. Compare qualitatively a boundary layer that has just traversed an unfavorable pressure gradient and is now traversing a favorable gradient with one for which the gradient has been favorable from the leading edge. How many inflection points are there in the velocity distributions in each case?

Section 15.5

1. Show that corresponding to the velocity field described by Eq. 21, the pressure variation is

$$\frac{\partial p}{\partial x} = -m\rho u_1^2 x^{2m-1}$$

Thus favorable pressure gradients are represented by positive, and adverse pressure gradients by negative values of m.

2. Verify that upon substitution from Eqs. 22, Eq. 14.3, with $\mu = $ constant, becomes an ordinary differential equation having the form of Eq. 23. Show that the boundary

conditions after transformation are

$$\text{at } \xi = 0: \quad f = f' = 0$$

$$\text{at } \xi = \infty: \quad f' = 1$$

Section 15.6

1. Consider the steady flow along a flat plate parallel to the flow. At a particular station where $du_e/dx = 0$, the velocity profile is given by

$$u = 4 \times 10^4 y - 2 \times 10^6 y^2$$

Show that the numerical value of $d\theta/dx$ is the same as that of the kinematic viscosity of the fluid.

2. Imagine the flat plate to be warped in such a way that we still have at some point the velocity profile given in Problem 15.6.1. Show that the value of du_e/dx necessary to make $d\theta/dx = 0$ at that point is $10^5 \, \nu/3$.

Section 15.7

1. Assume that the velocity distribution for steady flow in the boundary layer is given by

$$\frac{u}{u_e} = \frac{y}{\delta}$$

Substitute this expression in Eq. 33, and obtain the differential equation

$$\frac{d}{dx}(\delta^2) + \frac{10}{u_e}\frac{du_e}{dx}\delta^2 = 12\frac{\nu}{u_e}$$

the solution of which is

$$\delta^2 = \frac{12\nu}{u_e^{10}}\int_0^x u_e^9 \, dx$$

Assume u_e constant ($\partial p/\partial x = 0$), check the following relations, and compare with the Blasius solution.

$$\delta = \sqrt{\frac{12\nu x}{u_e}}, \quad c_f = 0.577\sqrt{\frac{\nu}{u_e x}}, \quad C_f = 1.154\sqrt{\frac{\nu}{u_e l}}$$

Section 16.5

1. The surface temperature of a flat plate flying at a Mach number M is higher than T_e, the temperature outside the boundary layer. By using Eq. 28, show that under a steady state when there is no heat transfer between the plate and air, and

under the assumption that $Pr = 1$, the surface temperature is

$$T_w = T_e \left[1 + \tfrac{1}{2}(\gamma - 1)M^2\right]$$

which is the stagnation temperature of the free stream. Based on this, show that the surface temperature of a flat plate flying at $M = 3$ at an altitude of 15 km is approximately $606°K$.

What is the surface temperature if a normal shock is produced ahead of the plate?

Section 16.6

1. The plane Couette flow is a flow between two infinite parallel plates, one of which is sliding relative to the other in a direction parallel to itself. Suppose the plate at $y = 0$ is stationary, and the one at $y = h$ is moving at a velocity U_1 and is maintained at temperature T_1.

(a) For an insulated lower plate, show that the temperature distribution is

$$T = T_1 + \frac{\mu U_1^2}{2k}\left(1 - \frac{y^2}{h^2}\right)$$

(b) If the lower plate is not insulated but is maintained at temperature T_1, show that the temperature distribution is

$$T = T_1 + \frac{\mu U_1^2}{2k}\frac{y}{h}\left(1 - \frac{y}{h}\right)$$

(c) Show that in case (a) the temperature recovery factor at the lower plate is equal to the Prandtl number.

2. Can a steady temperature distribution be obtained for the Couette flow of Problem 16.6.1 if both plates are insulated? If not, explain the reason from a physical point of view. What is the correct differential equation which governs the temperature distribution in this case?

3. An airplane is flying through sea level air at a Mach number of 0.8. Show that according to Pohlhausen's analysis, the reading on a flat-plate thermometer protruding ahead of the airplane is $46°C$.

4. Plot the distribution of stagnation temperature through the boundary layer as predicted by the Pohlhausen analysis of the flow past an insulated flat plate. Take $M_e = 2$ and $T_e = 288°K$. Use the values for θ given in Fig. 2 and the values for u/u_e given in Fig. 15.5.

Section 16.7

1. Fully developed flow in a tube is defined in Section 15.2. Show that for fully developed incompressible flow between two parallel plates separated by a distance

$2h$,

$$u = u_m \left(1 - \frac{y^2}{h^2}\right)$$

where u_m is the velocity at $y = 0$, the center of the channel. Assuming that temperature differences are so small that incompressibility is not seriously violated and that heat is generated by viscous dissipation, show that the temperature at the center, T_m, in terms of that at the walls, T_w, is $T_m = T_w + \mu u_m^2/3k$.

Based on the temperature difference $T_m - T_w$ across a characteristic length h, show that the Nusselt number at the plates is equal to 4.

Section 16.8

1. Verify Eq. 55. Use the Sutherland formula, Eq. 54, and make use of the constancy of the pressure through the boundary layer to calculate $C_{fw}\sqrt{Re_w}/C_f\sqrt{Re}$ in terms of T_w/T_e; multiply these values by the ordinates for $T_w/T_e = 0.25, 3$, and 5 at $M_e = 5$ in Fig. 6 and divide by 1.328 to find the error involved in neglecting compressibility.

2. Show that if Eqs. 36 and 37 (which neglect viscous dissipation) are nondimensionalized (as in Section 14.6) by dividing all velocities by u_e and x and y by the length L, and if we further introduce β from Eq. 38 and assume $Pr = 1$, that is, $k = \mu c_p$, these equations reduce to

$$u'u'_{x'} + v'u'_{y'} = \left(\frac{1}{Re}\right)u'_{y'y'} \tag{A}$$

$$u'_{x'} + v'_{y'} = 0 \tag{B}$$

$$u'\beta_{x'} + v'\beta_{y'} = \left[\frac{1}{PrRe}\right]\beta_{y'y'} \tag{C}$$

where subscripts designate partial derivatives, and $\beta = (T_w - T)/(T_w - T_e)$ as in Eq. 38. The boundary conditions are (for zero heat transfer at the plate)

$$\begin{array}{ll} \text{at } y = 0: & u' = v' = \beta = 0 \\ \text{at } y = \infty: & u' = 1, \beta = 1 \end{array} \tag{D}$$

Thus the equations and boundary conditions governing β and u' are identical for adiabatic flow of a fluid with $Pr = 1$ along a flat plate as well as with those of Section 15.3 for u'. Thus u' and β are identical functions of x' and y' and are, in turn, identical with the Blasius distributions of Section 15.3. Furthermore Eqs. 49 and 53 with $Pr = 1$ are good first approximations for adiabatic compressible flow of air ($Pr = 0.73$) well into the supersonic range of Mach numbers, if fluid properties at the wall are used, as described in Section 16.8. Plot the temperature distribution through the boundary layer.

Section 17.4

1. In Tollmien-type laminar instability the "disturbance stream function" is given by (see Fig. 6)

$$\psi' = \varphi(y) \exp i\alpha(x - ct)$$

Where α is real and positive, $c = c_r + ic_i$. The velocity components in the boundary layer are given respectively by $U(y) + \partial\psi'/\partial y$ and $-\partial\psi'/\partial x$. Give the physical meanings of α, c_r, and c_i. How do the perturbations vary with time if $c_i > 0$?

Section 17.5

1. A gas is flowing past a heated flat plate. The dependence of the coefficient of viscosity of the gas on temperature is described approximately by Eq. 16.54a. Consider a station on the plate where the surface temperature is T_w, shear stress is τ_0, and the rate at which heat is transferred into the fluid is q_0. Show that in order to remove the inflection point from the velocity profile at that station, a favorable pressure gradient of the minimum magnitude of $\omega\tau_0 q_0/kT_w$ is required locally.

2. If at a station on the flat plate of Problem 17.5.1, there is a local adverse pressure gradient of magnitude dp/dx, show that no inflection point will appear in the velocity profile at that station if heat is conducted from the fluid to the plate at a rate equal to or greater than $(kT_w/\omega\tau_0)\,dp/dx$.

Section 17.6

1. In Taylor-Goertler type of boundary layer instability the disturbance is described by (see Fig. 18)

$$u_1 = u'(y)\, e^{\beta t} \cos \alpha z$$

$$v_1 = v'(y)\, e^{\beta t} \cos \alpha z$$

$$w_1 = w'(y)\, e^{\beta t} \sin \alpha z$$

where the total velocity components are, respectively, $U(y) + u_1, v_1, w_1$. The boundary layer profile without the instability is $U(y)$. Give the physical meanings of α and β. How do the velocity perturbations vary with time if $\beta > 0$?

Section 17.7

1. With the assumption that the variation of density of air with height is given by Eq. 1.8, show that the Richardson number can be expressed as

$$\frac{\alpha g}{T_0}\left(\frac{g}{R\alpha} - 1\right)\left(1 - \frac{\alpha y}{T_0}\right)^{-1}\left(\frac{du}{dy}\right)^{-2}$$

Figure 19 indicates that the atmospheric flow is stable if the Richardson number is greater than 0.0417. From this show that for stability at the sea level, the increase in wind speed must be less than 0.15 m/sec per meter increase in height.

Section 18.3

1. Calculate the shearing stresses $-\rho\overline{u_1 v_1}$ for the disturbance described in Problem 17.4.1.

2. Show that $-\rho\overline{u_1^2 v_1}$ represents a rate of energy transfer.

Section 18.9

1. The following velocity profile for zero pressure gradient was measured at $x = 5.33$ m by Schubauer and Klebanoff (1951). Show that $H = \delta^*/\theta = 1.35$ and $Re_\theta = U_e\theta/\nu = 19150$. Calculate c_f from Eq. 43 and compare with Fig. 14. Plot U/U_e versus y/δ on a log-log plot to check Eq. 41. Find, from Eq. 42, the "apparent" origin for a boundary layer with zero pressure gradient and turbulent boundary layer from the leading edge. Since the actual pressure gradient is favorable what does the comparison of the apparent origin compared with the actual indicate about the effect of a favorable pressure gradient on the rate of growth of the boundary layer.

y meters	U m/sec
0	0
2.5(−4)	24.1
5.1(−4)	24.8
2.0(−3)	30.5
4.8(−3)	34.0
9.1(−3)	36.6
1.44(−2)	39.1
2.51(−2)	42.8
4.3(−2)	46.7
6.4(−2)	48.6
8.4(−2)	48.8

Note: $8.4(-2) \equiv .084$ etc.

2. Consider a rectangular plate 3 m long, 1 m wide parallel to an airflow of 100 m/sec. Transition from laminar to turbulent boundary layer occurs at $x = 1.5$ m. Calculate the drag of one surface. *Answer:* 25 N.

3. Show that if we use the $\frac{1}{7}$th power velocity distribution of Eq. 41, the displacement and momentum thickness of the turbulent boundary layer (Eqs. 15.29 and 15.30) are given by $\delta^* = \delta/8$ and $\theta = 7\delta/72$ so that, for a flat plate ($\partial p/\partial x = 0$), $H = \delta^*/\theta \cong 1.3$.

Section 18.11

1. Following are measurements of a velocity profile for a turbulent boundary layer in an adverse velocity (pressure) gradient, $U \sim x^{-0.255}$ along with the corre-

y meters	U m/sec	γ_S
0	0	
8.15(−5)	5.29	1.0
16.3(−5)	7.96	1.0
31.6(−5)	10.38	1.0
65.2(−5)	12.4	1.0
13.0(−4)	13.4	.99
17.9(−4)	14.3	.99
40.7(−4)	16.2	.99
81.4(−4)	17.7	.98
12.2(−3)	18.3	.98
17.1(−3)	19.4	.97
48.9(−3)	25.67	.64
81.4(−3)	31.8	.21
10.6(−2)	34.3	.03
11.4(−2)	34.6	0

sponding values of γ_S from Fig. 7. (The profile, in which (a) $\equiv 10^a$, is estimated from results reported by Bradshaw, 1967.) The local skin friction coefficient $c_f = 0.00131$, and the boundary layer thickness $\delta = 0.114$ m. Plot U^+ versus $\log_{10} y^+$ for air under standard conditions and compare with the distribution given by Eq. 33 in which the various regions are defined by Eqs. 34–36 and Fig. 7. Show first that $U_\tau = 0.884$ m/sec.

Section 18.12

1. Calculate the approximate correction for compressibility to skin friction by assuming that the incompressible formula (use Eq. 40) is valid for the wall properties. Take $C_{fi} = \tau_w / \frac{1}{2}\rho_w V_e^2$ and $Re = V_e l/\nu_w$, express ρ_w and ν_w in terms of the ratio T_w/T_e (take $\mu_w/\mu_e = T_w/T_e$) for an insulated wall in a $Pr = 1$ flow. For the compressible C_f take $C_f = \tau_w / \frac{1}{2}\rho_e V_e^2$ and $Re = V_e l/\nu_e$ and calculate C_f/C_{fi} for $C_{fi} = 0.004$. Compare with curves of Fig. 12 (von Kármán (1935) used instead Eq. 39 but the calculation is easier with Eq. 40).

Tables

Table 1
Conversion Factors Between SI and British Units

   SI (Système Internationale) units are used throughout the book; "Absolute" units and the "derived" unit N (newton), listed below, are used throughout. Other derived units (joule, watt, pascal) are given in parentheses. Conversion factors to and from the FPSR system are given; the numbers in parentheses in the conversion factors indicate multiplication by powers of 10; that is, $6.852 \, (-2) \equiv 6.852 \times 10^{-2}$, etc; see Section 8.1 for thermodynamic symbols. Reference: *International Standard, ISO, 1000;* available from National Standards Institute, 1430 Broadway, New York, NY 10018.

Quantity	SI Units	Multiply by	To Obtain FPSR Units	Multiply by	To Obtain SI Units
Mass (M)	kg	6.852 (−2)	slug	1.459 (+1)	kg
Length (L)	m	3.281	ft	3.048 (−1)	m
Density (ρ)	kg/m^3	1.940 (−3)	$slug/ft^3$	5.155 (+2)	kg/m^3
Temperature (T)	°C + 273 °K	1.8	°F + 460 °R	5.556 (−1)	°C + 273 °K
Velocity (V)	m/sec km/hr	3.281 6.214 (−1)	ft/sec mi/hr	3.048 (−1) 1.609	m/sec km/hr
Force (F)	N $kg \, m/sec^2$	2.248 (−1)	lb $slug \, ft/sec^2$	4.448	N $kg \, m/sec^2$
Work Energy (J)	Nm (joule, J)	7.376 (−1)	$slug \, ft^2/sec^2$ Btu	1.356	Nm (joule)
Power (W)	Nm/sec (watt, W)	7.376 (−1) 1.341 (−3)	$slug \, ft^2/sec^3$ hp (550 ft lb/sec)	1.356 7.456 (+2)	Nm/sec (watt)
Pressure (p)	N/m^2 (pascal, Pa)	2.088 (−2)	$slug/ft \, sec^2$ lb/ft^2	4.788 (+1)	N/m^2 (pascal)
Specific Energy, etc	Nm/kg	1.076 (+1)	ft lb/slug	9.290 (−2)	Nm/kg
Gas Constant	Nm/kg °K	5.981	ft lb/slug °R	1.672 (−1)	Nm/kg °K
Coef. of Viscosity (μ)	kg/m sec	2.088 (−2)	slug/ft sec	4.788 (+1)	kg/m sec
Kinematic Viscosity (ν)	m^2/sec	1.076 (+1)	ft^2/sec	9.290 (−2)	m^2/sec
Thermal Conductivity (k)	N/sec °K	1.249 (−1)	lb/sec °R	8.007	N/sec °K
Heat Transfer Coefficient	N/m sec °K	3.807 (−2)	lb/ft sec °R	2.627 (+1)	N/m sec °K

Table 2
Properties of Air and Water

Air at $p = 1.0132 \times 10^5$ N/m^2 (2116 lb/ft^2)							
T		ρ		$\mu \times 10^6$		$\nu \times 10^6$	
°C	°F	$\dfrac{kg}{m^3}$	$\dfrac{slug}{ft^3}$	$\dfrac{kg}{m\ sec}$	$\dfrac{slug}{ft\ sec}$	$\dfrac{m^2}{sec}$	$\dfrac{ft^2}{sec}$
−20	−4	1.39	.00270	15.6	0.326	11.2	121
−10	14	1.35	.00261	16.2	.338	12.0	130
0	32	1.29	.00251	16.8	.350	13.0	139
10	50	1.25	.00242	17.3	.362	13.9	150
[a]15	59	1.23	.00238	17.8	.372	14.4	155
20	68	1.21	.00234	18.0	.375	14.9	160
40	104	1.12	.00217	19.1	.399	17.1	184
60	140	1.06	.00206	20.3	.424	19.2	206
80	176	.99	.00192	21.5	.449	21.7	234
100	212	.94	.00183	22.8	.477	24.3	261
Water							
−20	−4						
−10	14						
0	32	1000	1.939	1787	37.5	1.80	19.3
10	50	1000	1.939	1307	27.2	1.31	14.0
[a]15	59	999	1.937	1154	24.1	1.16	12.4
20	68	997	1.935	1002	21.1	1.01	10.9
40	104	992	1.924	653	13.7	0.66	7.1
60	140	983	1.907	467	9.9	0.48	5.2
80	176	972	1.886	355	7.5	0.37	4.0
100	212	959	1.861	282	5.9	0.30	3.2

[a]Standard conditions.

Table 3

Properties of the Standard Atmosphere

h (km)	T (°C)	a (m/sec)	$p \times 10^{-4}$ (N/m^2) (pascals)	ρ (kg/m^3)	$\mu \times 10^5$ (kg/m sec)
0	15.0	340	10.132	1.226	1.780
1	8.5	336	8.987	1.112	1.749
2	2.0	332	7.948	1.007	1.717
3	−4.5	329	7.010	.909	1.684
4	−11.0	325	6.163	.820	1.652
5	−17.5	320	5.400	.737	1.619
6	−24.0	316	4.717	.660	1.586
7	−30.5	312	4.104	.589	1.552
8	−37.0	308	3.558	.526	1.517
9	−43.5	304	3.073	.467	1.482
10	−50.0	299	2.642	.413	1.447
11	−56.5	295	2.261	.364	1.418
12	−56.5	295	1.932	.311	1.418
13	−56.5	295	1.650	.265	1.418
14	−56.5	295	1.409	.227	1.418
15	−56.5	295	1.203	.194	1.418
16	−56.5	295	1.027	.163	1.418
17	−56.5	295	.785	.141	1.418
18	−56.5	295	.749	.121	1.418
19	−56.5	295	.640	.103	1.418
20	−56.5	295	.546	.088	1.418
30	−56.5	295	.117	.019	1.418
45	40.0	355	.017	.002	1.912
60	70.8	372	.003	3.9×10^{-4}	2.047
75	−10.0	325	.0006	8×10^{-5}	1.667

Table 4
Flow Parameters versus M for Subsonic Flow and $\gamma = 1.4$

M	p/p_0	ρ/ρ_0	T/T_0	a/a_0	A^*/A
.00	1.0000	1.0000	1.0000	1.0000	.00000
.01	.9999	1.0000	1.0000	1.0000	.01728
.02	.9997	.9998	.9999	1.0000	.03455
.03	.9994	.9996	.9998	.9999	.05181
.04	.9989	.9992	.9997	.9998	.06905
.05	.9983	.9988	.9995	.9998	.08627
.06	.9975	.9982	.9993	.9996	.1035
.07	.9966	.9976	.9990	.9995	.1206
.08	.9955	.9968	.9987	.9994	.1377
.09	.9944	.9960	.9984	.9992	.1548
.10	.9930	.9950	.9980	.9990	.1718
.11	.9916	.9940	.9976	.9988	.1887
.12	.9900	.9928	.9971	.9986	.2056
.13	.9883	.9916	.9966	.9983	.2224
.14	.9864	.9903	.9961	.9980	.2391
.15	.9844	.9888	.9955	.9978	.2557
.16	.9823	.9873	.9949	.9974	.2723
.17	.9800	.9857	.9943	.9971	.2887
.18	.9776	.9840	.9936	.9968	.3051
.19	.9751	.9822	.9928	.9964	.3213
.20	.9725	.9803	.9921	.9960	.3374
.21	.9697	.9783	.9913	.9956	.3534
.22	.9668	.9762	.9904	.9952	.3693
.23	.9638	.9740	.9895	.9948	.3851
.24	.9607	.9718	.9886	.9943	.4007
.25	.9575	.9694	.9877	.9938	.4162
.26	.9541	.9670	.9867	.9933	.4315
.27	.9506	.9645	.9856	.9928	.4467
.28	.9470	.9619	.9846	.9923	.4618
.29	.9433	.9592	.9835	.9917	.4767
.30	.9395	.9564	.9823	.9911	.4914
.31	.9355	.9535	.9811	.9905	.5059
.32	.9315	.9506	.9799	.9899	.5203
.33	.9274	.9476	.9787	.9893	.5345
.34	.9231	.9445	.9774	.9886	.5486
.35	.9188	.9413	.9761	.9880	.5624
.36	.9143	.9380	.9747	.9873	.5761
.37	.9098	.9347	.9733	.9866	.5896

Table 4 (*Continued*)

Flow Parameters versus *M* for Subsonic Flow and $\gamma = 1.4$

M	p/p_0	ρ/ρ_0	T/T_0	a/a_0	A^*/A
.38	.9052	.9313	.9719	.9859	.6029
.39	.9004	.9278	.9705	.9851	.6160
.40	.8956	.9243	.9690	.9844	.6289
.41	.8907	.9207	.9675	.9836	.6416
.42	.8857	.9170	.9659	.9828	.6541
.43	.8807	.9132	.9643	.9820	.6663
.44	.8755	.9094	.9627	.9812	.6784
.45	.8703	.9055	.9611	.9803	.6903
.46	.8650	.9016	.9594	.9795	.7019
.47	.8596	.8976	.9577	.9786	.7134
.48	.8541	.8935	.9560	.9777	.7246
.49	.8486	.8894	.9542	.9768	.7356
.50	.8430	.8852	.9524	.9759	.7464
.51	.8374	.8809	.9506	.9750	.7569
.52	.8317	.8766	.9487	.9740	.7672
.53	.8259	.8723	.9468	.9730	.7773
.54	.8201	.8679	.9449	.9721	.7872
.55	.8142	.8634	.9430	.9711	.7968
.56	.8082	.8589	.9410	.9701	.8063
.57	.8022	.8544	.9390	.9690	.8155
.58	.7962	.8498	.9370	.9680	.8244
.59	.7901	.8451	.9349	.9669	.8331
.60	.7840	.8405	.9328	.9658	.8416
.61	.7778	.8357	.9307	.9647	.8499
.62	.7716	.8310	.9286	.9636	.8579
.63	.7654	.8262	.9265	.9625	.8657
.64	.7591	.8213	.9243	.9614	.8732
.65	.7528	.8164	.9221	.9603	.8806
.66	.7465	.8115	.9199	.9591	.8877
.67	.7401	.8066	.9176	.9579	.8945
.68	.7338	.8016	.9153	.9567	.9012
.69	.7274	.7966	.9131	.9555	.9076
.70	.7209	.7916	.9107	.9543	.9138
.71	.7145	.7865	.9084	.9531	.9197
.72	.7080	.7814	.9061	.9519	.9254
.73	.7016	.7763	.9037	.9506	.9309
.74	.6951	.7712	.9013	.9494	.9362
.75	.6886	.7660	.8989	.9481	.9412
.76	.6821	.7609	.8964	.9468	.9461

Table 4 (*Continued*)
Flow Parameters versus *M* for Subsonic Flow and $\gamma = 1.4$

M	p/p_0	ρ/ρ_0	T/T_0	a/a_0	A^*/A
.77	.6756	.7557	.8940	.9455	.9507
.78	.6690	.7505	.8915	.9442	.9551
.79	.6625	.7452	.8890	.9429	.9592
.80	.6560	.7400	.8865	.9416	.9632
.81	.6495	.7347	.8840	.9402	.9669
.82	.6430	.7295	.8815	.9389	.9704
.83	.6365	.7242	.8789	.9375	.9737
.84	.6300	.7189	.8763	.9361	.9769
.85	.6235	.7136	.8737	.9347	.9797
.86	.6170	.7083	.8711	.9333	.9824
.87	.6106	.7030	.8685	.9319	.9849
.88	.6041	.6977	.8659	.9305	.9872
.89	.5977	.6924	.8632	.9291	.9893
.90	.5913	.6870	.8606	.9277	.9912
.91	.5849	.6817	.8579	.9262	.9929
.92	.5785	.6764	.8552	.9248	.9944
.93	.5721	.6711	.8525	.9233	.9958
.94	.5658	.6658	.8498	.9218	.9969
.95	.5595	.6604	.8471	.9204	.9979
.96	.5532	.6551	.8444	.9189	.9986
.97	.5469	.6498	.8416	.9174	.9992
.98	.5407	.6445	.8389	.9159	.9997
.99	.5345	.6392	.8361	.9144	.9999
1.00	.5283	.6339	.8333	.9129	1.0000

Numerical values taken from NACA TN 1428, courtesy of the National Advisory Committee for Aeronautics.

Table 5
Flow Parameters versus M for Supersonic Flow and $\gamma = 1.4$

M	$\dfrac{p}{p_0}$	$\dfrac{\rho}{\rho_0}$	$\dfrac{T}{T_0}$	$\dfrac{a}{a_0}$	$\dfrac{A^*}{A}$	$\dfrac{\dfrac{\rho}{2}V^2}{p_0}$	θ (deg)
1.00	.5283	.6339	.8333	.9129	1.0000	.3698	0
1.01	.5221	.6287	.8306	.9113	.9999	.3728	.04473
1.02	.5160	.6234	.8278	.9098	.9997	.3758	.1257
1.03	.5099	.6181	.8250	.9083	.9993	.3787	.2294
1.04	.5039	.6129	.8222	.9067	.9987	.3815	.3510
1.05	.4979	.6077	.8193	.9052	.9980	.3842	.4874
1.06	.4919	.6024	.8165	.9036	.9971	.3869	.6367
1.07	.4860	.5972	.8137	.9020	.9961	.3895	.7973
1.08	.4800	.5920	.8108	.9005	.9949	.3919	.9680
1.09	.4742	.5869	.8080	.8989	.9936	.3944	1.148
1.10	.4684	.5817	.8052	.8973	.9921	.3967	1.336
1.11	.4626	.5766	.8023	.8957	.9905	.3990	1.532
1.12	.4568	.5714	.7994	.8941	.9888	.4011	1.735
1.13	.4511	.5663	.7966	.8925	.9870	.4032	1.944
1.14	.4455	.5612	.7937	.8909	.9850	.4052	2.160
1.15	.4398	.5562	.7908	.8893	.9828	.4072	2.381
1.16	.4343	.5511	.7879	.8877	.9806	.4090	2.607
1.17	.4287	.5461	.7851	.8860	.9782	.4108	2.839
1.18	.4232	.5411	.7822	.8844	.9758	.4125	3.074
1.19	.4178	.5361	.7793	.8828	.9732	.4141	3.314
1.20	.4124	.5311	.7764	.8811	.9705	.4157	3.558
1.21	.4070	.5262	.7735	.8795	.9676	.4171	3.806
1.22	.4017	.5213	.7706	.8778	.9647	.4185	4.057
1.23	.3964	.5164	.7677	.8762	.9617	.4198	4.312
1.24	.3912	.5115	.7648	.8745	.9586	.4211	4.569
1.25	.3861	.5067	.7619	.8729	.9553	.4223	4.830
1.26	.3809	.5019	.7590	.8712	.9520	.4233	5.093
1.27	.3759	.4971	.7561	.8695	.9486	.4244	5.359
1.28	.3708	.4923	.7532	.8679	.9451	.4253	5.627
1.29	.3658	.4876	.7503	.8662	.9415	.4262	5.898
1.30	.3609	.4829	.7474	.8645	.9378	.4270	6.170
1.31	.3560	.4782	.7445	.8628	.9341	.4277	6.445
1.32	.3512	.4736	.7416	.8611	.9302	.4283	6.721
1.33	.3464	.4690	.7387	.8595	.9263	.4289	7.000
1.34	.3417	.4644	.7358	.8578	.9223	.4294	7.279
1.35	.3370	.4598	.7329	.8561	.9182	.4299	7.561
1.36	.3323	.4553	.7300	.8544	.9141	.4303	7.844

Table 5 (*Continued*)

Flow Parameters versus *M* for Supersonic Flow and γ = 1.4

M	$\dfrac{p}{p_0}$	$\dfrac{\rho}{\rho_0}$	$\dfrac{T}{T_0}$	$\dfrac{a}{a_0}$	$\dfrac{A^*}{A}$	$\dfrac{\frac{\rho}{2}V^2}{p_0}$	θ (deg)
1.37	.3277	.4508	.7271	.8527	.9099	.4306	8.128
1.38	.3232	.4463	.7242	.8510	.9056	4308	8.413
1.39	.3187	.4418	.7213	.8493	.9013	.4310	8.699
1.40	3142	.4374	.7184	.8476	.8969	.4311	8.987
1.41	.3098	.4330	.7155	.8459	.8925	.4312	9.276
1.42	.3055	.4287	.7126	.8442	.8880	.4312	9.565
1.43	.3012	.4244	.7097	.8425	.8834	.4311	9.855
1.44	.2969	.4201	.7069	.8407	.8788	.4310	10.15
1.45	.2927	.4158	.7040	.8390	.8742	.4308	10.44
1.46	.2886	.4116	.7011	.8373	.8695	.4306	10.73
1.47	.2845	.4074	.6982	.8356	.8647	.4303	11.02
1.48	.2804	.4032	.6954	.8339	.8599	.4299	11.32
1.49	.2764	.3991	.6925	.8322	.8551	.4295	11.61
1.50	.2724	.3950	.6897	.8305	.8502	.4290	11.91
1.51	.2685	.3909	.6868	.8287	.8453	.4285	12.20
1.52	.2646	.3869	.6840	.8270	.8404	.4279	12.49
1.53	.2608	.3829	.6811	.8253	.8354	.4273	12.79
1.54	.2570	.3789	.6783	.8236	.8304	.4266	13.09
1.55	.2533	.3750	.6754	.8219	.8254	.4259	13.38
1.56	.2496	.3710	.6726	.8201	.8203	.4252	13.68
1.57	.2459	.3672	.6698	.8184	.8152	.4243	13.97
1.58	.2423	.3633	.6670	.8167	.8101	.4235	14.27
1.59	.2388	.3595	.6642	.8150	.8050	.4226	14.56
1.60	.2353	.3557	.6614	.8133	.7998	.4216	14.86
1.61	.2318	.3520	.6586	.8115	.7947	.4206	15.16
1.62	.2284	.3483	.6558	.8098	.7895	.4196	15.45
1.63	.2250	.3446	.6530	.8081	.7843	.4185	15.75
1.64	.2217	.3409	.6502	.8064	.7791	.4174	16.04
1.65	.2184	.3373	.6475	.8046	.7739	.4162	16.34
1.66	.2151	.3337	.6447	.8029	.7686	.4150	16.63
1.67	.2119	.3302	.6419	.8012	.7634	.4138	16.93
1.68	.2088	.3266	.6392	.7995	.7581	.4125	17.22
1.69	.2057	.3232	.6364	.7978	.7529	.4112	17.52
1.70	.2026	.3197	.6337	.7961	.7476	.4098	17.81
1.71	.1996	.3163	.6310	.7943	.7423	.4086	18.10
1.72	.1966	.3129	.6283	.7926	.7371	.4071	18.40
1.73	.1936	.3095	.6256	.7909	.7318	.4056	18.69
1.74	.1907	.3062	.6229	.7892	.7265	.4041	18.98

Table 5 (*Continued*)

Flow Parameters versus *M* for Supersonic Flow and $\gamma = 1.4$

M	$\dfrac{p}{p_0}$	$\dfrac{\rho}{\rho_0}$	$\dfrac{T}{T_0}$	$\dfrac{a}{a_0}$	$\dfrac{A^*}{A}$	$\dfrac{\frac{\rho}{2}V^2}{p_0}$	θ (deg)
1.75	.1878	.3029	.6202	.7875	.7212	.4026	19.27
1.76	.1850	.2996	.6175	.7858	.7160	.4011	19.56
1.77	.1822	.2964	.6148	.7841	.7107	.3996	19.86
1.78	.1794	.2932	.6121	.7824	.7054	.3980	20.15
1.79	.1767	.2900	.6095	.7807	.7002	.3964	20.44
1.80	.1740	.2868	.6068	.7790	.6949	.3947	20.73
1.81	.1714	.2837	.6041	.7773	.6897	.3931	21.01
1.82	.1688	.2806	.6015	.7756	.6845	.3914	21.30
1.83	.1662	.2776	.5989	.7739	.6792	.3897	21.59
1.84	.1637	.2745	.5963	.7722	.6740	.3879	21.88
1.85	.1612	.2715	.5936	.7705	.6688	.3862	22.16
1.86	.1587	.2686	.5910	.7688	.6636	.3844	22.45
1.87	.1563	.2656	.5884	.7671	.6584	.3826	22.73
1.88	.1539	.2627	.5859	.7654	.6533	.3808	23.02
1.89	.1516	.2598	.5833	.7637	.6481	.3790	23.30
1.90	.1492	.2570	.5807	.7620	.6430	.3771	23.59
1.91	.1470	.2542	.5782	.7604	.6379	.3753	23.87
1.92	.1447	.2514	.5756	.7587	.6328	.3734	24.15
1.93	.1425	.2486	.5731	.7570	.6277	.3715	24.43
1.94	.1403	.2459	.5705	.7553	.6226	.3696	24.71
1.95	.1381	.2432	.5680	.7537	.6175	.3677	24.99
1.96	.1360	.2405	.5655	.7520	.6125	.3657	25.27
1.97	.1339	.2378	.5630	.7503	.6075	.3638	25.55
1.98	.1318	.2352	.5605	.7487	.6025	.3618	25.83
1.99	.1298	.2326	.5580	.7470	.5975	.3598	26.10
2.00	.1278	.2300	.5556	.7454	.5926	.3579	26.38
2.01	.1258	.2275	.5531	.7437	.5877	.3559	26.66
2.02	.1239	.2250	.5506	.7420	.5828	.3539	26.93
2.03	.1220	.2225	.5482	.7404	.5779	.3518	27.20
2.04	.1201	.2200	.5458	.7388	.5730	.3498	27.48
2.05	.1182	.2176	.5433	.7371	.5682	.3478	27.75
2.06	.1164	.2152	.5409	.7355	.5634	.3458	28.02
2.07	.1146	.2128	.5385	.7338	.5586	.3437	28.29
2.08	.1128	.2104	.5361	.7322	.5538	.3417	28.56
2.09	.1111	.2081	.5337	.7306	.5491	.3396	28.83
2.10	.1094	.2058	.5313	.7289	.5444	.3376	29.10
2.11	.1077	.2035	.5290	.7273	.5397	.3355	29.36

Table 5 (*Continued*)

Flow Parameters versus *M* for Supersonic Flow and γ = 1.4

M	$\dfrac{p}{p_0}$	$\dfrac{\rho}{\rho_0}$	$\dfrac{T}{T_0}$	$\dfrac{a}{a_0}$	$\dfrac{A^*}{A}$	$\dfrac{\frac{\rho}{2}V^2}{p_0}$	θ (deg)
2.12	.1060	.2013	.5266	.7257	.5350	.3334	29.63
2.13	.1043	.1990	.5243	.7241	.5304	.3314	29.90
2.14	.1027	.1968	.5219	.7225	.5258	.3293	30.16
2.15	.1011	.1946	.5196	.7208	.5212	.3272	30.43
2.16	.09956	.1925	.5173	.7192	.5167	.3252	30.69
2.17	.09802	.1903	.5150	.7176	.5122	.3231	30.95
2.18	.09650	.1882	.5127	.7160	.5077	.3210	31.21
2.19	.09500	.1861	.5104	.7144	.5032	.3189	31.47
2.20	.09352	.1841	.5081	.7128	.4988	.3169	31.73
2.21	.09207	.1820	.5059	.7112	.4944	.3148	31.99
2.22	.09064	.1800	.5036	.7097	.4900	.3127	32.25
2.23	.08923	.1780	.5014	.7081	.4856	.3106	32.51
2.24	.08785	.1760	.4991	.7065	.4813	.3085	32.76
2.25	.08648	.1740	.4969	.7049	.4770	.3065	33.02
2.26	.08514	.1721	.4947	.7033	.4727	.3044	33.27
2.27	.08382	.1702	.4925	.7018	.4685	.3023	33.53
2.28	.08252	.1683	.4903	.7002	.4643	.3003	33.78
2.29	.08123	.1664	.4881	.6986	.4601	.2982	34.03
2.30	.07997	.1646	.4859	.6971	.4560	.2961	34.28
2.31	.07873	.1628	.4837	.6955	.4519	.2941	34.53
2.32	.07751	.1609	.4816	.6940	.4478	.2920	34.78
2.33	.07631	.1592	.4794	.6924	.4437	.2900	35.03
2.34	.07512	.1574	.4773	.6909	.4397	.2879	35.28
2.35	.07396	.1556	.4752	.6893	.4357	.2859	35.53
2.36	.07281	.1539	.4731	.6878	.4317	.2839	35.77
2.37	.07168	.1522	.4709	.6863	.4278	.2818	36.02
2.38	.07057	.1505	.4688	.6847	.4239	.2798	36.26
2.39	.06948	.1488	.4668	.6832	.4200	.2778	36.50
2.40	.06840	.1472	.4647	.6817	.4161	.2758	36.75
2.41	.06734	.1456	.4626	.6802	.4123	.2738	36.99
2.42	.06630	.1439	.4606	.6786	.4085	.2718	37.23
2.43	.06527	.1424	.4585	.6771	.4048	.2698	37.47
2.44	.06426	.1408	.4565	.6756	.4010	.2678	37.71
2.45	.06327	.1392	.4544	.6741	.3973	.2658	37.95
2.46	.06229	.1377	.4524	.6726	.3937	.2639	38.18
2.47	.06133	.1362	.4504	.6711	.3900	.2619	38.42
2.48	.06038	.1347	.4484	.6696	.3864	.2599	38.66
2.49	.05945	.1332	.4464	.6681	.3828	.2580	38.89

Table 5 (*Continued*)

Flow Parameters versus M for Supersonic Flow and $\gamma = 1.4$

M	$\dfrac{p}{p_0}$	$\dfrac{\rho}{\rho_0}$	$\dfrac{T}{T_0}$	$\dfrac{a}{a_0}$	$\dfrac{A^*}{A}$	$\dfrac{\frac{\rho}{2}V^2}{p_0}$	θ (deg)
2.50	.05853	.1317	.4444	.6667	.3793	.2561	39.12
2.51	.05762	.1302	.4425	.6652	.3757	.2541	39.36
2.52	.05674	.1288	.4405	.6637	.3722	.2522	39.59
2.53	.05586	.1274	.4386	.6622	.3688	.2503	39.82
2.54	.05500	.1260	.4366	.6608	.3653	.2484	40.05
2.55	.05415	.1246	.4347	.6593	.3619	.2465	40.28
2.56	.05332	.1232	.4328	.6579	.3585	.2446	40.51
2.57	.05250	.1218	.4309	.6564	.3552	.2427	40.75
2.58	.05169	.1205	.4289	.6549	.3519	.2409	40.96
2.59	.05090	.1192	.4271	.6535	.3486	.2390	41.19
2.60	.05012	.1179	.4252	.6521	.3453	.2371	41.41
2.61	.04935	.1166	.4233	.6506	.3421	.2353	41.64
2.62	.04859	.1153	.4214	.6492	.3389	.2335	41.86
2.63	.04784	.1140	.4196	.6477	.3357	.2317	42.09
2.64	.04711	.1128	.4177	.6463	.3325	.2298	42.31
2.65	.04639	.1115	.4159	.6449	.3294	.2280	42.53
2.66	.04568	.1103	.4141	.6435	.3263	.2262	42.75
2.67	.04498	.1091	.4122	.6421	.3232	.2245	42.97
2.68	.04429	.1079	.4104	.6406	.3202	.2227	43.19
2.69	.04362	.1067	.4086	.6392	.3172	.2209	43.40
2.70	.04295	.1056	.4068	.6378	.3142	.2192	43.62
2.71	.04229	.1044	.4051	.6364	.3112	.2174	43.84
2.72	.04165	.1033	.4033	.6350	.3083	.2157	44.05
2.73	.04102	.1022	.4015	.6337	.3054	.2140	44.27
2.74	.04039	.1010	.3998	.6323	.3025	.2123	44.48
2.75	.03978	.09994	.3980	.6309	.2996	.2106	44.69
2.76	.03917	.09885	.3963	.6295	.2968	.2089	44.91
2.77	.03858	.09778	.3945	.6281	.2940	.2072	45.12
2.78	.03799	.09671	.3928	.6268	.2912	.2055	45.33
2.79	.03742	.09566	.3911	.6254	.2884	.2039	45.54
2.80	.03685	.09463	.3894	.6240	.2857	.2022	45.75
2.81	.03629	.09360	.3877	.6227	.2830	.2006	45.95
2.82	.03574	.09259	.3860	.6213	.2803	.1990	46.16
2.83	.03520	.09158	.3844	.6200	.2777	.1973	46.37
2.84	.03467	.09059	.3827	.6186	.2750	.1957	46.57
2.85	.03415	.08962	.3810	.6173	.2724	.1941	46.78
2.86	.03363	.08865	.3794	.6159	.2698	.1926	46.98

Table 5 (*Continued*)
Flow Parameters versus M for Supersonic Flow and $\gamma = 1.4$

M	$\dfrac{p}{p_0}$	$\dfrac{\rho}{\rho_0}$	$\dfrac{T}{T_0}$	$\dfrac{a}{a_0}$	$\dfrac{A^*}{A}$	$\dfrac{\frac{\rho}{2}V^2}{p_0}$	θ (deg)
2.87	.03312	.08769	.3777	.6146	.2673	.1910	47.19
2.88	.03263	.08675	.3761	.6133	.2648	.1894	47.39
2.89	.03213	.08581	.3745	.6119	.2622	.1879	47.59
2.90	.03165	.08489	.3729	.6106	.2598	.1863	47.79
2.91	.03118	.08398	.3712	.6093	.2573	.1848	47.99
2.92	.03071	.08307	.3696	.6080	.2549	.1833	48.19
2.93	.03025	.08218	.3681	.6067	.2524	.1818	48.39
2.94	.02980	.08130	.3665	.6054	.2500	.1803	48.59
2.95	.02935	.08043	.3649	.6041	.2477	.1788	48.78
2.96	.02891	.07957	.3633	.6028	.2453	.1773	48.98
2.97	.02848	.07872	.3618	.6015	.2430	.1758	49.18
2.98	.02805	.07788	.3602	.6002	.2407	.1744	49.37
2.99	.02764	.97705	.3587	.5989	.2384	.1729	49.56
3.00	.02722	.07623	.3571	.5976	.2362	.1715	49.76
3.01	.02682	.07541	.3556	.5963	.2339	.1701	49.95
3.02	.02642	.07461	.3541	.5951	.2317	.1687	50.14
3.03	.02603	.07382	.3526	.5938	.2295	.1673	50.33
3.04	.02564	.07303	.3511	.5925	.2273	.1659	50.52
3.05	.02526	.07226	.3496	.5913	.2252	.1645	50.71
3.06	.02489	.07149	.3481	.5900	.2230	.1631	50.90
3.07	.02452	.07074	.3466	.5887	.2209	.1618	51.09
3.08	.02416	.06999	.3452	.5875	.2188	.1604	51.28
3.09	.02380	.06925	.3437	.5862	.2168	.1591	51.46
3.10	.02345	.06852	.3422	.5850	.2147	.1577	51.65
3.11	.02310	.06779	.3408	.5838	.2127	.1564	51.84
3.12	.02276	.06708	.3393	.5825	.2107	.1551	52.02
3.13	.02243	.06637	.3379	.5813	.2087	.1538	52.20
3.14	.02210	.06568	.3365	.5801	.2067	.1525	52.39
3.15	.02177	.06499	.3351	.5788	.2048	.1512	52.57
3.16	.02146	.06430	.3337	.5776	.2028	.1500	52.75
3.17	.02114	.06363	.3323	.5764	.2009	.1487	52.93
3.18	.02083	.06296	.3309	.5752	.1990	.1475	53.11
3.19	.02053	.06231	.3295	.5740	.1971	.1462	53.29
3.20	.02023	.06165	.3281	.5728	.1953	.1450	53.47
3.21	.01993	.06101	.3267	.5716	.1934	.1438	53.65
3.22	.01964	.06037	.3253	.5704	.1916	.1426	53.83
3.23	.01936	.05975	.3240	.5692	.1898	.1414	54.00
3.24	.01908	.05912	.3226	.5680	.1880	.1402	54.18

Table 5 (*Continued*)

Flow Parameters versus M for Supersonic Flow and $\gamma = 1.4$

M	$\dfrac{p}{p_0}$	$\dfrac{\rho}{\rho_0}$	$\dfrac{T}{T_0}$	$\dfrac{a}{a_0}$	$\dfrac{A^*}{A}$	$\dfrac{\frac{\rho}{2}V^2}{p_0}$	θ (deg)
3.25	.01880	.05851	.3213	.5668	.1863	.1390	54.35
3.26	.01853	.05790	.3199	.5656	.1845	.1378	54.53
3.27	.01826	.05730	.3186	.5645	.1828	.1367	54.71
3.28	.01799	.05671	.3173	.5633	.1810	.1355	54.88
3.29	.01773	.05612	.3160	.5621	.1793	.1344	55.05
3.30	.01748	.05554	.3147	.5609	.1777	.1332	55.22
3.31	.01722	.05497	.3134	.5598	.1760	.1321	55.39
3.32	.01698	.05440	.3121	.5586	.1743	.1310	55.56
3.33	.01673	.05384	.3108	.5575	.1727	.1299	55.73
3.34	.01649	.05329	.3095	.5563	.1711	.1288	55.90
3.35	.01625	.05274	.3082	.5552	.1695	.1277	56.07
3.36	.01602	.05220	.3069	.5540	.1679	.1266	56.24
3.37	.01579	.05166	.3057	.5529	.1663	.1255	56.41
3.38	.01557	.05113	.3044	.5517	.1648	.1245	56.58
3.39	.01534	.05061	.3032	.5506	.1632	.1234	56.75
3.40	.01513	.05009	.3019	.5495	.1617	.1224	56.91
3.41	.01491	.04958	.3007	.5484	.1602	.1214	57.07
3.42	.01470	.04908	.2995	.5472	.1587	.1203	57.24
3.43	.01449	.04858	.2982	.5461	.1572	.1193	57.40
3.44	.01428	.04808	.2970	.5450	.1558	.1183	57.56
3.45	.01408	.04759	.2958	.5439	.1543	.1173	57.73
3.46	.01388	.04711	.2946	.5428	.1529	.1163	57.89
3.47	.01368	.04663	.2934	.5417	.1515	.1153	58.05
3.48	.01349	.04616	.2922	.5406	.1501	.1144	58.21
3.49	.01330	.04569	.2910	.5395	.1487	.1134	58.37
3.50	.01311	.04523	.2899	.5384	.1473	.1124	58.53
3.60	.01138	.04089	.2784	.5276	.1342	.1033	60.09
3.70	9.903×10^{-3}	.03702	.2675	.5172	.1224	.09490	61.60
3.80	8.629×10^{-3}	.03355	.2572	.5072	.1117	.08722	63.04
3.90	7.532×10^{-3}	.03044	.2474	.4974	.1021	.08019	64.44
4.00	6.586×10^{-3}	.02766	.2381	.4880	.09329	.07376	65.78
4.10	5.769×10^{-3}	.02516	.2293	.4788	.08536	.06788	67.08
4.20	5.062×10^{-3}	.02292	.2208	.4699	.07818	.06251	68.33

Table 5 (*Continued*)

Flow Parameters versus M for Supersonic Flow and $\gamma = 1.4$

M	$\dfrac{p}{p_0}$	$\dfrac{\rho}{\rho_0}$	$\dfrac{T}{T_0}$	$\dfrac{a}{a_0}$	$\dfrac{A^*}{A}$	$\dfrac{\frac{\rho}{2}V^2}{p_0}$	θ (deg)
4.30	4.449×10^{-3}	.02090	.2129	.4614	.07166	.05759	69.54
4.40	3.918×10^{-3}	.01909	.2053	.4531	.06575	.05309	70.71
4.50	3.455×10^{-3}	.01745	.1980	.4450	.06038	.04898	71.83
4.60	3.053×10^{-3}	.01597	.1911	.4372	.05550	.04521	72.92
4.70	2.701×10^{-3}	.01464	.1846	.4296	.05107	.04177	73.97
4.80	2.394×10^{-3}	.01343	.1783	.4223	.04703	.03861	74.99
4.90	2.126×10^{-3}	.01233	.1724	.4152	.04335	.03572	75.97
5.00	1.890×10^{-3}	.01134	.1667	.4082	.04000	.03308	76.92
6.00	6.334×10^{-4}	5.194×10^{-3}	.1220	.3492	.01880	.01596	84.96
7.00	2.416×10^{-4}	2.609×10^{-3}	.09259	.3043	9.602×10^{-3}	8.285×10^{-3}	90.97
8.00	1.024×10^{-4}	1.414×10^{-3}	.07246	.2692	5.260×10^{-3}	4.589×10^{-3}	95.62
9.00	4.739×10^{-5}	8.150×10^{-4}	.05814	.2411	3.056×10^{-3}	2.687×10^{-3}	99.32
10.00	2.356×10^{-5}	4.948×10^{-4}	.04762	.2182	1.866×10^{-3}	1.649×10^{-3}	102.3
100.00	2.790×10^{-12}	5.583×10^{-9}	4.998×10^{-4}	.02236	2.157×10^{-8}	1.953×10^{-8}	127.6
∞	0	0	0	0	0	0	130.5

Numerical values taken from NACA TN 1428, courtesy of the National Advisory Committee for Aeronautics.

Table 6
Parameters for Shock Flow ($\gamma = 1.4$)

M_{1n}	p_2/p_1	ρ_2/ρ_1	T_2/T_1	a_2/a_1	p_2^0/p_1^0	M_2 for Normal Shocks Only
1.00	1.000	1.000	1.000	1.000	1.0000	1.0000
1.01	1.023	1.017	1.007	1.003	1.0000	.9901
1.02	1.047	1.033	1.013	1.007	1.0000	.9805
1.03	1.071	1.050	1.020	1.010	1.0000	.9712
1.04	1.095	1.067	1.026	1.013	.9999	.9620
1.05	1.120	1.084	1.033	1.016	.9999	.9531
1.06	1.144	1.101	1.039	1.019	.9998	.9444
1.07	1.169	1.118	1.046	1.023	.9996	.9360
1.08	1.194	1.135	1.052	1.026	.9994	.9277
1.09	1.219	1.152	1.059	1.029	.9992	.9196
1.10	1.245	1.169	1.065	1.032	.9989	.9118
1.11	1.271	1.186	1.071	1.035	.9986	.9041
1.12	1.297	1.203	1.078	1.038	.9982	.8966
1.13	1.323	1.221	1.084	1.041	.9978	.8892
1.14	1.350	1.238	1.090	1.044	.9973	.8820
1.15	1.376	1.255	1.097	1.047	.9967	.8750
1.16	1.403	1.272	1.103	1.050	.9961	.8682
1.17	1.430	1.290	1.109	1.053	.9953	.8615
1.18	1.458	1.307	1.115	1.056	.9946	.8549
1.19	1.485	1.324	1.122	1.059	.9937	.8485
1.20	1.513	1.342	1.128	1.062	.9928	.8422
1.21	1.541	1.359	1.134	1.065	.9918	.8360
1.22	1.570	1.376	1.141	1.068	.9907	.8300
1.23	1.598	1.394	1.147	1.071	.9896	.8241
1.24	1.627	1.411	1.153	1.074	.9884	.8183
1.25	1.656	1.429	1.159	1.077	.9871	.8126
1.26	1.686	1.446	1.166	1.080	.9857	.8071
1.27	1.715	1.463	1.172	1.083	.9842	.8016
1.28	1.745	1.481	1.178	1.085	.9827	.7963
1.29	1.775	1.498	1.185	1.088	.9811	.7911
1.30	1.805	1.516	1.191	1.091	.9794	.7860
1.31	1.835	1.533	1.197	1.094	.9776	.7809
1.32	1.866	1.551	1.204	1.097	.9758	.7760
1.33	1.897	1.568	1.210	1.100	.9738	.7712
1.34	1.928	1.585	1.216	1.103	.9718	.7664
1.35	1.960	1.603	1.223	1.106	.9697	.7618
1.36	1.991	1.620	1.229	1.109	.9676	.7572

Table 6 (*Continued*)
Parameters for Shock Flow ($\gamma = 1.4$)

M_{1n}	p_2/p_1	ρ_2/ρ_1	T_2/T_1	a_2/a_1	p_2^0/p_1^0	M_2 for Normal Shocks Only
1.37	2.023	1.638	1.235	1.111	.9653	.7527
1.38	2.055	1.655	1.242	1.114	.9630	.7483
1.39	2.087	1.672	1.248	1.117	.9606	.7440
1.40	2.120	1.690	1.255	1.120	.9582	.7397
1.41	2.153	1.707	1.261	1.123	.9557	.7355
1.42	2.186	1.724	1.268	1.126	.9531	.7314
1.43	2.219	1.742	1.274	1.129	.9504	.7274
1.44	2.253	1.759	1.281	1.132	.9476	.7235
1.45	2.286	1.776	1.287	1.135	.9448	.7196
1.46	2.320	1.793	1.294	1.137	.9420	.7157
1.47	2.354	1.811	1.300	1.140	.9390	.7120
1.48	2.389	1.828	1.307	1.143	.9360	.7083
1.49	2.423	1.845	1.314	1.146	.9329	.7047
1.50	2.458	1.862	1.320	1.149	.9298	.7011
1.51	2.493	1.879	1.327	1.152	.9266	.6976
1.52	2.529	1.896	1.334	1.155	.9233	.6941
1.53	2.564	1.913	1.340	1.158	.9200	.6907
1.54	2.600	1.930	1.347	1.161	.9166	.6874
1.55	2.636	1.947	1.354	1.164	.9132	.6841
1.56	2.673	1.964	1.361	1.166	.9097	.6809
1.57	2.709	1.981	1.367	1.169	.9061	.6777
1.58	2.746	1.998	1.374	1.172	.9026	.6746
1.59	2.783	2.015	1.381	1.175	.8989	.6715
1.60	2.820	2.032	1.388	1.178	.8952	.6684
1.61	2.857	2.049	1.395	1.181	.8914	.6655
1.62	2.895	2.065	1.402	1.184	.8877	.6625
1.63	2.933	2.082	1.409	1.187	.8838	.6596
1.64	2.971	2.099	1.416	1.190	.8799	.6568
1.65	3.010	2.115	1.423	1.193	.8760	.6540
1.66	3.048	2.132	1.430	1.196	.8720	.6512
1.67	3.087	2.148	1.437	1.199	.8680	.6485
1.68	3.126	2.165	1.444	1.202	.8640	.6458
1.69	3.165	2.181	1.451	1.205	.8599	.6431
1.70	3.205	2.198	1.458	1.208	.8557	.6405
1.71	3.245	2.214	1.466	1.211	.8516	.6380
1.72	3.285	2.230	1.473	1.214	.8474	.6355
1.73	3.325	2.247	1.480	1.217	.8431	.6330
1.74	3.366	2.263	1.487	1.220	.8389	.6305

Table 6 (*Continued*)

Parameters for Shock Flow ($\gamma = 1.4$)

M_{1n}	p_2/p_1	ρ_2/ρ_1	T_2/T_1	a_2/a_1	p_2^0/p_1^0	M_2 for Normal Shocks Only
1.75	3.406	2.279	1.495	1.223	.8346	.6281
1.76	3.447	2.295	1.502	1.226	.8302	.6257
1.77	3.488	2.311	1.509	1.229	.8259	.6234
1.78	3.530	2.327	1.517	1.232	.8215	.6210
1.79	3.571	2.343	1.524	1.235	.8171	.6188
1.80	3.613	2.359	1.532	1.238	.8127	.6165
1.81	3.655	2.375	1.539	1.241	.8082	.6143
1.82	3.698	2.391	1.547	1.244	.8038	.6121
1.83	3.740	2.407	1.554	1.247	.7993	.6099
1.84	3.783	2.422	1.562	1.250	.7948	.6078
1.85	3.826	2.438	1.569	1.253	.7902	.6057
1.86	3.870	2.454	1.577	1.256	.7857	.6036
1.87	3.913	2.469	1.585	1.259	.7811	.6016
1.88	3.957	2.485	1.592	1.262	.7765	.5996
1.89	4.001	2.500	1.600	1.265	.7720	.5976
1.90	4.045	2.516	1.608	1.268	.7674	.5956
1.91	4.089	2.531	1.616	1.271	.7628	.5937
1.92	4.134	2.546	1.624	1.274	.7581	.5918
1.93	4.179	2.562	1.631	1.277	.7535	.5899
1.94	4.224	2.577	1.639	1.280	.7488	.5889
1.95	4.270	2.592	1.647	1.283	.7442	.5862
1.96	4.315	2.607	1.655	1.287	.7395	.5844
1.97	4.361	2.622	1.663	1.290	.7349	.5826
1.98	4.407	2.637	1.671	1.293	.7302	.5808
1.99	4.453	2.652	1.679	1.296	.7255	.5791
2.00	4.500	2.667	1.688	1.299	.7209	.5773
2.01	4.547	2.681	1.696	1.302	.7162	.5757
2.02	4.594	2.696	1.704	1.305	.7115	.5740
2.03	4.641	2.711	1.712	1.308	.7069	.5723
2.04	4.689	2.725	1.720	1.312	.7022	.5707
2.05	4.736	2.740	1.729	1.315	.6975	.5691
2.06	4.784	2.755	1.737	1.318	.6928	.5675
2.07	4.832	2.769	1.745	1.321	.6882	.5659
2.08	4.881	2.783	1.754	1.324	.6835	.5643
2.09	4.929	2.798	1.762	1.327	.6789	.5628
2.10	4.978	2.812	1.770	1.331	.6742	.5613
2.11	5.027	2.826	1.779	1.334	.6696	.5598

Table 6 (*Continued*)
Parameters for Shock Flow ($\gamma = 1.4$)

M_{1n}	p_2/p_1	ρ_2/ρ_1	T_2/T_1	a_2/a_1	p_2^0/p_1^0	M_2 for Normal Shocks Only
2.87	9.443	3.734	2.529	1.590	.3670	.4833
2.88	9.510	3.743	2.540	1.594	.3639	.4827
2.89	9.577	3.753	2.552	1.597	.3608	.4820
2.90	9.645	3.763	2.563	1.601	.3577	.4814
2.91	9.713	3.773	2.575	1.605	.3547	.4807
2.92	9.781	3.782	2.586	1.608	.3517	.4801
2.93	9.849	3.792	2.598	1.612	.3487	.4795
2.94	9.918	3.801	2.609	1.615	.3457	.4788
2.95	9.986	3.811	2.621	1.619	.3428	.4782
2.96	10.06	3.820	2.632	1.622	.3398	.4776
2.97	10.12	3.829	2.644	1.626	.3369	.4770
2.98	10.19	3.839	2.656	1.630	.3340	.4764
2.99	10.26	3.848	2.667	1.633	.3312	.4758
3.00	10.33	3.857	2.679	1.637	.3283	.4752
3.10	11.05	3.947	2.799	1.673	.3012	.4695
3.20	11.78	4.031	2.922	1.709	.2762	.4643
3.30	12.54	4.112	3.049	1.746	.2533	.4596
3.40	13.32	4.188	3.180	1.783	.2322	.4552
3.50	14.13	4.261	3.315	1.821	.2129	.4512
3.60	14.95	4.330	3.454	1.858	.1953	.4474
3.70	15.80	4.395	3.596	1.896	.1792	.4439
3.80	16.68	4.457	3.743	1.935	.1645	.4407
3.90	17.58	4.516	3.893	1.973	.1510	.4377
4.00	18.50	4.571	4.047	2.012	.1388	.4350
5.00	29.00	5.000	5.800	2.408	.06172	.4152
6.00	41.83	5.268	7.941	2.818	.02965	.4042
7.00	57.00	5.444	10.47	3.236	.01535	.3947
8.00	74.50	5.565	13.39	3.659	8.488×10^{-3}	:3929
9.00	94.33	5.651	16.69	4.086	4.964×10^{-3}	.3898
10.00	116.5	5.714	20.39	4.515	3.045×10^{-3}	.3876
100.00	11,666.5	5.997	1945.4	44.11	3.593×10^{-8}	.3781
∞	∞	6	∞	∞	0	.3780

Data taken from NACA TN 1428, courtesy of National Advisory Committee of Aeronautics.

Table 7

Thermal Properties of Air ($p = 1$ atmosphere)

°R	Pr	c_p/R	$z = \dfrac{p}{\rho RT}$	γ
200	.768	3.56	0.985	1.420
400	.732	3.51	0.998	1.405
600	.701	3.51	1.000	1.400
800	.684	3.55	1.000	1.393
1000	.680	3.63	1.000	1.381
1200	.682	3.73	1.000	1.368
1400	.688	3.81	1.000	1.356
1600	.698	3.90	1.000	1.346
1800	.702	3.98	1.000	1.336
2000		4.05	1.000	1.329
3000		4.41	1.000	1.294
4000		4.99	1.001	1.261
5000		7.66	1.011	1.198

Data from Tables of *Thermal Properties of Gases*, Dept. of Commerce, National Bureau of Standards Circular 564, U.S. Government Printing Office, 1955.

OBLIQUE SHOCK CHART
(From Liepmann and Puckett, 1947.)

Bibliography

Abbott, I. H., and von Doenhoff, A. E. (1949), *Theory of Wing Sections, Including a Summary of Airfoil Data*, McGraw-Hill, New York. Paperback edition, Dover, New York, 1959. [116, 136, 357, 359]

Adamson, T. J., Jr. (1964), "The Structure of the Rocket Plume without Reactions," from *Supersonic Flow, Chemical Processes and Radiative Transfer*, Pergamon, New York. [224]

Allen, H. J. (1938), *Calculation of the Chordwise Load Distribution over Airfoil Sections with Plain, Split, or Serially-Hinged Trailing Edge Flaps*, NACA Report 634. [137]

Ashley, H., and Landahl, M. (1965), *Aerodynamics of Wings and Bodies*, Addison-Wesley, Reading, Mass. [171]

Back, L. H., Cuffel, R. F., and Messier, P. F. (1969), "Laminarization of a Turbulent Boundary Layer in Nozzle Flow," *AIAA J.* 7, 730–733. [359]

Bailey, N. P. (1944), "Thermodynamics of Air at High Velocity," *J. Aero. Sci.* 11, 227–238. [215]

Bekofske, K., and Liu, V.-C. (1972), "Internal Gravity Wave-Atmospheric Wind Interaction: A Cause of Clear Air Turbulence," *Science* 178, 1089–1092. [369]

Belle, G., see Blick, E. F.

Beushausen, W., see Lippisch, A.

Blick, E. F., Watson, W., Belle, G., Chu, H. (1975), "Bird Aerodynamics Experiments," *Proceedings of Symposium on Swimming and Flying in Nature*, Plenum Press, New York. [414]

Boison, J. C., see Van Driest, E. R.

Borst, H. V., see Hoerner, S. F.

Bradshaw, P. (1964), *Experimental Fluid Mechanics*, Macmillan, New York.

Bradshaw, P. (1975), *Introduction to Turbulence and Its Measurement*, Pergamon Press, New York. [239, 349, 371, 380, 485]

Bridgewater, J., see Lock, R. C.

Burgers, J. M., see von Kármán, T.

Carlson, H. W., and Harris, R. V., Jr. (1969), "A Unified System of Supersonic Aerodynamic Analysis," *Analytic Methods in Aircraft Aerodynamics*, NASA SP 228, pp. 639–658. [293, 294, 296]

Cebeci, T., Mosinskis, G. J., and Smith, A. M. O. (1972), "Predicted Separation points at positive pressure gradients in incompressible flow," *J. Aircraft* **9**, 618–624. [407]

Cebeci, T., and Smith, A. M. O. (1974), *Analysis of Turbulent Boundary Layers*, Academic Press, New York. [378, 396, 402, 406, 415]

Chapman, D. R., and Rubesin, M. W. (1949), "Temperature and Velocity Profiles in the Compressible Laminar Boundary Layer with Arbitrary Distribution of Surface Temperature," *J. Aero. Sci.* **16**, 547–565. [341]

Chu, H., see Blick, E. F.

Clauser, F. (1956), "The Turbulent Boundary Layer," in *Advances in Applied Mechanics*, vol. 4, T. von Kármán, ed., Academic Press, New York. [388]

Coles, D. R., and Hurst, E. A. (1968), *Proceedings of the Stanford Conference on Turbulent Boundary Layer Prediction*, Vol. 2, AFOSR-IFP, University Press, Stanford, Calif. [396]

Colwell, G. T., see Goradia, S. H.

Cuffel, R. F., see Back, L. H.

Dhawan, S. (1953), *Direct Measurement of Skin Friction*, NACA Report 1121. [314]

Dhawan, S., see Liepmann, H. W.

Diaconis, W. S., see Jack, J. R.

von Doenhoff, A. E., and Tetervin, N. (1943), *Determination of General Relations for Behavior of Turbulent Boundary Layers*, NACA Report 772. [404]

von Doenhoff, A. E., see Abbott, I. H.

Dryden, H. L. (1959), "Transition from Laminar to Turbulent Flow," in *High Speed Aerodynamics and Jet Propulsion*, vol. 5, Princeton University Press, Princeton, N.J. [365, 375]

Durand, W. F. (1943), "Mathematical Aids," in *Aerodynamic Theory*, vol. 1, W. F. Durand, ed., Durand Reprinting Committee, California Institute of Technology, Pasadena, Calif. [422]

Emmons, H. W. (1951), "Laminar Turbulent Transition in a Boundary Layer," Part I, *J. Aero. Sci.* **18**, 490–498; Part II, *Proc. of 1st Nat. Conf. for Appl. Mech.*, Edwards Brothers, Ann Arbor, Mich., 1952. [355]

Englar, R. J. (1975), "Circulation Control for High-Lift and Drag Generation on STOL Aircraft," *J. Aircraft*, **12**, (5), 457–464. [417, 419]

Falkner, V. M., and Skan, S. W. (1930), *Some Approximate Solutions of the Boundary Layer Equations*, British Aeronautical Research Committee, reports and memo 1314. [316]

Fejer, A. A., see Loehrke, R. I.

Ferri, A. (1939), *Experimental Results with Airfoils Tested in the High Speed Tunnel at Guidonia*, NACA TM 946. [272]

Fink, M. R., see Paterson, R. W.

Froessel, W. (1938), *Flow in Smooth Straight Pipes at Velocities Above and Below Sound Velocity*, NACA TM 844. [215]

Gadd, G. E. (1961), *Interactions of Normal Shock Waves and Turbulent Boundary Layers*, British Aeronautical Research Council, A.R.C. 22,559; F.M. 3051. [347]

Galbraith, R. A., McD., and Head, M. R. (1975), "Eddy Viscosity and Mixing Length from Measured Boundary Layer Developments." *Aeronautical Quarterly*, **26**, (2), 133–154. [397, 405]

Giacomelli, R. (1943), "Historical Sketch," in *Aerodynamic Theory*, vol. 1, W. F. Durand, ed., Durand Reprinting Committee, California Institute of Technology, Pasadena, Calif. [116]

Glauert, H. (1924), *A Theory of Thin Airfoils*, British Aeronautical Research Committee, Rep. & Memo. 910. [134]

Glauert, H. (1927), *Theoretical Relationships for an Airfoil with Hinged Flap*, British Aeronautical Research Committee, Rep. & Memo. 1095. [134]

Glauert, H. (1937), *Elements of Airfoil and Airscrew Theory*, Cambridge University Press, Cambridge. [123, 156, 158]

Goertler, H., and Hammerlin, G. (1955), *50 Jahre Grenzschichtforschung*, H. Goertler and W. Tollmien, eds., Vieweg & Sohn, Braunschweig. [367]

Goldstein, S., ed. (1938), *Modern Developments in Fluid Dynamics*, Clarendon Press, Oxford. Paperback edition, Dover, New York, 1965. [76, 98, 298, 306, 311, 375]

Goodier, J. N., see Timoshenko, S.

Goodmanson, L. T., and Gratzer, L. B. (1973), "Recent Advances in Aerodynamics for Transport Aircraft—Part 1," *Aero. Astro.* **11**, (12), 30–45; Part 2, *Aero. and Astro.* **12**, No. 1, 52–60, 1974. [286, 289, 418]

Goradia, S. H., and Colwell, G. T. (1975), "Analysis of High-Lift Wing Systems," *Aeronautical Quarterly* **26**, (2), 88–108. [415, 417]

Goranson, R. F. (1944), *Ground Effect in Aircraft Characteristics*, NACA Wartime Report WR L95. [173]

Gratzer, L. B., see Goodmanson, L. T.

Hammerlin, G., see Goertler, H.

Harris, R. V., Jr., see Carlson, H. W.

Hayes, W. D. (1947), *Linearized Supersonic Flow*, Report AL 222, North American Aviation, Inc. [283]

Head, M. R. (1960), *Entrainment in Turbulent Boundary Layers*, British Aeronautical Research Committee, Rep. & Memo. 3152. [404]

Head, M. R., see Galbraith, R. A., McD.

Heaslet, M. A., and Lomax, H. (1954), "Supersonic and Transonic Small Perturbation Theory," in *High Speed Aerodynamics and Jet Propulsion*, vol. 6, W. R. Sears, ed., Princeton University Press, Princeton, N.J. [253, 283, 295]

Hess, J. L. (1971), *Numerical Solution of the Integral Equation for the Neumann Problem with Application to Aircraft and Ships*, presented at SIAM Fall Meeting, Oct. 1971, Douglas Aircraft Company Engineering Paper 5987. [142, 144, 173]

Hess, J. L. (1972), *Calculation of Potential Flow about Arbitrary Three-Dimensional Lifting Bodies*, McDonnell Douglas Report No. MDC J5679-01. [173]

Hess, J. L., and Smith, A. M. O. (1967), "Calculation of Potential Flow about Arbitrary Bodies," in *Progress in Aeronautical Sciences*, vol. 8, D. Küchemann, ed., Pergamon Press, New York. [110, 107, 173]

Hinze, J. O., (1959), *Turbulence*, McGraw-Hill, New York. [388, 409]

Hoerner, S. F. (1965), *Fluid Dynamic Drag*, published by the author. Box 342, Brick Town, N.J., 08723. [298]

Hoerner, S. F., and Borst, H. B., (1975). *Fluid Dynamic Lift*, Hoerner Fluid Dynamics, Box 342, Brick Town, N.J. 08723. [410]

Howarth, L., ed. (1953), *Modern Developments in Fluid Dynamics—High Speed Flow*, Clarendon Press, Oxford. [261, 328]

Hurst, E. A., see Coles, D. R.

Jack, J. R., and Diaconis, W. S. (1955), *Variation of Boundary-Layer Transition with Heat Transfer at Mach Number 3.12*, NACA TN 3562. [362]

Jones, B. M. (1938), "Flight Experiments on the Boundary Layer," *J. Aero. Sci.* **5**, 81–94. [371]

Jones, R. T. (1956), *Theory of Wing-Body Drag at Supersonic Speeds*, NACA Report 1284. (Supersedes NACA RM A53H18a, 1953.) [293]

Jones, R. T. (1972), "Reduction of Wave Drag by Antisymmetric Arrangement of Wings and Bodies," *Aero. Astro.* **10**, (2), 171–176. [289]

Jones, R. T., and Nisbet, J. W. (1974), "Transonic-Transport Wings—Oblique or Swept?" *Aero. Astro.* **12**, (1), 40–47. [289]

Karamcheti, K. (1966), *Principles of Ideal-Fluid Aerodynamics*, Wiley, New York. [32]

von Kármán, T. (1921), "Uber Laminare und turbulente Reibung," *Z. angew. Math. Mech.* **1**, 233–252. Translated as NACA TM 1092, 1946. [318]

von Kármán, T. (1930), "Mechanische Ähnlichkeit und Turbulenz," *Proc. 3rd Internat. Cong. Appl. Mech.*, Stockholm.

von Kármán, T. (1934), "Turbulence and Skin Friction," *J. Aero. Sci.* **1**, 1–20.

von Kármán, T. (1935), "The Problem of Resistance in Compressible Fluids," *Atti del V Convegno della Fondazione Volta*, Rome, 222–276. See von Kármán, *Collected Works*, vol. 3, Butterworth, London, 1956, p. 179. [290, 401]

von Kármán, T. (1947), "Supersonic Aerodynamics—Principles and Applications," *J. Aero. Sci.* **14**, 373–402. [220]

von Kármán, T. (1954), *Aerodynamics, Selected Topics*, Cornell University Press, Ithaca, N.Y.

von Kármán, T., and Burgers, J. M. (1943), "General Aerodynamic Theory—Perfect Fluids," in *Aerodynamic Theory*, vol. 2, W. F. Durand, ed., Durand Reprinting Committee, California Institute of Technology, Pasadena, Calif. [123]

Keenan, J. H., and Neumann, E. P. (1945), *Friction in Pipes at Subsonic and Supersonic Velocities*, NACA TN 963. [215]

Klebanoff, P. S. (1955), *Characteristics of Turbulence in a Boundary Layer with Zero Pressure Gradient*, NACA Rep. 1247. [395]

Klebanoff, P. S., see Schubauer, G. B.

Kovasznay, L. S. G. (1940), "Hot-Wire Investigation of the Wake behind Cylinders at Low Reynolds Numbers," *Proc. R. Soc., London*, A, **198**, 174–190. [375]

Kruger, C., see Vincenti, W.

Kuethe, A. M., and Schetzer, J. D., (1959), *Foundations of Aerodynamics*, 2nd edition, Wiley, New York. [v]

Kuethe, A. M. (1972), "Effect of Streamwise Vortices on Wake Properties Associated with Sound Generation," *J. Aircraft* **9**, 715–719. [420]

Lachmann, G. V., ed. (1961), *Boundary Layer and Flow Control*, Pergamon Press, New York. [415]

Landahl, M., see Ashley, H.

Laufer, J. (1950), "Some Recent Measurements in a Two-Dimensional Channel," *J. Aero. Sci.* **17**, 277–288. [391]

Legendre, R. (1966), "Vortex Sheets Rolling up along Leading Edges of Delta Wings," in *Progress in Aeronautical Sciences*, vol. 7, D. Küchemann, ed., Pergamon Press, New York. [414]

Liepmann, H. W., and Dhawan, S. (1951), "Direct Measurement of Local Skin Friction in Low-Speed and High-Speed Flow," *Proc. 1st U.S. Nat. Cong. Appl. Mech.* [314]

Liepmann, H. W., and Puckett, A. E. (1947), *Introduction to Aerodynamics of a Compressible Fluid*, Wiley, New York. [507]

Liepmann, H. W., and Roshko, A. (1957), *Elements of Gasdynamics*, Wiley, New York. [224, 226, 246, 254, 267, 290]

Liepmann, H. W., Roshko, A., and Dhawan, S. (1952), *On Reflection of Shock Waves from Boundary Layers*, NACA Report 1100. [345, 347]

Lighthill, M. J. (1954), "Higher Approximations," in *High Speed Aerodynamics and Jet Propulsion*, vol. 6, W. R. Sears, ed., Princeton University Press, Princeton, N.J. [274]

Lighthill, M. J. (1961), "Sound Generated Aerodynamically" (The Bakerian Lecture, 1961), *Proc. R. Soc. London*, A, **267**, 148–182. [420]

Lin, C. C. (1945), "On the Stability of Two-Dimensional Parallel Flows," Parts I, II, III, *Quart. Appl. Math.* **3**, 117–142, 218–234, 277–301, respectively.

Lin, C. C. (1955), *Hydrodynamic Stability*, Cambridge University Press, Cambridge. [352, 354, 367]

Lippisch, A., and Beushausen, W. (1946), *Pressure Distribution Measurement at High Speed and Oblique Incidence of Flow*, Translation No. F-TS-634-EW, Air Matériel Command. [281]

Liu, V.-C., see Bekofske, K.

Lock, R. C., and Bridgewater, J. (1967), "Theory and Aerodynamic Design for Swept-winged Aircraft at Transonic and Supersonic Speeds," in *Progress in Aeronautical Sciences*, vol. 8, D. Küchemann, ed., Pergamon Press, New York. [286, 288, 415]

Loehrke, R. I., Morkovin, M. V., and Fejer, A. A. (1975), "Review—Transition in Nonreversing Oscillating Boundary Layers," *Jour. Fluids Engg.* **97**, 534–549. [364]

Lomax, H., and Heaslet, M. A. (1956), "Recent Developments in the Theory of Wing-Body Wave Drag," *J. Aero. Sci.* **23**, 1061–1074.

Lomax, H., see Heaslet, M. A.

Maccoll, J. W. (1937), "The Conical Shock Wave Formed by a Cone Moving at a High Speed," *Proc. R. Soc. London*, A, **159**, 459–472. [239, 240]

Maccoll, J. W., see Taylor, G. I.

Mack, L. M. (1954), Report 20-80, Jet Propulsion Laboratory, California Institute of Technology, Pasadena, Calif. [338]

Mack, L. M. (1972), "Boundary Layer Stability Theory," Tape cassettes of lectures and copies of figures available through AIAA. [356]

Mack, L. M. (1975), "Linear Theory and the Problem of Supersonic Boundary Layer Transition," *AIAA J.* **13**, 278–290. [363]

McCormick, B. W. (1967), *Aerodynamics of V-STOL Flight*, Academic Press, New York.

Messier, P. F., see Back, L. H.

von Mises, R. (1945), *Theory of Flight*, McGraw-Hill, New York. Paperback edition, Dover, New York, 1959. [123]

Moore, F. K. (1956), "Three-Dimensional Boundary Layer Theory," in *Advances in Applied Mechanics*, vol. 4, H. L. Dryden and T. von Kármán, eds., Academic Press, New York. [326]

Morkovin, M. V. (1972), "Critical Evaluation of Transition," Tape cassettes of lectures and copies of figures available through AIAA. [356]

Morkovin, M. V., see Loehrke, R. I.

Mosinskis, G. J., see Cebeci, T.

Munch, C. L., see Paterson, R. W.

National Committee for Fluid Mechanics Films (1972), *Illustrated Experiments in Fluid Mechanics*. MIT Press, Cambridge, Mass. and London. [v]

Neumann, E. P., see Keenan, J. H.

Niewland, G. Y., and Spee, B. M., (1973), "Transonic Airfoils: Recent Developments in Theory, Experiment, and Design," in *Annual Review of Fluid Mechanics*, vol. 5, M. Van Dyke and W. G. Vincenti, eds., Annual Reviews Inc., Palo Alto, Calif. [267]

Nikuradse, J. (1932), "Gesetzmässigkeiten der turbulenten Strömung in glatten Rohren," *Forschungsheft* 356, *Ver. deutsch. Ing.* [313]

Nikuradse, J. (1942), "Laminare Reibungsschichten an der längsangeströmten Platte," Monograph, *Zentrale f. wiss. Berichtswesen*, Berlin. [392]

Nisbet, J. W., see Jones, R. T.

Pai, S. I. (1956), *Viscous Flow Theory*, Van Nostrand, New York. [328]

Paterson, R. W., Vogt, P. G., Fink, M. R., and Munch, C. L. (1973), "Vortex Noise of Isolated Airfoils," *J. Aircraft* **10**, 296–302. [420]

Peirce, B. O. (1929), *Short Table of Integrals*, 3rd ed., Ginn, Boston.

Pfenninger, W. (1965), *Summary Report about the Investigation of a 10-ft Chord 33° Swept Low Drag Suction Wing*, Northrup Report. [360]

Pfenninger, W., and Reed, V. D. (1966), "Laminar Flow; Research and Experiments," *Aero. Astro.* **4**, (7), 44–50. [415]

Pinkerton, R. M. (1936), *Calculated and Measured Pressure Distributions over the Midspan Section of the NACA 4412 Airfoil*, NACA Report 563. [411]

Pohlhausen, E. (1921), "Der Wärmeaustausch zwischen festen Körpern und Flüssigkeiten mit kleiner Reibung und kleiner Wärmeleitung," *Z. angew. Math. Mech.* **1**, 115. [336]

Prandtl, L. (1921), *Applications of Modern Hydrodynamics to Aeronautics*, NACA Report 116. [160]

Prandtl, L. (1943), "The Mechanics of Viscous Fluids," in *Aerodynamics Theory*, vol. 3, W. F. Durand, ed., Durand Reprinting Committee, California Institute of Technology, Pasadena, Calif. [399]

Prandtl, L. (1952), *Essentials of Fluid Dynamics*, Hafner, New York. [369]

Prandtl, L., and Tietjens, O. G. (1931), *Applied Hydro- and Aeromechanics*, McGraw-Hill, New York. Paperback edition, Dover, New York, 1957. [160]

Puckett, A. E., see Liepmann, H. W.

Reed, V. D., see Pfenninger, W.

Reynolds, O. (1883), "An Experimental Investigation of the Circumstances Which Determine Whether the Motion of Water Shall Be Direct of Sinuous, and of the Law of Resistance in Parallel Channels," *Phil. Trans. R. Soc. London* **174**, 935–982.

Ribner, H. S. (1964), "The Generation of Sound by Turbulent Jets," in *Advances in Applied Mechanics*, vol. 8, H. L. Dryden and T. von Kármán, eds., Academic Press, New York. [351]

Ribner, H. S. (1968), "Jets and Noise" (J. Rupert Turnbull Lecture) AFOSR-

UTIAS *Symposium on Aerodynamic Noise*, University of Toronto, Toronto, pp. 1–42. Reprinted from *Canad. Appl. Sci. J.*, October 1968. [420]

Roshko, A. (1954), *On the Development of Turbulent Wakes from Vortex Streets*, NACA Report 1191. [375]

Roshko, A., see Liepmann, H. W.

Rubbert, P. E., and Saaris, G. R. (1969), "3-D Potential Flow Method Predicts V/STOL Aerodynamics," *SAE J.* **77**, 44–51. [173]

Rubbert, P. E., and Saaris, G. R. (1972), *Review and Evaluation of a Three-Dimensional Lifting Potential Flow Analysis Method for Arbitrary Configurations*, AIAA Paper No. 72-188. [173]

Rubesin, M. W. (1953), *A Modified Reynolds Analogy*, NACA TN 2917. [408]

Rubesin, M. W., see Chapman, D. R.

Saaris, G. R., see Rubbert, P. E.

Schetzer, J. D., see Kuethe, A. M.

Schlichting, H. (1968), *Boundary Layer Theory*, J. Kestin, trans., 6th ed., McGraw-Hill, New York. [317, 323, 328, 352, 356, 358, 368, 391, 394, 409]

Schlichting, H., and Ulrich, A. (1940), "Zur Berechnung des Umschlages laminar-turbulent," Memorial Lecture 1940 of the Lilienthalgesellschaft für Luft-fahrtforschung, Flugzeugbau. Complete text in Bericht S 10 of the Lilienthal-Gesellschaft. [352]

Schoenherr, K. E. (1932), "Resistance of Flat Plates Moving Through a Fluid," *Trans. Soc. Nav. Arch. Mar. Eng.* **40**, 279–313. [400]

Schrenk, O. (1940), *Simple Approximation Method for Obtaining Spanwise Lift Distribution*, NACA TM 948. [165]

Schubauer, G. B. (1954), "Turbulent Process as Observed in Boundary Layer and Pipe," *J. Appl. Physics* **25**, 188–196. [395]

Schubauer, G. B., and Klebanoff, P. S. (1951), *Investigation of Separation of the Turbulent Boundary Layer*, NACA Report 1030. [388, 402, 406, 484]

Schubauer, G. B., and Klebanoff, P. S. (1955), *Contributions to the Mechanics of Boundary Layer Transition*, NACA TN 3489. [355, 406]

Schubauer, G. B., and Skramstad, H. K. (1947), "Laminar Boundary-Layer Oscillations and Stability of Laminar Flow," *J. Aero. Sci.* **14**, 69–78. [352, 353, 357, 363]

Sears, W. R. (1947), "On Projectiles of Minimum Wave Drag," *Quart. Appl. Math.* **4**, 361–366. (Reprinted in *Sears Collected Papers through 1973*, pp. 52–57, Cornell University Press, Ithaca, N.Y., 1974.) [292]

Sears, W. R., ed. (1954), *General Theory of High Speed Aerodynamics*, Vol. 6 of *High Speed Aerodynamics and Jet Propulsion*, Princeton University Press, Princeton, N.J. [278]

Shapiro, A. (1961), *Shape and Flow, The Fluid Mechanics of Drag.* Doubleday, Garden City, N.Y. [v]

Sichel, M. (1971) "Two-Dimensional Shock Structure in Transonic and Hypersonic flow" in *Advances in Applied Mechanics*, vol. 11, C.-S. Yih, ed., Academic Press, New York. [267]

Skan, S. W., see Falkner, V. M.

Skramstad, H. K., see Schubauer, G. B.

Smith, A. M. O. (1956), "Transition, Pressure Gradient and Stability Theory," *Proc. 9th Internat. Cong. Appl. Mech.*, Brussels, **4**, 234–244. [370]

Smith, A. M. O. (1975), "High-Lift Aerodynamics," 37th Wright Brothers Lecture AIAA, *J. Aircraft* **12**, (6), 501–531. [410, 415]

Smith, A. M. O., see Cebeci, T., Hess, H. L.

Soderman, P. T. (1973), *Leading Edge Serrations Which Reduce Noise of Low Speed Rotors*, NASA TN D-7371. [414, 420]

Sokolnikoff, I. S. (1939), *Advanced Calculus*, 1st ed. McGraw-Hill, New York, p. 347. [120]

Spangler, J. G., and Wells, C. S., Jr. (1968), "Effects of Free-Stream Disturbances on Boundary Layer Transition," *AIAA J.* **6**, 543–545. [364]

Spee, B. M., see Niewland, G. Y.

Sternberg, J. (1954), *The Transition from a Turbulent to a Laminar Boundary Layer*, Ballistics Research Laboratory Report 908. [359]

Stuart, J. T. (1963), in *Laminar Boundary Layers*, L. Rosenhead, ed., Clarendon Press, Oxford, chap. 9. [361, 409]

Stuart, J. T. (1971), "Nonlinear Stability Theory," in *Annual Review of Fluid Mechanics*, vol. 3, M. Van Dyke and W. G. Vincenti, eds., Annual Reviews Inc., Palo Alto, Calif. [356, 368]

Taylor, G. I., and Maccoll, J. W. (1933), "Air Pressure on a Cone Moving at High Speeds," *Proc. R. Soc. London*, A, **139**, 279–311. [239, 367]

Tetervin, N., see von Doenhoff, A. E.

Thompson, B. G. J. (1967), "A New Two-Parameter Family of Mean Velocity Profiles for Incompressible Boundary Layers on Smooth Walls." Brit. Aero. Res. Com. Rep. Mem. No. 3463. [396]

Thwaites, B., ed. (1960), *Incompressible Aerodynamics*, Clarendon Press, Oxford. [396]

Tietjens, O. G., see Prandtl, L.

Tifford, A. N. (1950), "On Surface Effects of a Compressible Laminar Boundary Layer," *J. Aero. Sci.* **17**, 187–188. [343]

Timoshenko, S., and Goodier, J. N., (1951), *Theory of Elasticity*, McGraw-Hill, New York. [432]

Townsend, A. A. (1956), *The Structure of Turbulent Flows*, Cambridge University Press, Cambridge. [409]

Ulrich, A., see Schlichting, H.

Van Driest, E. R. (1952), *Investigation of Laminar Boundary Layer in Compressible Fluids Using the Crocco Method*, NACA TN 2597. [338]

Van Driest, E. R., and Boison, J. C. (1957), "Experiments on Boundary Layer Transition at Supersonic Speeds," *J. Aero. Sci.* **24**, 885–899. [343]

Vincenti, W., and Kruger, C. (1965), *Introduction to Physical Gas Dynamics*, Wiley, New York. [435]

Vogt, P. G., see Paterson, R. W.

Watson, W., see Blick, E. F.

Weaver, J. H. (1948), "A Method of Wind Tunnel Testing through the Transonic Range," *J. Aero. Sci.* **15**, 28–34. [265]

Wells, C. S., Jr., see Spangler, J. G.

Whitcomb, R. T. (1956), *A Study of the Zero-Lift Drag-Rise Characteristics of Wing-Body Combinations Near the Speed of Sound*, NACA Report 1273. [285]

Whitcomb, R. T. (1974), Paper No. 74-10, *9th Cong. of the Internat. Council of the Aeronautical Sci.*, Haifa, Israel. [266]

Yih, C.-S. (1969), *Fluid Mechanics*, McGraw-Hill, New York. [368]

Zemansky, M. W. (1943), *Heat and Thermodynamics*, McGraw-Hill, New York. [183]

Subject Index